P9-CJR-798

# STRUCTURAL AND FUNCTIONAL ORGANIZATION OF THE NUCLEAR MATRIX

**VOLUME 162B**

International Review of **Cytology** A Survey of **Cell Biology**

Edited by

**Ronald Berezney**
Department of Biological Sciences
State University of New York at Buffalo
Buffalo, New York

**Kwang W. Jeon**
Department of Zoology
University of Tennessee
Knoxville, Tennessee

# STRUCTURAL AND FUNCTIONAL ORGANIZATION OF THE NUCLEAR MATRIX

**VOLUME 162B**

**Academic Press**
San Diego New York Boston London Sydney Tokyo Toronto

This book is printed on acid-free paper. ♾



Academic Press, Inc.
A Division of Harcourt Brace & Company
525 B Street, Suite 1900, San Diego, California 92101-4495

*United Kingdom Edition published by*
Academic Press Limited
24-28 Oval Road, London NW1 7DX

International Standard Serial Number: 0074-7696

International Standard Book Number: 0-12-364566-2

PRINTED IN THE UNITED STATES OF AMERICA
95 96 97 98 99 00 EB 9 8 7 6 5 4 3 2 1

# CONTENTS



## Nuclear Matrix Proteins as Structural and Functional Components of the Mitotic Apparatus

D. He, C. Zeng, and B. R. Brinkley



## Nuclear Matrix Isolated from Plant Cells

Susana Moreno Díaz de la Espina

## The Dynamic Properties and Possible Functions of Nuclear Lamins

Robert D. Moir, Timothy P. Spann, and Robert D. Goldman



## Intracellular Structure and Nucleocytoplasmic Transport

Paul S. Agutter



## Toward a Molecular Understanding of the Structure and Function of the Nuclear Pore Complex

Nelly Panté and Ueli Aebi



## Nuclear Pore Complex Proteins

Ricardo Bastos, Nelly Panté, and Brian Burke

## Targeting and Association of Proteins with Functional Domains in the Nucleus: The Insoluble Solution

Heinrich Leonhardt and M. Cristina Cardoso



## Nuclear Matrix Acceptor Binding Sites for Steroid Hormone Receptors: A Candidate Nuclear Matrix Acceptor Protein

Andrea Lauber, Nicole P. Sandhu, Mark Schuchard, M. Subramaniam, and Thomas C. Spelsberg

## Nuclear Matrix and the Cell Cycle

Peter Loidl and Anton Eberharter



## Specificity and Functional Significance of DNA Interaction with the Nuclear Matrix: New Approaches to Clarify the Old Questions

Sergey V. Razin, Irina I. Gromova, and Olga V. Iarovaia

# CONTRIBUTORS

Numbers in parentheses indicate the pages on which the authors' contributions begin.

Ueli Aebi (225), *Department of Cell Biology and Anatomy, The Johns Hopkins University School of Medicine, Baltimore, Maryland 21205*

Paul S. Agutter (183), *Department of Biological Sciences, Napier University, Edinburgh, Scotland EH14 1DJ, United Kingdom*

Ricardo Bastos (257), *Department of Cell Biology, Harvard Medical School, Boston, Massachusetts 02115*

B. R. Brinkley (1), *Department of Cell Biology, Baylor College of Medicine, Houston, Texas 77030*

Brian Burke (257), *Department of Cell Biology, Harvard Medical School, Boston, Massachusetts 02115*

M. Cristina Cardoso (303), *Department of Nephrology, Hypertension, and Genetics, Freie Universität Berlin, Franz-Volhard-Klinik am Max-Delbrück-Centrum für Molekulare Medizin, 13122 Berlin, Germany*

Anton Eberharter (377), *Department of Microbiology, University of Innsbruck-Medical School, A-6020 Innsbruck, Austria*

Robert D. Goldman (141), *Department of Cell and Molecular Biology, Northwestern University Medical School, Chicago, Illinois 60611*

Irina I. Gromova (405), *Institute of Gene Biology of the Russian Academy of Sciences, 117342 Moscow, Russia and Department of Molecular Biology, University of Aarhus, Aarhus C, Denmark*

D. He (1), *Department of Cell Biology, Baylor College of Medicine, Houston, Texas 77030*

Olga V. Iarovaia (405), *Institute of Gene Biology of the Russian Academy of Sciences, 117342 Moscow, Russia and International Centre for Genetic Engineering and Biotechnology, I-34012 Trieste, Italy*

Andrea Lauber (337), *Department of Biochemistry and Molecular Biology, Mayo Clinic and Graduate School of Medicine, Rochester, Minnesota 55904*

Heinrich Leonhardt (303), *Department of Nephrology, Hypertension, and Genetics, Freie Universität Berlin, Franz-Volhard-Klinik am MaxDelbrück-Centrum für Molekulare Medizin, 13122 Berlin, Germany*

Peter Loidl (377), *Department of Microbiology, University of Innsbruck-Medical School, A-6020 Innsbruck, Austria*

Robert D. Moir (141), *Department of Cell and Molecular Biology, Northwestern University Medical School, Chicago, Illinois 60611*

Susana Moreno Díaz de la Espina (75), *Laboratorio de Biología Celular y Molecular Vegetal, Departamento de Biologíe de Plantas, Centro de Investigaciones Biólogicas, 28006 Madrid, Spain*

Nelly Panté (225,257), *M.E. Müller Institute for Microscopy, Biozentrum, University of Basel, CH-4056 Basel, Switzerland*

Sergey V. Razin (405), *Institute of Gene Biology of the Russian Academy of Sciences, 117342 Moscow, Russia and International Center for Genetic Engineering and Biotechnology, I-34012 Trieste, Italy*

Nicole P. Sandhu (337), *Department of Biochemistry and Molecular Biology, Mayo Clinic and Graduate School of Medicine, Rochester, Minnesota 55904*

Mark Schuchard (337), *Department of Biochemistry and Molecular Biology, Mayo Clinic and Graduate School of Medicine, Rochester, Minnesota 55904*

Timothy P. Spann (141), *Department of Cell and Molecular Biology, Northwestern University Medical School, Chicago, Illinois 60611*

Thomas C. Spelsberg (337), *Department of Biochemistry and Molecular Biology, Mayo Clinic and Graduate School of Medicine, Rochester, Minnesota 55904*

M. Subramaniam (337), *Department of Biochemistry and Molecular Biology, Mayo Clinic and Graduate School of Medicine, Rochester, Minnesota 55904*

C. Zeng (1), *Verna and Marrs McLean Department of Biochemistry, Baylor College of Medicine, Houston, Texas 77030*

# PREFACE

Research on the nuclear matrix has grown enormously since Berezney and Coffey first reported its isolation and initial characterization in 1974. Since then, more than 1000 papers have been published on the subject by numerous workers around the world, yet there has been no book devoted to reviewing the major developments in this growing field. One of our aims in producing Volume 162, Parts A and B, has been to fill this gap.

We invited many of the world's leading experts in the field to contribute to these volumes, and we were delighted with their positive responses. The chapters cover a variety of topics including isolation of the nuclear matrix, its morphology and correlation with the nuclear structure observed *in situ,* structural and functional domains of the nuclear matrix and its components, and biochemistry and molecular biology of the matrix proteins and associated DNA and RNA. Chapters also discuss functional properties associated with the nuclear matrix such as DNA replication, transcription, RNA splicing, transcription regulation, intranuclear, and nucleocytoplasmic transport and targeting, cell cycle regulation, mitotic regulation steroid hormone receptors, and viral and cancer drug associations. While these chapters deal with a broad range of topics, they are all interrelated in considering the genome organization, function, and regulation in relation to the nuclear architectural parameters. Thus, there are overlaps among some chapters, especially at the conceptual level where authors have been encouraged to "speak their peace" about the nuclear matrix and the progress made in understanding the functional organization of the cell nucleus. This is healthy for a field still in its relatively early stage of development but which shows every sign of entering a stage of explosive growth.

Among the more exciting developments is the recent application of high-resolution three-dimensional microscopy and computer-based image analysis in association with fluorescence techniques to label sites of genomic function inside the cell nucleus. Another development is the recent breakthrough in cloning and identifying genes for nuclear matrix-associated proteins. Combination of these approaches will give us insight into understanding how individual proteins are arranged and interact with each other and in association with the genetic information and expression. This could lead to important advances in correlating molecular details of the genome and its expression with the hierarchy of organization and function characteristic of all living organisms. We hope that these volumes will not only be informative but will also stimulate the interest of many to study the exciting subject of the nuclear matrix.

We sincerely thank our authors, whose willingness to write accounts of recent and cutting-edge developments in the nuclear matrix field in a timely and scholarly fashion has made this project successful. We are also very grateful to the editors and production staff members at Academic Press for their understanding, patience, skill, and devotion to producing the high-quality text and illustrations including many color plates.

Ronald Berezney
Kwang W. Jeon

# LIST OF ABBREVIATIONS FOR THE VOLUME

**a.a.** amino acid
**ADP** adenosine diphosphate
**AFB** anticarrot fibrillar bundle antibody
**AFM** atomic force microscopy
**AR** rat androgen
**araATP** arabinofuranosyladenosine 5′-triphosphate
**araCTP** arabinofuranosylcytosine 5′-triphosphate
**ARBP** attachment region-binding protein
**ARS** autonomously replicating sequences
**ASTP** ATP-stimulated translocation promoter
**BAA** bromoacetaldehyde
**BrdU** bromodeoxyuridine
**BrUTP** 5-bromouridine-5-triphosphate
**CAA** chloroacetaldehyde
**CB** coiled body
**CCNU** chloroethyl-cyclohexl-nitrosourea
**CHAT** choline acetyltransferase gene
**CHO** Chinese hamster cell
**DAPI** 4,6-diamidino-2-phenylindole
**DBSF** DNA binding stimulatory factor
**DFC** dense fibrillar component
**DHFR** dihydrofolate reductase
**DHRR** dihydrofolate reductase
**DHS** DNase-hypersensitive site
**DNase I-HS** DNase Ihypersensitive site
**DRB** 5,6-dichloro-(B-D-ribofuranosyl)-benzimidazole
**dsDNA** double-stranded DNA
**dsRNA** double-stranded RNA
**DTT** dithiothreitol
**DUE** DNA-unwinding element
**EBNA1** Epstein-Barr virus nuclear antigen 1
**EDTA** ethylenediaminetetracetic acid
**EM** electron microscopy
**ER** endoplasmic reticulum

**ERc** estrogen receptor
**FA** actin filaments
**FaraATP** arabinofuranosyl-2-fluroadosine 5′-triphosphate
**FC** fibrillar centers
**FISH** fluorescence in situ hybridization
**FRT** FLP recognition target sites
**GAPDH** glyceraldehyde-3-phosphate dehydrogenase
**GC** granular component
**GlcNac** N-acetylglucosamine
**GR** glucocorticoid receptor
**GRE** glucocorticoid responsive elements
**HD** homeodomain
**HDase** histone deacetylase
**HMG** high mobility group
**HnRNP** heterogeneous nuclear ribonucleoprotein
**huIFN-β** human interferon beta
**ICG** interchromatin granule
**ICS** intermediate Cairns structures
**IF** intermediate filaments
**IFA** anti-intermediate filament antibody
**IFN** interferon
**Ig** immunoglobulin
**IG** interchromatin granule
**IMPDH** IMP dehydrogenase
**INCENP** inner centromere protein
**IR** inverted repeat
**LAP** lamin-associated protein
**LBR** lamin B receptor
**LCR** locus control region
**LCS** latest Cairns structures
**LIS** lithium 3,5-diiodosalicylate
**MAb** monoclonal antibody
**M-AP** mitogen-activated protein
**MAP** microtubule-associated protein
**MAR** matrix attachment region
**MDa** megadalton = $10^6$ daltons
**MKLP** mitotic kinesin-like protein
**MMTV** mouse mammary tumor virus
**MNase** micrococcal nuclease
**MPE** methidiumpropyl-EDTA-iron(II)
**MPF** matrix protein filaments
**$M_r$** molecular radius
**MT** microtubule
**MTOC** MT organizing centers
**MVM** minute virus of mice
**N/A** neutral/alkaline
**NAP** nucleotide protein
**NaTT** sodium tetrathionate
**NE** nuclear envelope
**NEL** nuclear envelope lattice
**NEM** N-ethylmaleimide
**NLS** nuclear localization signal
**NM** nuclcar matrix
**nmDNA** nuclear matrix-associated DNA

**N/N** neutral/neutral
**NOR** nucleolar organizer region
**NPC** nuclear pore complex
**NU** in situ nuclei
**NuM** nucleolar-matrix
**NuMA** nuclear matrix mitotic apparatus
**nup** nucleoporin
**OBR** origin of bidirectional replication
**OC** osteocalcin
**O-glcNAc** O-linked *N*-acetyglucosamine moieties
**ORC** origin recognition complex
**ORI** origin of replication
**PCN** perichromonucleolin
**PCR** polymerase chain reaction
**PF** perichromatin fibril
**PFGE** pulse field gel electrophoresis
**PG** perichromatin granules
**PKA** protein kinase A
**PKC** protein kinase C
**PMSF** phenylmethylsulfonyl-fluoride
**PNB** prenucleolar body
**PPB** preprophase band
**PPIase** proline isomerase
**PR** progesterone receptor
**pRb** retinoblastoma protein
**PRE** progesterone response element
**R** receptor
**RAF** receptor accessory factor
**RANBP1** Ran-binding protein-1
**RAP** repressor-activator binding protein
**Rb** retinoblastoma
**RBF** receptor binding factor
**rDNA** ribosomal DNA
**RNM** RNA-binding motif
**RNP** ribonucleoprotein
**RP** multidrug resistance-associated protein
**RPA** replication protein A
**RRE** Rev-responsive element
**rRNA** ribosomal RNA
**R-WGA** rhodamine-conjugated wheat germ agglutinin
**SAR** scaffold attachment region
**SAF** scaffold attachment factor
**SAF-A** scaffold attachment factor A
**SATB1** special AT-rich sequence binding protein 1
**scs** special chromatin structures
**SDS–PAGE** SDS–polyacrylamide gel electrophoresis
**SMI1** suppress MAR inhibition
**snRNP** small nuclear ribonucleoprotein
**snoRNA** small nucleolar RNA
**SMC** stability of microchromosome
**S/MAR** scaffold/matrix-attached regions
**SR** steroid-receptor
**ssDNA** single-stranded DNA

**SSB** single-stranded binding protein
**STEM** scanning transmission EM
**SV40** simian virus 40
**T-antigen** large tumor antigen
**TEM** transmission electron microscopy
**TF** transcription factor
**Topo II** topoisomerase II
**TP** terminal protein
**Tpr** translocated promoter region
**TRAP** T3 receptor auxiliary protein
**TRE** thyroid hormone response element
**TSA** trichostatin A
**VDRE** vitamin D responsive element
**VZV** varicella-zoster virus
**WAP** whey acidic protein
**WGA** wheat germ agglutinin

# Nuclear Matrix Proteins as Structural and Functional Components of the Mitotic Apparatus

D. He,* C. Zeng,† and B. R. Brinkley*
* Department of Cell Biology and † Verna and Marrs McLean Department of Biochemistry, Baylor College of Medicine, Houston, Texas 77030

The eukaryotic nucleus is a membrane-enclosed compartment containing the genome and associated organelles supported by a complex matrix of nonhistone proteins. Identified as the nuclear matrix, this component maintains spatial order and provides the structural framework needed for DNA replication, RNA synthesis and processing, nuclear transport, and steroid hormone action. During mitosis, the nucleoskeleton and associated chromatin is efficiently dismantled, packaged, partitioned, and subsequently reassembled into daughter nuclei. The dramatic dissolution of the nucleus is accompanied by the assembly of a mitotic apparatus required to facilitate the complex events associated with nuclear division. Until recently, little was known about the fate or disposition of nuclear matrix proteins during mitosis. The availability of specific molecular probes and imaging techniques, including confocal microscopy and improved immunoelectron microscopy using resinless sections and related procedures, has enabled investigators to identify and map the distribution of nuclear matrix proteins throughout the cell cycle. This chapter will review the structure, function, and distribution of the protein NuMA (nuclear matrix mitotic apparatus) and other nuclear matrix proteins that depart the nucleus during the interphase/mitosis transition to become structural and functional components within specific domains of the mitotic apparatus.

**KEY WORDS:** Nuclear matrix, Nucleoskeleton, Core filaments, Chromosome scaffold, Mitotic apparatus, NuMA, Centromere, Kinetochore.

## I. Introduction

The eukaryotic nucleus is a highly integrated compartment containing the genome and a complex assortment of structural and regulatory factors required for DNA replication, RNA synthesis and processing, nuclear transport, and steroid hormone action (Nickerson *et al.,* 1990; Spector, 1993). Although originally thought to be composed largely of chromatin, the nucleolus, and amorphous nucleoplasm, the nucleus is now believed to be highly structured with a complex skeletal lattice or nuclear matrix (Berezney and Coffey, 1974, 1977) on which chromatin and numerous other factors are spatially arranged to maintain an integrated three-dimensional organization (Nickerson and Penman, 1992). A distinct nucleoskeleton has been difficult to identify in intact nuclei by conventional microscopic techniques due, in part, to masking by a dense array of chromatin fibers. However, if chromatin is extracted by extensive nuclease digestion and subsequent salt extraction, an underlying anastomosing network of 9- to 13-nm core filaments can be demonstrated (He *et al.,* 1990). In many ways, this unique filamentous system resembles the cytoplasmic intermediate filament complex, but it is confined entirely within the nucleus where it intersects with the nuclear lamina and connects the nucleolus, forming an integral lattice of fibers collectively called the nuclear matrix. Although the molecular composition of the nucleoskeleton remains largely uncharted, its integrity apparently requires ribonucleoprotein (RNP) and an array of nuclear matrix proteins (Fey *et al.,* 1986).

As a biochemical concept, the nuclear matrix is thought to be an essential solid-state integrator of most of the metabolic events associated with the function of the interphase nucleus. The small fraction of DNA remaining with the nuclear matrix after extraction appears to be enriched in actively transcribing genes indicating a close structural relationship between the matrix and transcriptionally active chromatin (Robinson *et al.,* 1982; Ciejek *et al.,* 1983; Hentzen *et al.,* 1984; Small *et al.,* 1985; Thorburn *et al.,* 1988; Nickerson *et al.,* 1992).

Although the interphase nucleus is structurally and functionally complex, the entire framework and associated chromatin can be efficiently dismantled, packaged, and partitioned equally to daughter cells and subsequently reassembled into functional nuclei during each mitotic cycle. The dramatic dissolution of the nucleus is accompanied by the assembly of a complex mitotic apparatus that facilitates the alignment, transport, and partitioning of the replicated genome to daughter cells. We have known very little about the fate of the nuclear matrix during mitosis because only recently have probes become available for identifying and tracing its components. It has been generally assumed that most of the nucleus dissolves at the time of nuclear envelope breakdown, except for the chromosomes that become attached to the mitotic spindle. Indeed some nucleoskeletal proteins such as the lamins are solubilized and released into the cytoplasm at mitosis. Many nuclear matrix proteins, however, persist and relocate to specific domains of the mitotic apparatus where they form a new matrix binding to, and in some cases, encompassing the mitotic spindle. Certain intermediate filaments of the cytoskeleton are known to form a cage around the mitotic spindle (Zieve *et al.*, 1980), but the spindle matrix described in this review is apart and distinct from these elements and appears to be derived exclusively from the nuclear matrix.

In this review, we attempt to identify and account for those nuclear matrix proteins that depart the nucleus at the onset of mitosis and become structural and functional components of the mitotic apparatus. We define the nuclear matrix-derived proteins of the mitotic apparatus as those initially identified by immunofluorescence in the nucleus after extensive nuclease digestion and extraction in high salts.

Foremost among this list is a high-molecular-weight coiled-coil protein, NuMA (nuclear-mitotic apparatus protein), identified initially by Lydersen and Pettijohn (1980) as a nonhistone protein that departed the nucleus at mitosis and became associated with the poles of the mitotic spindle. Independently rediscovered a decade later in several laboratories and assigned new names such as centrophilin (Tousson *et al.*, 1991), SPN (Kallajoki *et al.*, 1991), and SP-H (Maekawa *et al.*, 1991), NuMA's full-length amino acid sequence was determined (Compton *et al.*, 1992; Yang *et al.*, 1992) and various isoforms have been identified (Tang *et al.*, 1993). Since this protein is likely the best known example of a nuclear matrix protein that functions in mitosis and has been reasonably well characterized in our laboratory and elsewhere, we present it here in considerable detail.

A survey of the current literature indicates that, like NuMA, numerous other nuclear matrix proteins become targeted to the mitotic apparatus after dissolution of the nucleus, where they assume vital functions in the mitotic process. Most of these components eventually make their way directly or indirectly to specific sites including the chromosome scaffold, perichromosome region, centromere–kinetochore complex, midzone, midbody, and spindle poles. Although some of these may be "chromosomal passengers" (Earnshaw and Bernat, 1990), using the mitotic apparatus for conveyance to daughter nuclei, evidence will be presented that many nuclear matrix-derived proteins likely play vital roles in the mitotic process.

The molecular basis of mitosis and chromosome segregation is, as yet, poorly understood, as many pieces of the puzzle are largely unaccounted for, especially those involved in regulatory processes or checkpoints. Perhaps some of the missing parts can be found among the nuclear matrix-derived proteins, heretofore neglected as functional components of the mitotic apparatus.

## II. Nuclear Mitotic Apparatus Protein

Nuclear mitotic apparatus protein (NuMA) was initially identified in 1980 as a high-molecular-weight nonhistone protein (Lydersen *et al.*, 1980). The name NuMA reflects the distinct pattern of translocation of this protein from the interphase nucleus to the spindle poles of the mitotic apparatus during the cell cycle (Lydersen and Pettijohn, 1980). Since the early 1990s, a number of proteins with a similar cell cycle-dependent redistribution have been described, including centrophilin, mitotin, SPN, SP-H, and proteins identified by various antibodies such as fA12, 1F1/1H1, 5E3, CC-3, W1, H1B2, and B1C8 (Compton *et al.,* 1991; He *et al.,* 1991; Kallajoki *et al.,* 1991; Maekawa *et al.,* 1991; Thibodeau and Vincent, 1991; Todorov *et al.,* 1992; Tousson *et al.,* 1991; Nickerson *et al.,* 1992; Tang *et al.,* 1993; Wan *et al.,* 1994). cDNA analysis has indicated that centrophilin, SPN, SP-H, and the antigens recognized by 1F1/1H1, and W1 all represent the same protein now uniformly termed NuMA (Compton *et al.,* 1992; Kallajoki *et al.,* 1993; Maekawa and Kuriyama, 1993; Tang *et al.,* 1993), while the others are proteins distinct from NuMA. The following sections review the properties and functions of NuMA in proliferating cells.

### A. Structural Properties of NuMA Protein

NuMA is a common protein of 200–240 kDa in mammalian cells as shown by a variety of probes (Price and Pettijohn, 1986; Kallajoki *et al.,* 1991; Tang *et al.,* 1993). Most of the monoclonal anti-NuMA antibodies raised against human cell extracts are specific for primate cells or react with only a few other mammalian cell lines (Kallajoki *et al.,* 1991; Tousson *et al.,* 1991; Compton *et al.,* 1992), suggesting significant sequence diversity in the NuMA protein among species. Although NuMA is expected to also exist in lower species, NuMA protein has not been reported in *Xenopus, Drosophila,* or sea urchin, probably due to the lack of suitable probes. In K562 erythroleukemic cells, NuMA is measured at a concentration of $\sim 2 \times 10^5$ copies of protein per cell (Compton *et al.,* 1992). The human gene for NuMA has been localized to chromosome 11 at band q13 by *in situ* hybridization (Sparks *et al.,* 1993). cDNA clones of human NuMA have been obtained and sequenced (Compton *et al.,* 1992; Yang *et al.,* 1992; Tang *et al.,* 1993).

A unique feature of the primary structure of NuMA protein is its unusually long central $\alpha$-helical region of approximately 1500 amino acids flanked by globular terminal domains (Fig. 1a). The $\alpha$-helix contains seven segments of heptad periodicity with predominantly hydrophobic amino acids in positions **a** and **d** of the heptad (Compton *et al.,* 1992; Yang *et al.,* 1992). These hydrophobic residues could potentially generate a spine along one side of the helix that is capable of wrapping around the hydrophobic spine of another helix to form a coiled-coil rod (Cohen and Parry, 1986, 1990). In addition, the central domain has sequences similar to other structural pro-

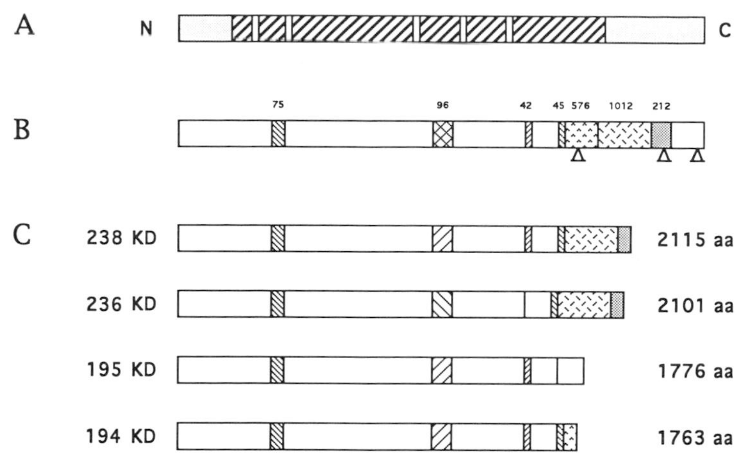


FIG. 1 Structure and isoforms of NuMA protein. A, The predicted secondary structure of NuMA. Interruptions of the helix by pairs of proline residues are indicated. B, Identified sequence blocks involved in the alternative splicing of NuMA pre-mRNA. Numbers above each box indicate the size of sequence block. C, NuMA isoforms identified by cDNA and RT-PCR analysis.

teins that have been shown or predicted to be coiled-coils including myosins, cytokeratins, and nuclear lamins (Yang *et al.*, 1992; Tang *et al.*, 1993). Therefore, the primary structure of NuMA predicts that it is organized into a coiled-coil structure through the interactions along the hydrophobic faces of two or more NuMA proteins. Purified recombinant NuMA protein has been observed to form a starlike oligomer structure (Compton and Cleveland, 1993b). Whether this nonlinear shape represents a natural assembly for NuMA is not known; however, it suggests a considerable oligomerization potential of NuMA protein.

No functional domains have been identified in the central helix of NuMA other than leucine zipper sequences (Maekawa and Kuriyama, 1993; Tang *et al.*, 1993). In contrast, several lines of evidence suggest that the carboxyl-terminal region of NuMA protein contains important domains responsible for the distribution and functions of NuMA during the cell cycle. First, the C-terminus is essential for the localization of NuMA in the interphase nucleus and mitotic spindle poles. Although no known nuclear localization sequence or microtubule (MT) binding sites are found in NuMA cDNA (Compton *et al.*, 1992; Yang *et al.*, 1992), expressed truncated NuMA peptides lacking the C-terminus fail to import to the nucleus or bind to the spindle poles (Compton and Cleveland, 1993a; Tang *et al.*, 1994; see also Section II,B). Second, NuMA appears to be a target for several cell cycle-related protein kinases and most of the phosphorylation sites are found in the C-terminal domain. A number of consensus kinase recognition sequences are present in NuMA, including sites for cAMP-dependent kinase (×2), PKC (×2), CDC2 (×4), and $Ca^{2+}$/calmodulin kinases (×16 sites) (Yang *et al.,* 1992), indicating the potential for complicated regulation of NuMA protein during the cell cycle. A recent study involving site-directed mutagenesis of NuMA at the CDC2 binding domains has directly demonstrated that at least the phosphorylation on one site by CDC2 kinase is required for NuMA's role in the organization of the mitotic spindle (Compton and Luo, 1995). Finally, the C-terminal domain of NuMA also contains multiple alternative splicing sites responsible for the generation of isoforms of NuMA protein.

Recently, detailed analysis of cDNA clones and the 3′ ends of transcripts obtained from RT–PCR assays indicated that human mRNAs of NuMA possess at least four forms (Maekawa and Kuriyama, 1993; Tang *et al.,* 1993). By examining human cDNA clones from various expression libraries, Tang and co-workers (1993) identified six sequence blocks present in some cDNA fragments, but not in others (Fig. 1b). Two alternatively included exons of 42 and 75 bp reside in the central $\alpha$-helical rod. Four more sequence blocks of 45, 576, 1012, and 212 bp are contiguously located in the carboxyl-terminal domain, and each may be composed of more than one exon. Three translation termination codons are identified in the C-terminal region suggesting that complicated differential splicing is used to generate NuMA proteins with different carboxyl termini. Direct evidence for alternative splicing of NuMA comes from the analysis of two full-length NuMA sequences with 2115 and 2101 amino acid (aa) residues, respectively, in which one of the cDNAs contains the 42-bp exon in $\alpha$-helical region but the other does not (Compton *et al.,* 1992; Yang *et al.,* 1992). However, the total number of NuMA isoforms and the distribution of the alternative sequence blocks is not yet known. At least for the present, a form with the spliced 75-bp block has only been found in a cDNA fragment but not the full-length gene product. RT-PCR analysis of human glioma mRNA has revealed three types of 3′ end variants of NuMA with deduced protein length of 2115, 1776, and 1763 residues (assuming exon inclusion of both 75- and 42-bp differential sequence blocks in the central rod) and calculated molecular masses of 238, 195, and 194 kDa, respectively (Fig. 1c) (Tang *et al.,* 1993). Therefore, although sequences of NuMA isoforms are incomplete, available data suggest a minimum of four possible proteins of 2115, 2101, 1776, and 1763 amino acids. On Western blots using NuMA antibodies, a doublet at 220–230 kDa and a second band at 200 kDa are observed (Kallajoki *et al.,* 1991; Maekawa *et al.,* 1991; Tang *et al.,* 1993; Zeng *et al.,* 1994b). It seems likely that the upper band corresponds to the 2115- and/or 2101-aa isoforms and that the lower band corresponds to the 1776- and 1763-aa isoforms. Furthermore, NuMA also appears to contain mutually exclusive exons of the same size (one of a pair of exons is always included in the mRNA, but never both) (C. W. Smith *et al.,* 1989). Within the $\alpha$-helical region (amino acids 1268–1299), two 96-bp segments of different sequence but with the same flanking sequence have been reported (Compton *et al.,* 1992; Yang *et al.,* 1992; Maekawa and Kuriyama, 1993), indicating the mutually exclusive selection of 96-bp exons during splicing. Interestingly, mutually exclusive alternative splicing is common in other structural proteins, including both $\alpha$ and $\beta$ tropomyosin (Helfman *et al.,* 1986; Ruiz-Opazo and Nadal-Ginard, 1987), troponin-T (Medford *et al.,* 1984; Breitbart and Nadal-Ginard, 1987), and myosin light chain (Periasamy *et al.,* 1984; Strehler *et al.,* 1985). The usual interpretation of such splicing is that it permits different forms of the protein to evolve for different purposes. Mechanisms and factors involved in polymorphic splicing patterns of NuMA pre-mRNA remain unknown.

Much of the variation in NuMA cDNA sequences has been detected using human cDNA libraries produced from different established cell lines, implicating possible tissue specificity in alternative splicing of NuMA. However, three isoforms of NuMA with different carboxyl termini have been obtained from a single cell line (Tang *et al.,* 1993). Moreover, immunoblotting patterns of cell extracts suggest that different isoforms may also exist in HeLa and CHO cells (Kallajoki *et al.,* 1991; Maekawa *et al.,* 1991; Zeng *et al.,* 1994b). Therefore, multiple NuMA proteins could indeed exist within one cell type, suggesting multiple functions or control of functions for NuMA within a single cell.

Isoforms of NuMA have been named NuMA-1, NuMA-m, NuMA-s, referring to large, medium, and small NuMAs of 238, 195, and 194 kDa, and NuMA-x where the 42-bp exon has been excluded (Tang *et al.,* 1994). Since additional alternatively spliced products may yet be discovered, we suggest that a different nomenclature be adopted to include $\alpha$ to describe the large isoforms around 230 kDa, and $\beta$, for isoforms around 200 kDa. Thus, NuMA $\alpha 1$ represents the NuMA with 2115 amino acids (Yang *et al.,* 1992), NuMA $\alpha 2$ refers to the isoform of 2101 amino acids (42-bp exon exclusion) (Compton *et al.,* 1992), and $\beta 1$ and $\beta 2$ would represent the 194- and 195-kDa proteins, respectively (Tang *et al.,* 1993). Known isoforms and their respective sizes are summarized in Table I.

The patterns of alternative splicing observed for NuMA generate a variety of NuMA isoforms, each of which shares extensive regions of homology, but vary from each other in specific domains, thus allowing for the fine modulation of NuMA functions. Delineation of cellular distribution and possible roles played by each isoform await further characterization. Notably, the globular C-terminal tail of NuMA contains the most intensive diversity due to extensive alternative splicing and phosphorylation. Such a

TABLE I
Isoforms of NuMA Protein

| Suggested name | Original name | Amino acid | Deduced $M_r$ (kDa) | Apparent $M_r$ (kDa) | Reference |
|---|---|---|---|---|---|
| $\alpha1$ | NuMA-l | 2115 | 238 | 220–230 | Yang *et al.* (1992) |
| $\alpha2$ | NuMA-lx | 2101 | 236 | 220–230 | Compton *et al.* (1992) |
| $\beta1$ | NuMA-m | 1776 | 195 | ~200 | Tang *et al.* (1993) |
| $\beta2$ | NuMA-s | 1763 | 194 | ~200 | Tang *et al.* (1993) |

high level of structural polymorphism probably accounts for reports of diverse function for NuMA in eukaryotic cells.

In summary, NuMA is an unusually long coiled-coil protein, sharing certain structural similarities to several cytoskeletal proteins. Extensive post-transcriptional and post-translational modifications, especially in the carboxyl-terminal region, generate multiple forms of NuMA protein that potentially could provide different functions. Currently, four isoforms of NuMA have been identified, two of approximately 230 kDa (we suggest denoting these isoforms as $\alpha1$ and $\alpha2$) and two of approximately 200 kDa (we suggest denoting these as $\beta1$ and $\beta2$).

## B. The Distribution of NuMA in Proliferating Cells

Two nuclear distribution patterns for NuMA during interphase have been identified with different antibodies, in contrast to the consistent polar localization of NuMA in mitotic cells. Most anti-NuMA antibodies produce either a diffuse or uniformly punctate pattern in the interphase nucleus as shown by immunofluorescence (van Ness and Pettijohn, 1983; Kallajoki *et al.,* 1991; Maekawa *et al.,* 1991; Tousson *et al.,* 1991; Compton *et al.,* 1992; Yang *et al.,* 1992). Some antibodies, however, including the original NuMA antiserum (Lydersen and Pettijohn, 1980) and the monoclonal antibodies 8.22 (Zeng *et al.,* 1994a) and W1 (Tang *et al.,* 1993, 1994), have revealed a pattern of mainly dots and speckles along with light diffuse staining in the nucleus (Fig. 2). Some NuMA antibodies stain the nucleus uniformly when examined by conventional epifluorescence, but reveal discrete spots when visualized by laser scanning confocal microscopy (Compton *et al.,* 1992). The actual molecular basis for the different nuclear staining patterns is not clear. Most likely it is due to the recognition of differentially spliced or phosphorylated NuMA proteins by various antibodies. In addition, other types of post-translational modifications, or the associations of NuMA with

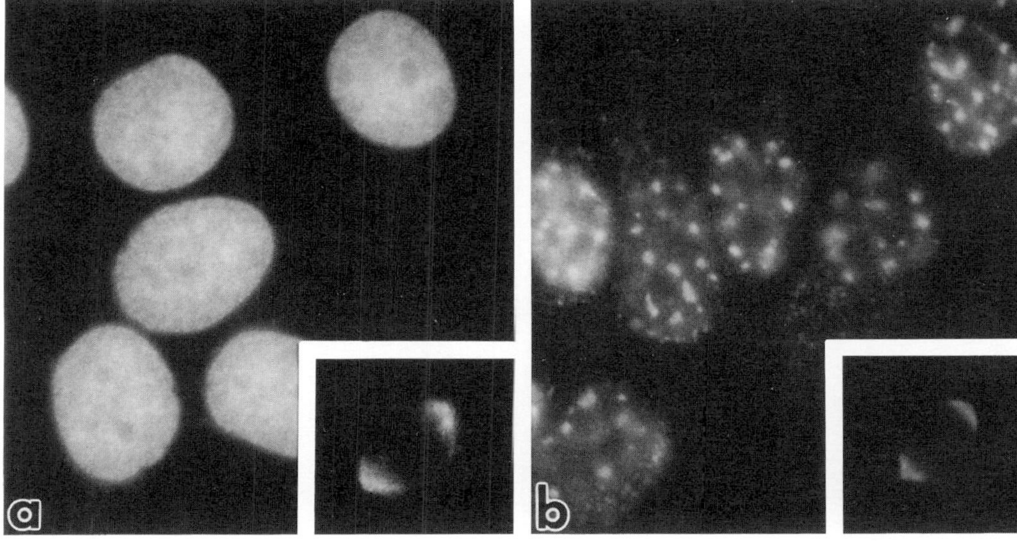


FIG. 2 Staining patterns of NuMA antibodies. a, Diffuse nuclear staining by mAb 2D3. b, Speckled nuclear staining by mAb 8.22. Inserts, staining of mitotic cells with the same antibody as in a or b.

different nuclear components within discrete domains, could contribute to different staining patterns.

A recent study of the expression of human NuMA cDNAs in hamster cells has shown that only the large size of NuMA ($\alpha$ type, 230 kDa) was able to locate to the nucleus. Two smaller isoforms ($\beta$1 and $\beta$2, 200 kDa) expressed in CHOP cells showed localization to the cytoplasm, mainly around the centrosomal region after 6 hr of transfection, although they did locate to the mitotic spindle poles along with endogenous hamster NuMA. Analysis of the C-terminus of NuMA has revealed a sequence motif, required for the entrance to the nucleus, which is absent in smaller isoforms. This sequence feature probably accounts for the cytoplasmic accumulation of small NuMA proteins (Tang *et al.,* 1994). Whether the natural smaller NuMA isoforms distribute in the same fashion as the expressed ones by transfected cDNAs awaits further identification with more specific probes. Interestingly, in the nucleoskeleton of human cells, two isoforms of NuMA have been identified that are most likely the large and small types ($\alpha$ and $\beta$) of NuMA based on their Western blotting patterns (Zeng *et al.,* 1994b). Thus far, studies involving immunofluorescent analysis of normal cells have all indicated that NuMA proteins are localized exclusively in the nucleus. Therefore, the ratio and role for a cytoplasmic form of NuMA remain unclear.

Descriptions of the localization of NuMA at different stages of mitosis vary slightly among individual studies and have resulted in different postulated functions for NuMA during mitosis (Price and Pettijohn, 1986; Kallajoki *et al.,* 1991; Tousson *et al.,* 1991; Compton *et al.,* 1992). Such variation may be attributed to the differences among cell lines and fixation protocols as well as the specificity of utilized antibodies. In general, translocation of NuMA at mitosis begins with the aggregation of NuMA within the interchromosomal space in early prophase. Progressively, NuMA transfers to the two centrosomes prior to the formation of spindle poles. The process of aggregation and migration of NuMA coincides with chromosome condensation and the disassembly of the nuclear lamina. Although pole-derived spindle fibers are occasionally stained with the NuMA antibodies at metaphase (Compton *et al.,* 1992), the majority of NuMA is localized at the spindle poles in a crescent shape. The association of NuMA with the spindle poles is of sufficient strength that it is resistant to extraction in 2 *M* salt (Zeng *et al.,* 1994b). Under certain conditions, NuMA has also been detected in the centromere/kinetochore regions by immunofluorescence (Compton *et al.,* 1991; Tousson *et al.,* 1991). Since this pattern was most often observed in either isolated chromosomes or in mitotically arrested cells, it may be due to the redistribution of NuMA from the pericentriolar region (Compton *et al.,* 1992).

NuMA gradually diminishes from the mitotic poles in anaphase and begins to reappear in the daughter nuclei regions at late telophase. Several studies have indicated that the carboxyl terminus is important for the transition of NuMA distribution between the mitotic and interphase state. A truncated large NuMA lacking its C-terminal domain failed to either associate with the spindle poles at metaphase or relocate in the daughter nuclei after mitosis (Compton and Cleveland, 1993a). Similarly, expression of a chimeric polypeptide of $\beta$-gal and the entire C-terminus of large NuMA in hamster cells demonstrated localization of the fusion protein in interphase nuclei and mitotic spindle poles. Detailed analysis with a series of fusion peptides indicated that amino acids 1975–2115 of $\alpha$-NuMA are imperative for nuclear localization, and the entire C-terminus is required for spindle association (Tang *et al.,* 1994). A dramatic conformational change possibly occurs to bring about the dissociation of NuMA with the mitotic poles and its relocation in the nucleus. This may elucidate why most antibodies fail to detect the majority of NuMA at late anaphase through early telophase.

Whether NuMA relocates to daughter nuclei before or after nuclear envelope assembly remains disputed. Microinjection of rhodamine-conjugated wheat germ agglutinin (R-WGA) into the prometaphase cell has been shown to block the relocation of NuMA into the nuclei after 2 hr (Compton *et al.*, 1992). Since WGA can impair nuclear pore permeability and block the later steps of nuclear assembly (Finlay *et al.,* 1987; Yoneda *et al.,* 1987; Newmeyer and Forbes, 1988), the cytoplasmic accumulation of NuMA in this experiment seems to suggest that the postmitotic nuclear import of NuMA may occur via transit through the nuclear pores (Compton *et al.,* 1992). This proposal, however, is inconsistent with the majority of immunofluorescence results in which NuMA commences to locate to chromosome regions that are still condensed in telophase prior to the complete formation of daughter nuclei (Price and Pettijohn, 1986; Tousson *et al.,* 1991; Yang *et al.,* 1992). Particularly, double-label immunofluorescence with anti-NuMA and anti-lamin antibodies has shown that NuMA associates directly with condensed chromosomes earlier than the association of lamins with chromosomes (Yang *et al.,* 1992). Considering the 2-hr incubations used in the study, the consequent cytoplasmic accumulation of NuMA by WGA may represent newly synthesized NuMA protein in the cytoplasm of daughter cells. Nevertheless, the half-life and metabolism of NuMA remain uncertain at the present time, and how newly synthesized NuMA enters the nucleus requires additional study.

### C. Mitotic Function of NuMA

During mitosis, at least two functions for NuMA have been proposed. In light of its localization at the minus ends of MTs in the pericentriolar regions, NuMA may be involved in spindle MT assembly (Kallajoki *et al.*, 1991; Maekawa *et al.*, 1991; Tousson *et al.*, 1991). Alternatively, NuMA may play a role in nuclear reassembly at the end of mitosis (Price and Pettijohn, 1986; Compton *et al.*, 1992).

### 1. NuMA and Spindle Organization

***a. Is NuMA a Specific MAP in the Mitotic MT Organization Centers?*** Several studies involving the processes of disassembly and reassembly of spindle MTs have revealed an intimate association of NuMA with the minus ends of mitotic MTs (Kallajoki *et al.*, 1991, 1992; Maekawa *et al.*, 1991; Maekawa and Kuriyama, 1991; Tousson *et al.*, 1991). Treatments with MT inhibitors result in the dissolution of both spindle MTs and immunovisualized polar crescents of NuMA into numerous discrete foci containing both tubulin and NuMA dispersed throughout the cytoplasm. After removal of disassembly agents, NuMA invariably becomes a component of multiple small MT organizing centers (MTOCs) in the cytoplasm. Nascent MTs emanate radially from these centers. Multiple NuMA containing foci gradually consolidate into polar crescents, and MT asters elongate and converge to form a bipolar spindle.

Strong evidence showing a key role of NuMA as a component of MTOCs in spindle organization has been obtained from Taxol-treated cells (Kallajoki *et al.*, 1992). Taxol induces MTs of mitotic PtK2 cells to form short bundles or asters that relocate at the cell periphery. Microinjecting of NuMA antibody has been shown to result in the disassembly of these MT bundles and asters and the formation of disorganized MT arrays. In addition, Taxol treatment also disperses NuMA foci that are associated with the minus end of MT bundles, as shown by the method of tubulin hook decoration (Maekawa *et al.*, 1991). Other proteins, including a constituent centrosomal component 5051 and a kinesin-like MT motor protein, CHO1, have also been detected in the pericentriolar regions (Calarco-Gillam *et al.*, 1983; Clayton *et al.*, 1985; Kuriyama and Sellitto, 1989). Interestingly, 5051 protein is not dispersed from the centrosome by Taxol treatment, and CHO1 protein distributes as small cytoplasmic spots that fail to colocalize with the MT foci (Maekawa and Kuriyama, 1991). Thus, NuMA appears to be one of perhaps a few proteins in mitotic cells that invariably localizes at multiple MT nucleation centers during MT regrowth following mitotic arrest. Indeed, phosphorylation analysis has directly shown that mutagenesis of the CDC2 binding sites of NuMA abolishes its association with the spindle pole and disorganizes the spindle, thereby blocking mitosis (Compton and Luo, 1995). In light of its coiled-coil structure, NuMA may form oligomers or short filaments around centrosomes serving as a structural lattice to help initiate and/or stabilize spindle organization, or to support sliding of MTs (Fig. 3, color plate 1) (Yang and Snyder, 1992). Indirect evidence from studies of the nuclear matrix suggests a higher order of NuMA at the polar region where the intensity of polar NuMA staining in mitotic cells remains as strong as the nucleoskeleton staining of NuMA in the nucleus after extensive cellular extraction (Zeng *et al.*, 1994b).

A major argument against NuMA being a mitotic microtubule-associated protein (MAP) is the fact that its sequence contains no known MT binding motifs (Compton *et al.*, 1992; Yang *et al.*, 1992), such as those identified in MAP1B (Noble *et al.*, 1989), MAP2 (Lewis *et al.*, 1988), and tau (Himmler *et al.*, 1989). It should be noted that these MAPs bind stoichiometrically to the wall of MTs (Borisy *et al.*, 1975; Berkowitz *et al.*, 1977; Scheele and Borisy, 1979). Due to its specific minus-end association, NuMA could possibly bind to MTs via a different mechanism than that used by conventionally defined MAPs. Although *in vivo* studies remain to be attempted, MT cosedimentation data indicate that NuMA extracted from mitotic cells binds to Taxol-stabilized brain MT assemblies (Maekawa *et al.*, 1991; Kallajoki *et al.*, 1992). In addition, the remarkable crescent shape of NuMA staining at the spindle poles strongly suggests that the majority of NuMA molecules in the spindle have high affinity only to the minus ends of kinetochore MTs, but rarely to the ends of astral MTs. Therefore, being a mitotically transient pericentriolar material, NuMA appears to function preferentially in the nucleation and/or maintenance of chromosome-associated spindle fibers.

***b. Effects of Injected NuMA Antibodies between Species*** Microinjections of certain purified NuMA antibodies into proliferating cells have strongly supported a key role of NuMA during mitosis. The effects of injected antibodies on cells, however, are surprisingly different among cell lines from different species. Using interphase cells, ~50% of injected HeLa cells showed mitotic arrest in prometaphase after 24 hr, whereas ~70% of injected PtK2 rat kangaroo cells formed micronuclei (Kallajoki *et al.*, 1991, 1993). Injection of antibodies into mitotic cells from primates and marsupials also induced distinguishable interruption patterns on the spindle function. Approximately 80% of African green monkey (CV-1) or HeLa cells displayed a prometaphase-like arrest with disrupted spindles if antibodies were introduced into cells from prophase to metaphase (Kallajoki *et al.*, 1991; Yang and Snyder, 1992). In contrast, injections of PtK2 cells at the same mitotic stages appeared not to interfere with progression through anaphase but caused the formation of micronuclei in ~80% of the daughter cells (Kallajoki *et al.*, 1993). A strikingly consistent observation was that antibodies interfered with the mitotic process only if introduced prior to the onset of anaphase. Both primate and PtK2 cells injected during anaphase were able to successfully complete mitosis and yield daughter cells with apparently normal nuclei.

Detailed time course analysis on cells injected at metaphase further revealed that different mechanisms may be involved in the antibody's action between species. In primate cells, antibodies initially caused the collapse of the mitotic apparatus with spindle fibers becoming disheveled. Over time, MTs gradually merged to form an abnormal single microtubule array and NuMA became localized at the putative centrosome region (Yang and Snyder, 1992). Surprisingly, such a dramatic microtubule disorganization and rearrangement failed to occur when PtK2 cells were injected. These cells displayed a morphologically normal spindle, followed by midbody formation and the appearance of a cytoplasmic MT complex in micronucleated interphase cells (Kallajoki *et al.,* 1993). Collectively, microinjection of NuMA antibodies into primate cells results in disruption of the spindle structure. Thus, in these cells, NuMA appears to be most important in the establishment and stabilization of the mitotic spindle as already discussed. In marsupial cells, injected antibodies appear to interfere with an unidentified mitotic event that ultimately leads to micronucleation. Two striking features, however, implicate a distortion of spindle function in PtK2 cells. First, micronucleation itself usually suggests a defect in the mechanism of chromosome segregation. Second, the antibody effect is mitotic stage-dependent and fails to induce micronucleation if injected after the onset of anaphase (Yang and Snyder, 1992; Kallajoki *et al.,* 1993). In addition, despite the apparent normal morphology of spindles in injected PtK2 cells, no evidence has shown that the kinetics of the mitotic apparatus remain the same as in the untreated cells. Thus, despite the absence of visible mitotic arrest in injected PtK2 cells, the interrupted function of NuMA by antibodies still appears to involve the earlier stages of mitosis rather than the later process of nuclear reformation.

Interestingly, the mitotic defects induced by injections of NuMA antibodies are surprisingly parallel to the well-known spectrum of responses of mammalian cell lines to MT inhibitors (Brinkley *et al.,* 1967; Jensen *et al.,* 1987; Kung *et al.,* 1990; Rieder and Palazzo, 1992). For instance, human cells such as HeLa S3 cells, representing primate types, remain arrested in mitosis by Colcemid throughout the course of treatment up to 72 hr or until cell detachment, whereas Chinese hamster ovary cells, characteristic of rodent types, are only transiently blocked and consequently resume an interphase morphology in the presence of Colcemid or related mitotic inhibitors. The fate of cells that escape mitotic block without chromosome segregation is to become micronucleated (Kung *et al.,* 1990; Rieder and Palazzo, 1992). Indeed a striking difference has also been observed between PtK2 and HeLa cells after Colcemid and Taxol treatment (Kallajoki *et al.,* 1993). Apparently marsupials are more like rodents in their response to

MT inhibitors. Comparing the phenotypes between anti-NuMA antibody injection and MT inhibition, there is considerable similarity in the outcomes of the two events. Both treatments lead to a high incidence of mitotic arrest in primate cells, and an alternatively high incidence of micronuclei formation in marsupial cells (Yang and Snyder, 1992; Kallajoki *et al.,* 1993). Therefore, the inability of PtK2 cells to be arrested by anti-NuMA antibody injection possibly relates to a similar process occurring in MT inhibition. Such a correlation hints that NuMA antibody-induced micronuclei formation may involve some aspects of MT perturbation, rather than nuclear assembly.

The mechanism responsible for the escape from mitotic arrest is not clear. However, it appears to be more complicated than merely the disturbance of the spindle alone, perhaps involving a mitotic checkpoint. The disruption of spindle assembly in HeLa cells has been shown to correlate with abnormal maintenance of elevated cyclin B levels and continued activation of $p34^{cdc2}$, whereas the oscillation of cyclin B level and $p34^{cdc2}$ activity is only transiently inhibited in CHO cells (Kung *et al.,* 1990). A checkpoint control serving to couple cell cycle progression to other cellular events has been proposed (Murray and Kirschner, 1989). The stringency of the control mechanism may vary among cell species, thereby causing different cellular responses to the mitotic block (Kallajoki *et al.,* 1993).

In summary, NuMA appears to be a minus-end specific MAP as shown by MT inhibition and MT cosedimentation. Despite the different cellular effects induced by antibody microinjections, NuMA is a component of mitotic MTOC and essential for spindle function. The mechanism responsible for the different phenotypes of antibody injection awaits further characterization. However, it appears to be correlated with species-dependent cellular responses to MT inhibition, a complex process that is yet poorly understood. Considering its coiled-coil structure and lack of an ATP binding site, NuMA is unlikely to serve as an MT motor but could play a structural role in nucleation, stabilization, or supporting of spindle MTs.

### 2. NuMA and Postmitotic Reassembly

In view of its cell cycle-dependent localization and association with the nuclear matrix (Lydersen and Pettijohn, 1980), NuMA has been proposed to function in postmitotic nuclear assembly (Price and Pettijohn, 1986; Compton *et al.,* 1992). Supporting evidence for this hypothesis comes from experiments involving *in vivo* overexpression of human wild-type and mutant NuMA cDNAs in hamster cells by introduction of constructed plasmids carrying various coding sequences of NuMA lacking the amino-terminal head or the carboxyl-terminal tail. Regardless of the different cellular localizations of the truncated NuMAs, the terminal phenotypes were similar.

Both headless and tailless forms induced the formation of micronuclei without distortion of mitotic spindles (Compton and Cleveland, 1993a).

A temperature-sensitive hamster cell line tsBN2 (Nishimoto *et al.,* 1978) shares certain similarities to the mitotic phenotype induced by the expression of truncated NuMA. Shifting to the restrictive temperature leads to premature entry of tsBN2 cells into mitosis and later formation of micronuclei. A missense mutation in the RCC1 gene that encodes a highly conserved chromatin binding protein is responsible for the temperature-sensitive phenotype (Kai *et al.,* 1986; Uchida *et al.,* 1990; Nishitani *et al.,* 1991). The endogenous NuMA in tsBN2 associates with spindle poles in the abnormal mitosis at the restrictive temperature but fails to reenter interphase micronuclei. Expression of wild-type human NuMA suppressed the postmitotic micronucleation in tsBN2 cells without affecting the RCC1-dependent phenotype of premature entry into mitosis (Compton and Cleveland, 1993a).

A function of NuMA in terminal phases of chromosome separation and/or nuclear reassembly has been suggested based on two facts of *in vivo* overexpression. First, the expression of NuMA mutants in normal BHK cells induced micronucleation without disruption of the mitotic apparatus. Second, the expression of wild-type NuMA can suppress the abnormal micronucleation in the mutant tsBN2 cells without interfering with the original phenotype of premature mitosis (Compton and Cleveland, 1993a). Since both the normal and mutant host cells are rodent, however, the micronucleation induced by truncated NuMAs or restrictive temperature may also be a species-specific phenotype related to the escape of rodent cells from mitotic block as already noted. In particular, micronucleation strongly suggests an abnormal chromosome segregation. The dislocation of endogenous NuMA to the nucleus in tsBN2 cells may be attributed to the failure of chromosomes to reach the poles. Thus, the introduced NuMAs appear to be involved in chromosome segregation that occurs prior to nuclear formation. Moreover, exogenous NuMAs in these experiments were expressed at interphase due to the way the microinjection or transfection was carried out (Compton and Cleveland, 1993a), making it impossible to distinguish the stage of mitosis (e.g., prior to or after onset of anaphase), at which the expressed NuMA proteins commenced their function and competed or interfered with endogenous NuMA. Consequently, of the two hypotheses regarding NuMA's role in either spindle stability or in postmitotic reassembly, the *in vivo* overexpression data appear to be consistent with both but may actually favor the former.

In summary, most experimental data involving mitotic cells widely support the hypothesis that NuMA is a mitosis-specific, minus-end MAP that plays a structural role in the organization and function of spindles. The antibody injection and *in vivo* expression experiments also argue that

NuMA is essential in early stages of mitosis related to chromosome segregation. Moreover, although the data are somewhat more indirect, evidence also implies that NuMA may be involved in postmitotic reassembly of daughter nuclei.

## D. The Structural Role of NuMA in Interphase

During the longest portion of the cell cycle, interphase, NuMA resides within the nucleus, implying a nuclear function of NuMA protein. Moreover, different nuclear staining patterns have been seen with various antibodies, suggesting a complex function for NuMA or NuMA forms in interphase. Such complexity may be due to expression of various isoforms and/or other regulatory pathways such as phosphorylation. The localization of NuMA in the nucleoskeleton and associated RNA splicing apparatus (Zeng *et al.,* 1994a,b) argues for a more global role for NuMA in nuclear organization and function. Unfortunately, most studies of NuMA have been confined to mitotic cells. Obviously, considerably more information is needed before the role of NuMA in the interphase nucleus can be fully understood.

### 1. Localization of NuMA in Nucleoskeletal Core Filaments

NuMA was first identified by an antibody raised against a fraction of the nuclear matrix (Lydersen and Pettijohn, 1980), a network that remains after removal of chromatin by nuclease treatment and salt extraction (Berezney and Coffey, 1974, 1977; Nickerson *et al.,* 1990; Berezney, 1991; van Driel *et al.,* 1991). Electron microscopy of sequentially extracted cell nuclei has provided excellent morphological support for the presence of a filamentous scaffold associated with granules and amorphous materials (Fey *et al.,* 1986; Jackson and Cook, 1988). A 9- to 13-nm, fibrous network consisting of highly branched RNA containing core filaments has been identified following gradient extraction of nuclei (He *et al.,* 1990). This intermediate filament-like system may represent a true nucleoskeleton that serves as a scaffold to which other proteins associate to form the complete nuclear matrix. It has been shown by immunofluorescence that the nuclear staining of NuMA is resistant to DNA digestion and salt extraction up to 2 *M* (Kallajoki *et al.,* 1991; Zeng *et al.,* 1994a), indicating coexistence of NuMA and nuclear matrix as defined phenomenologically. Moreover, NuMA has been characterized as a stable component of a subclass of nuclear core filaments by immunogold-labeled electron microscopy using the monoclonal antibody, 2D3 (Zeng *et al.,* 1994b). In addition, immunoblots of fractions of core filament preparations have indicated that isoforms of NuMA contribute to the formation of core filaments (Zeng *et al.,* 1994b).

Unlike the cytoskeleton, nuclear core filaments are composed of heterogenous proteins and hnRNAs involved in various nuclear activities (Fey *et al.,* 1986; Nickerson *et al.,* 1989; He *et al.,* 1990). The molecular definition of core filaments is yet incomplete and thus far only a few associated proteins have been identified (He *et al.,* 1991; Nickerson *et al.,* 1992). The immunogold localization of NuMA provides initial evidence for a specific protein component in a subset of fibers in this nucleoskeletal system. The structural similarities of NuMA to lamins and intermediate filaments (Yang *et al.,* 1992) also strongly favor the identification of NuMA as a component of the nucleoskeleton. In this regard, core filaments share several features with intermediate filaments including size, morphology, and solubility properties (He *et al.,* 1991). In light of its long coiled-coil configuration with hydrophobic heptad repeats, NuMA could very well form heterooligomers or polymers as subunits of a portion of nuclear core filaments (Zeng *et al.,* 1994b).

## 2. Association of NuMA with the Spliceosomes

RNA splicing takes place in spliceosomes that are macromolecular complexes consisting of pre-mRNA, small nuclear ribonucleoprotein particles (snRNPs), and other splicing factors (Krainer and Maniatis, 1988; Bindereif and Green, 1990; Smith, 1992; Moore *et al.,* 1993). The snRNPs consist of various species of U snRNAs, respectively, and common or unique polypeptides binding to each snRNA (Maniatis and Reed, 1987; Lührmann *et al.,* 1990). At the cellular level, the spliceosome components are specifically recognized by anti-snRNP antibodies as dots and speckles throughout the nucleus (Reuter *et al.,* 1984, 1990; Spector, 1984, 1990; Nyman *et al.,* 1986). Three NuMA antibodies have been shown to produce a speckled pattern of staining in interphase nuclei (Lydersen and Pettijohn, 1980; Tang *et al.,* 1993; Zeng *et al.,* 1994a) similar to the pattern produced with anti-snRNP antibodies. The colocalization of NuMA antigens and snRNPs within speckles has been confirmed by double-label immunofluorescence with a snRNP autoantibody ($\alpha$Sm) and one anti-NuMA antibody, mAB 8.22 (Zeng *et al.,* 1994a). The colocalization persists even in nuclear core filament preparations, or after RNase treatment that deforms the speckles to amorphous patches. More direct evidence for the association of NuMA with snRNPs has been obtained by coimmunoprecipitation of NuMA isoforms with anti-snRNP antibodies. Moreover, NuMA protein appears to associate with active spliceosomes based on the ability of anti-NuMA antibodies to immunoprecipitate spliceosomes assembled on an exogenous substrate pre-mRNA in *in vitro* splicing extracts (Zeng *et al.,* 1994a). Spliceosomes are assembled through a series of specific recognition and association steps between consensus sequences in the RNA precursor and snRNPs and other splicing factors (Lührmann *et al.,* 1990; Moore *et al.,* 1993).

Coprecipitation results suggest that one function of NuMA may be to associate with pre-mRNAs undergoing processing, at least in the initial stages of formation of spliceosomes. Given its filamentous feature, NuMA most likely plays a structural rather than catalytic role in splicing, similar to its function in mitosis (Zeng *et al.,* 1994a).

Although little is known about how spliceosomes are organized within discrete nuclear domains, different studies have noticed a preferential association of pre-mRNAs, snRNPs, and other splicing factors with the nuclear matrix (Spector *et al.,* 1983; Reuter *et al.,* 1984; Zeitlin *et al.,* 1987; H. C. Smith *et al.,* 1989). Recent microscopic studies have suggested that pre-mRNAs are both transcribed and processed in a discrete nuclear "track" that might correspond to nuclear architecture (Huang and Spector, 1991; Carter *et al.,* 1993; Jiménez-Garcia and Spector, 1993; Xing *et al.,* 1993). As a coiled-coil component of nuclear core filaments, NuMA could form a structural interface between the splicing apparatus and the nucleoskeleton and play a key role in the 3-D organization of splicing factors into nuclear domains (Fig. 3). Obviously, a more detailed analysis of NuMA and its associated components is needed to fully understand how this protein functions in the formation and organization of spliceosomes.

In contrast to the three NuMA antibodies that produce a speckled pattern of interphase nuclear staining, most NuMA antibodies reported in the literature produce a rather uniform fluorescent staining pattern in the nucleus. Therefore, it is likely that only a portion of the NuMA protein, perhaps a type of isoform or specific phosphorylated state of NuMA, is involved in RNA processing. A more common function of NuMA in the interphase nucleus may be to link other nuclear components to the nucleoskeleton. A recent DNA binding study suggested that NuMA, as well as the intermediate filament protein desmin, has a low affinity for the AT-rich MAR (matrix-attached region) sequences (Ludérus *et al.,* 1994). The functional role of NuMA in the nucleus is still in an initial stage of investigation. The way in which NuMA molecules are assembled into nucleoskeletal filaments and how these integrate with other nuclear domains are key questions yet to be answered.

## E. Multiple Functions of NuMA Protein during the Cell Cycle

Significant progress has been made in understanding the structure and functions of the NuMA protein. cDNA analysis has revealed several structural features of NuMA and the heterogeneous composition of the NuMA family. Various evidence has indicated that NuMA plays multiple roles in both mitosis and interphase. The mitotic function of NuMA appears to be mainly involved in the nucleation, stabilization, or supporting of spindle

MTs. Additionally, NuMA may be also involved in the postmitotic reformation of the nucleus. Thus far, studies of NuMA have enhanced our understanding of the mitotic apparatus, but less has been learned about NuMA's role in the interphase nucleus. Its presence as a coiled-coil component of core filaments argues for a structural role in the nucleoskeletal organization. The evidence for the association of NuMA with RNA splicing, in conjunction with its localization on the nucleoskeleton, may provide an initial step toward understanding of the structural integration of spliceosomes with nuclear architecture in eukaryotic cells.

It is perhaps worthwhile to note that NuMA can be classified either as a nucleoskeletal or cytoskeletal protein, depending on the stage of the cell cycle. The dynamic translocation of NuMA from nucleoskeleton in interphase to the cytoskeleton of the mitotic apparatus suggests a unique bifunctional role in two major molecular events in the cell cycle (Fig. 3). Proteins such as NuMA that play multiple roles in cells would seem to have a selective advantage in evolution and may indeed be more common in eukaryotic cells than anticipated. It is likely, therefore, that NuMA is one member of a class of nuclear matrix proteins that resides in both the nucleus and mitotic apparatus, forming a molecular fabric that embraces the genome throughout the life of a cell. As a component of the nuclear core filaments, NuMA may signal the continuity of the nucleoskeleton throughout the cell cycle, even in mitosis when the nuclear envelope is disrupted. Indeed, studies have suggested that a number of nuclear matrix components persist as filamentous structures throughout mitosis (He *et al.,* 1991; Nickerson *et al.,* 1992; Wan *et al.,* 1994). Thus, the mitotic apparatus accentuated by MTs appears to be an essential blend of the nucleoskeleton and cytoskeleton required for the accurate partitioning of the replicated genome. This interesting concept, so elegantly demonstrated by NuMA, will be further documented with other examples in the following sections of this review.

## III. Targeting and Relocation of Nuclear Matrix Proteins to the Mitotic Apparatus

In addition to NuMA, many nuclear matrix-derived proteins leave the nucleus at the time of nuclear envelope dissolution and make their way to various zones within the mitotic apparatus. Although little is known about the mechanism of how these proteins become targeted to various spindle sites, most end up in one of several zones including the chromosome scaffold, perichromosome region, centromere/kinetochore region, midzone, midbody, and spindle poles (Fig. 4).

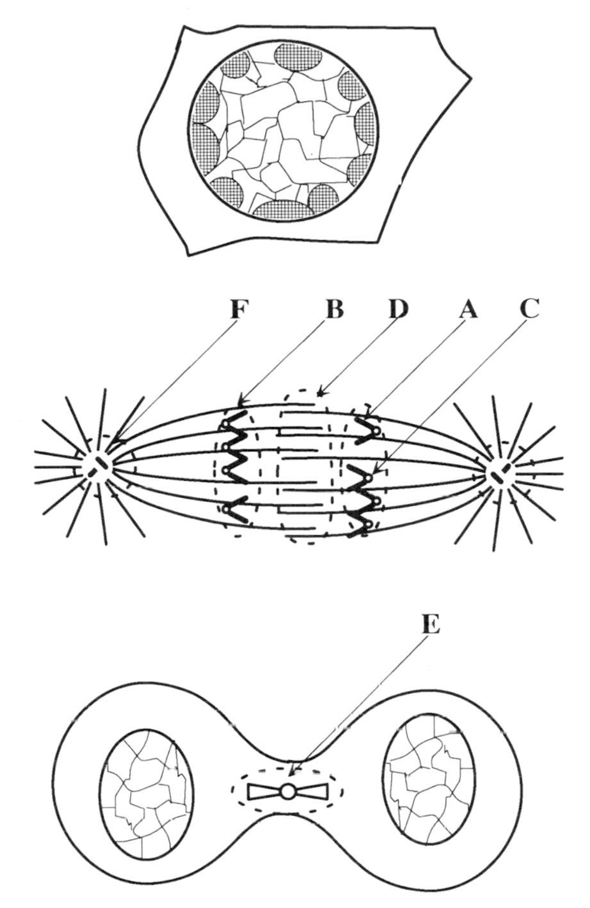


FIG. 4 Diagram showing zones of nuclear matrix proteins. During prophase (top diagram), chromosome scaffold forms adjacent to the nuclear envelope. In anaphase (middle diagram), nuclear matrix-derived proteins occupy one or more specific zones of the mitotic apparatus including, A, the chromosome scaffold, B, perichromosome region, C, centromere region, D, spindle midzone and F, spindle pole. During late telophase (bottom diagram), the midbody, E, can be seen in a cytoplasmic bridge connecting two daughter cells.

## A. Chromosome Scaffold

The chromosome scaffold is a nonhistone core framework extending along the axis of the chromosome arms. With structural features similar to the nuclear matrix, the scaffold is usually demonstrated by sequentially extracting mitotic chromosomes with detergent, nuclease, and salt (Adolph *et al.,* 1977b; Lewis and Laemmli, 1982). After extraction, the chromosome scaffold can be resolved as a dense filamentous network that accounts for 3–4% of the total chromosome mass (Lewis and Laemmli, 1982). Unlike the nuclear matrix, the chromosome scaffold can be visualized in intact chromosomes by silver staining (Howell and Hsu, 1979). Recently, using the highly AT-specific fluorochrome daunomycin, a coiled or folded AT-rich core was observed in native chromosomes and thought to correspond to the chromosome scaffold where the GC-rich DNA loops protrude (Saitoh and Laemmli, 1994). Although some morphological aspects of the chromosome scaffold, such as thickness, length, and helical intensity, appear to differ in various studies due to individual extraction protocols, the composition of scaffolds are quite similar based on electrophoresis analysis (Lewis and Laemmli, 1982). Detailed characterization of individual components of the chromosome scaffold, however, has been quite limited.

Since both the nuclear matrix and chromosome scaffold are comparable frameworks involved in genome structure and organization and are demonstrated by similar biochemical protocols, it is plausible that they are derived from similar structural components in the nucleus. Indeed, using embedding-free EM techniques (D. He and B. R. Brinkley, unpublished), the chromosome scaffold of the early condensing chromosome adjacent to the nuclear envelope was still shown to be a specified domain of the nuclear matrix (Fig. 5). Perhaps one of the best documented components found in both the chromosome scaffold and the nuclear matrix is DNA topoisomerase type II or topo II (Adolph *et al.,* 1977a; Lewis and Laemmli, 1982; Berrios *et al.,* 1985). Other components identified in isolated chromosome scaffolds are SCII and the centromere proteins, especially CENP-B (Earnshaw *et al.,* 1985). The identification and characterization of DNA sequence elements known as MARs or SARs (scaffold-attached regions) and their respective binding proteins represent significant steps in defining specific DNA interactions with the nuclear matrix and chromosome scaffold. Among all types of nuclear matrix proteins, MAR-binding proteins appear to display the least change in organization and function when the cells progress from interphase to mitosis. Therefore, these regions may be the most consistent and reliable markers for chromosome scaffolds. Unfortunately, work on the MAR-binding proteins has been limited to *in vitro* studies and little is known about their association with the nuclear matrix *in vivo* and how they become altered during the cell cycle.

Like the nuclear matrix, chromosome scaffolds function to maintain a high level of structure and organization in the chromosomes. During mitosis, chromosome scaffolds appear to be involved in chromosome condensation/decondensation and structural reinforcement at the centromere/kinetochore region during segregation of sister chromatids (Uemura *et al.,* 1987). The following sections will briefly review some nuclear proteins that appear to relocate from the nuclear matrix to the chromosome scaffold during mitosis.

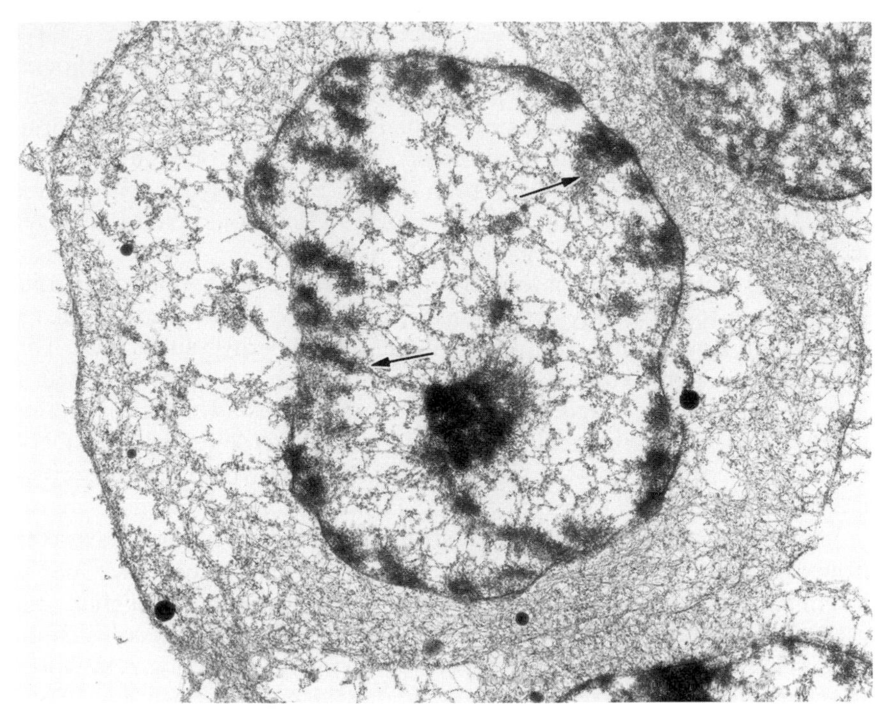

FIG. 5 Embedding-free EM section of an extracted HeLa cell at prophase. Note that the chromosomes begin to condense adjacent to the nuclear envelope. At this stage, the chromosome scaffold (arrows) can be seen as a specified domain of the nuclear matrix.

### 1. DNA Topoisomerase Type II (Topo II)

Early studies identified two major proteins in the chromosome scaffold, SCI (170 kDa) and SCII (135 kDa) from *Drosophila* (Lewis and Laemmli, 1982). The most abundant protein of the chromosome scaffold, SCI, was later identified as topo II (Earnshaw *et al.,* 1985) but the second most abundant protein, SCII, had not been as extensively studied. A recent analysis has shown that SCII may belong to the SMC1 (stability of microchromosome) family of proteins (Saitoh *et al.,* 1994). SMC1 appears to be required for accurate chromosome segregation in budding yeast and is likely a novel chromosomal ATPase (Strunnikov *et al.,* 1993). SCII colocalizes with topo II on the chromosome scaffold but appears not to be associated with the interphase nuclear matrix (Saitoh *et al.,* 1994).

Topo II can be recovered at very high yields (~70%) in fractions enriched in chromosome scaffolds (Earnshaw *et al.,* 1985) and is recognized as a major protein of interphase nuclei, where up to 60% can be recovered in the nuclear matrix fraction of most vertebrate cells (Halligan *et al.,* 1985). In HeLa, $3 \times 10^5$ copies per cell of topo II protein have been reported corresponding to about 3 copies per 70-kb DNA loop (Gasser and Laemmli, 1986). Recent work has demonstrated the existence of two topo II isoforms in mammalian cells (Chung *et al.,* 1989). The 170 kDa of topo II$\alpha$ is the major form, while the 180 kDa of topo II$\beta$ represents only 20–25% of total topo II protein (Drake *et al.,* 1987). *In vivo,* topo II is thought to be arranged as homodimers.

The enzymatic function of topo II is to catalyze the relaxation of superhelical turns in topologically constrained DNA by a double-strand breakage and rejoining mechanism that is dependent on the hydrolysis of ATP. This essential and highly conserved protein has been shown to be involved in many aspects of gene regulation and DNA metabolism through its ability to alter DNA topology, including initiation of DNA replication, DNA repair, transcription, and the separation of the intertwined daughter chromatids at the end of DNA replication (Fisher *et al.,* 1992).

Immunofluorescence and cellular fractionation experiments have confirmed that topo II is present in both metaphase chromosomes and in interphase nuclei. At interphase, topo II$\alpha$ is distributed throughout the nucleoplasm as slightly diffused spots excluded from the nucleolus, while topo II$\beta$ is found exclusively in association with the nucleolar remnant in the nuclear matrix preparation (Zini *et al.,* 1984; Heck and Earnshaw, 1986). *In vitro* DNA binding analysis has shown that topo II preferentially binds and aggregates MAR-containing fragments by a cooperative interaction, but the affinity is lower than MAR-binding proteins per se (Adachi *et al.,* 1989). In mitotic cells, topo II is mainly localized to a longitudinal central (core) region of metaphase chromosomes (Earnshaw *et al.,* 1985; Taagepera *et al.,* 1993). A helical staining pattern of topo II is seen in the most compact chromosomes, where the scaffold appears to be folded into a coil (Boy de la Tour and Laemmli, 1988). This localization, in conjunction with the observation that topo II preferentially binds and aggregates MAR-containing fragments by a cooperative interaction (Adachi *et al.,* 1989), has led to the model that topo II resides at the base of the chromatin loops to anchor the DNA responsible for the radial loop organization of chromatin in mitotic chromosomes. This view, however, has been challenged by several other observations. First, the cellular content of this enzyme varies widely with different stages of proliferation and cell cycle. The amount of topo II decreases rapidly and dramatically in cells that cease to proliferate. For instance, there are $7 \times 10^4$ copies of topo II per chicken erythroblast, whereas less than 300 copies are found in nonproliferating erythrocytes during erythropoiesis (Heck and Earnshaw, 1986). Also, the spotted staining pattern seen in interphase nuclei appears to be cell cycle dependent, since a high ratio of $G_1$ cells are negatively stained by topo II antibody (Earnshaw *et al.,* 1985). These results argue that topo II may not be an essential DNA anchoring protein as previously proposed. Second, when topo II is extracted from metaphase chromosomes under mild conditions, the morphology of the chromosome remains intact (Hirano and Mitchison, 1993), suggesting that this component is not indispensable in maintaining the structural integrity of mitotic chromosomes. Third, the appearance of topo II in chromosome "cores" largely depends on the compositions of the spreading buffers used for preparation (Ohnuki, 1968; Boy de la Tour and Laemmli, 1988). In fact, in some recent studies, topo II has been reported to distribute uniformly throughout the chromosome arms and is not restricted to the axis as shown by immunofluorescence (Hirano and Mitchison, 1993) or by measurement of microinjected rhodamine-conjugated topo II (Swedlow *et al.,* 1993). It is possible, however, that the distribution pattern of injected topo II may be due to the influence of an abnormally high content of exogenous protein. Collectively, therefore, the function of topo II is not only the general anchoring of DNA loops but may also function in proliferation and related cell cycle events, such as initiation of DNA replication, separation of replicated chromatin strands, and reorganization of the genome at the transitions between interphase and mitosis.

Additional evidence for the function of topo II in chromosome formation comes from the complement experiments in a topo II immunodepleted system (Adachi *et al.,* 1991). HeLa nuclei, which have a high content of endogenous topo II, are able to convert to mitotic chromosomes in a topo II-depleted extract prepared from *Xenopus* egg. Chicken erythrocyte nuclei, on the other hand, which have a very low amount of topo II, failed to form condensed chromosomes in the same topo II-depleted mitotic extract. If purified topo II is added to the system, however, then chicken chromosomes can undergo normal condensation. Moreover, the requirement of topo II for chromosome condensation appears to be dosage-dependent (Adachi *et al.,* 1991). Using a totally different cell-free system, Newport and Forbes (1987) also found topo II to be important in chromosome condensation.

Topo II also appear to be essential for chromosome segregation for both yeast and mammalian cells, although the exact mechanism remains unknown (Uemura *et al.,* 1987; Clarke *et al.,* 1993). In this regard, topo II may also be concentrated at the centromeric region as shown by immunofluorescent staining (Taagepera *et al.,* 1993). Its presence at the centromere may be to facilitate a more condensed and rigid chromatin organization needed to withstand mechanical stresses applied by spindle MTs during chromosome movement.

The function of topo II during cell cycle transitions appears to be regulated by a kinase activity. Immunoblot and immunoprecipitation experiments have shown that both $\alpha$ and $\beta$ isoforms of topo II are phosphorylated as detected by a monoclonal antibody MPM-2, that recognizes a specific group of phosphorylated epitopes (Woessner *et al.,* 1990, 1991). This has been confirmed by the colocalization of MPM-2 and anti-topo II immunostaining in various mammalian cells including mouse, Indian muntjac, and hamster (Taagepera *et al.,* 1993). Moreover, topo II has also been shown to be the target of cdc2 and various other kinases (Sayhoun *et al.,* 1986; Cardenas *et al.,* 1992). The multiple levels of phosphorylation of topo II reached during the $G_2$–M transition may be required for the new function of topo II in mitosis (Taagepera *et al.,* 1993).

In summary, topo II appears to be an essential protein of the nuclear matrix and chromosome scaffold that functions largely in the organization of chromatin. Specifically, topo II catalizes changes in DNA topology at stages in the cell cycle, requiring dramatic remodeling of the genome including initiation of DNA replication, transcription, and chromosome condensation/decondensation. Its function, if any, during mitosis remains less clear.

## 2. MAR/SAR Binding Proteins

DNA in eukaryotic cells is believed to be organized into loop domains in both interphase nuclei and mitotic chromosomes (Gasser and Laemmli, 1987). These loops are thought to be anchored to a proteinaceous framework (nuclear matrix/chromosome scaffold) via specific evolutionarily conserved DNA elements at the basis of the loops known as matrix-attached regions (MARs), or scaffold-attached regions (SARs) (Cockerill and Garrard, 1986; Gasser and Laemmli, 1986; Gasser *et al.,* 1989). For convenience and consistency, we will only use the term of MAR throughout this section. MARs have been characterized as stretches of 300–1000 bp of DNA, possessing typically about 70% AT-rich sequences containing several topo II-susceptive motifs (Adachi *et al.,* 1989; Käs and Laemmli, 1992). Most MARs are localized at the boundaries of transcription units, coincident with enhancers, promoters, or other functionally important sequences (Gasser and Laemmli, 1986). It is reasonable to propose that specific proteins (MAR-binding proteins) may interact or associate with MARs in both the nuclear matrix and chromosome scaffold.

MAR-binding proteins are usually identified by Southwestern blotting by incubating nitrocellulose filters containing electrophoretically separated proteins with labeled MAR sequences. The MAR-binding proteins are then visualized as individual bands. Using this procedure, several MAR-binding proteins have been characterized including a 95-kDa ARBP (attachment region binding protein) from chickens (von Kries *et al.,* 1991), a 120-kDa SAF-A (scaffold attachment factor A), a double-strand DNA-binding protein from HeLa nuclei (Romig *et al.,* 1992), SP120 from rat brain (Tsutsui *et al.,* 1993), 116-kDa RAP-1 (repressor–activator binding protein) from yeast (Hofmann *et al.,* 1989), SATB1 (special AT-rich-binding protein 1) (Dickinson *et al.,* 1992), lamins (Ludérus *et al.,* 1992, 1994), and topoisomerase II (Adachi *et al.,* 1989). These proteins all show high affinity to MARs in binding assays *in vitro.* Although there is no consensus sequence found in MARs, they appear to bind to proteins via specific sites. At the sequence level, multiple AT-rich stretches throughout the MARs appear to have significantly higher protein-binding affinity as shown by *in vitro* binding assays (Hakes and Berezney, 1991). At higher structural levels, however, MARs are thought to bind to proteins by way of two possible domains: (1) via a single-stranded DNA region (Kay and Bode, 1994) or alternatively, a strong potential for extensive base unpairing under superhelical strain (Probst and Herzog, 1985; Kohwi-Shigematsu and Kohwi, 1990; Bode *et al.,* 1992). Specific sequence motifs of MARs that function as unwinding nucleation sites have also been identified. Mutants at the unwinding nucleation site of two MARs have led to resistance to unwinding and loss of high affinity binding to the nuclear scaffold (Bode *et al.,* 1992); or (2) via the narrow minor groove of the DNA double helix in the MARs region, as revealed by deletion or competition experiments (Adachi *et al.,* 1989; Nakagomi *et al.,* 1994). Occasionally, some MAR-binding proteins, i.e., lamin proteins, can recognize both types of binding domains (Ludérus *et al.,* 1992). Interestingly, specific structural features of proteins with affinities for the single strand or minor groove regions of MARs have yet to be found.

Binding of MARs to their target proteins results in the formation of DNA loops at 30- to 100-kb intervals along the DNA strand. These loops are thought to insulate genetic units *in vivo.* Experimentally, such looped structures have been successfully obtained by incubating purified proteins with the DNA fragments containing selected MARs. For instance, the loops induced by MAR-binding protein RAP-1 were shown to be sequence specific when incubated with heterologous DNA. Not only the size, but also the base sequence of the loop were coincident to MARs that were mapped in the scaffold binding assay (Hofmann *et al.,* 1989). In addition, another MAR-binding protein, SAF-A, formed approximately 35-nm filamentous aggregates from which the DNA molecules containing MARs protruded to form a specific looped structure resembling a matrix-attached DNA spreading sample (Romig *et al.,* 1992). Although more complicated regulation probably occurs in living cells, such interactions very likely represent the basic anchoring condition of chromatin in both interphase nuclei and mitotic chromosomes. As we mentioned earlier, such reactions may vary with cell cycle progression and proliferation states (e.g., topo II) (Heck and Earnshaw, 1986) or with cell differentiation (e.g., SATB1) (Dickinson *et al.,* 1992; Filatova and Zbarskii, 1993).

Binding between MARs and their target proteins appears to be evolutionarily conserved, since MAR-binding proteins can bind to MARs from a variety of tissues, and MARs usually can bind to proteins from different species as well (Cockerill and Garrarad, 1986; Izaurralde *et al.,* 1988; Phi-Van and Strätling, 1990). For instance, a MAR sequence from the chicken lysozyme locus can react with the nuclear matrix proteins of human, porcine, *Drosophila,* and *Xenopus* oocyte, and recognize a single major band on Southwestern assays from each species (von Kries *et al.,* 1991).

Since most MARs were originally identified in nuclear matrix or scaffold preparations, we have very limited knowledge about the *in situ* distribution pattern of MAR-binding proteins. A monoclonal antibody against SP120 was found to uniformly stain human HEp2 interphase nuclei, leaving nucleoli only weakly stained; occasionally, fluorescence preferentially accumulates at the nuclear periphery. Rat brain cells, however, displayed a spotted, cell-dependent, nuclear staining pattern using the same antibody (Tsutsui *et al.,* 1993). In addition, chicken MAR-binding protein ARBP is specifically localized to the inner nuclear matrix. When nuclear matrix preparation was further extracted to make an empty shell of nuclear pore–lamina complex (Kaufmann *et al.,* 1983), all ARBP was released (von Kries *et al.,* 1991). In this regard, it would be useful to determine whether MAR-binding proteins represent major components of the nucleoskeletal core filaments. As yet, however, EM immunolocalization experiments involving MAR-binding proteins have not been attempted.

With the exception of the lamins, distribution of MAR-binding proteins in mitotic cells has rarely been tested. However, we predict that many will be localized in the chromosome scaffold based on their capability to form DNA loops. Evidence for this hypothesis comes from the analysis of chromosome structure, where DNA loops have been found with the same sizes as those in the nucleus (Gasser *et al.,* 1989). It seems unlikely that such a looped organization would result from a different set of DNA–protein interactions unique to mitosis. Therefore, most MAR-binding proteins in interphase may distribute to the scaffold of metaphase chromosomes. On the other hand, perhaps caution should be taken when making such predictions since some DNA-binding proteins may have a redistribution pattern similar to lamins, a specific MAR-binding protein localized primarily in the periphery of the nucleus (Ludérus *et al.,* 1994). During mitosis, hyperdephosphorylated lamins disassociate from chromatin and become dispersed throughout the cytoplasm. At late telophase, lamin B reassociates with the surface of the merged chromosomal mass where the new nuclear envelope will later form. This type of distribution may represent a few special MAR-binding proteins that are not scaffold components. Indeed, several nuclear matrix proteins may also become solubilized when cells progress into mitosis (He *et al.,* 1991).

In addition to their role in the formation of DNA loops of chromatin, many MAR-binding proteins also function in the regulation of gene expression. For instance, enhanced expression of transfected genes can be demonstrated in constructs linked to MAR sequences (Stief *et al.,* 1989). In a more recent finding, a new MAR-binding protein was specifically found in breast tumor nuclei but not in the normal breast tissue (Kohwi-Shigematsu, personal communication), indicating that MAR-binding proteins may be involved in tumorigenesis or tumor-specific metabolism. During mitosis, however, gene expression is halted and the function of MAR-binding proteins within the chromosome scaffold probably includes the maintenance of chromosome structure, especially that required to counter the pulling forces associated with chromosome alignment and segregation. MAR-binding proteins may also play active roles in chromosome condensation/decondensation and in sister chromatid partitioning.

### 3. Other DNA Binding Proteins

In addition to its well-known function in encoding various RNAs, the eukaryotic genome has a much larger coding capacity for DNA structure itself and the interactions of DNA with other nuclear components, via protein binding codes (Vogt, 1992). In this regard, MARs are not the only "matrix-attached regions" through which genomic DNA binds to proteinaceous substructure. One line of evidence implicates the highly repetitive DNA segments located at the centromere and telomere regions and interspersed throughout the genome. In addition, centromeric and telomeric DNA contain a variety of unique protein binding motifs. For example, the CENP-B box is a 17-bp DNA sequence located in human $\alpha$-satellite DNA that binds centromere protein B (CENP-B) throughout the cell cycle (Masumoto *et al.,* 1989). Moreover, highly repetitive DNA sequences interspersed throughout the genome are often found to be coincident with the attachments to the nuclear matrix (Vogt, 1992). Recently, a sequence-directed DNA bend-determinant has been identified with a binding affinity for a specific nuclear scaffold protein, P130, in rat cells (Hibino *et al.,* 1993). Several other DNA structural features involved in specific binding to matrix proteins have been reported (Wahls *et al.,* 1991; Grady *et al.,* 1992). In addition, a few MAR-binding proteins, including SAF-A (Romig *et al.,* 1992), topo II (Adachi *et al.,* 1989), SP120 (Tsutsui *et al.,* 1993), and lamins (Ludérus *et al.,* 1994), have been shown to have affinity not only for MARs, but for non-MAR, AT-rich, or single-stranded DNA elements. These DNA elements may also have a general affinity for proteins other than MAR-binding proteins. Therefore, it seems likely that nuclear binding reactions other than the MAR-mediated types are common to the eukaryotic nucleus. Indeed, when 12 nuclear matrix proteins from rat were identified by non-equilibrium 2-D electrophoresis, as many as 7 were found to be specific DNA-binding proteins (Hakes and Berezney, 1991; Nakayasu and Berezney, 1991). These included lamin A and C, and five novel matrix proteins termed matrins D, E, G, F, and 4. The latter have been localized to the nuclear interior by immunofluorescence and bind to DNA in a saturable, temperature-dependent, salt-resistant manner. These observations suggest that MARs only represent one subset of DNA-binding sites. Moreover, the non-MAR types of binding between DNA and the nuclear matrix may play a greater role in maintaining structure than MAR-type binding. For instance, the centromere and telomere are important structural elements of the chromosome where non-MAR-binding sequences abound (Vogt, 1992). From these observations, it is plausible that MAR-binding proteins may function primarily in chromatin organization and gene expression in interphase nuclei, whereas the non-MAR-mediated DNA-binding proteins could have a greater involvement in maintaining structural organization in mitotic chromosomes.

### 4. Rb Protein

Another example of matrix-associated, DNA-binding protein is the product of the retinoblastoma susceptibility gene (Rb). Rb is a 110 kD nuclear phosphoprotein that binds double stranded DNA nonspecifically (Lee *et al.,* 1987; Goodrich and Lee, 1993; Riley *et al.,* 1994). Rb protein has been identified as a prototypical tumor suppressor and cell cycle-regulating factor (Goodrich and Lee, 1993). The mechanism by which Rb influences the cell cycle is not well understood; however, its action is presumed to involve interaction with many proteins. Through a variety of approaches, multiple cellular proteins have been shown to interact with the hypophosphorylated form of Rb (see Riley *et al.,* 1994), which is found primarily from late M through most of $G_1$ (DeCaprio *et al.,* 1989; Buchkovich *et al.,* 1989; Chen *et al.,* 1989; Ludlow, *et al.,* 1993). Some Rb-binding proteins are themselves associated with the nuclear matrix, including several "inactivating" viral oncoproteins (SV40 Large T antigen, adenovirus E1a protein, and the human papilloma E7 protein; Goodrich and Lee, 1993), lamin A/C (Mancini *et al.,* 1994), and p84 (Durfee *et al.,* 1994).

Initial observations of cultured cells exposed to low-salt and detergent demonstrated hypophosphorylated Rb is capable of "tethering to the nuclear structure" in a cell cycle-dependent ($G_1$) fashion (Mittnatch and Weinberg, 1991). Upon further examination, phosphorylated Rb and naturally occurring Rb mutants found in many tumors (not interacting with known Rb-binding proteins), also did not tether (Mittnatch and Weinberg, 1991; Templeton *et al.,* 1991).

Core filament matrix preparations were also examined to determine the nature of Rb interaction with the nucleus. Immunoblotting experiments showed that during early $G_1$, a significant amount of hypophosphorylated

Rb protein associates with the core filament matrix fraction (Mancini *et al.,* 1994). As shown by immunofluorescence, a spotted pattern is displayed throughout the matrix, particularly concentrated at the nuclear periphery and in nucleolar remnants. Rb is one of only a few nuclear matrix proteins that has been localized using the resinless embedding technique for electron microscopy. In such perparations, distinct fibrogranular masses and nucleolar remnants were labeled with gold-conjugated antibody, while the core filaments themselves remain unlabeled (Mancini *et al.,* 1994).

These observations suggest Rb is a facultative component of the nuclear matrix and is not essential for the maintenance of constitutive nuclear structure. Perhaps the nuclear matrix-associated Rb can sequester growth regulatory proteins during $G_1$ and the concomitant phosphorylation and disassociation of Rb from the nuclear matrix controls signals driving cell cycle progression (Mancini *et al.,* Riley *et al.,* 1994). As determined by subcellular fractionation and immunoblotting, similar to other matrix-bound proteins, hypophosphorylated Rb is present in the chromosome scaffold (M. A. Mancini and W. H. Lee, personal communication). Although the function of chromosome-bound Rb is unknown, identification of protein phosphatase type 1 catalytic subunit (PP1) as an Rb-binding protein (Durfee *et al.,* 1993) may imply regional dephosphorylation occurs during M phase. Indeed, PP1 has been shown to be specifically chromosome-bound during mitosis (Fernandez *et al.,* 1992). The marked dephosphorylation during the late stages of mitosis, perhaps in part through PP1 action, may suggest Rb is required for mitotic regulation (Ludlow *et al.,* 1993; Riley *et al.,* 1994).

## B. Perichromosome Region

The perichromosomal region represents a uniquely differentiated compartment of mitotic cells, and indeed, many proteins derived from the nuclear matrix appear to be targeted to this region (Hernandez-Verdun and Gautier, 1994). Early EM studies (Brinkley, 1965; Hsu *et al.,* 1986; Stevens, 1965; Goessens and LePoint, 1974; Paweletz and Risue-no, 1982; Ochs *et al.,* 1983) identified a distinct sheath on the surface of most eukaryotic chromosomes and distinct clusters of ribosome-like particles at the perichromosome region, possibly derived from the nucleolus (Brinkley, 1965). Utilizing an EDTA regressive staining procedure specific for ribonuclear proteins (Bernhard, 1969), RNP particles were detected at the periphery of CHO metaphase chromosomes (Okada and Comings, 1980). Cryosection EM, used to minimize the artifact induced by the resin embedding, also showed clustered particles involving 11- to 16-nm granules linked by filaments along the periphery of chromosomes (Gautier *et al.,* 1992c). In addition, scanning EM has provided 3-D images of granules attached to fibers surrounding and connected directly to chromosomes (Zee, 1992). In studies using embedding-free, whole mount preparation of PtK2 cells, granular filaments were found attached to condensed chromosomes that were thought to represent the previous nuclear matrix elements (Capco and Penman, 1983). Using resinless section EM, we recently observed a clear region surrounding each metaphase chromosome that contained a filamentous granular network resembling thc nuclear matrix, while the filamentous network observed elsewhere in the cytoplasm was less dense and smoother, resembling the intermediate filament cytoskeleton (Fig. 6) (He and Brink-

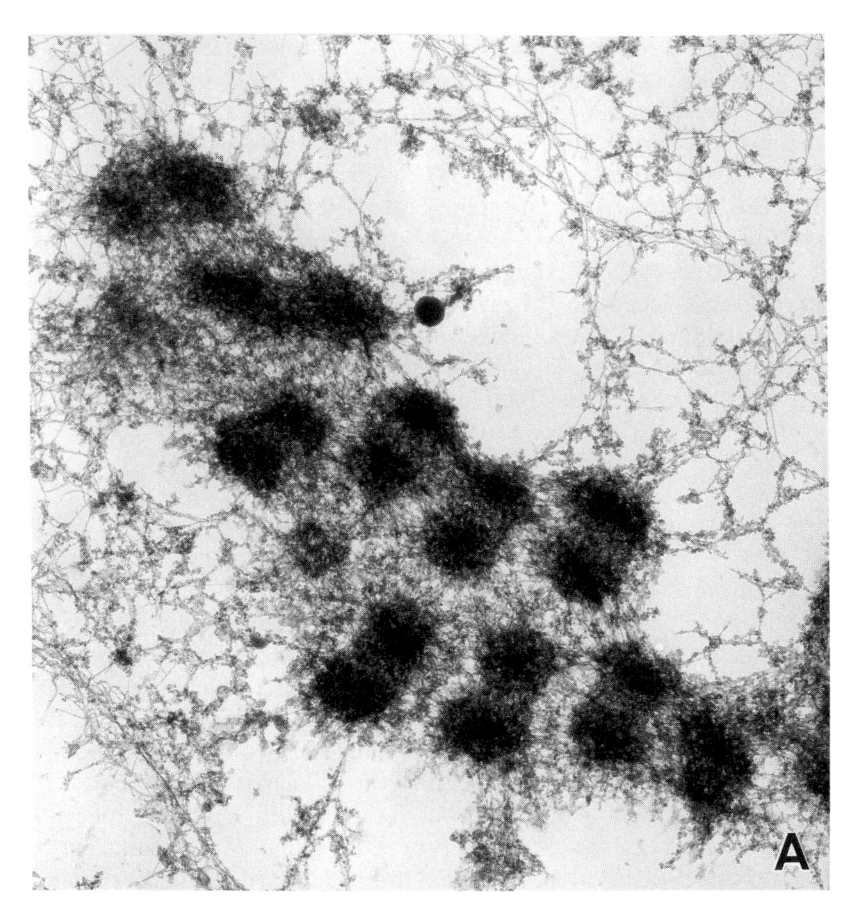


FIG. 6. The perichromosome region is clearly shown in differentially extracted, embedding free EM preparations. The metaphase plate is sectioned parallel to the spindle axis. A, low magnification showing chromosomes connected by a dense network. B, higher magnification showing the fibrogranular connections.

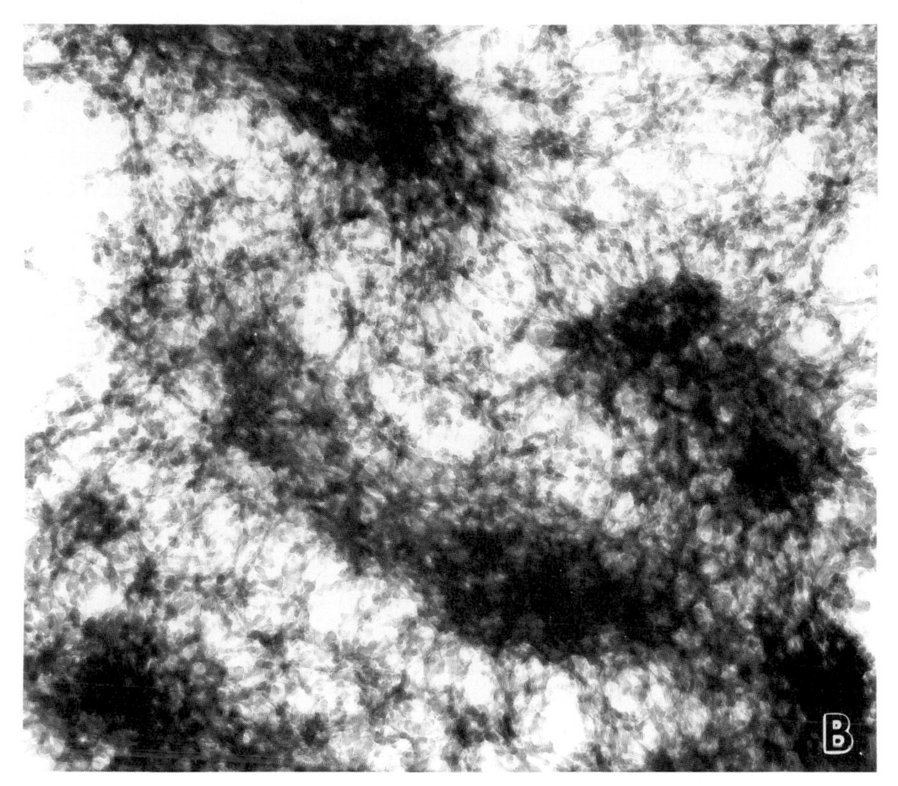


FIG. 6 (Continued).

ley, 1995). Collectively, these observations not only support the existence of a differentiated perichromosome region or compartment, they also indicate a distinct spatial relationship to RNP and the nuclear matrix.

The hnRNP antibody, fA12, is known to react with a group of 33- to 40-kDa hnRNP core polypeptides (Leser *et al.*, 1989) and the distribution and fate of fA12 antigen in mitotic cells has been traced by immunofluorescence and in resinless EM preparations (He *et al.*, 1991). During interphase, fA12 identifies a fine pattern of granules, uniformly extending throughout the nucleus with the exception of nucleoli. Immunogold labeling of core filaments in resinless EM sections showed that fA12 decorated fibrogranular material but not the core filaments themselves. This is consistent with previous EM studies that used embedded sections with both polyclonal and monoclonal antibodies to hnRNP core polypeptides and showed localization to the perichromatin fibrils (Fakan *et al.*, 1984, 1986; Puvion *et al.*, 1984). During mitosis, fA12 antigen becomes redistributed to the perichromosome region as well as to the spindle pole regions. Again, fA12 decorated similar fibrogranular material that attached to the filaments surrounding and interconnecting chromosomes and to similar material that surrounds the centrioles. At the end of telophase, fA12 antigens became relocated to the dense midbody leaving a central unstained spot. It is likely that when cells progressed into mitosis, some of the filaments that connected chromosomes in interphase were retained and formed perichromosome fibrils representing a persistent form of the nuclear matrix.

Other nuclear matrix proteins also display a similar redistribution pattern from the nucleus to the perichromosome region. snRNPs have been found to associate with the nuclear matrix in interphase (Miller *et al.,* 1978; Vogelstein and Hung, 1982; Spector *et al.,* 1983). A monoclonal antibody that recognizes a 28-kDa snRNP protein has been used to trace the distribution of snRNPs throughout the cell cycle (Spector and Smith, 1986). During interphase, the antibody stains a speckled pattern leaving the nucleoli unstained. Discrete intranuclear protein clusters were identified at these sites as shown by immunogold EM. At the onset of mitosis, however, snRNPs become redistributed to a region between condensing chromosomes. By metaphase and anaphase, some snRNPs remain associated with the surface of the chromosome (perichromosome region) while others become dispersed into the cytoplasm. During telophase, reformation of nuclear protein clusters occurs in association with the chromosome surface. Monoclonal antibody H1B2 recognizes a 250-kDa nuclear matrix protein, and its pattern of distribution has been studied using resinless EM gold labeling (Nickerson *et al.,* 1992). During interphase, this antigen appears to be masked and can only be visualized by antibodies applied to nuclear matrix preparations following chromatin extraction. Interestingly, the antigen begins to unmask at a discrete site at prophase. The stained area is gradually enlarged, and from prometaphase to telophase it displays a dramatic redistribution resembling the fA12 antigen staining pattern. Again, 9- to 13-nm filaments were found to surround both the chromosomes and the spindle poles and labeled fibrogranular material remains associated with these filaments. Whether this antigen is related to RNPs remains unknown.

Another interesting group of nuclear matrix proteins includes those that redistribute from the nucleoli in the interphase nucleus to the perichromosome regions of mitotic chromosome (Hernandez-Verdun *et al.,* 1993). Among 36 nucleolar autoantigens identified, 10 were found to be relocated to the perichromosome region in mitosis (Gautier *et al.,* 1992b). PCN (perichromonucleolin) (Shi *et al.,* 1987), fibrillarin (Ochs *et al.,* 1985; Yasuda and Maul, 1990), Ki-67 polypeptides (Verheijen *et al.,* 1989; Gerdes *et al.,* 1991), B23 protein (Ochs *et al.,* 1983; Tawfic *et al.,* 1993), and G04 antigen (Gautier *et al.,* 1992a,c) are all nuclear matrix proteins associated with nucleoli during interphase but become relocated to the perichromosome sheath at mitosis. Their pathways for the relocation from nucleoli to the perichromosomes, however, appear to vary slightly. For example, while PCN was found to fuse into large patches during prophase (Shi *et al.,* 1987), the Ki-67 displayed an irregular meshwork throughout the nucleoplasm (Bading *et al.,* 1989; Verheijen *et al.,* 1989).

In addition to the proteins found principally in the inner region of the nuclear matrix and nucleoli, some nuclear matrix proteins redistribute from the periphery of the interphase nucleus to the perichromosome region of mitotic chromosomes. These proteins include perichromin (33 kDa) (McKeon *et al.,* 1984), lamin B (Gerace and Blobel, 1980), and P1 antigen (Chaly *et al.,* 1984), which was later named peripherin (Chaly and Brown, 1988). Since the name peripherin has also been used for an unrelated intermediate filament type III polypeptide (Molday *et al.,* 1987), we prefer to use the name P1 here and throughout. Perichromin has been detected in HeLa, CHO, and *Drosophila* cells, where it is localized at the periphery of the interphase nucleus. It is resistant to NP-40, micrococcal nuclease, and 0.5 *M* NaCl extraction but when 2 *M* NaCl was applied, a portion of the antigen was found to disperse to the outer surface of the nucleus. During mitosis, perichromin was found to coat chromosomes from metaphase to telophase. On the other hand, P1 and lamin B occupy a different perichromosome location. P1 is distributed to the nuclear periphery during interphase but at metaphase it coats a zone at the periphery of the metaphase plate. From anaphase and even later stages, it forms a coat that enshrouds the entire chromosomal mass. Similarly, lamin B forms a coat around the chromosome mass during late telophase as chromosomes become more tightly associated (Burke and Gerace, 1986). However, at earlier stages of mitosis, lamin B is largely dispersed throughout the cytoplasm where it associates with vesicular organelles (Gerace and Blobel, 1980). Another nucleolar protein, fibrillarin, is a component of the nuclear matrix but becomes associated with the surface of individual chromosomes during metaphase, except for the centromere regions. At the onset of anaphase, however, it becomes redistributed from the surfaces of individual chromosomes to the surfaces of the entire chromosome mass. Thus, these nuclear matrix proteins appear to be targeted very specifically to a special zone along the chromosome surface that may represent future sites for the formation of the nuclear envelope. However, an alternative model has been proposed where nuclear envelope reassembly is lamin-independent with lamins entering the nucleus through the nuclear pores (Newport *et al.,* 1990).

Those nuclear matrix proteins that become uniquely relocated to the perichromosome region may provide a function in both interphase and mitotic cells. Interestingly, many members of this particular protein group are associated in some way with RNA metabolism. It is not surprising, therefore, that they associate with the core filaments in interphase, since this nucleoskeletal system contains mostly proteins, a small fraction of

DNA, but a large amount of total nuclear RNA (He *et al.,* 1990). Both RNA transcription and processing are thought to take place on the nuclear matrix (He *et al.,* 1990, 1991). Moreover, it is well known that rRNA is synthesized and processed in the nucleoli. For example, fibrillarin has been shown to be associated with the early processing of rRNA (Benavente *et al.,* 1989), while protein B23 is involved in the later stage of ribosome assembly (Tawfic *et al.,* 1993). In addition to their well-established function in the RNA metabolism, these proteins also appear to have a significant influence on nuclear structure. Interrupting nuclear RNA, either by inhibiting synthesis or by digesting isolated nuclei with RNase, leads to dramatic collapse of the nuclear matrix. Since RNPs are localized in fibrogranular material that is enmeshed in the core filament network but not a part of the filaments themselves, they may form a nexus with core filaments thereby maintaining the integrity of the network (He *et al.,* 1991).

Nuclear matrix proteins that become targeted to the perichromosome region may be important in maintaining structure of the mitotic apparatus. Capco and Penman (1983) described a non-MT fibrous network in mitotic cells with a morphologic similarity to interphase core filaments (He *et al.,* 1990). It is well known that diassembly of spindle MTs has little effect on the position of chromosomes within the spindle. Indeed, a fibrous matrix persists in the mitotic apparatus even when MTs have been completely disassembled. In an early study, mitotic chromosomes were found to be released in clusters when dissected from living mitotic cells, suggesting the presence of an interchromosomal matrix (Hoskins, 1968). As will be described (Section III,C,5), nonrandom distribution and alignment of centromeres in interphase nuclei and mitotic cells suggests a continuous association with an underlying matrix. Indeed, the spindle matrix may serve to interconnect chromosomes and keep them in a defined order and organization. In this regard, interchromosomal connecting filaments derived from the interphase nuclear matrix may provide a continuous network around the genome. The perichromosome sheath, therefore, may somehow confine genetic material and prevent it from being dispersed during mitosis. In this regard, the chromosome sheath may be analogous to the nuclear envelope in interphase (Yasuda and Maul, 1990; Gautier *et al.,* 1992a,c). Another possible function of the nuclear matrix-derived chromsome sheath might be to retain the spatial order and integrity of transcription and splicing machinery needed in the newly formed daughter nuclei when chromosomes undergo decondensation.

The perichromosome region also appears to be a site where nucleolar proteins specifically accumulate during mitosis. This is consistent with the finding that 45S and 32S rRNA precursors remain associated with the chromosome periphery throughout mitosis (Fan and Penman, 1971). These proteins along with others such as Ki-67 and B23 may also play roles in the initiation of new rRNA metabolism in daughter nuclei. Indeed, we cannot rule out the possibility that the distribution of these complexes to the chromosome surface during mitosis assures the delivery of long-lived RNA and associated proteins needed in the newly formed daughter nuclei (Rattner, 1992).

## C. Centromere/Kinetochore Region

### 1. Centromere/Kinetochore–Prekinetochore

The centromere is also a site for the accumulation of nuclear matrix-derived proteins during mitosis. The terms kinetochore and centromere have long been used interchangeably in the cell biology and genetic literature to identify the site where spindle microtubules interface with chromosomes. In fact, this complex locus contains several unique structural domains including the trilaminar kinetochore, a central domain and a pairing domain packaged to facilitate spindle microtubule capture and attachment, chromosome movement, and pairing of sister chromatids (Brinkley *et al.,* 1989, 1992; Brinkley, 1990). The centromere is composed of a large block of nontranscribing, heterochromatic DNA that accounts for up to 5% of the total genomic DNA in most eukaryotic cells (Willard and Waye, 1987; Wong and Rattner, 1988; Clarke, 1990).

A discussion of the centromere/kinetochore complex and its counterpart in the interphase nucleus, the prekinetochore, is appropriate in this chapter because in addition to its constitutive structures and well-known function associated with mitotic chromosome movements, it appears to represent a unique site for the transient appearance of various facultative proteins whose origins include the interphase nuclear matrix and whose functions remains somewhat enigmatic.

A consistent feature of the centromere of most eukaryotic chromosomes is repetition. All domains of the centromere appear to contain blocks of DNA with sequences that are, for the most part, a few hundred base pairs long. These sequences are present as tandem repeats that extend for several mb along the centromere. A structural model has been proposed whereby short repetitive segments containing microtubule-binding proteins are interspersed along a linear 30-nm chromosome fiber that can be folded to form the kinetochore plates and the entire centromere/kinetochore complex (Zinkowski *et al.,* 1991). Evidence that specific kinetochore proteins associate with repetitive centromeric DNA units periodically is shown by antikinetochore antibody staining. According to this model, the platelike kinetochore may be formed by a specific pattern of folding of the 30-nm chromosomal fiber. Although there is little evidence that DNA directly binds to the outer or middle plates (Cooke *et al.,* 1993), DNA is most certainly located at or near the inner plate and likely specifies their formation. Most of the known proteins in the centromere region have been identified using autoimmune sera from the patients with scleroderma CREST (calcinosis, Raynaud's phenomenon, esophageal dysmotility, sclerodactyly, and telangiectasia) syndrome. Antibodies from these patients serve as specific probes for detecting the centromere/kinetochore complex (Moroi *et al.,* 1980). In the $G_1$ nucleus, these antisera stain discrete "spots," the number of which corresponds to the diploid chromosome numbers, while at $G_2$ the number of fluorescent spots is exactly doubled (Brenner *et al.,* 1981). These intranuclear foci, termed prekinetochores, have recently been found to persist after extensive extraction of nuclei with nuclease and high salt and are, therefore, stable components of the nucleoskeleton and chromosome scaffold (D. He and B. R. Brinkley, unpublished). In extracted cellular preparations of HeLa and Indian muntjac nuclei, the centromere/prekinetochore region represents most of the discernible organelles other than the nucleolar remnants. A dynamic pattern of structural organization within this locus has been revealed by immunofluorescence and immunogold electron microscopy, which strongly suggests that the centromere/prekinetochore region in agreement with the model of Zinkowski *et al.* (1991) undergoes a programmed unfolding–refolding, associated with centromeric DNA replication. This is followed by postreplication segregation of the two sister halves, all of which appear to involve intimate association with the nuclear matrix (He and Brinkley, 1995).

### 2. Constitutive and Facultative Proteins of the Centromere

The centromere/kinetochore contains two classes of proteins, constitutive and facultative. CENP-A (17–19 kDa), CENP-B (80 kDa), CENP-C (140 kDa), and CENP-D (50 kDa) are present at or near the kinetochores throughout the entire cell cycle, with the possible exception of brief disassociation during the replication of the corresponding DNA sequence (Haaf and Ward, 1994). Since these proteins are present at both mitosis and interphase and appear to be permanent structural components, they have been classified as constitutive centromere proteins (Brinkley *et al.,* 1992). As mentioned previously (Section III,C,1), they can be detected in all phases of the cell cycle by immunofluorescence or Western blotting with human CREST autoantibodies (Earnshaw and Rothfield, 1985; Valdivia and Brinkley, 1985). Among these, CENP-A has been identified as a centromere-specific, H3-like core nucleosomal histone (Palmer *et al.,* 1991; Sullivan *et al.,* 1994). CENP-D appears to be a homolog to the RCC1 protein, which is thought to be involved in the regulation of chromosome condensation (Kingwell and Rattner, 1987; Bischoff *et al.,* 1990). CENP-E, another centromere protein, has been identified as a kinesin-like motor associated with the outer kinetochore surface (Yen *et al.,* 1991, 1992). CENP-A, CENP-D, and CENP-E can be easily released from the nucleus after treatment with low salt and thus are not thought to be part of the nuclear matrix proteins (Earnshaw *et al.,* 1985; Valdivia and Brinkley, 1985; Ohtsubo *et al.,* 1989; Yen *et al.,* 1991), whereas CENP-B and CENP-C resist extraction and remain with the scaffold or nuclear matrix (Balczon and Brinkley, 1987).

The second category of centromere/kinetochore proteins, i.e., facultative proteins (Brinkley *et al.,* 1992), represents the components temporally residing at this region only at certain periods of the cell cycle. Indeed, some of the latter appear to be derived from the nuclear matrix of the interphase nucleus in a cell cycle-dependent fashion. For instance, P330 or CENP-F, INCENPs, and Mitosin/CENP-F begin to accumulate in the nucleus in S or $G_2$ phase but gradually concentrate at the centromere during prophase–prometaphase. These proteins associate with the centromere/kinetochore region transiently, but at the onset of anaphase are transferred to either the midzone or the poles of the mitotic spindle and remain there until the end of mitosis (Earnshaw and Bernat, 1990). Facultative proteins are thought to regulate kinetochore functions in some way as discussed in a subsequent section.

### 3. The Role of Chromatin Anchoring

The centromere/kinetochore complex not only attaches the chromosomes to spindle fibers to bring about segregation of sister chromatids during anaphase of cell division, it represents the only locus that links sister chromatids after DNA replication. The nuclear matrix proteins in this region could be involved in one of several kinetochore functions including capturing and stabilizing spindle MTs, pairing and separation of sister chromatids, or attachment of centromere proteins to DNA. Centromeres are known to contain a large amount of highly repetitive DNA (Willard and Waye, 1987; Werrick and Willard, 1991), and a number of DNA sequence elements have been identified with high affinity for certain proteins at the centromere/kinetochore region. For instance, a highly conserved sequence unit (GGAAT)n has been identified at the centromere of human chromosomes that specifically bind nuclear proteins of HeLa cells *in vitro* (Grady *et al.,* 1992). This sequence is similar to CDE III, a critical DNA element of yeast chromosomes that also binds specific nuclear proteins from HeLa nuclear extract with high affinity (Lechner and Carbon, 1991). Another example is a sequence-directed DNA bend protein that binds to centromere protein P130 in rat (Hibino *et al.,* 1993). Perhaps the best documented sequence motif is the previously mentioned 17-bp "CENP-B box" found in human

$\alpha$ satellite DNA that specifically binds CENP-B (Masumoto *et al.,* 1989). A similar sequence motif has also been found in mouse minor satellite DNA (Wong and Rattner, 1988). Considering the large quantity of centromere DNA in the total genome (5% for human and 10% for mouse, for example) (Willard and Waye, 1987; Wong and Rattner, 1988), centromeric DNA may represent DNA–matrix binding elements that are stronger and in larger scale than ordinary MARs. In addition to the possible functions mentioned earlier, centromere DNA binding proteins could play an important role in high-order chromosome structure and positioning in both interphase and mitosis.

### 4. Role in High Order Chromatin Organization

An important function of constitutive centromere protein during interphase may be the modulation of unfolding and refolding of 30-nm DNA fibers associated with centromere replication. In recent studies, we have identified a chracteristic pattern of movement and organization of the centromere/prekinetochore complex during the interphase nuclear cycles of Indian muntjac, Chinese muntjac, and HeLa cells by immunofluorescence and immunogold labeling (He and Brinkley, 1995). During early $G_1$, prekinetochores appear as round patches with unstained electron-dense cores. At late $G_1$, these loci unfold and become greatly elongated. In Indian muntjac cells, for example, a single prekinetochore can lengthen up to 100 times its metaphase length. In all cell lines, unfolded prekinetochores display numerous repetitive subunits resembling beads on a string. A similar elongation of the kinetochores of mitotic chromosomes can be induced by hypotonic stretching (Zinkowski *et al.,* 1991). The natural unfolding and refolding of the prekinetochore prior to and following replication may be the best example of high-order structural changes seen within chromatin segments in an intact nucleus. Significantly, morphological changes such as those described above can also be seen in nuclei after chromatin has been extracted by nuclear matrix preparation procedures. Furthermore, embedding-free immunolabeling EM revealed that throughout interphase, prekinetochores are firmly associated with the nuclear core filament network (He and Brinkley, 1995; see also Fig. 9).

### 5. Role in Interchromosomal Arrangement

Centromeric proteins appear to be involved in specific chromosomal positioning in the interphase nucleus. It is well known that arrays of prekinetochores assume nonrandom distributions within nuclei of many cell lines, and the distribution patterns are often associated with cell differentiation and/or tumorigenesis (Manuelidis, 1990; Brinkley *et al.,* 1986; Haaf and

Schmid, 1989; Spector, 1993). Of particular interest are Chinese muntjac cells, whose centromeres become rearranged within the nucleus from a random pattern of distribution to a series of linear arrays (Brinkley *et al.*, 1984) prior to the start of replication (D. He and B. R. Brinkley, unpublished data). A somewhat similar phenomenon has also been seen in mouse 3T3 cells. In mitosis, the centromeres are also arranged in a specific pattern on the metaphase plate reflecting the strict organization of chromosomes on the mitotic spindle (Hoskins, 1968). These observations suggest that centromeres, perhaps in association with the nucleoskeleton or a spindle matrix, display a dynamic pattern of interchromosomal organization throughout the division cycle. When an antiserum recognizing CENP-B and CENP-C is used in immunogold EM localization studies, prekinetochores are heavily labeled and represent distinct assemblies associated with the nucleoskeletal matrix network (He and Brinkley, 1995). These observations support the hypothesis that interchromosomal arrangements are mediated by intercentromere connections to the nuclear matrix proteins.

### 6. Structural Role during the $G_2$/M Transition

During the transition from interphase to mitosis, chromatin condensation appears to initiate at the centromere regions, which appear to represent the initial condensation center for the chromosome arms (Fig. 5) (He and Brinkley, 1995). Immunogold EM localization using CREST autoantibodies shows that at early prophase, each chromosome displays an initial focus of condensation at the centromere consisting of an area of about 1 $\mu m^2$ associated with the lamin fibers. Each condensation focus contains a pair of intensely labeled prekinetochores (Fig. 7). As prophase progresses, more chromatin condensation occurs in the flanking region of these initial foci resulting in further migration of the condensation band along the chromosome axis extending perpendicular to the connection between the paired prekinetochores. Another interesting feature of the interphase/mitosis transition in the centromere region is the development and differentiation of the kinetochore plates. The first indication that structural differentiation of the plates is taking place is at early prophase where a dense chromatin mass can be seen by electron microscopy at the primary constriction (Brinkley and Stubblefield, 1970; Roos, 1973). As condensation continues, mature kinetochore plates appear at about the same time that nuclear envelope disruption occurs. We have carefully examined this region by immunogold EM using antikinetochore CREST antibodies and found that the pattern of labeling of the kinetochore region is species-specific, with strikingly different patterns seen between chromosomes of primates and those of nonprimate species. In human and other primate chromosomes, the CREST label is distributed throughout the centromere region forming a band that

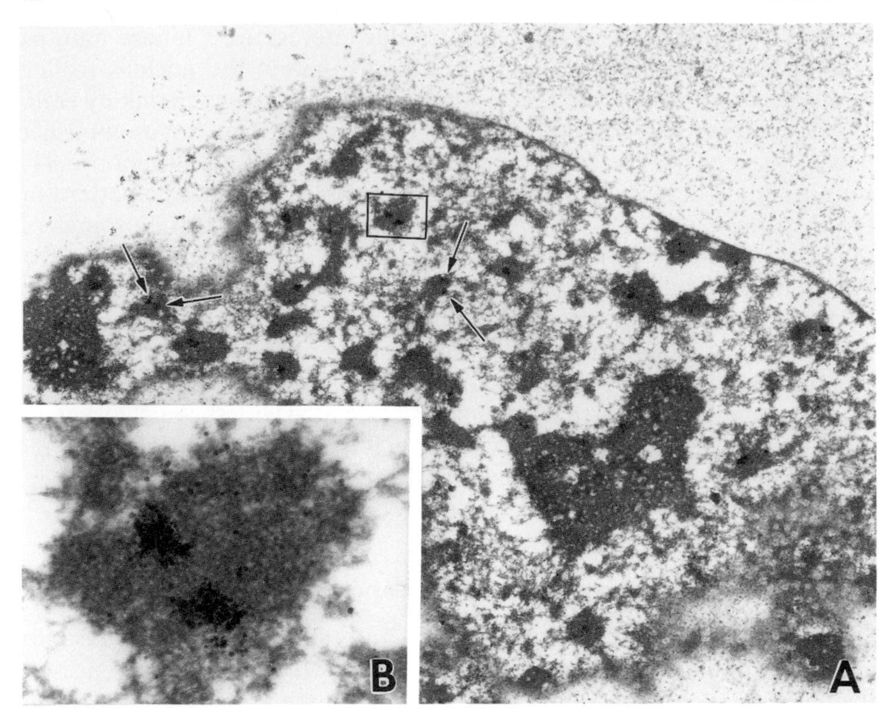


FIG. 7 Early prophase of Chinese muntjac cell, immunogold labeled with antikinetochore antibodies. A, low magnification. Chromosome condensation is initiated at the centromere (box). Arrows show specific labeling at the centromeres. B, higher magnification showing duplicated centromeres.

extends from the inner plate of one sister kinetochore to the inner plate of the opposite sister with the greatest concentrations just under each plate (Cooke *et al.,* 1990; see also Fig. 8). In chromosomes of nonprimates including CHO, PtK1, Indian and Chinese muntjac cells, however, the label is confined entirely to a narrow zone just under each of the inner kinetochore plates. We interpret these differences in labeling pattern to reflect variations in the binding of CREST antigens along centromeric DNA. In human and other primates, binding appears to be distributed over very long stretches

---

FIG. 8 Immunogold-labeled EM micrographs showing antikinetochore antibodies. In HeLa cells (A), the CENPs are distributed continually between two kinetochores (arrows), while in Indian muntjac cells (B), the labeling is distributed on or adjacent to the inner kinetochore plate (arrows).

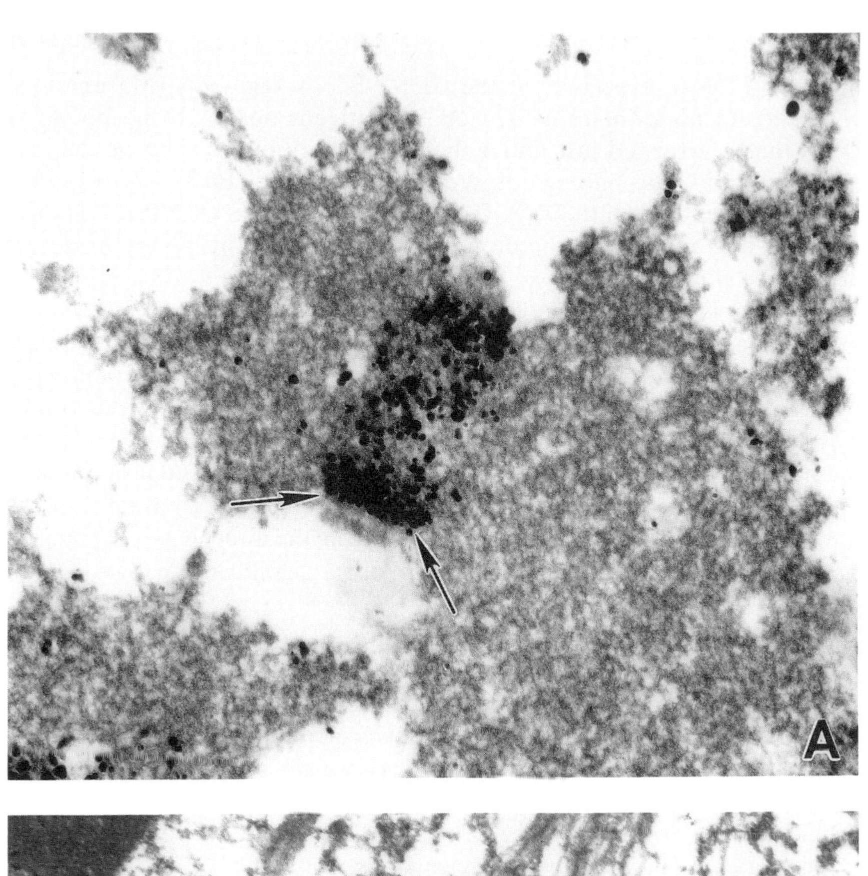
A

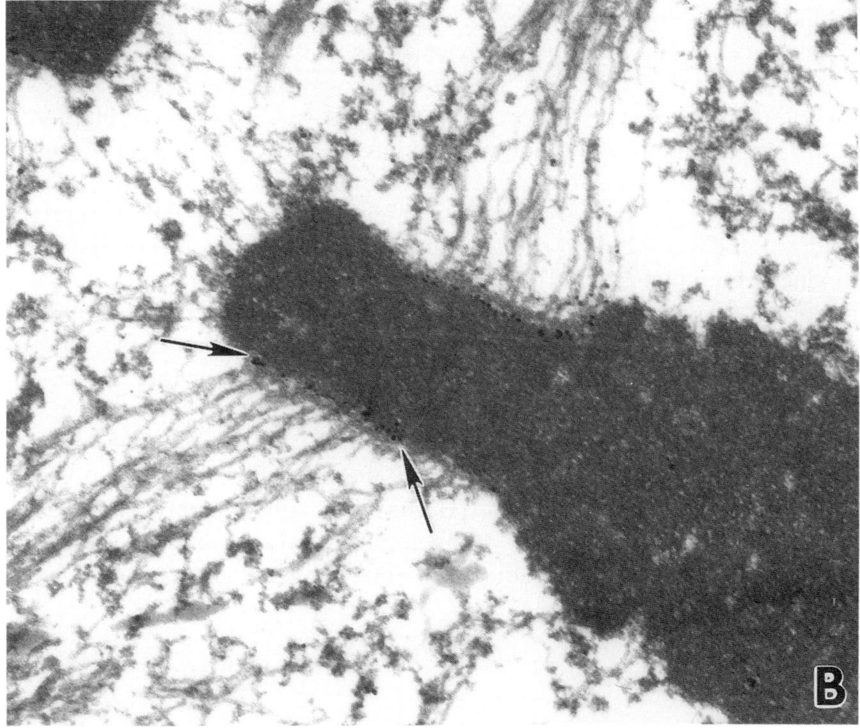
B

of alphoid DNA, especially at the CENP-B box segments (Masumoto *et al.*, 1989). In nonprimate cells, CREST antigens appear to be bound to DNA that is arranged just under the kinetochore plates. This model is in agreement with the findings of Wong and Rattner (1988) who reported that minor-band satellite DNA (containing the mouse CENP-B box) was confined to a zone immediately under the kinetochore plates of mouse metaphase chromosomes.

Although kinetochore plates can no longer be seen on prekinetochores in interphase nuclei, the pattern of CREST staining seen on metaphase chromosomes is conserved in prekinetochores. After each centromere duplicates in $G_2$, two separate and distinctly stained dots can be seen separated by a distance of about 0.8 $\mu$m in nonprimate nuclei; whereas in primates the stained dots appear to be connected by a fluorescent band just as it does on metaphase chromosomes, suggesting that the underlying framework of the centromere region has been transposed from the interphase ($G_2$) nuclear matrix to the chromosome scaffold without a significant change in morphology.

Even though such immunogold EM images may indicate basic differences in the organization of centromeric DNA in mammalian chromosomes, they by no means contradict the notion that prekinetochores are associated with the chromosome scaffold and that centromeric–DNA anchors the prekinetochore to the nucleoskeleton in interphase nuclei. Indeed, our recent immunogold EM images from resinless sections clearly show the prekinetochore to be intimately associated with the core filaments in the nucleoskeleton in a variety of species (Figs. 7 and 9).

Prekinetochores appear to be associated with nucleoskeletal elements; however, it is not yet clear how nuclear matrix-bound proteins relate, if at all, to the formation of kinetochore plates. Indeed, at this stage of our knowledge, the composition of the trilaminar plates of mitotic chromosomes remains a mystery. In this regard, when antikinetochore CREST immunogold staining is used, most of the gold particles appear not to bind directly to kinetochore plates. CENP-B and -C, however, appear to be located near the plates (Cooke *et al.*, 1990; Saitoh *et al.*, 1992), and these antigens resist extraction in high salts, making them candidates for nuclear matrix proteins and possibly involving them in the assembly of kinetochore plates and function of the kinetochore in chromosome movement. This notion is supported by microinjection experiments where CREST immunoglobulin injected into interphase nuclei resulted in aberrant chromosome movement and alignment in the next mitosis (Bernat *et al.*, 1990; Simerly *et al.*, 1990). Moreover injection of CENP-C monoclonal antibody resulted in the assembly of kinetochore plates of reduced size and altered organization (Tomkiel *et al.*, 1994).

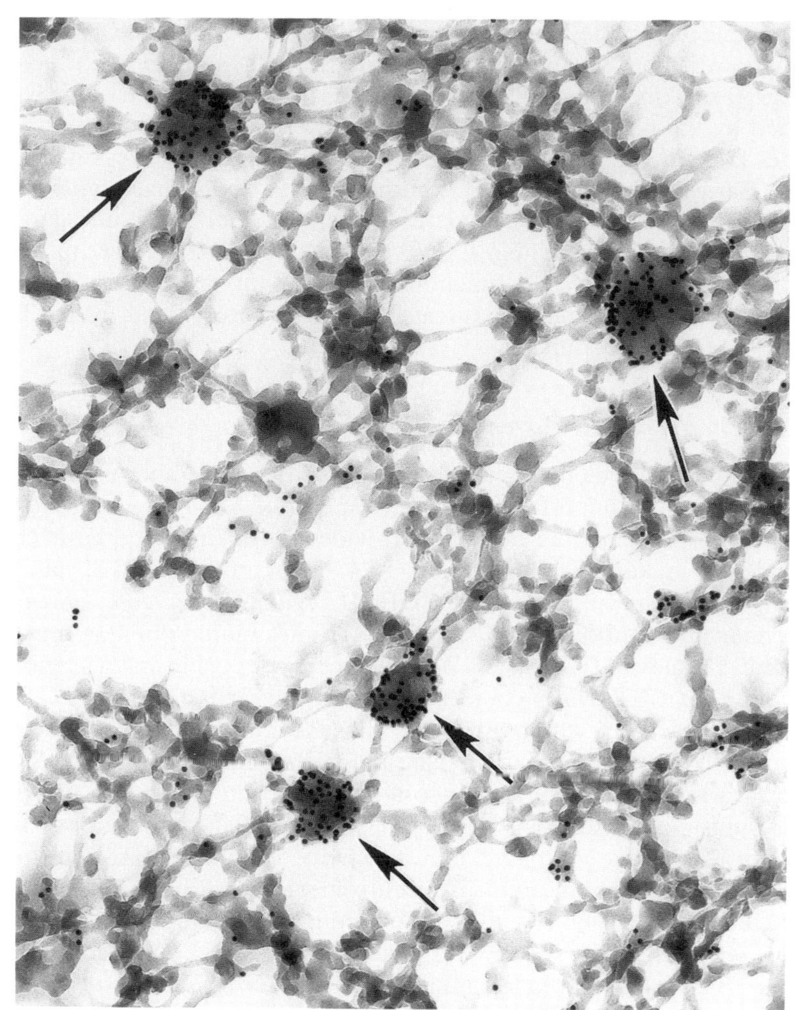

FIG. 9 Immunogold label with antikinetochore antibodies on embedding-free section of HeLa nuclei at $G_1$ phase. In this preparation, prekinetochores are tightly associated with core filaments. The label is confined to the periphery and each prekinetochore contains an unlabeled central core (arrows).

The relationship of CENP-B to kinetochore plates is further complicated by the observation that human Y chromosomes appear to lack CENP-B and have no CENP-B box (Earnshaw *et al.,* 1989), yet these chromosomes display functional kinetochores with fully developed trilaminar plate assemblies. Although it is possible that a small amount of CENP-B is present but undetected in the Y chromosome centromere, its absence as shown by immunocytochemical staining may argue that this antigen is not actually required for kinetochore differentiation and function. Indeed, this is supported by the observation that inactive centromeres in dicentric chromosomes contain CENP-B but are unable to bind and stabilize kinetochore MTs (Earnshaw *et al.,* 1989). On the other hand, CENP-C has been reported to be absent in inactive centromeres. Taken together, these immunocytochemical studies suggest that CENP-C, but not CENP-B, is required for kinetochore assembly and function. Proof for this hypothesis, however, must await more definitive data such as might be obtained from CENP-gene knockout experiments.

## 7. Role in MT-Based Movement

Nuclear matrix proteins located at or near the centromere/kinetochore complex may be important for some aspects of MT-based chromosome movements. Perhaps the most direct evidence for this hypothesis comes from lower eukaryotic cells where a 240-kDa protein, CBF3, has been found to bind specifically to the CDE III region of the yeast centromere (Lechner and Carbon, 1991). The high affinity binding of this multiunit protein complex to the yeast CEN sequence is so specific that a single base mutation on CEN DNA could abolish the interaction (Ng and Carbon, 1987). Significantly, affinity-purified CBF3 contains an ATP-dependent motor activity that is able to mediate the movement of CEN DNA-coated microbeads along MTs in a plus-to-minus direction (Hyman *et al.,* 1992). Thus, most likely CBF3 is a protein responsible for centromere DNA binding as well as MT-based motor activity. Another yeast centromere protein, CBF5p, has low affinity for centromere DNA (Jiang *et al.,* 1993a,b) and contains the same repetitive MT-binding motif as found in MAP 1A and MAP 1B (Noble *et al.,* 1989; Langkopf *et al.,* 1992). Moreover, CBF5p has also been shown to bind to bovine MTs *in vitro* (Jiang *et al.,* 1993a,b). On the other hand, the evidence for involvement of centromere DNA-binding proteins in MT-based movement in higher eukaryotic cells is somewhat indirect. Microinjection of CREST antiserum recognizing antigens CENP-A, -B, and -C in interphase cells prevented chromosomes from following normal prometaphase movements (Bernat *et al.,* 1990; Simerly *et al.,* 1990). Microinjection of PtK2 cells with CENP-B, -C, and -D decreases chromosome velocity to near zero (Wise and Bhattacharjee, 1992). In HeLa cells, CENP-B was found to exist adjacent to tubulin *in vitro* as shown by the chemical crosslinking experiments (Balczon and Brinkley, 1987). In addition, CENP-B can bind to the C-terminal region of tubulin as do the MAPs (Armas-Portela *et al.,* 1992). Interestingly, a subunit of yeast CBF3 protein contains a short stretch of serines similar to that found in CENP-

B (Lechner and Carbon, 1991; Doheny *et al.*, 1993). Unlike CBF3, however, CENP-B appears not to be a motor protein. It should be noted that yeast chromosomes have no platelike kinetochore structure and the MTs appear to attach directly to centromere DNA. Moreover, yeast appear not to have developed a specific motor protein homologous to dynein and kinesin. Although the functional concept of the centromere appears to be evolutionarily conserved in all eukaryotic species, the structure, DNA sequence, and protein composition, including MT motors, vary widely throughout the plant and animal kingdoms. Although we know relatively little about nuclear matrix proteins in yeast and lower eukaryotes, it is clear that nuclear matrix proteins such as P330, INCENPs, AP-10, and CENP-F and antigens of mAB37A5 and mAB154 transiently associate with the kinetochore during prophase or prometaphase and become associated with the surface of the outer layer. Such facultative association suggests that these proteins may play an important role in checkpoints or related regulatory events at the centromere/kinetochore region.

## 8. Role in Signaling

Kinetochore function, like many other cell cycle-dependent processes, appears to be regulated in part by control surveillance or mitotic checkpoints (Murray, 1992). For example, if kinetochores fail to align properly at the metaphase plate or lose their attachment to the spindle MTs, mitotic progression is delayed (Zirkle, 1970; Nicklas and Arana, 1992). Thus, some proteins of the kinetochore may serve in signal transduction associated with checkpoint regulation during the mitotic process. Indirect evidence for this hypothesis comes from the arrest of yeast cells by mutation or depletion of a centromere protein CBF5. The truncated CBF5 gene product with reduced copies of MT-binding motifs has been shown to delay cell cycle progression at $G_2$/M transition, while depletion of CBF5 protein arrests most cells at the $G_1$/S position (Jiang *et al.*, 1993b). In mammalian cells, microinjection of purified CREST antibody in $G_2$ phase impairs the stability of the kinetochore and arrests cells in metaphase (Simerly *et al.*, 1990; Bernat *et al.*, 1991). Once the chromosomes begin to separate at anaphase, microinjection of CREST antibodies fails to interfere with segregation because the metaphase checkpoint has been completed. Recently, a class of phosphoepitopes recognized by mAB3F3/2 has been found to express asymmetrically at each of the sister kinetochores in prometaphase chromosomes while congressing to the metaphase plate (Gorbsky and Ricketts, 1993). It was noted that the sister kinetochores that most recently become stably attached to the MTs were labeled more intensely than the others. More significantly, kinetochores of chromosomes that have become properly aligned on the metaphase plate cease to display this unique phosphoepitope staining feature. Therefore, the phosphorylation and dephosphorylation of 3F3/2 antigen appears to serve as a signal responsive to the kinetochore stability, chromosome positioning, and the MT-binding state. This observation suggests that kinases and phosphatases are directly involved in the metaphase checkpoint. An interchromosoml signaling pathway involving the metaphase checkpoint appears to assure that all chromosomes are properly aligned at the metaphase plate prior to the onset of anaphase (Gorbsky and Ricketts, 1993). Although such a provocative kinetochore-based checkpoint model remains largely unproven, the key players in the signaling pathway could be facultative proteins associated with this locus, including those derived in part from the nuclear matrix.

In summary, two classes of nuclear matrix-derived proteins, constitutive and facultative, appear to be involved in centromere/kinetochore activities. Some constitutive centromere proteins, including CENP-B and C, appear to be bound to the scaffold or nucleoskeleton. These proteins, perhaps in conjunction with the other matrix proteins, may also be involved in the organization of the prekinetochore, including DNA binding and unfolding/refolding associated with centromere DNA replication. Other important kinetochore functions, including capturing and stabilizing MTs, may also involve constitutive nuclear matrix proteins, but the extent and specificity of such involvement remains untested. Some facultative proteins of the centromere are also localized in the nuclear matrix throughout interphase, but the functions of these proteins also remain undetermined. Perhaps the most intriguing question relates to the transient association of some nuclear matrix-derived proteins with the kinetochore prior to the onset of anaphase. It will be important to characterize these proteins and determine their role in checkpoint signaling and other functions required for chromosome alignment and sister chromatid separation.

## D. Spindle Midzone during Metaphase to Anaphase Transition

The midzone and interzone, a compartment of the mitotic spindle midway between the spindle poles, is identified by an antiparallel arrangement of nonkinetochore MTs extending perpendicular to the metaphase plate. This zone becomes detectable at early anaphase when sister chromosomes separate and migrate to opposite poles, but as will be described later in this section, the midzone may actually begin to form at metaphase. MTs in the midzone will ultimately constitute part of a cytoplasmic bridge, or midbody, connecting daughter cells at the end of telophase. Several important mitotic events involve the midzone, including chromosome disjunction, spindle elongation at anaphase B, and formation of the cleavage furrow. Moreover, many protein components of this region appear to be of the insoluble variety usually identified with the interphase nuclear matrix. Although the pathway for the redistribution of the nuclear matrix proteins to the midzone is unclear, it appears to involve chromosomes as early as prophase, particularly near the centromere/kinetochore regions. At the onset of anaphase, however, many of these proteins leave the chromosome and become associated with the spindle midzone. Proteins that transiently associate with chromosomes in this manner have been termed "chromosome passengers" (Earnshaw and Bernat, 1990). As such, they appear to use the chromosomes as a means of conveyance to the cell equator where they are thought to carry out important, but as yet unknown, functions. It is tempting to speculate that their association with centromeres at prophase may activate or regulate the kinetochores during this phase of mitosis. The migration of these proteins from the nuclear matrix to the chromosome, and then onto the midzone, implies an even more involved role in the complex process of mitosis.

A growing number of proteins in this group have been described, including 135- and 150-kDa INCENPs (inner centromere proteins) (Cooke *et al.*, 1987), mAB 37A5 antigen (Pankov *et al.*, 1990), 400-kDa CENP-F (Rattner *et al.*, 1993), and P330, which may be the same protein as CENP-F (Casiano *et al.*, 1993) (Table II). In addition to these proteins, the 95- and 105-kDa CHO1 antigen or MKLP-1 (mitotic kinesin-like protein 1; Sellito and Kuriyama, 1988) most likely belong to the same group. They display the same transient distribution behaviors and their mitotic immunofluorescent staining is resistant to RNase and DNase treatments. Whether or not staining persists in the interphase nucleus after extensive extraction is yet to be determined (Sellito and Kuriyama, 1988; Nislow *et al.*, 1992). The precise distribution patterns of some of these proteins have been defined by confocal microscopy and immunogold staining electron microscopy (Earnshaw and Cooke, 1991; Casiano *et al.*, 1993; Rattner *et al.*, 1993).

P330, INCENPs, and CENP-F distribute in either a punctate or uniform staining pattern throughout the nucleus in a cell cycle-dependent manner. The immunofluorescent staining of INCENPs is intensified from late $G_1$, while the staining of CENP-F/P330 is obviously increased from late S or $G_2$ phase, and both P330 and CENP-F reach peak levels prior to prophase (Casiano *et al.*, 1993; Rattner *et al.*, 1993). For the most of the remaining proteins, there appears to be little significant staining until prometaphase, although immunoblotting has shown that at least some are present in the interphase nucleus (Pankov *et al.*, 1990). Possibly, their lack of staining in interphase is due to some form of masking by associated chromatin. As cells progress into mitosis, their epitopes appear to become unmasked. From prophase or prometaphase, these proteins gradually concentrate at the centromeres in distinct pathways. P330 first moves through the narrow space between chromosomes while INCENPs migrate directly to both the centromere and telomere. At early metaphase both proteins associate specifically with the centromere/kinetochore (Cooke *et al.,* 1987).

Interestingly, many proteins in this class are found to associate with specific domains of the centromere/kinetochore. CENP-F/P330 and an antigen of mAB 37A5 are localized specifically at the surface of the outer kinetochore plate (Pankov *et al.,* 1990; Casiano *et al.,* 1993; Rattner *et al.,* 1993) while INCENPs are specifically concentrated between sister chromatids along the inner centromere as indicated by their names. It should be noted that INCENPs were originally identified in colcemid-blocked cells (Cooke *et al.,* 1987). Later it was reported that in nonblocked cells, INCENPs appeared dispersed throughout the entire centromere/kinetochore region instead of the inner centromere (Earnshaw and Cooke, 1991), perhaps due to the disruption of MTs altering the organization of centromere/kinetochore proteins.

The remarkable detachment and migration of proteins from the centromere to the midzone at the onset of anaphase deserves additional investigation. Immunofluorescence studies with most antibodies specific for these proteins have shown well-defined parallel strings in the inner part of the midzone, and this pattern is maintained until telophase as the strings shorten progressively (Cooke *et al.,* 1987; Nislow *et al.,* 1992). As shown in late metaphase, by immunogold EM labeling, the INCENPs form thin streaks parallel to the spindle axis that traverse the metaphase plate between the chromosomes. During early anaphase, INCENPs are localized on fibers (probably microtubules) that are organized into a highly ordered hollow cylinder. In this regard, a 42-aa domain in INCENPs has been found to be required for their translocation from the chromosome to the midzone. This migration pattern appears to be evolutionarily conserved because when chicken INCENPs are expressed in human and pig cells, the characteristic pattern of distribution is displayed during the cell cycle (Mackay *et al.,* 1993). Additionally, a portion of INCENPs moves to the cortex where the cleavage furrow will form during late anaphase. By telophase, all of these proteins are found concentrated at the midbody between two daughter cells, where they are most likely discarded (Nislow *et al.,* 1992).

Several other nuclear proteins redistribute to the midzone of the spindle during mitosis using pathways somewhat different from those described above. For example, a ~205-kDa protein identified by mAB3G3 antibody has been shown to display a spotted pattern throughout the interphase nucleus (Compton *et al.,* 1991). In mitotic cells, this protein is localized at the midzone as well as the pole region of the spindle (Compton *et al.,* 1991). The mitotic kinesin-like protein 1, MKLP-1, recognized by CHO1 antibody, is localized in both the nucleus and centrosome in interphase. At metaphase it concentrates principally at the spindle pole regions leaving spindle fibers only faintly stained. With further progression into mitosis, it concentrates at the midzone and organizes into discrete short lines along the spindle axis and eventually aggregates in the midbody of the bridge between two daughter cells (Nislow *et al.,* 1992). Therefore, this particular group of proteins demonstrates an alternate redistribution pathway, whereby nuclear proteins reach the midzone not by way of chromosomes, but by way of the centrosomes.

The role of spindle-associated nuclear matrix proteins in the mitotic process is not well understood. Their relocation from the nuclear matrix to the mitotic apparatus and persistence throughout mitosis suggests that they are indeed essential for some aspect of M phase. It seems unlikely that they are passive "passengers." The initial localizations of many of these proteins at the centromere/kinetochore region at prometaphase implies that they may be important factors in the early movement and alignment of chromosomes. For example, CENP-F translocates to the centromeres as early as prophase even before the nuclear envelope breaks down (Rattner *et al.,* 1993). Such specific targeting would suggest a possible role in the maturation of the centromere, or the establishment of the kinetochore, a key event in the mitotic process.

Nuclear matrix proteins localized in the midzone migrate to this region of the spindle very early, implying that they may be essential in some way for the metaphase to anaphase transition. Indeed, microinjection of purified CHO1 monoclonal antibody interfered with the metaphase–anaphase transition in PtK1 cells (Nislow *et al.,* 1990). As mentioned earlier, it is certainly possible that proteins located at the centromere/kinetochore region at prometaphase or early metaphase may play a role in capturing and stabilizing the plus-ends of MTs in the kinetochore at metaphase (Pankov *et al.,* 1990). The subsequent translocation of these proteins to the midzone could possibly inactivate MT stabilization, thereby initiating depolymerization at the kinetochore end, causing MTs to shorten (Gorbsky *et al.,* 1987; Mitchison, 1988) at the onset of anaphase. Any or all of these events could be involved in the mitotic checkpoint process associated with kinetochores as described earlier (see also Gorbsky and Ricketts, 1993). It seems unlikely that this group of proteins would have motor functions required for anaphase chromosome movement, since they abruptly leave chromosomes prior to the anaphase event. However, their abrupt departure from the centromere and association with the polar end of MTs could stabilize this region as part of the spindle elongation process of anaphase B.

Finally, we cannot rule out the possibility that these proteins simply constitute part of a complex mitotic matrix that connects and integrates spindle function in mitosis. The most predominant components of the midzone are antiparallel MT bundles extending from opposing poles. It is widely accepted that the elongation of the spindle in anaphase B depends

TABLE II

Nuclear Matrix Proteins in the Mitotic Apparatus

| Protein or antibody | MW (kDa) | Interphase | Mitosis | Reference |
|---|---|---|---|---|
| Topo II (Scl) | $\alpha$: 170<br>$\beta$: 180 | Nuclear matrix (S phase), proliferation-related | Chromosome scaffold | Berrios *et al.* (1985) |
| ARBP | 95 | Nuclear matrix | Chromosome scaffold | von Kries *et al.* (1991) |
| SAF-A | 120 | Nuclear matrix | Chromosome scaffold | Romig *et al.* (1992) |
| SP120 | 120 | Nuclear matrix (stained diffusely or as spots) | Chromosome scaffold | Tsutsui *et al.* (1993) |
| RAP-1 (TUF or GRF1) | 116 | Nuclear matrix | Chromosome scaffold | Hofmann *et al.* (1989) |
| SATB1 | 90 | Nuclear matrix | Chromosome scaffold | Dickinson *et al.* (1992) |
| Lamins | A: 70<br>B: 67<br>C: 60 | Nuclear periphery | Cytoplasm, and late telophase at perichromosome | Gerace and Blobel (1980); Ludérus *et al.* (1992) |
| P130 | 130 | Centromere | NA[a] | Hibino *et al.* (1993) |
| Matrin 4 | 105 | Nuclear matrix | NA | Nakayasu and Berezney (1991) |
| Matrins D, E, F, G | 60–75 | Nuclear matrix | NA | Nakayasu and Berezney (1991) |
| Rb protein | 110 | Nuclear matrix in early G1 | Chromosome scaffold and cytoplasm | Mancini *et al.* (1994) |
| CENP-B | 80 | Centromere (nuclear matrix) | Kinetochore inner layer or inner centromere | Earnshaw and Rothfield (1985) |
| CENP-C | 110 | Centromere (nuclear matrix) | Kinetochore inner layer | Earnshaw and Rothfield (1985) |
| CBF-3 (yeast) | 240 | Centromere | Centromere | Lechner and Carbon (1991) |

| | | | | |
|---|---|---|---|---|
| MPM2 | NA | Nuclear matrix | Centromere, spindle poles, and midzone | Binarova *et al.* (1993) |
| 28 kDa Sm antigen | 28 | Speckles in nuclear matrix | Perichromosome and cytoplasm | Spector and Smith (1986) |
| fA12 | 33, 40 | Speckles in nuclear matrix | Perichromosome and poles | He *et al.* (1991) |
| H1B2 | 250 | Speckles in nuclear matrix | Perichromosome and poles | Nickerson *et al.* (1992) |
| PCN | 30, 36 | Nucleolus | Perichromosome | Shi *et al.* (1987) |
| Fibrillarin | 34 | Nucleolus | Perichromosome | Ochs *et al.* (1985) |
| Ki-67 | 345, 395 | Nucleolus | Perichromosome | Gerdes *et al.* (1991) |
| B23 | 38 | Nucleolus | Perichromosome | Ochs *et al.* (1983) |
| G04 | 94 | Nucleolus | Perichromosome | Gautier *et al.* (1992b) |
| Perichromin | 33 | Nucleolus | Perichromosome | McKeon *et al.* (1984) |
| P1 or peripherin | 27–31 | Nuclear periphery | Perichromosome | Chaly *et al.* (1984) |
| INCENPs | 135, 150 | Nuclear matrix | Kinetochore to midzone | Cooke *et al.* (1987) |
| 37A5 | 140, 155 | Nuclear matrix | Kinetochore to midzone | Pankov *et al.* (1990) |
| CHO1 (MKLP-1) | 95, 105 | Nucleus and centrosome | Spindle pole to midzone | Sellitto and Kuriyama (1988) |
| CENP-F | 400 | Nuclear matrix | Kinetochore to midzone | Rattner *et al.* (1993) |
| P330 | 330 | Nuclear matrix | Kinetochore to midzone | Casiano *et al.* (1993) |
| NuMA | 200–240 | Nuclear matrix (diffuse or speckles) | Spindle poles | Lydersen and Pettijohn (1980) |
| B1C8 | 160, 180 | Speckles in nuclear matrix | Spindle poles and midbody | Wan *et al.* (1994) |
| Mitotin | 125 | Speckles (proliferation related) | Concentrated in poles, also in spindle | Todorov *et al.* (1992) |
| Mitosin/CENP-F | 350 | Nuclear speckles | Kinetochore to poles | Zhu *et al.* (1995) |
| 5E3 | 205 | Nuclear punctate | Spindle poles | Compton *et al.* (1991) |
| 3G3 | 205 | Nuclear matrix | Spindle poles and midzone | Compton *et al.* (1991) |

[a] NA, not available.

on the MT plus end stabilization and the sliding of antiparallel MTs from the adjacent matrix (McDonald *et al.,* 1979; Wordeman *et al.,* 1989). Such sliding is reflected by the shortening of the MT overlapping region in the direction of the spindle axis. Since many nuclear matrix proteins become relocated to the inner portion of the midzone, they may be essential for the structure and function of the spindle in anaphase. In this regard, it has been shown that the expressed CHO1 antigen can form a cross-bridge connecting antiparallel MTs *in vitro.* In the presence of MT-ATP, these MTs slide in a fashion reminiscent of MT movements in anaphase B (Nislow *et al.,* 1992).

It should also be noted that the midzone is the region where the two half-spindles are held together, a feature that might require an extensive matrix needed to cross-link the two half-spindles. In this regard, it is well known that the spindle midzone contains an electron-dense, osmiophilic matrix that coats antiparallel MTs (Buck and Tisdale, 1962; McIntosh and Landis, 1971). In conventional EM, this matrix can be seen only as a dense "cloud" similar to that found in the perichromosomal or pericentriolar area. Using extracted cells and the embedding-free EM technique, a nonmicrotubular network or mitotic matrix is resolved in this area (Capco and Penman, 1983; Nickerson and Penman, 1992). In the midzone and midbody, these filaments are found to connect to the bundled MTs. Therefore, given their characteristic immunofluorescence and immuno-EM staining patterns, most of the nuclear matrix proteins discussed in this section are likely to be associated with or part of this dense fibrogranular mitotic matrix. Indeed, sequence analysis has demonstrated that a number of these proteins may contain a coiled-coil structure. For example, CHO1 antigen contains a 200-amino-acid domain that may form a coiled-coil rod much like myosin and kinesin heavy chains (Nislow *et al.,* 1992). INCENPs have a >200 aa domain that has primary sequence similarity to the coiled-coil domain of myosin, tropomyosin, and keratins (Mackay *et al.,* 1993). Thus, their association with this matrix network may play a major role in structural reinforcement of the two half-spindles, thereby mediating MT movements associated with anaphase B.

Not to be overlooked is the important observation that INCENPs appear to be involved in determining the location of cleavage furrow formation. Indeed, the targeting of a portion of INCENPs to the cortex, where the furrow will form, may signal one of the earliest events in this important process (Earnshaw and Cooke, 1991).

In summary, this unique group of matrix proteins displays a dynamic translocation from the nucleus to the centromere/kinetochore regions in the early phases of mitosis and ultimately come to rest at the midzone at anaphase forming a fibrogranular mitotic matrix network. During their brief association with the chromosome in early mitosis, it is tempting to speculate that they may play some function in the maturation of the centromere/kinetochore complex and be involved in the checkpoint regulation of chromosome alignment at the metaphase plate. In the midzone, they appear to enshroud antiparallel MTs and possibly function in the events of spindle elongation at anaphase B.

## E. The Midbody

The midbody is the last remnant of the mitotic spindle representing a thin, sometimes elongated cytoplasmic bridge that connects daughter cells at the end of mitosis. Composed of tight bundles of antiparallel MTs, the midbody forms in late anaphase and persists into late telophase or to $G_1$ phase. Electron microscopic studies have demonstrated that the amorphous matrix loosely surrounding bundles of microtubules in the midzone in anaphase gradually condenses into a rigid midbody. Also, electrophoretic analysis has indicated that isolated midbodies have a very similar protein composition to that of mitotic spindles with over 35 bands identified on SDS gels (Mullins and McIntosh, 1982). As mentioned earlier, most nuclear matrix proteins found in the midzone persist throughout anaphase and remain in the midbody at the end of telophase. Proteins from other mitotic regions also appear to relocate to midbody such as H1B2, fA12 (He *et al.*, 1991; Nickerson *et al.*, 1992). The midbody is a cellular organelle with well-defined ultrastructure that has only recently been analyzed by resinless EM sections (Nickerson *et al.*, 1992). The distribution of individual proteins in the midbody has been difficult to study due to the high density and opacity at this site. The antitubulin staining pattern is well known, consisting of a densely stained bridge with a central unstained region. There is little doubt that the unstained region is where MTs are overlapped and surrounded by a dense matrix blocking the tubulin epitope (Mullins and Biesele, 1977). The immunofluorescent staining patterns observed with antibodies against various nuclear matrix proteins have revealed a variety of patterns. For example, antibodies of INCENPs display a pattern similar to tubulin, whereas CENP-F antibody stains a sharp, narrow band spanning the central region. CHO1 staining, however, is confined exclusively to a central spot (Nislow *et al.*, 1992). Some antibodies such as the anti-P330 brightly stain the whole bridge but leave unstained two narrow bands at each side of the central region. Based on these reports it appears that the midbody is compartmentalized into at least three domains: the central region, the narrow bands, and the distal region. Each of these domains probably has a different protein composition and perhaps a different ultrastructural organization. EM micrographs have also shown a dense "cloud" surrounding parallel microtubules (Mullins and McIntosh, 1982). It is significant that the midbody can be imaged by the same procedures used to resolve nuclear core filaments using embedding-free electron microscopy (Nickerson and Penman, 1992). This observation provides strong evidence that compartments of the midbody share considerable homology with the nucleoskeleton. Parallel MT bundles are laterally connected by a filamentous network that also links MTs to cytoplasmic membranes. These filaments have various diameters and lengths and are associated with granular masses. It appears that the network is coincident with the dense matrix seen by conventional EM. An antibody raised against nuclear matrix protein H1B2 has been shown to decorate some, but not all, short 12- to 17-nm thin filaments that connect MT bundles, as well as other fibers in the filamentous network (Nickerson *et al.*, 1992). Collectively, these observations strongly support the notion that the midbody is a heterogeneous web composed of fibers that are morphologically and biochemically similar to those of the nucleoskeleton.

Although the function of the midbody is unknown, it probably assures further partitioning of the replicated genomes and maintains cytoplasmic continuity between the newly formed daughter cells. Failure of the midbody to form results in failure to maintain proper separation and can lead to the fusion of daughter nuclei or formation of binucleate cells. The rigid skeletal-like organization at this critical site is needed to stabilize the mitotic products. Once it completes its function, the midbody is discarded, taking with it proteins no longer needed. This feature has led some to refer to this discarded appendage as the cell's "garbage disposer."

### F. Spindle Poles

At the onset of mitosis, duplicated sets of centrosomes are partitioned to opposite ends of the cell and become the poles of the spindle. The actual molecular events leading to this important process, however, are complex and poorly understood. The spindle poles become enlarged in size at prophase with the accumulation of dense-staining pericentriolar material surrounding each centriole pair (Rieder and Borisy, 1982). At prophase, the spindle poles also serve as sites for the nucleation of many more individual microtubules than are assembled around the interphase centrosome. Indeed, *in vitro* microtubule assembly assays revealed that spindle poles can nucleate about five times more microtubules than their interphase counterparts (Kuriyama and Borisy, 1981). All these changes appear to occur in concert with the recruitment of new proteins around spindle poles during prophase (Leslie, 1990). Some proteins that migrate to the polar regions during prometaphase have been clearly identified as nuclear matrix proteins that become associated with spindle poles following the disruption of the nucleus at prophase. These include NuMA (see Section II), B1C8

(Wan *et al.*, 1994), mitotin (Todorov *et al.*, 1992; Montag, 1992), Mitosin/CENP-F (Zhu *et al.*, 1995), and mAB5E3 antigen (Compton *et al.*, 1991; see also Table II). In addition to those proteins confined exclusively to the poles, some nuclear matrix proteins are present at both the spindle poles as well as at other regions of the mitotic cell throughout mitosis. For example, as mentioned above, fA12 and H1B2 antigens appear to relocate to both the perichromosome and the pole regions, while mAB3G3 relocates to both the midzone and the spindle poles.

During interphase, all nuclear matrix proteins that were identified to target to the spindle poles ultimately demonstrate a speckled or diffused inner nuclear distribution pattern. At least two of them display a cell cycle-dependent immunostaining pattern. Mitotin is a proliferation-related nuclear matrix protein detected in a variety of cell lines but not in quiescent cells or in *in vitro* differentiated HL60 cells (Philipova *et al.*, 1987, 1991), while another protein, Mitosin/CENP-F, has been identified as one of the Rb-binding proteins. Both proteins display a marked increase in the intensity of staining during $G_2$/M and a decrease at the end of mitosis. It has been suggested that the $G_2$/M increase of mitotin is regulated by phosphorylation since it is synthesized throughout the cell cycle at a constant level (Zhelev *et al.*, 1990). The ultrastructural localization of several nuclear matrix-derived spindle pole components has been studied by immunogold EM labeling on resinless sections in core filament preparations. While B1C8, H1B2, and fA12 all decorated dense fibrogranular assemblies enmeshed in a filamentous network, the filaments themselves remained unstained (He *et al.*, 1991; Nickerson *et al.*, 1992; Wan *et al.*, 1994). The anti-NuMA antibody 2D3 decorated a subset of filaments along with some granular material (Zeng *et al.*, 1994b). Although these proteins are all associated with the spindle pole regions during mitosis, they arrive at the spindle poles at different times and through different pathways. NuMA and B1C8 antigen were found to be associated with the spindle poles at the time of nuclear envelope disassociation. Moreover, fA12 and H1B2 antigens are initially relocated to both perichromosome and pericentriolar regions at approximately the same time. In contrast, mitotin and Mitosin/CENP-F redistribute to the spindle poles at much later stages. At prophase, mitotin is relocated from interphase speckles to a region between the condensing chromosomes, but by the time cells reach metaphase, mitotin migrates to the pericentriolar area. While the spindle poles remain intensively stained for the rest of metaphase and anaphase, the whole spindle is diffusely stained. During telophase, immunostaining is concentrated to the dense midbody along with some large spots in the cytoplasm that ultimately become integrated into daughter nuclei (Montag, 1992; Todorov *et al.*, 1992). Mitosin/CENP-F, on the other hand, displays a totally different redistribution pathway. During prophase, it specifically concentrates at the centromeres where EM gold labeling shows the proteins to be located at the outer layer of the kinetochore or in the corona. After metaphase, kinetochore staining disappears but staining is noted in the pericentriolar dense matrix (not the centrioles) and some labeled assemblies are found associated with the MT bundles (Zhu *et al.,* 1995). The mitotic distribution of NuMA and B1C8 antigen have also been examined in detail using the resinless EM technique. B1C8 showed a transient cytoplasmic speckle pattern as described for mitotin. During metaphase and anaphase, the antigen is mostly concentrated at fibers surrounding the centrioles and as a small mass associated with MT bundles (Wan *et al.,* 1994). NuMA, on the other hand, is visualized at the spindle pole regions with different antibodies where it displays a crescent shape throughout mitosis. Unlike the other pericentriolar nuclear matrix proteins, mAB5E3 detects an antigen that is primarily concentrated and maintained at the spindle poles throughout mitosis (Compton *et al.,* 1991).

The spindle poles appear to play a major role in the mitotic spindle organization and chromosome segregation process, and it is likely that the nuclear matrix-derived proteins associated with this region help mediate these critical functions. Because the processes that occur in the nucleus and those that take place in mitosis are so divergent, the function of nuclear matrix-derived proteins in mitosis is likely to be considerably different from that in interphase. For example, the speckled distribution of B1C8 and NuMA in interphase nuclei corresponds to regions known to be rich in RNA splicing factors. Moreover, B1C8 antibody and some NuMA antibodies have been found to immunoprecipitate splicing factors *in vitro* (Blencowe *et al.,* 1994; Zeng *et al.,* 1994a). Like B1C8 and some NuMA antibodies, mitotin antibodies produce a speckled staining pattern in the nucleus and colocalize with components of the spliceosome. Therefore, it is likely that a major role of these proteins in the matrix of the intact nucleus involves RNA processing. Since NuMA is identified as a component of the nuclear core filaments, it may serve to anchor spliceosomal assemblies and other nuclear structures to the nucleoskeleton (Zeng *et al.,* 1994b). The nucleus is disrupted during mitosis in most eukaryotic cells and, of course, all RNA processing ceases. As cells progress from interphase to mitosis, the structure we envision as the interphase nuclear matrix obviously undergoes considerable transformation and redistribution. As we have noted throughout this chapter, many proteins of the nucleoskeleton make their way either directly or circuitously to strategic sites in the spindle. Since eukaryotic cells rarely tolerate and maintain "excess baggage," it seems reasonable that those large protein assemblies (fibrils and granules) of the nucleoskeleton that redistribute to mitotic cells would assume significant new function in mitosis.

The spindle pole is a likely place to identify and explore possible functions of nuclear matrix-derived proteins. It is well known that the pericentriolar material contains factors needed in the anchorage, nucleation, and assembly of spindle microtubules. Therefore, it is logical to correlate the nuclear matrix-derived components with these important functions. Indeed, previous EM studies using conventional fixation and embedding techniques have identified a cloud of fibrous, electron-dense material surrounding the centrioles. More recently, studies using resinless-section EM procedures have revealed a pericentriolar fibrogranular network that is not too unlike networks observed within the nucleoskeleton (Capco and Penman, 1983; Nickerson *et al.,* 1992). Indeed, the placement of nuclear matrix-derived proteins at the spindle poles argues strongly for a role in anchorage, assembly, and organization of MTs and in the overall maintenance of spindle structure. In this regard, it is interesting to note that these matrix proteins often associate with MT bundles that extend between the poles and the kinetochore.

Double immunofluorescent staining with anti-NuMA and antitubulin antibodies strongly suggests that NuMA may be essential for the nucleation of spindle MTs following recovery from mitotic inhibitors as described earlier (Kallajoki *et al.,* 1991; Tousson *et al.,* 1991). Involvement with microtubules is supported by microinjection experiments where either aberrant assembly of spindles or spindle collapse (Kallajoki *et al.,* 1991, 1993; Yang and Snyder, 1992) was produced, depending on the stage of the cell cycle when the antibodies were microinjected. Recent structural information on NuMA adds considerable support to its role in MT stabilization. Domains that target NuMA molecules to the spindle and back to the nucleus have been identified (see Section II,B). Collectively, therefore, NuMA and perhaps other nuclear matrix-derived proteins associated with spindle poles may play important roles in MT assembly and spindle formation.

Just as the core filaments of the nucleoskeleton can provide a solid-state substrate for important functions in interphase nuclei, the fibrogranular pericentriolar material could provide substrate for molecules involved in regulating mitosis. The switch between interphase nuclear matrix activity and that of the mitotic apparatus seems to be regulated in part by protein phosphorylation. In addition to the well-known example of lamins, NuMA, B1C8, mitotin, and AP-10 are all phosphoproteins regulated by kinases and phosphatases during progression into and out of mitosis. In this regard, it is noteworthy that MPM-2, an antibody that recognizes phosphoepitopes in mitotic cells, heavily stains spindle pole midbodies and kinetochores (Vandre *et al.,* 1984; Binarova *et al.,* 1993). Moreover, MPM-2 also blocks the nucleation of MTs from isolated centrosomes (Centouze *et al.,* 1986). Thus, it seems likely that a three-dimensional protein matrix located at the poles and at other sites within the spindle provides a lattice for regulating complex molecular signals that control the progression of mitosis in the same conceptual way that the nucleoskeleton is thought to control the progression of nuclear functions. According to this concept, the nuclear matrix may provide a structural and regulatory continuum for the cell's genome throughout the cell cycle as it relocates from the interphase nucleus to the mitotic apparatus and back to the nucleus again.

## IV. Concluding Remarks

Throughout the cell cycle, the genome of eukaryotic cells is contained within one of two structurally complex compartments, the mitotic apparatus or the nucleus. When mitosis ends and interphase begins, the entire chromosome complement becomes decondensed and rearranged within a space that is considerably different in structure and design from the mitotic spindle. Nevertheless, the nucleoplasm contains many of the same proteins found in the mitotic apparatus. The recycling of nuclear matrix proteins between interphase and mitosis suggests that chromatin is constantly embraced, and perhaps regulated by many of the same nonhistone proteins throughout all phases of the cell cycle. Some proteins, such as the scaffold proteins, probably play a role in chromosome condensation–decondensation as well as anchorage of DNA and reinforcement of structure. More than just providing a continuous matrix, however, many nuclear matrix proteins are apparently bifunctional and take on totally new roles in mitosis. For example, some may function in RNA processing in the nucleus, but become targeted to the spindle poles to associate with the minus ends of microtubules and function in chromosome movement and postmitotic nuclear assembly. One such protein, NuMA, displays great versatility due to its molecular design. As a large coiled-coil protein, it contains no apparent functional domains in its long central $\alpha$-helix except leucine zippers, but appears to be targeted to and from the nucleus by domains in the carboxyl terminus. NuMA also appears to be regulated by a number of kinases targeted to multiple phosphorylation sites in the C-terminal domain. In addition, several isoforms of NuMA proteins exist.

The targeting of various nuclear matrix-derived proteins to specific domains in the mitotic apparatus provides further evidence for their multifunctional roles in the two phases of the cell cycle. Although fluorescent antibodies enable their localization in the spindle, we can only speculate as to their possible functions based on what is known about specific regions of the spindle. Some proteins may simply be "passengers," being transported to daughter nuclei to facilitate critical nuclear functions in early $G_1$ phase, while others may form a stable matrix about the spindle. Of particular interest to students of mitosis is the possibility that spindle matrix proteins function in signaling. The molecular basis of mitotic checkpoints is essentially unknown, but interchromosomal signal transduction of the type envisioned at the metaphase–anaphase checkpoint might involve transmission of signals along the matrix fibers in the spindle. Obviously, additional studies are required before the complete significance of the relationship between the nuclear matrix and spindle matrix can be fully understood.

## Acknowledgments

Appreciation is extended to our colleagues Susan Berget, Ilia Ouspenski, and Dwayne Wise for helpful advice and consultation. We are especially grateful to Betty Ledlie for administrative assistance, proofreading, and typing of the chapter and to Kevin Brinkley for proofreading and editorial assistance. This research was funded in part by NIH Grant CA-41424 to B.R.B. Support and facilities were also provided through a grant from the W. M. Keck Foundation.

## References

Adachi, Y., Käs, E., and Laemmli, U. K. (1989). Preferential, cooperative binding of DNA topoisomerase II to scaffold-associated regions. *EMBO J.* **8,** 3997–4006.

Adachi, Y., Luke, M., and Laemmli, U. K. (1991). Chromosome assembly *in vitro:* Topoisomerase II is required for condensation. *Cell* **64,** 137–148.

Adolph, K. W., Cheng, S. M., and Laemmli, U. K. (1977a). Role of nonhistone proteins in metaphase chromosome structure. *Cell* **12,** 805–816.

Adolph, K. W., Cheng, S. M., Paulson, J. R., and Laemmli, U. K. (1977b). Isolation of a protein scaffold from mitotic HeLa cell chromosomes. *Proc. Natl. Acad. Sci. U.S.A.* **11,** 4937–4941.

Armas-Portela, R., Kremer, L., and Avila, J. (1992). The centromere protein CENP-B behaves as a microtubule-associated protein. *Acta Histochem., Suppl.* **41,** 37–43.

Bading, H., Rauterberg, E. W., and Moelling, K. (1989). Distribution of c-*myc,* c-*myb* and Ki-67 antigens in interphase and mitotic human cells evidenced by immuno-fluorescence staining technique. *Exp. Cell Res.* **185,** 50–59.

Balczon, R. D., and Brinkley, B. R. (1987). Tubulin interaction with kinetochore proteins: Analysis by *in vitro* assembly and chemical cross-linking. *J. Cell Biol.* **105,** 855–862.

Benavente, R., Dabauvalle, M.-C., Scheer, U., and Chaly, N. (1989). Functional role of newly formed pore complexes in postmitotic nuclear reorganization. *Chromosome* **98,** 233–241.

Berezney, R. (1991). The nuclear matrix, a heuristic model for investigating genomic organization and function in the cell nucleus. *J. Cell. Biochem.* **47,** 109–123.

Berezney, R., and Coffey, D. S. (1974). Identification of a nuclear protein matrix. *Biochem. Biophys. Res. Commun.* **60,** 1410–1417.

Berezney, R., and Coffey, D. S. (1977). Nuclear matrix: Isolation and characterization of a framework structure from rat liver nuclei. *J. Cell Biol.* **73,** 616–637.

Berkowitz, S. A., Katagiri, J., Binder, H. K., and Williams, R. C., Jr. (1977). Separation and characterization of microtubule proteins from calf brain. *Biochemistry* **16,** 5610–5617.

Bernat, R. L., Borisy, G. G., Rothfield, N. F., and Earnshaw, W. C. (1990). Injection of anticentromere antibodies in interphase disrupts events required for chromosome movement at mitosis. *J. Cell Biol.* **111,** 1519–1533.

Bernat, R. L., Delannoy, M. R., Rothfield, N. F., and Earnshaw, W. C. (1991). Disruption of centromere assembly during interphase inhibits kinetochore morphogenesis and function in mitosis. *Cell* **66,** 1229–1238.

Bernhard, W. (1969). A new staining procedure for electron microscopical cytology. *J. Ultrastruct. Res.* **27,** 250–265.

Berrios, M., Osheroff, N., and Fisher, P. A. (1985). *In situ* localization of DNA topoisomerase II, a major polypeptide component of the *Drosophila* nuclear matrix fraction. *Proc. Natl. Acad. Sci. U.S.A.* **82,** 4142–4146.

Binarova, P., Cihalikova, J., and Dolezel, J. (1993). Localization of MPM-2 recognized phosphoproteins and tubulin during cell progression in root meristem cells. *Cell Biol. Int.* **17,** 847–856.

Bindereif, A., and Green, M. (1990). Identification and functional analysis of mammalian splicing factors. *Genet. Eng.* **12,** 201–224.

Bischoff, F. R., Maier, G., Tilz, G., and Ponstingl, H. (1990). A 47-kDa human nuclear protein recognized by antikinetochore autoimmune sera is homologous with the protein encoded by RCC1, a gene implicated in onset of chromosome condensation. *Proc. Natl. Acad. Sci. U.S.A.* **87,** 8617–8621.

Blencowe, B. J., Nickerson, J. A., Issner, R., Penman, S., and Sharp, P. A. (1994). Association of nuclear matrix antigens with exon-containing splicing complexes. *J. Cell Biol.* **127,** 593–607.

Bode, J., Kohwi, Y., Dickinson, L., Joh, T., Klehr, D., Mielke, C., and Kohwi-Shigematsu, T. (1992). Biological significance of unwinding capability of nuclear matrix-associating DNAs. *Science* **255,** 195–197.

Borisy, G. G., Marcum, J. M., Olmsted, D. B., Murphy, D. B., and Johnson, K. A. (1975). Purification of tubulin and associated high molecular weight proteins from porcine brain and characterization of microtubule assembly *in vitro. Ann. N.Y. Acad. Sci.* **253,** 107–132.

Boy de la Tour, E., and Laemmli, U. K. (1988). The metaphase scaffold is helically folded: Sister chromatids have predominantly opposite helical handedness. *Cell* **55,** 937–944.

Breitbart, R. E., and Nadal-Ginard, B. (1987). Developmentally induced, muscle specific trans factors control the differential splicing of alternative and constitutive troponin-T exons. *Cell* **49,** 793–803.

Brenner, S., Pepper, D., Berns, M. W., Tan, E., and Brinkley, B. R. (1981). Kinetochore structure, duplication and distribution in mammalian cells: Analysis by human autoantibodies from schleroderma patients. *J. Cell Biol.* **91,** 95–102.

Brinkley, B. R. (1965). The fine structure of the nucleolus in mitotic divisions of Chinese hamster cells *in vitro. J. Cell Biol.* **27,** 411–422.

Brinkley, B. R. (1990). Centromeres and kinetochores: Integrated domains on eukaryotic chromosomes. *Curr. Opin. Cell Biol.* **2,** 446–452.

Brinkley, B. R., and Stubblefield, E. (1970). Ultrastructural and interaction of the kinetochore and centriole in mitosis and meiosis. *In* "Advances in Cell biology" (D. M. Prescott, L. Goldstein, and E. McConkey, Eds.), pp. 119–185. Appleton–Century–Crofts, New York.

Brinkley, B. R., Stubblefield, E., and Hsu, T. C. (1967). The effects of colcemid inhibition and reversal on the fine structure of the mitotic apparatus of Chinese hamster cells *in vivo. J. Ultrastruct. Res.* **19,** 1–18.

Brinkley, B. R., Valdivia, M. M., Tousson, A., and Brenner, S. L. (1984). Compound kinetochores of the Indian muntjac: Evolution by linear fusion of unit kinetochores. *Chromosoma* **91,** 1–11.

Brinkley, B. R., Brenner, S. L., Hall, J. M., Tousson, A., Balczon, R. D., and Valdivia, M. M. (1986). Arrangements of kinetochores in mouse cells during meiosis and spermiogenesis. *Chromosoma* **94,** 309–317.

Brinkley, B. R., Valdivia, M. M., Tousson, A., and Balczon, R. D. (1989). The Kinetochore: Structure and Molecular Organization. *In* "Mitosis: Molecules and Mechanisms" (J. S. Hyams and B. R. Brinkley, eds.), pp. 77–118. Academic Press, NY.

Brinkley, B. R., Ouspenski, I., and Zinkowski, R. P. (1992). Structure and molecular organization of the centromere-kinetochore complex. *Trends Cell Biol.* **2,** 15–21.

Buchkovich, K., Duffy, L. A., and Harlow, E. (1989). The retinoblastoma protein is phosphorylated during specific phases of the cell cycle. *Cell* **58,** 1097–1105.

Buck, R. C., and Tisdale, J. M. (1962). The fine structure of the mid-body of the rat erythroblast. *J. Cell Biol.* **13,** 109–115.

Burke, B., and Gerace, L. (1986). A cell-free system to study reassembly of the nuclear envelope at the end of mitosis. *Cell* **44,** 636–652.

Calarco-Gillam, P. D., Siebert, M. C., Hubble, R., Mitchison, T., and Kirschner, M. (1983). Centrosome development in early mouse embryo as defined by an auto-antibody against pericentriolar material. *Cell* **35,** 621–629.

Capco, D. G., and Penman, S. (1983). Mitotic architecture of the cell: The filament networks of the nucleus and cytoplasm. *J. Cell Biol.* **96,** 896–906.

Cardenas, M. E., Dang, Q., Glover, C. V., and Gasser, S. M. (1992). Casein kinase II phosphorylates the eukaryote-specific C-terminal domain of topoisomerase II *in vivo. EMBO J.* **11,** 1785–1796.

Carter, K. C., Bowman, D., Carrington, W., Fogarty, K., McNeil, J. A., Fay, F. S., and Lawrence, J. B. (1993). A three dimensional view of precursor messenger RNA metabolism within the mammalian nucleus. *Science* **259,** 1331–1335.

Casiano, C. A., Landberg, G., Ochs, R., and Tan, E. M. (1993). Autoantibodies to a novel cell cycle-regulated protein that accumulates in the nuclear matrix during S phase and is localized in the kinetochores and spindle midzone during mitosis. *J. Cell Sci.* **106,** 1045–1056.

Centouze, V. E., Vandre, D. D., and Borisy, G. G. (1986). Growth of microtubules on mitotic centrosomes is blocked by MPM-2. *J. Cell Biol.* **103,** 412a.

Chaly, N., and Brown, D. L. (1988). The prometaphase configuration and chromosome order in early mitosis. *J. Cell Sci.* **91,** 325–335.

Chaly, N., Bladon, T., Setterfield, G., Little, J. E., Kaplan, J. G., and Brown, D. L. (1984). Changes in distribution of nuclear matrix antigens during the mitotic cell cycle. *J. Cell Biol.* **99,** 661–671.

Chen, P. L., Scully, P., Shew, J. Y., Wang, J. Y., and Lee, W. H. (1989). Phosphorylation of the retinoblastoma gene product is modulated during the cell cycle and cell differentiation. *Cell* **58,** 1193–1198.

Chung, T. D., Drake, F. H., Tan, K. B., Per, S. R., Crooke, S. T., and Mirabelli, C. K. (1989). Characterization and immunological identification of cDNA clones encoding two human DNA topoisomerase II isozymes. *Proc. Natl. Acad. Sci. U.S.A.* **86,** 9431–9435.

Ciejek, E. M., Tsai, M. J., and O'Malley, B. W. (1983). Actively transcribed genes are associated with the nuclear matrix. *Nature* **306,** 607–609.

Clarke, D. J., Johnson, R. T., and Downes, C. S. (1993). Topoisomerase II inhibition prevents anaphase chromatid segregation in mammalian cells independently of the generation of DNA strand breaks. *J. Cell Sci.* **105,** 563–569.

Clarke, L. (1990). Centromeres of budding and fission yeasts. *Trends Genet.* **6,** 150–154.

Clayton, L., Black, C. M., and Lloyd, C. W. (1985). Microtubule nucleating sites in higher plant cells identified by an autoantibody against pericentriolar material. *J. Cell Biol.* **101,** 319–324.

Cockerill, P. N., and Garrarad, W. T. (1986). Chromosomal loop anchorage of the kappa immunoglobulin gene occurs next to the enhancer in a region containing topoisomerase II sites. *Cell (Cambridge, Mass.)* **44,** 273–282.

Cohen, C., and Parry, D. A. D. (1986). $\alpha$-helical coiled coils: A widespread motif in proteins. *Trends Biochem. Sci.* **11,** 245–248.

Cohen, C., and Parry, D. A. D. (1990). $\alpha$-helical coiled coils and bundles: How to design an $\alpha$-helical proteins. *Proteins: Struct., Funct., Genet.* **7,** 1–5.

Compton, D. A., and Cleveland, D. W. (1993a). NuMA is required for the proper completion of mitosis. *J. Cell Biol.* **120,** 947–957.

Compton, D. A., and Cleveland, D. W. (1993b). Assembly of star-like oligomers of NuMA supports an anchoring role for NuMA in the interphase nucleus and at mitotic spindle poles. 43rd Annu. Meet. Am. Soc. Cell Biol., Abstr. 464. *Mol. Biol. Cell* Suppl. **4,** 80a.

Compton, D. A., and Luo, C. (1995). Mutation of the predicted p34cdc2 phosphorylation sites in NuMA impair the assembly of the mitotic spindle and block mitosis. *J. Cell Sci.* **108,** 621–633.

Compton, D. A., Yen, T. J., and Cleveland, D. W. (1991). Identification of novel centromere/kinetochore-associated proteins using monoclonal antibodies generated against human mitotic chromsome scaffolds. *J. Cell Biol.* **112,** 1083–1097.

Compton, D. A., Szilak, I., and Cleveland, D. W. (1992). Primary structure of NuMA, an intranuclear protein that defines a novel pathway for segregation of proteins at mitosis. *J. Cell Biol.* **16,** 1395–1408.

Cooke, C. A., Heck, M. M. S., and Earnshaw, W. C. (1987). The INCENP antigens: Movement from the inner centromere to the midbody during mitosis. *J. Cell Biol.* **105,** 2053–2067.

Cooke, C. A., Bernat, R. L., and Earnshaw, W. C. (1990). CENP-B: A major human centromere protein located beneath the kinetochore. *J. Cell Biol.* **110,** 1475–1488.

Cooke, C. A., Bazett-Jones, D. P., Earnshaw, W. C., and Rattner, J. B. (1993). Mapping DNA within the mammalian kinetochore. *J. Cell Biol.* **120,** 1083–1091.

DeCaprio, J. A., Ludlow, J. W., Lynch, D., Furukawa, Y., Griffin, J., Piwnica-Worms, H., Huang, C.-M., and Livingston, D. M. (1989). The product of the retinoblastoma susceptibility gene has properties of a cell cycle regulatory element. *Cell* **58,** 1085–1095.

Dickinson, L. A., Joh, T., Kohwi, Y., and Kohwi-Shigematsu, T. (1992). A tissue-specific MAR/SAR DNA-binding protein with unusual binding site recognition. *Cell* **70,** 631–645.

Doheny, K. F., Sorger, P. K., Hyman, A. A., Tugendreich, S., Spencer, F., and Hieter, P. (1993). Identification of essential components of the *S. cerevisiae* kinetochore. *Cell* **73,** 761–774.

Drake, F. H., Zimmerman, J. P., McCabe, F. L., Bartus, H. F., Per, S. R., Sullivan, D. M., Ross, W. E., Mattern, M. R., Johnson, R. K., Crooke, S. T., *et al.* (1987). Purification of topoisomerase II from amsacrine-resistant P388 leukemia cells. Evidence for two forms of the enzyme. *J. Biol. Chem.* **262,** 16739–16747.

Durfee, T., Beecher, K., Chen, P. L., Yeh, S.-H., Yang, Y., Kilburn, A. E., Lee, W.-H., and Elledge, S. J. (1993). The retinoblastoma protein associates with the protein phosphatase type 1 catalytic subunit. *Genes Dev.* **7,** 555–569.

Durfee, T., Mancini, M. A., Jones, D., Elledge, S. J., and Lee, W.-H. (1994). The amino-terminal region of the retinoblastoma gene product binds a novel nuclear matrix proteins that co-localizaes to centers for RNA processing. *J. Cell Biol.* **127,** 609–622.

Earnshaw, W. C., and Bernat, R. L. (1990). Chromosomal passengers: Towards an integrated view of mitosis. *Chomosoma* **100,** 139–146.

Earnshaw, W. C., and Cooke, C. A. (1991). Analysis of the distribution of the INCENPs throughout mitosis reveals the existence of a pathway of structural changes in the chromosomes during metaphase, and early events in cleavage furrow formation. *J. Cell Sci.* **98,** 443–461.

Earnshaw, W. C., and Rothfield, N. (1985). The identification of a family of human centromere proteins using autoimmune sera from patients with scleroderma. *Chromosome* **91,** 313–319.

Earnshaw, W. C., Halligan, B., Cooke, C. A., Heck, M. M. S., and Liu, L. F. (1985). Topoisomerase II is a structural component of mitotic chromosome scaffolds. *J. Cell Biol.* **100,** 1706–1715.

Earnshaw, W. C., Ratrie, H., and Stetten, G. (1989). Visualization of centromere protein CENP-B and CENP-C on a stable dicentric chromosome in cytological spreads. *Chromosoma* **98,** 1–12.

Fakan, S., Leser, G., and Martin, T. E. (1984). Ultrastructural distribution of nuclear ribonucleoproteins as visualized by immunocytochemistry on thin sections. *J. Cell Biol.* **98,** 358–363.

Fakan, S., Leser, G., and Martin, T. E. (1986). Immunoelectron microscope visualization of nuclear ribonucleoprotein antigens within spread transcription complexes. *J. Cell Biol.* **103,** 1153–1157.

Fan, H., and Penman, S. (1971). Regulation of synthesis and processing of nucleolar components in metaphase-arrested cells. *J. Mol. Biol.* **59,** 27–42.

Fernandez, A., Brautigan, D. L., and Lamb, N. J. C. (1992). Protein phosphatase type 1 in mammalian cell mitosis: chromosomal localization and involvement in mitotic exit. *J. Cell Biol.* **116,** 1421–1430.

Fey, E. G., Krochmalnic, G., and Penman, S. (1986). The nonchromatin substructures of the nucleus: The ribonucleoprotein (RNP)-containing and RNP-depleted matrices analyzed by sequential fractionation and resinless section microscopy. *J. Cell Biol.* **102,** 1654–1665.

Filatova, L. S., and Zbarskii, I. B. (1993). The immunochemical demonstration of high-molecular proteins in the nuclear matrix of tumor cells. *Byull. Eksp. Biol. Med.* **116,** 418–421.

Finlay, D. R., Newmeyer, D. D., Price, T. M., and Forbes, D. J. (1987). Inhibition of *in vitro* nuclear transport by a lectin that binds to nuclear pores. *J. Cell Biol.* **104,** 189–200.

Fisher, L. M., Austin, C. A., Hopewell, R., Margerrison, E. E., Oram, M., Patel, S., Plummer, K., Sng, J. H., and Sreedharan, S. (1992). *Philos. Trans. R. Soc. London, Ser. B* **336,** 83–91.

Gasser, S. M., and Laemmli, U. K. (1986). Cohabitation of scaffold binding regions with upstream-enhancer elements of three developmentally regulated genes. *D. melanogaster. Cell* **46,** 521–530.

Gasser, S. M., and Laemmli, U. K. (1987). Improved methods for the isolation of individual and clustered mitotic chromosomes. *Exp. Cell Res.* **173,** 85–98.

Gasser, S. M., Amati, B. B., Cardenas, M. E., and Hofmann, J. F. X. (1989). Studies on scaffold attachment sites and their relation to genome function. *Int. Rev. Cytol.* **119,** 57–96.

Gautier, T., Robert-Nicoud, M., Guilly, M.-N., and Hernandez-Verdun, D. (1992a). Relocation of nucleolar proteins around chromosomes at mitosis. A study by confocal laser scanning microscopy. *J. Cell Sci.* **102,** 729–737.

Gautier, T., Dauphin-Villemant, C., André, C., Masson, C., Arnoult, J., and Hernandez-Verdun, D. (1992b). Identification and characterization of a new set of nucleolar ribonucleoproteins which line the chromosomes during mitosis. *Exp. Cell Res.* **200,** 5–15.

Gautier, T., Masson, C., Quintana, C., Arnoult, J., and Hernandez-Verdun, D. (1992c). The ultrastructure of the chromosome periphery in human cell lines: An *in situ* study using cryomethods in electron microscopy. *Chromosoma* **101,** 502–510.

Gerace, L., and Blobel, G. (1980). The nuclear envelope lamina is reversibly depolymerized during mitosis. *Cell* **19,** 277–287.

Gerdes, J., Li, L., Schlueter, C., Duchrow, M., Wohlemberg, C., Gerlach, C., Stahmer, I., Kloth, S., Brandt, E., and Flad, H.-D. (1991). Immunobiochemical and molecular biologic characterization of the cell proliferation-associated nuclear antigen that is defined by monoclonal antibody Ki-67. *Am. J. Pathol.* **138,** 867–873.

Goessens, G., and LePoint, A. (1974). The fine structure of the nucleolus during interphase and mitosis in Ehrlich tumor cells cultivated *in vitro. Exp. Cell Res.* **87,** 63–72.

Goodrich, D. W., and Lee, W. H. (1993). Molecular characterization of the retinoblastoma susceptibility gene. *Biochim. Biophys. Acta* **1155,** 43–61.

Gorbsky, G. J., and Ricketts, W. A. (1993). Differential expression of a phosphoepitope at the kinetochores of moving chromosomes. *J. Cell Biol.* **122,** 1311–1321.

Gorbsky, G. J., Sammak, P. J., and Borisy, G. G. (1987). Chromosomes move poleward in anaphase along stationary microtubules that coordinately disassemble from their kinetochore ends. *J. Cell Biol.* **104,** 9–18.

Grady, D. L., Rattliff, R. L., Robinson, D. L., McCanlies, E. C., Meyne, J., and Moyzis, R. K. (1992). Highly conserved repetitive DNA sequences are present at human centromeres. *Proc. Natl. Acad. Sci. U.S.A.* **89,** 1695–1699.

Haaf, T., and Schmid, M. (1989). Centromeric association and non-random distribution of centromeres in human tumor cells. *Hum. Genet.* **81,** 137–143.

Haff, T., and Ward, D. C. (1994). Structural analysis of alpha-satellite DNA and centromere proteins using extended chromatin and chromosomes. *Hum. Mol. Genet.* **3,** 697–709.

Hakes, D. J., and Berezney, R. (1991). DNA binding properties of the nuclear matrix and individual nuclear matrix proteins. Evidence for salt-resistant DNA binding sites. *J. Biol. Chem.* **266,** 11131–11140.

Halligan, B. D., Edwards, K. A., and Liu, L. F. (1985). Purification and characterization of a type II DNA topoisomerase from bovine calf thymus. *J. Biol. Chem.* **260,** 2475–2482.

He, D., and Brinkley, B. R. (1995). In preparation.

He, D., Nickerson, J. A., and Penman, S. (1990). Core filaments of the nuclear matrix. *J. Cell Biol.* **110,** 569–580.

He, D., Martin, T., and Penman, S. (1991). Localization of heterogeneous nuclear ribonucleoprotein in the interphase nuclear matrix core filaments and on perichromosomal filaments at mitosis. *Proc. Natl. Acad. Sci. U.S.A.* **88,** 7469–7473.

Heck, M. M. S., and Earnshaw, W. C. (1986). Topoisomerase II: A specific marker for proliferating cells. *J. Cell Biol.* **103,** 2569–2581.

Helfman, D. M., Cheley, S., Kuismanen, E., Finn, L. A., and Yamawaki-Kataoka, Y. (1986). Nonmuscle and muscle tropomyosin isoforms are expressed from a single gene by alternative RNA splicing and polyadenylation. *Mol. Cell. Biol.* **6,** 3582–3595.

Hentzen, P. C., Rho, J. H., and Bekhor, I. (1984). Nuclear matrix DNA from chicken erythrocytes contains b-globin gene sequences. *Proc. Natl. Acad. Sci. U.S.A.* **81,** 304–307.

Hernandez-Verdun, D., and Gautier, T. (1994). The chromosome periphery during mitosis. *BioEssays* **16,** 179–185.

Hernandez-Verdun, D., Roussel, P., and Gautier, T. (1993). Nucleolar proteins during mitosis. *Chromosomes Today* **11,** 79–90.

Hibino, Y., Tsukada, S., and Sugano, N. (1993). Properties of a DNA-binding protein from rat nuclear scaffold fraction. *Biochem. Biophys. Res. Commun.* **197,** 336–342.

Himmler, A., Prechsel, D., Kirschner, M. W., and Martin, D. W. (1989). Tau consists of a set of proteins with repeated C-terminal microtubule-binding domains and variable N-terminal domains. *Mol. Cell. Biol.* **9,** 1381–1388.

Hirano, T., and Mitchison, T. J. (1993). Topoisomerase II does not play a scaffolding role in the organization of mitotic chromosomes assembled in *Xenopus* egg extracts. *J. Cell Biol.* **120,** 601–612.

Hofmann, J. F.-X., Laroche, T., Brand, A. H., and Gasser, S. M. (1989). RAP-1 factor is necessary for DNA loop formation *in vitro* at the silent mating type locus HML. *Cell* **57,** 725–737.

Hoskins, G. C. (1968). Sensitivity of micrurgically removed chromosomal spindle fibers to enzyme disruption. *Nature* (*London*) **217,** 748–750.

Howell, W. M., and Hsu, T. C. (1979). Chromosome core structure revealed by silver staining. *Chromosoma* **73,** 61–66.

Hsu, T. C., Arrighi, F. E., Klevecz, R. R., and Brinkley, B. R. (1965). The nucleoli in mitotic divisions of mammalian cells *in vitro*. *J. Cell Biol.* **26,** 539–553.

Huang, S., and Spector, D. L. (1991). Nascent pre-mRNA transcripts are associated with nuclear regions enriched in splicing factors. *Genes Dev.* **5,** 2288–2302.

Hyman, A. A., Middleton, K., Centola, M., Mitchison, T. J., and Carbon, J. (1992). Microtubule-motor activity of a yeast centromere-binding protein complex. *Nature* **359,** 533–536.

Izaurralde, E., Mirkovitch, J., and Laemmli, U. K. (1988). Interaction of DNA with nuclear scaffolds *in vitro*. *J. Mol. Biol.* **200,** 111–125.

Jackson, D. A., and Cook, P. R. (1988). Visualization of a filamentous nucleoskeleton with a 23-nm axial repeat. *EMBO J.* **7,** 3667–3678.

Jensen, C. G., Davison, E. A., Bowser, S. S., and Rieder, C. L. (1987). Primary cilia cycle in PtK1 cells: Effects of colcemid and taxol on cilia formation and resorption. *Cell Motil. Cytoskel.* **7,** 187–197.

Jiang, W., Lechner, J., and Carbon, J. (1993a). Isolation and characterization of a gene (CBF2) specifying a protein component of the budding yeast kinetochore. *J. Cell Biol.* **121,** 513–519.

Jiang, W., Middleton, K., Yoon, H., Fouquet, C., and Carbon, J. (1993b). An essential yeast protein, $CBF5_p$, binds *in vitro* to centromeres and microtubules. *Mol. Cell. Biol.* **13,** 4884–4893.

Jiménez-Garcia, L. F., and Spector, D. L. (1993). *In vivo* evidence that transcription and splicing are coordinated by a recruiting mechanism. *Cell* **73,** 47–59.

Kai, R., Ohtsubo, M., Sekiguchi, T., and Nishimoto, T. (1986). Molecular cloning of a human gene that regulates chromosome condensation and is essential for cell proliferation. *Mol. Cell. Biol.* **6,** 2027–2032.

Kallajoki, M., Weber, K., and Osborn, M. (1991). A 210 KD nuclear matrix protein is a functional part of the mitotic spindle: A microinjection study using SPN monoclonal antibodies. *EMBO J.* **10,** 3351–3362.

Kallajoki, M., Weber, K., and Osborn, M. (1992). Ability to organize microtubules in taxol-treated mitotic $PtK_2$ cells goes with the SPN antigen and not with the centrosome. *J. Cell Sci.* **102,** 91–102.

Kallajoki, M., Harborth, J., Weber, K., and Osborn, M. (1993). Microinjection of a monoclonal antibody against SPN antigen, now identified by peptide sequences as the NuMA protein, induces micronuclei in PtK2 cells. *J. Cell Sci.* **104,** 139–150.

Käs, E., and Laemmli, U. K. (1992). *In vivo* topoisomerase II cleavage of the *Drosophila* histone and satellite III repeats: DNA sequence and structure characteristics. *EMBO J.* **11,** 705–716.

Kaufmann, S. H., Gibson, W., and Shaper, J. H. (1983). Characterization of the major polypeptides of the rat liver nuclear envelope. *J. Biol. Chem.* **258,** 2710–2719.

Kay, V., and Bode, J. (1994). Binding specificity of a nuclear scaffold: Supercoiled, single-stranded, and scaffold-attached-region DNA. *Biochemistry* **33,** 367–374.

Kingwell, B., and Rattner, J. B. (1987). Mammalian centromere/kinetochore composition: A 50 kDa antigen is present in the mammalian centromere/kinetochore. *Chromosoma* **95,** 403–407.

Kohwi-Shigematsu, T., and Kohwi, Y. (1990). Torsional stress stabilizes extended base unpairing in suppresser sites flanking immunoglobulin heavy chain enhancer. *Biochemistry* **29,** 9551–9560.

Krainer, A. R., and Maniatis, T. (1988). RNA splicing. *In* "Frontiers in Molecular Biology: Transcription and Splicing" (B. D. Hames and D. M. Glover, eds.), pp. 131–296. IRL Press, Washington, DC.

Kung, A. L., Sherwood, S. W., and Schimke, R. T. (1990). Cell line-specific differences in the control of cell cycle progression in the absence of mitosis. *Proc. Natl. Acad. Sci. U.S.A.* **87,** 9553–9557.

Kuriyama, R., and Borisy, G. G. (1981). Microtubule-nucleating activity of centrosomes in Chinese hamster ovary cells is independent of the centriole cycle but coupled to the mitotic cycle. *J. Cell Biol.* **91,** 822–826.

Kuriyama, R., and Sellitto, C. (1989). A 225 kD phosphoprotein associated with the mitotic centrosome in sea urchin eggs. *Cell Motil. Cytoskel.* **12,** 90–103.

Langkopf, A., Hammerback, J. A., Mueller, R., Vallee, R. B., and Garner, C. C. (1992). Microtubule-associated proteins 1A and LC2. *J. Biol. Chem.* **267,** 16561–16566.

Lechner, J., and Carbon, J. (1991). A 240 kD multisubunit protein complex, CBF3, is a major component of the budding yeast centromere. *Cell* **64,** 717–725.

Lee, W.-H., Shew, J. Y., Hong, F. D., Sery, T. W., Donoso, L. A., Young, L. J., Bookstein, R., and Lee, E. Y.-H. P. (1987). The retinoblastoma susceptibility gene encodes a nuclear phosphoprotein associated with DNA binding activity. *Nature* (*London*) **329,** 642–645.

Leser, G. P., Fakan, S., and Martin, T. E. (1989). Ultrastructural distribution of ribonucleoprotein complexes during mitosis. snRNP antigens are contained in mitotic granule clusters. *Eur. J. Cell Biol.* **50,** 376–389.

Leslie, R. J. (1990). Recruitment: The ins and outs of spindle pole formation. *Cell Motil. Cytoskel.* **16,** 225–228.

Lewis, C. D., and Laemmli, U. K. (1982). Higher order metaphase chromosome structure: Evidence for metalloprotein interactions. *Cell* **29,** 171–181.

Lewis, S. A., Wang, D., and Cowan, N. J. (1988). Microtubule associated protein MAP2 shares a microtubule-binding motif with tau protein. *Science* **242,** 936–939.

Ludérus, M. E. E., de Graaf, A., Mattia, E., den Blaauwea, J. L., Grande, M. A., de Jong, L., and van Driel, R. (1992). Binding of matrix attachment regions to lamin $B_I$. *Cell* **70,** 949–959.

Ludérus, M. E. E., den Blaauwen, J. L., de Smit, C. B., Compton, D. A., and van Driel, R. (1994). Binding of matrix attachment regions to lamin polymers involves single-stranded regions and the minor groove. *Mol. Cell. Biol.* **14,** 6297–6305.

Ludlow, J. W., Glendening, C. L., Livingston, D. M., and DeCaprio, J. A. (1993). Specific enzymatic dephosphorylation of the retinoblastoma protein. *Mol. Cell. Biol.* **13,** 367–372.

Lührmann, R., Kastner, B., and Bach, M. (1990). Structure of spliceosomal snRNPs and their role in pre-mRNA splicing. *Biochim. Biophys. Acta* **1087,** 265–292.

Lydersen, B., and Pettijohn, D. (1980). Human specific nuclear protein that associates with the polar region of the mitotic apparatus: Distribution in a human/hamster hybrid cell. *Cell* **22,** 489–499.

Lydersen, B., Kao, F. T., and Pettijohn, D. (1980). Expression of genes coding for non-histone chromosomal proteins in human/Chinese hamster cell hybrids, an electrophoretic analysis. *J. Biol. Chem.* **255,** 3002–3007.

Mackay, A. M., Eckley, D. M., Chue, C., and Earnshaw, W. C. (1993). Molecular analysis of the INCENPs (inner centromere proteins): Separate domains are required for association with microtubules during interphase and with the central spindle during anaphase. *J. Cell Biol.* **123,** 373–385.

Maekawa, T., and Kuriyama, R. (1991). Differential pathways of recruitment for centrosomal antigens to the mitotic poles during bipolar spindle formation. *J. Cell Sci.* **100,** 533–540.

Maekawa, T., and Kuriyama, R. (1993). Primary structure and microtubule-interacting domain of the SP-H antigen, mitotic map located at the spindle pole and characterized as a homologous protein to NuMA. *J. Cell Sci.* **105,** 589–600.

Maekawa, T., Leslie, R., and Kuriyama, R. (1991). Identification of a minus end-specific microtubule-associated protein located in the mitotic poles in cultured mammalian cells. *Eur. J. Cell Biol.* **54,** 255–267.

Mancini, M. A., Bei, S., Nickerson, J. A., Penman, S., and Lee, W.-H. (1994). The retinoblastoma gene product is a cell cycle-dependent, nuclear matrix-associated protein. *Proc. Natl. Acad. Sci. U.S.A.,* **91** 418–422.

Maniatis, T., and Reed, R. (1987). The role of small nuclear ribonucleotprotein particles in pre-mRNA splicing. *Nature* (*London*) **325,** 673–678.

Manuelidis, L. (1984). Different central nervous system cell types display distinct nonrandom arrangements of satellite DNA sequences. *Proc. Natl. Acad. Sci. U.S.A.* **81,** 3123–3127.

Manuelidis, L. (1990). A view of interphase chromosomes. *Science* **250,** 1533–1540.

Masumoto, H., Masukata, H., Muro, Y., Nozaki, N., and Okazaki, T. (1989). A human centromere antigen (CENP-B) interacts with a short specific sequence in alphoid DNA, a human centromeric satellite. *J. Cell Biol.* **109,** 1963–1973.

McDonald, K. L., Edwards, M. K., and McIntosh, J. R. (1979). Cross-sectional structure of the central mitotic spindle of *Diatoma vulgare.* Evidence for specific interactions between antiparallel microtubules. *J. Cell Biol.* **83,** 443–461.

McKeon, F. D., Tuffanelli, D. L., Kobayashi, S., and Kirschner, M. W. (1984). The redistribution of a conserved nuclear envelope protein during the cell cycle suggests a pathway for chromosome condensation. *Cell* **36,** 83–92.

McIntosh, J. R., and Landis, S. C. (1971). The distribution of spindle microtubules during mitosis in cultured human cells. *J. Cell. Biol.* **49,** 468–497.

Medford, R. M., Nguyen, H. T., Destree, A. T., Summers, E., and Nadal-Ginard, B. (1984). A novel mechanism of alternative RNA splicing for the developmentally regulated generation of toponin-T isoforms from a single gene. *Cell* **38,** 409–421.

Miller, T. E., Huang, C. Y., and Pogo, A. O. (1978). Rat liver nuclear skeleton and ribonucleoprotein complexes containing HnRNA. *J. Cell Biol.* **76,** 675–691.

Mitchison, T. J. (1988). Microtubule dynamics and kinetochore function in mitosis. *Annu. Rev. Cell Biol.* **4,** 527–549.

Mittnacht, S., and Weinberg, R. A. (1991). G1/S phosphorylation of the retinoblastoma protein is associated with an altered affinity for the nuclear compartment. *Cell* **65,** 381–393.

Molday, R. S., Hicks, D., and Molday, L. (1987). Peripherin: A rim-specific membrane proterod outer segment discs. *Invest. Ophthalmol. Visual Sci.* **28,** 50–61.

Montag, M. (1992). Localization of the nuclear matrix protein mitotin in mouse cells with a mitotic or endomitotic cell cycle. *Exp. Cell Res.* **202,** 207–210.

Moore, M. J., Query, C. C., and Sharp, P. A. (1993). Splicing of precursors to messenger RNAs by the spliceosome. *In* "The RNA World" (R. F. Gesteland and J. F. Atkins, eds.), pp. 303–357. Cold Spring Harbor Lab., Cold Spring Harbor, NY.

Moroi, Y., Peebles, C., Fritzler, M. J., Steigerwald, J., and Tan, E. M. (1980). Autoantibody to centromere (kinetochore) in scleroderma sera. *Proc. Natl. Acad. Sci. U.S.A.* **77,** 1627–1631.

Mullins, J. M., and Biesele, J. J. (1977). Terminal phase of cytokinesis in D-98S cells. *J. Cell Biol.* **73,** 627–684.

Mullins, J. M., and McIntosh, J. R. (1982). Isolation and initial characterization of the mammalian midbody. *J. Cell Biol.* **94,** 654–661.

Murray, A. W. (1992). Creative blocks: Cell-cycle checkpoints and feedback controls. *Nature (London)* **359,** 599–604.

Murray, A. W., and Kirschner, M. W. (1989). Dominoes and clocks: The union of two views of the cell cycle. *Science* **246,** 614–621.

Nakagomi, K., Kohwi, Y., Dickinson, L. A., and Kohwi-Shigematsu, T. (1994). A novel DNA-binding motif in the nuclear matrix attachment DNA-binding protein SATB1. *Mol. Cell. Biol.* **14,** 1852–1860.

Nakayasu, H., and Berezney, R. (1991). Nuclear matrins: Identification of the major nuclear matrix proteins. *Proc. Natl. Acad. Sci. U.S.A.* **88,** 10312–10316.

Newmeyer, D. D., and Forbes, D. J. (1988). Nuclear import can be separated into distinct steps *in vitro:* Nuclear pore binding and translocation. *Cell* **52,** 641–653.

Newport, J. W., and Forbes, D. J. (1987). The nucleus: Structure, function and dynamics. *Annu. Rev. Biochem.* **56,** 535–565.

Newport, J. W., Wilson, K. L., and Dunphy, W. G. (1990). A lamin-independent pathway for nuclear envelope assembly. *J. Cell Biol.* **111,** 2247–2259.

Ng, R., and Carbon, J. (1987). Mutational and *in vitro* protein-binding studies on centromere DNA from *Saccharomyces cerevisiae. Mol. Cell. Biol.* **7,** 4522–4534.

Nickerson, J. A., and Penman, S. (1992). Localization of nuclear matrix core filament proteins at interphase and mitosis. *Cell Biol. Int. Rep.* **16,** 811–826.

Nickerson, J. A., Krochmalnic, G., Wan, K. M., and Penman, S. (1989). Chromatin architecture and nuclear RNA. *Proc. Natl. Acad. Sci. U.S.A.* **86,** 177–181.

Nickerson, J. A., He, D., Fey, E. G., and Penman, S. (1990). The nuclear matrix. *In* "The Eukaryotic Nucleus: Molecular Biochemistry and Macromolecular Assemblies" (P. R. Strauss and S. H. Wilson, eds.), Vol. 2, pp. 763–782. Telford Press, Caldwell, New Jersey.

Nickerson, J. A., Krockmalnic, G., Wan, K. M., Turner, C. D., and Penman, S. (1992). A normally masked nuclear matrix antigen that appears at mitosis on cytoskeleton filaments adjoining chromosomes, centrioles, and midbodies. *J. Cell. Biol.* **116,** 977–987.

Nicklas, R. B., and Arana, P. (1992). Evolution and the meaning of metaphase. *J. Cell Sci.* **102,** 681–690.

Nishimoto, T., Ellen, E., and Basilico, C. (1978). Premature chromosome condensation in a tsDNA mutant of BHK cells. *Cell* **15,** 475–483.

Nishitani, H., Ohtrubo, M., Yamashita, K., Iida, H., Pines, J., Yasudo, H., Shibata, Y., Hunter, T., and Nishimoto, T. (1991). Loss of RCC1, a nuclear DNA binding protein, uncouples the completion of DNA replication from the activation of cdc2 protein kinase and mitosis. *EMBO J.* **10,** 1555–1564.

Nislow, C., Sellitto, C., Kuriyama, R., and McIntosh, J. R. (1990). A monoclonal antibody to a mitotic microtubule-associated protein blocks mitotic progression. *J. Cell Biol.* **111,** 511–522.

Nislow, C., Lombillo, V. A., Kuriyama, R., and McIntosh, J. R. (1992). A plus-end-directed motor enzyme that moves antiparallel microtubules *in vitro* localizes to the interzone of mitotic spindles. *Nature* **359,** 543–547.

Noble, M., Lewis, S. A., and Cowan, N. J. (1989). The microtubule binding domain of MAP1B contains a repeated sequence motif unrelated to that of MAP2 and tau. *J. Cell Biol.* **109,** 3367–3376.

Nyman, U., Hallman, H., Hadlaczky, G., Pettersson, I., Sharp, G., and Ringertz, N. R. (1986). Intranuclear localization of snRP antigens. *J. Cell Biol.* **102,** 137–144.

Ochs, R., Lischwe, M., O'Leary, P., and Busch, H. (1983). Localization of nucleolar phosphoproteins B23 and C23 during mitosis. *Exp. Cell Res.* **146,** 139–149.

Ochs, R., Lischwe, M. A., Spohn, W. H., and Busch, H. (1985). Fibrillarin: A new protein of the nucleolus identified by autoimmune sera. *Biol. Cell.* **54,** 123–134.

Ohnuki, Y. (1968). Structure of chromosomes. I. Morphological studies of the spiral structure of human somatic chromosomes. *Chromosoma* **25,** 402–428.

Ohtsubo, M., Okazaki, H., and Nishimoto, T. (1989). The RCC1 protein, a regulator for the onset of chromosome condensation locates in the nucleus and binds to DNA. *J. Cell Biol.* **109,** 1389–1397.

Okada, T. A., and Comings, D. E. (1980). A search for protein chores in the chromosomes: Is the scaffold an artifact? *Am. J. Hum. Genet.* **32,** 814–832.

Palmer, D. K., O'Day, K., Trong, H. L., Charbonneau, H., and Margolis, R. L. (1991). Purification of the centromere-specific protein CENP-A and demonstration that it is a distinctive histone. *Proc. Natl. Acad. Sci. U.S.A.* **88,** 3734–3738.

Pankov, R., Lemieux, M., and Hancock, R. (1990). An antigen located in the kinetochore region in metaphase and on polar microtubule ends in the midbody region in anaphase, characterized using a monoclonal antibody. *Chromosoma* **99,** 95–101.

Paweletz, N., and Risue-no, M. C. (1982). Transmission electron microscopic studies on the mitotic cycle of nucleolar proteins impregnated with silver. *Chromosoma* **85,** 261–273.

Periasamy, M., Strehler, E. E., Garfinkel, L. I., Gubits, R. M., Ruiz-Opazo, N., and Nadal-Ginard, B. (1984). Fast skeleton muscle myosin light chains 1 and 3 are produced from a single gene by a combined process of differential RNA transcription and splicing. *J. Biol. Chem.* **259,** 13595–13604.

Philipova, R. N., Zhelev, N. Z., Todorov, I. T., and Hadjiolov, A. A. (1987). Monoclonal antibody against a nuclear matrix antigen in proliferating human cells. *Biol. Cell.* **60,** 1–8.

Philipova, R. N., Vassilev, A. P., Kaneva, R. P., Andreeva, P., Todorov, I. T., and Hadjiolov, A. A. (1991). Expression of the nuclear protein mitotin in differentiating *in vitro* HL 60 cells. *Biol. Cell.* **72,** 47–50.

Phi-Van, L., and Strätling, W. H. (1990). Association of DNA with nuclear matrix. *Prog. Mol. Subcell. Biol.* **11,** 1–11.

Price, C. M., and Pettijohn, D. E. (1986). Redistribution of the nuclear mitotic apparatus protein (NuMA) during mitosis and nuclear assembly. *Exp. Cell Res.* **166,** 295–311.

Probst, H., and Herzog, R. (1985). DNA regions associated with the nuclear matrix of *Ehrlich ascites* cells expose single-stranded sites after deproteinization. *Eur. J. Biochem.* **146,** 167–171.

Puvion, E., Viron, A., and Xu, F. X. (1984). High resolution autoradiographical detection of RNA in the interchromatin granules of DRB-treated cells. *Exp. Cell Res.* **152,** 357–367.

Rattner, J. B. (1992). Integrating chromosome structure with function. *Chromosoma* **101,** 259–264.

Rattner, J. B., Rao, A., Fritzler, M. J., Valencia, D. W., and Yen, T. J. (1993). CENP-F is a .ca 400 kDa kinetochore protein that exhibits a cell-cycle dependent localization. *Cell Motil. Cytoskel.* **26,** 214–226.

Reuter, R., Appel, B., Bringmann, P., Rinke, J., and Lührmann, R. (1984). 5′-terminal caps of snRNAs are reactive with antibodies specific for 2, 2, 7-trimethylguanosine in whole cells and nuclear matrices. *Exp. Cell Res.* **154,** 548–560.

Rieder, C. L., and Borisy, G. G. (1982). The centrosome cycle in PtK2 cells: Asymmetric distribution and structural changes in the pericentriolar material. *Biol. Cell.* **44,** 117–132.

Rieder, C. L., and Palazzo, R. E. (1992). Colcemid and the mitotic cycle. *J. Cell Sci.* **102,** 387–392.

Riley, D. J., Lee, E. Y., Lee, W.-H. (1994). The retinoblastoma protein: More than a tumor suppressor. *Annu. Rev. Cell Biol.* **10,** 1–29.

Robinson, S. I., Nelkin, B. D., and Vogelstein, B. (1982). The ovalbumin gene is associated with the nuclear matrix of chicken oviduct cells. *Cell* **28,** 99–106.

Romig, H., Fackelmayer, F. O., Renz, A., Ramsperger, U., and Richter, A. (1992). Characterization of SAF-A, a novel nuclear DNA binding protein from HeLa cells with high affinity for nuclear matrix/scaffold attachment DNA elements. *EMBO J.* **11,** 3431–3440.

Roos, U. P. (1973). Light and electron microscopy of rat kangaroo cells in mitosis. I. Formation and breakdown of the mitotic apparatus. *Chromosoma* **41,** 195–220.

Ruiz-Opazo, N., and Nadal-Ginard, B. (1987). A-tropomyosin gene organization. *J. Biol. Chem.* **261,** 4755–4765.

Saitoh, H., Tomkiel, J. E., Cooke, C. A., Ratrie, H. R., Maurer, M., Rothfield, N. F., and Earnshaw, W. C. (1992). CENP-C, an autoantigen in scleroderma, is a component of the human inner kinetochore plate. *Cell* **170,** 115–125.

Saitoh, N., Goldberg, I. G., Wood, E. R., and Earnshaw, W. C. (1994). ScII: An abundant chromosome scaffold protein is a member of a family of putative ATPases with an unusual predicted tertiary structure. *J. Cell Biol.* **127,** 303–318.

Saitoh, Y., and Laemmli, U. K. (1994). Metaphase chromosome structure: Bands arise from a differential folding path of the high AT-rich scaffold. *Cell* **76,** 609–622.

Sayhoun, N., Wolf, M., Besterman, J., Hsieh, T.-S., Sander, M., LeVine, H., Chang, K.-J., and Cuatrecasas, P. (1986). Protein kinase C phosphorylates topoisomerase II: Topoisomerase activation and its possible role in phorbol ester-induced differentiation of HL-60 cells. *Proc. Natl. Acad. Sci. U.S.A.* **83,** 1603–1607.

Scheele, R. D., and Borisy, G. G. (1979). *In vitro* assembly of microtubules. *In* "Microtubules" (K. Roberts and J. S. Hyams, Eds.), pp. 175–254. Academic Press, London.

Sellitto, C., and Kuriyama, R. (1988). Distribution of matrix component of the midbody during the cell cycle in Chinese hamster ovary cells. *J. Cell Biol.* **106,** 431–439.

Shi, L., Zumei, N., Shi, Z., Ge, W., and Yang, Y. (1987). Involvement of a nucleolar component, perichromonucleolin, in the condensation and deconstruction of chromosomes. *Proc. Natl. Acad. Sci. U.S.A.* **84,** 7953–7956.

Simerly, C., Balczon, R., Brinkley, B. R., and Schatten, G. (1990). Microinjected centromere kinetochore antibodies interfere with chromosome movement in meiotic and mitotic mouse oocytes. *J. Cell Biol.* **111,** 1491–1504.

Small, D., Nelkin, B., and Volgelstein, B. (1985). The association of transcribed genes with the nuclear matrix of *Drosophila* cells during heat shock. *Nucleic Acids Res.* **13,** 2413–2431.

Smith, C. W., Patton, J. G., and Nadal-Ginard, B. (1989). Alternative splicing in the control of gene expression. *Annu. Rev. Genet.* **23,** 527–577.

Smith, H. C. (1992). Organization of RNA splicing in the cell nucleus. *Curr. Top. Cell Regul.* **33,** 145–166.

Smith, H. C., Harris, S. G., Zillmann, M., and Berget, S. M. (1989). Evidence that a nuclear matrix protein participates in pre-messenger RNA splicing. *Exp. Cell Res.* **182,** 521–533.

Sparks, C. A., Bangs, P. L., McNeil, G. P., Lawrence, J. B., and Fey, E. (1993). Assignment of the nuclear mitotic apparatus protein NuMA gene to human chromosome 11q13. *Genomics* **17,** 222–224.

Spector, D. L. (1984). Colocalization of U1 and U2 small nuclear RNPs by immunocytochemistry. *Biol. Cell.* **51,** 109–112.

Spector, D. L. (1990). Higher order nuclear organization: Three-dimensional distribution of small nuclear ribonucleoprotein particles. *Proc. Natl. Acad. Sci. U.S.A.* **87,** 147–151.

Spector, D. L. (1993). Macromolecular domains within the cell nucleus. *Annu. Rev. Cell Biol.* **9,** 265–315.

Spector, D. L., and Smith, H. C. (1986). Redistribution of U-snRNPs during mitosis. *Exp. Cell Res.* **163,** 87–94.

Spector, D. L., Schrier, W. H., and Busch, H. (1983). Immunoelectron microscopic localization of snRNPs. *Biol. Cell.* **49,** 1–10.

Stevens, B. J. (1965). The fine structure of the nucleolus during mitosis in the grasshopper neuroblast cell. *J. Cell Biol.* **24,** 349–368.

Stief, A. D., Winger, M., Strátling, W. H., and Sippel, A. E. (1989). A nuclear element mediates elevated and position-independent gene activity. *Nature* **341,** 343–345.

Strehler, E. E., Periasamy, M., Strehler-Page, M. E., and Nadal-Ginard, B. (1985). Myosin light chain 1 and 3 gene has two structurally distinct and differentially regulated promoters evolving at different rates. *Mol. Cell. Biol.* **5,** 3168–3182.

Strunnikov, A. V., Larionoc, V. L., and Koshland, D. (1993). SMC1: An essential yeast gene encoding a putative head-rod-tail protein is required for nuclear division and defines a new ubiquitous protein family. *J. Cell Biol.* **123,** 1635–1648.

Sullivan, K. F., Hechenberger, M., and Masri, K. (1994). Human CENP-A contains a histone H3 related histone fold domain that is required for targeting to the centromere. *J. Cell Biol.* **127,** 581–592.

Swedlow, J. R., Sedat, J. W., and Agard, D. A. (1993). Multiple chromosomal populations of topoisomerase II detected *in vivo* by time-lapse, three-dimensional wide-field microscopy. *Cell* **73,** 97–108.

Taagepera, S., Rao, P. N., Drake, F. H., and Gorbsky, G. J. (1993). DNA topoisomerase II$\alpha$ is the major chromosome protein recognized by the mitotic phosphoprotein antibody MPM-2. *Proc. Natl. Acad. Sci. U.S.A.* **90,** 8407–8411.

Tang, T. K., Tang, C. C., Chen, Y., and Wu, C. (1993). Nuclear proteins of the bovine esophageal epithelium II: The NuMA gene gives rise to multiple mRNAs and gene products reactive with monoclonal antibody W1. *J. Cell Sci.* **104,** 249–260.

Tang, T. K., Tang, C. C., Chan, Y., and Wu, C. (1994). Nuclear mitotic apparatus protein (NuMA): Spindle association, nuclear targeting and differential subcellular localization of various NuMA isoforms. *J. Cell Sci.* **107,** 1389–1402.

Tawfic, S., Goueli, S. A., Olson, M. O., and Ahmed, K. (1993). Androgenic regulation of the expression and phosphorylation of prostatic nucleolar protein B23. *Cell. Mol. Biol. Res.* **39,** 43–51.

Templeton, D. J., Park, S.-H., Lanier, L., and Weinberg, R. A. (1991). Nonfunctional mutants of the retinoblastoma protein are characterized by defects in phosphorylation, viral oncoprotein association, and nuclear tethering. *Proc. Natl. Acad. Sci. U.S.A.* **88,** 3033–3037.

Thibodeau, A., and Vincent, M. (1991). Monoclonal antibody CC-3 recognizes phosphoproteins in interphase and mitotic cells. *Exp. Cell Res.* **195,** 145–153.

Thorburn, A., Moore, R., and Knowland, J. (1988). Attachment of transcriptionally active DNA sequences to the nucleoskeleton under isotonic conditions. *Nucleic Acids Res.* **16,** 7183–7184.

Todorov, I. T., Philipova, R. N., Joswig, G., Werner, D., and Ramaekers, F. C. S. (1992). Detection of the 125-kDa nuclear protein mitotin in centrosomes, the poles of the mitotic spindle, and the midbody. *Exp. Cell Res.* **199,** 398–401.

Tomkiel, J. E., Cooke, C. A., Saitoh, H., Bernat, R. L., and Earnshaw, W. C. (1994). CENP-C is required for maintaining proper kinetochore size and for a timely transition to anaphase. *J. Cell Biol.* **125,** 531–545.

Tousson, A., Zeng, C., Brinkley, B. R., and Valdivia, M. M. (1991). Centrophilin: A novel mitotic spindle protein involved in microtubule nucleation. *J. Cell Biol.* **112,** 427–440.

Tsutsui, K., Tsutsui, K., Okada, S., Waterai, S., Seki, S., Yasuda, T., and Shomori, T. (1993). Identification and characterization of a nuclear scaffold protein that binds the matrix attachment region DNA. *J. Biol. Chem.* **268,** 12886–12894.

Uchida, S., Sekiguchi, T., Nishitani, H., Miyauchi, K., Ohtsubo, M., and Nishimoto, T. (1990). Premature chromosome condensation is induced by a point mutation in the hamster RCC1 gene. *Mol. Cell. Biol.* **10,** 577–584.

Uemura, T., Ohkura, H., Adachi, Y., Morino, K., Shiozaki, K., and Yanagida, M. (1987). DNA topoisomerase II is required for condensation and separation of mitotic chromosomes in *S. pombe. Cell* **50,** 917–925.

Valdivia, M. M., and Brinkley, B. R. (1985). Fractionation and initial characterization of the kinetochore from mammalian metaphase chromosomes. *J. Cell Biol.* **101,** 1124–1134.

Vandre, D. D., Davis, F. M., Rao, P. N., and Borisy, G. G. (1984). Phosphoproteins are components of mitotic microtubule organizing centers. *Proc. Natl. Acad. Sci. U.S.A.* **81,** 4439–4443.

van Driel, R., Humbel, B., and de Jong, L. (1991). The nucleus: A black box being opened. *J. Cell. Biochem.* **47,** 311–316.

van Ness, J., and Pettijohn, D. E. (1983). Specific attachment of NuMA protein to metaphase chromosomes: Possible function in nuclear reassembly. *J. Mol. Biol.* **171,** 175–205.

Verheijen, R., Kuijpers, H. J. H., Schlingemann, R. O., Boehmer, A. L. M., van Driel, R., Brakenhoff, G. J., and Ramaekers, F. C. S. (1989). Ki-67 detects a nuclear matrix-associated proliferation-related antigen. I. Intracellular localization during interphase. *J. Cell Sci.* **92,** 123–130.

Vogelstein, B., and Hunt, B. F. (1982). A subset of small nuclear ribonucleoprotein particle antigens is a component of the nuclear matrix. *Biochem. Biophys. Res. Commun.* **105,** 1224–1232.

Vogt, P. (1992). Code domains in tandem repetitive DNA sequence structures. *Chromosoma* **101,** 585–589.

von Kries, J. P., Buhrmester, H., and Strätling, W. H. (1991). A matix/scaffold attachment region binding protein: Identification, purification, and mode of binding. *Cell* **64,** 123–135.

Wahls, W. P., Swenson, G., and Moore, P. D. (1991). Two hypervariable mini-satellite DNA binding proteins. *Nucleic Acids Res.* **19,** 3269–3274.

Wan, K. M., Nickerson, J. A., Krockmalnic, G., and Penman, S. (1994). The B1C8 protein is in the dense assemblies of the nuclear matrix and relocates to the spindle and pericentriolar filaments at mitosis. *Proc. Natl. Acad. Sci. U.S.A.* **91,** 594–598.

Werrick, R., and Willard, H. F. (1991). Physical map of the centromeric region of human chromosome 7: Relationship between two distinct alpha-satellite arrays. *Nucleic Acids Res.* **19,** 2295–2301.

Willard, H. F., and Waye, J. S. (1987). Hierarchical order in chromosome-specific human alpha satellite DNA. *Trends Genet.* **3,** 192–198.

Wise, D. A., and Bhattacharjee, L. (1992). Antikinetochore antibodies interfere with prometaphase but not anaphase chromosome movement in living PtK2 cells. *Cell Motil. Cytoskel.* **23,** 157–167.

Woessner, R. D., Chung, T. D. Y., Hofmann, G. A., Mattern, M. R., Mirabelli, C. K., Drake, F. H., and Johnson, R. K. (1990). Differences between normal and ras-transformed NIH-

3T3 cells in expression of the 170 kD and 180 kD forms of topoisomerase II. *Cancer Res.* **50,** 2901–2908.

Woessner, R. D., Mattern, M. R., Mirabelli, C. K., Johnson, R. K., and Drake, F. H. (1991). Proliferation and cell cycle-dependent differences in the expression of the 170 kD and 180 kD forms of topoisomerase II in NIH-3T3 cells. *Cell Growth Differ.* **4,** 209–214.

Wong, A. K., and Rattner, J. B. (1988). Sequence organization and cytological localization of the minor satellite of mouse. *Nucleic Acids Res.* **16,** 11645–11661.

Wordeman, L., Davis, F. M., Rao, P. N., and Cande, W. Z. (1989). Distribution of phosphorylated spindle-associated proteins in the diatom *Stephanopyxis turris. Cell Motil. Cytoskel.* **12,** 33–41.

Xing, Y., Johnson, C. V., Dobner, P. R., and Lawrence, J. B. (1993). Higher level organization of individual gene transcription and RNA splicing. *Science* **259,** 1326–1330.

Yang, C. H., and Snyder, M. (1992). The nuclear-mitotic apparatus protein (NuMA) spindle apparatus. *Mol. Biol. Cell.* **3,** 1259–1267.

Yang, C. H., Lambie, E. J., and Snyder, M. (1992). NuMA: An unusually long coiled-coil related protein in the mammalian nucleus. *J. Cell. Biol.* **116,** 1303–1317.

Yasuda, Y., and Maul, G. G. (1990). A nucleolar auto-antigen is part of a major chromosomal surface component. *Chromosoma* **99,** 152–160.

Yen, T. J., Compton, D. A., Wise, D., Zinkowski, R. P., Brinkley, B. R., Earnshaw, W. C., and Cleveland, D. W. (1991). CENP-E, a novel centromere-associated protein required for progression from metaphase to anaphase. *EMBO J.* **10,** 1245–1254.

Yen, T. J., Li, G., Schaar, B. T., Szilak, I., and Cleveland, D. W. (1992). CENP-E is a putative kinetochore motor that accumulates just before mitosis. *Nature* **359,** 536–539.

Yoneda, Y., Imamoto-Sonobe, N., Yamaizumi, M., and Ulchida, T. (1987). Reversible inhibition of protein import into the nucleus by wheat germ agglutinin injected into cultured cells. *Exp. Cell Res.* **173,** 586–595.

Zee, S. Y. (1992). Chromosome exoskeleton in plant cells visualized by scanning electron microscopy. *Protoplasma* **170,** 86–89.

Zeitlin, S., Parent, A., Silverstein, S., and Efstratiadis, A. (1987). Pre-mRNa splicing and the nuclear matrix. *Mol. Cell. Biol.* **7,** 111–120.

Zeng, C., He, D., Berget, S. M., and Brinkley, B. R. (1994a). Nuclear-mitotic apparatus protein, a structural protein interface between the nucleoskeleton and RNA splicing. *Proc. Natl. Acad. Sci. U.S.A.* **91,** 1505–1509.

Zeng, C., He, D., and Brinkley, B. R. (1994b). Localization of NuMA protein isoforms in the nuclear matrix of mammalian cells. *Cell Motil. Cytoskel.* **29,** 167–176.

Zhelev, N. Z., Todorov, I. T., Philipova, R. N., and Hadjiolov. A. A. (1990). Phosphorylation-related accumulation of the 125k nuclear matrix protein mitotin in human mitotic cells. *J. Cell Sci.* **95,** 59–64.

Zhu, X. L., Mancini, M. A., Chang, K. H., Liu, C. Y., Chen, C. F., Shan, B., Jones, D., Yang-Feng, T. L., Lee, W.-H. (1995). Characterization of a novel 350 kD nuclear phosphoprotein specifically involved in mitotic phase progression. *Molec. Cell Biol.* (in press).

Zieve, G. W., Hiedemann, S. R., and McIntosh, J. R. (1980). Isolation and partial characterization of a cage of filaments that surrounds the mammalian mitotic spindle. *J. Cell Biol.* **87,** 160–169.

Zini, N., Santi, S., Ognibene, A., Bavelloni, A., Neri, L. M., Valmori, A., Mariani, E., Negri, C., Astaldi-Ricotti, G. C., and Maraldi, N. M. (1984). Discrete localization of different DNA topoisomerases in HeLa and K562 cell nuclei and subnuclear fractions. *Exp. Cell Res.* **210,** 336–348.

Zinkowski, R. P., Meyne, J., and Brinkley, B. R. (1991). The centromere-kinetochore complex: A repeat subunit model. *J. Cell Biol.* **113,** 1091–1110.

Zirkle, R. E. (1970). Ultra-violet microbeam irradiation of newt-cell cytoplasm: Spindle destruction, false anaphase, and delay of true anaphase. *Radiat. Res.* **41,** 516–537.

# Nuclear Matrix Isolated from Plant Cells

Susana Moreno Díaz de la Espina
Laboratorio de Biología Celular y Molecular Vegetal, Departamento de Biología de Plantas, Centro de Investigaciones Biológicas, 28006 Madrid, Spain

Residual nuclear matrices can be successfully obtained from isolated nuclei of different monocot and dicot plant species using either high ionic or low ionic extraction protocols. The protein composition of isolated nuclear matrices depends on the details of isolation protocols. They are stable and present in all cases, a tripartite organization with a lamina, nucleolar matrix, and internal matrix network, and also maintain some of the basic architectural features of intact nuclei. *In situ* preparations demonstrate the continuity between the nuclear matrix and the plant cytoskeleton. Two-dimensional separation of isolated plant nuclear matrix proteins reveals a heterogeneous polypeptide composition corresponding rather to a complex multicomponent matrix than to a simple nucleoskeletal structure. Immunological identification of some plant nuclear matrix components such as A and B type lamins, topoisomerase II, and some components of the transcription and splicing machineries, internal intermediate filament proteins, and also specific nucleolar proteins like fibrillarin and nucleolin, which associate to specific matrix domains, establish a model of organization for the plant nuclear matrix similar to that of other eukaryotes. Components of the transcription, processing, and DNA-anchoring complexes are associated with a very stable nucleoskeleton. The plant matrix-attached regions share structural and functional characteristics with those of insects, vertebrates, and yeast, and some of them are active in animal cells. In conclusion, the available data support the view that the plant nuclear matrix is basically similar in animal and plant systems, and has been evolutionarily conserved in eukaryotes.

**KEY WORDS:** Nuclear matrix, Lamins, Intermediate filaments, Topoisomerase II, Nucleolar matrix, Fibrillarin, Nucleolar fibrillar centers, Interchromatin granules, P105 nuclear antigen, MARs.

## I. Introduction

With the great amount of information available about the molecular mechanisms of gene expression in eukaryotes, and the accumulating evidence from biochemistry, electron microscopy, and cell and molecular biology, we now know that replication, transcription, and RNA splicing and transport are coordinated complex processes, which require not only informative sequences but also a great number of inducible enzymes, factors, and cofactors organized in multimeric complexes that are unlikely to function coordinately in a soluble way (Hoffman, 1993). The nuclear matrix (NM), a network of insoluble protein fibers pervading the nucleus, has turned out to be the structure responsible for the topological organization and coordination of all these processes (de Jong *et al.,* 1990; Jackson, 1991).

In the past 20 years, an insoluble nuclease-resistant nuclear matrix has been isolated from different types of cells ranging from lower eukaryotes to insects and vertebrates (Berezney, 1984; Verheijen *et al.,* 1988). This structure is involved in main nuclear functions. The nuclear matrix is responsible for the higher value folding of chromatin (Getzenberger *et al.,* 1991). Replication and transcription factories are associated with the matrix (Berezney, 1991; Hozák *et al.,* 1993a; Jackson *et al.,* 1993). Splicing factors as well as small nuclear ribonucleoproteins (snRNPs) and hnRNPs also appear bound to the matrix (Verheijen *et al.,* 1988).

The nuclear matrix has a complex polypeptide composition that depends on the isolation method used, the species, and the cell type (Peters and Comings, 1980). Up to several hundred polypeptides can be detected by two-dimensional SDS–polyacrylamide gel electrophoresis (SDS–PAGE) in nuclear matrix preparations (Peters and Comings, 1980), some of which are conserved in all cell types studied and thus considered to form "the minimal matrix," serving housekeeping functions.

The best characterized proteins of the nuclear matrix are the lamins, which form the lamina, a highly dynamic peripheral meshwork of intermediate filament (IF) proteins connected to the cytoskeleton. On the contrary, the protein components of the internal framework of the matrix are poorly characterized. All the evidence accumulated in the past two decades closely involves the nuclear matrix in trafficking, signaling, and coordination of gene expression, and has positively contributed to settle the old controversy about the real existence of a nuclear matrix pervading the nucleus in eukaryotes, converting the studies on the nuclear matrix into one of the most exciting fields of present research in cell biology.

In contrast to the situation in vertebrates, the information on the nuclear matrix of plant cells is recent and scarce (see Table I). In fact, the development of procedures to prepare nuclear matrices from plant cells started a decade later than those for animal cells (Barthelemy and Moreno Díaz de la Espina, 1984; Stolyarov, 1984; Ghosh and Dey, 1986). That is so in part because of technical reasons: mass preparation of plant nuclei is difficult to achieve in good conditions; in addition, the technology of suspension culture in plants, which would facilitate these kind of experiments, is re-

TABLE I
Main Studies on the Plant Nuclear Matrix

| Species | Tissue | Type of study | Reference |
|---|---|---|---|
| *Allium cepa* | Root meristems | Ultrastructure and protein analysis | Barthelemy and Moreno Díaz de la Espina (1984) |
| | | | Moreno Díaz de la Espina *et al.* (1991) |
| | | Effects of different extraction procedures on the NM | Cerezuela and Moreno Díaz de la Espina (1990) |
| | | Characterization of plant lamins | Minguez and Moreno Díaz de la Espina (1993a) |
| | | Isolated nuclear matrix | Minguez and Moreno Díaz de la Espina (1993b) |
| | Roots and bulb cells | Cytological and protein analysis | Ghosh and Dey (1986) |
| *Cucurbita pepo* | Seedlings | Localization of an endonuclease in the NM | Rzepecki *et al.* (1992) |
| *Daucus carota* | Protoplasts | Intermediate filaments | Beven *et al.* (1991) |
| | | Identification of a nucleolar matrix protein | Corben *et al.* (1989) |
| | Suspension cultures | Intermediate filament antigens | Frederick *et al.* (1992) |
| | Somatic embryos | Ultrastructural and protein analysis | Masuda *et al.* (1993) |
| *Glycine hispida* | | MAR of heat shock genes | Schöffl *et al.* (1993) |
| *Nicotiana tabacum* | Protoplasts | SAR | Hall *et al.* (1991) |
| | | | Breyne *et al.* (1992) |
| *Petunia* | Transformed shoots | MAR | Dietz *et al.* (1994) |
| *Phaseolus vulgaris* | Root meristems | Ultrastructure and protein analysis | Galcheva-Gargova *et al.* (1988) |
| *Pisum sativum* | Seedlings | Lamin-like proteins and casein kinase II in the NM | Li and Roux (1992) |
| | | NT Pase | Tong *et al.* (1993) |
| | | SAR sequences | Slatter *et al.* (1991) |
| | Protoplasts | Lamins and IF | McNulty and Saunders (1992) |
| *Trillium kamtschaticum* | Microsporocytes | Ultrastructure and protein analysis | Ohyama *et al.* (1987) |
| *Zea mays* | Seedlings | Nuclear matrix proteins | Galcheva Gargova *et al.* (1988) |
| | | NM DNA | Stoilov *et al.* (1992); Lvandrenko and Avramanova 1992 |
| | | Nuclear matrix proteins | Ivanchenko *et al.* (1993) |
| | | SARs in the Adh 1 promoter | Paul and Ferl (1993) |
| | Seeds and roots | NM and transcriptional activity | Krachmarov *et al.* (1991) |
| | Leaves | SARs in the Adh 1 gene | Avramanova and Bennetzen (1993) |

stricted to a few dicotyledonous species, and *in situ* preparation methods avoiding isolation of nuclei, which have been successfully applied to cultured animal cells (Jackson *et al.,* 1988) and plasmodia (Waitz and Loidl, 1988), are not useful in plants because of the cell wall.

The abundance of proteases in plant vacuoles, which could contaminate the nuclear preparations, and the high DNA content per nucleus in most plant species are additional problems in the preparation of plant nuclear matrix proteins. However, the lack of information has been due not only to technical reasons but also to the fact that cellular and molecular studies on plants usually lag behind those on animal cells. As structural and functional studies on plant nuclei progressed, they revealed that except for some minor differences, the basic organization of the replication, transcription, and RNA splicing and processing in them follows the same pattern as in animal cells (Jordan *et al.,* 1980; Puvion-Dutilleul, 1983; Medina *et al.,* 1989; Moreno Díaz de la Espina *et al.,* 1992; Spector, 1993a). For this reason, the structure responsible for the topological organization and coordination of these processes should also be conserved in plant cells.

In the 1980s the development of protocols to obtain nuclear matrices isolated from plant cells started in our laboratory and also in other laboratories. They were mainly based on the "high salt" extraction procedure described by Berezney and included not only extractions with detergents and low and high ionic strength buffers, but also extensive extraction with nucleases. The early results confirmed the presence of an insoluble nuclease-resistant nuclear matrix in plant cells with a structural organization analogous to that described in animal systems. Later the adaptation of different extraction procedures to plant nuclei has allowed the analysis of the ultrastructural organization and protein composition of the nuclear matrix in several mono- and dicotyledonous species of plants (see Table I).

Although the studies on the plant nuclear matrix are still behind those on animal systems, the progress made in this field during the past few years has been very important. At present it has been proven that the nuclear matrix is a conserved nuclear structure in plants with a similar organization to that in animal cells (see Table I). The analysis of its protein components allowed the identification of some plant nuclear matrix proteins, from which lamins are the best characterized (Li and Roux, 1992; McNulty and Saunders, 1992; Minguez and Moreno Díaz de la Espina, 1993a,b). The plant matrices specifically bind matrix-attached region (MAR) sequences from different animal systems (Hall *et al.,* 1991; Breyne *et al.,* 1992; Avramanova and Bennetzen, 1993; Minguez *et al.* experiments in progress); endogenous MAR sequences also have been identified (Hall *et al.,* 1991) demonstrating that the recognition mechanisms are conserved during evolution.

In this article I will try to update the most relevant information about the isolated nuclear matrix of plant cells, placing special emphasis on the peculiarities of the plant matrix.

## II. Isolation of Plant Nuclear Matrices: Protein Composition and Ultrastructural Organization of the Corresponding Structures

The nuclear matrix can be obtained by sequential extractions of isolated nuclei to eliminate the nucleic acids and most of the soluble and loosely bound components. Various protocols have been developed for the isolation of nuclear matrices from different cell systems (de Jong *et al.,* 1990; Stuurman *et al.,* 1992a). They basically consist of a permeabilization of nuclei after DNA and occasionally RNA are digested, and eventually all soluble or loosely bound components are extracted. As expected, the composition, ultrastructure, and properties of the final matrix isolated greatly depend on the procedures used to achieve the extraction of nuclei. Some of the steps are especially important for the further processing of the extraction and also for the stability and composition of the final nuclear matrix, as is the case of the preparation of nuclei and the extraction of the residual structure with either low or high ionic strength media.

Two different procedures are the most commonly used to prepare nuclear matrices from plant species. The first, adapted from the original protocol of Berezney and Coffey (Berezney, 1984), consists basically in the digestion of nuclei with exogenous nucleases and further extraction with buffers of high ionic strength. The effects of various parameters of the isolation protocol on the final matrix, such as RNase digestion, the reversal of the high salt extraction, and digestion steps, stabilization by heat, or promotion of disulfides have been systematically analyzed in animal systems (Kaufman *et al.,* 1981; Belgrader *et al.,* 1991; Martelli *et al.,* 1991; Stuurman *et al.,* 1992a,b). These protocols have been used mainly in studies of protein composition and ultrastructural organization of the plant nuclear matrix.

The second method employs more physiological ionic strength to eliminate histones and other soluble proteins from the nucleus. After stabilization by heat, histones and other loosely bound proteins are extracted by the nonionic detergent lithium 3′, 5′-diiodosalycylate (LIS), and DNA is digested by restriction enzymes (Mirkovitch *et al.,* 1984). This method results in the loss of transcribed sequences of active genes and RNA polymerase II from the matrix (Mirkovitch *et al.,* 1984), but facilitates the characterization of MAR sequences attached to the matrix, and constitutes the choice protocol for these kinds of studies.

As expressed above, *in situ* preparation methods are not adequate for plant cells. For this reason, nuclei, prepared either from cultured cells obtained from protoplast or from plant tissues (see Table I) are the usual starting material for nuclear matrix preparation. As the protein composition and ultrastructural organization for the final nuclear matrix are strongly influenced by the procedures used for nuclear isolation, fractionation should be performed in an isolation medium containing millimolar amounts of magnesium ions to stabilize the nuclei, protease, and RNase inhibitors to protect that components of the matrix, and reducing agents to avoid the stabilization of neighboring proteins by disulfides (Moreno Díaz de la Espina *et al.,* 1991). Isolated nuclei should appear well preserved and of high purity by light and electron microscopy and maintain their general morphological characteristics with prominent well-contrasted nucleoli, condensed chromatin masses, interchromatin regions rich in fibrils and granules, and generally intact nuclear envelopes without contaminating cytoplasmic IF (Figs. 1A and 1C; Moreno Díaz de la Espina *et al.,* 1991).

## A. Experimental Procedures Based on High Ionic Strength Extraction

The first experimental procedure used for the preparation of onion nuclear matrices [adapted from the original method described by Berezney and Coffey for rat hepatocytes (Berezney, 1984)] consisted basically in the permeabilization of freshly purified nuclei with nonionic detergents, extraction with low and high salt buffers, and eventual digestion with DNase and RNase (see Table II). Observation of the nuclei by light microscope after each of the extraction steps showed a general relaxation of the nuclear

---

FIG. 1 Light and electron micrographs of isolated nuclei (A,C) and nuclear matrices (B,D,E,F,G) from onion meristematic root cells. (A) Light micrograph of a pellet of isolated nuclei prepared by differential centrifugation as described in Moreno Díaz de la Espina *et al.* (1991), and stained with methyl green pyronin solution. (B) Light micrograph of a pellet of isolated nuclear matrices prepared by procedure II as described in Table II, and later stained with red ponceau. The internal matrix shows a reticulate structure and the nucleolar matrix a lower contrast than the corresponding structures in isolated nuclei (arrowhead). (C) Electron micrograph of an isolated nuclei fixed with 2.5% glutaraldehyde and 2% osmium tetroxyde and processed as in Moreno Díaz de la Espina *et al.* (1991). The nucleus keeps its general morphology after the isolation (compare with the *in situ* nucleus shown in Fig. 5A), displaying an intact nuclear envelope (ne), big masses of condensed chromatin (Chr), a typical nucleolus (Nu), and interchromatin regions rich in granules (ir). (D–G) Electron micrographs of isolated nuclear matrices prepared by procedures I (D), II (E), III (F), and IV (G), respectively, as described in Table II. In all cases the resulting matrices are well organized and display the three typical domains: lamina (lam), nucleolar matrix (Num), and internal matrix (im). Slight morphological variations are observed in the internal matrix and nucleolar matrices obtained by different procedures (see text), while the lamina always presents a similar organization. Fibrillar centers within the nucleolar matrix are denoted by arrowheads. When not expressed, bars = 1 $\mu$m. A and B bars = 10 $\mu$m.

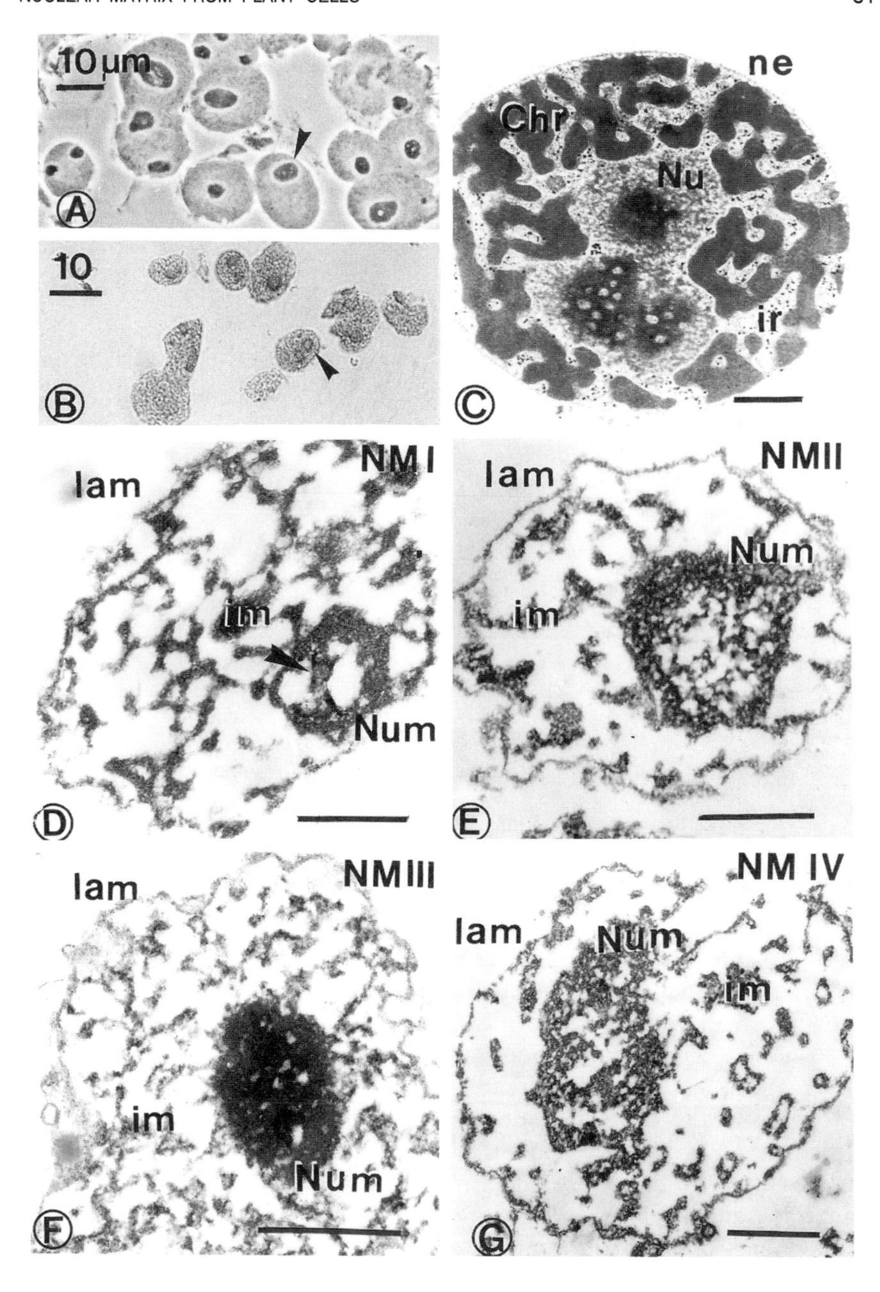

10 µm
A
10
B
ne
Chr
Nu
ir
C
lam
NM I
im
Num
D
lam
NMII
Num
im
E
lam
NMIII
im
Num
F
NM IV
lam
Num
im
G

TABLE II

Schematic Representation of the Procedures Used to Prepare Nuclear Matrices from Onion Meristematic Cells

| | Meristematic root tissue | | |
|---|---|---|---|
| | Freshly purified nuclei | | |
| 0.2 m$M$ $MgCl_2$ extraction[a] | 50 μg/ml DNase I[a] | 2 m$M$ TTNa[a] | 100 μg/ml DNase I[b] |
| 2 $M$ NaCl extraction + 0.2 m$M$ $MgCl_2$ | 0.5% Triton X-100 | 0.5% Triton X-100 | 25 m$M$ LIS[c] |
| 0.5% Triton X-100 | 100 μg/ml DNase I<br>100 μg/ml RNase A | 100 μg/ml DNase I<br>100 μg/ml RNase A | 200 μg/ml DNase I[b] |
| 100 μg/ml DNase I<br>100 μg/ml RNase A | 0.2 m$M$ $MgCl_2$ extraction | 0.2 m$M$ $MgCl_2$ extraction | **NM V** |
| | 2 $M$ NaCl extraction<br>**NM II** | 2 $M$ NaCl extraction<br>**NM III** | |
| **NM I** | 4 $M$ urea | 2 $M$ NaCl + 20 m$M$ DDT | |
| | Lamin-enriched<br>**NM II** | **NM IV** | |

[a] The corresponding solutions were prepared in 10 m$M$ Tris–HCl, pH 7.4. When not expressed $MgCl_2$ concentration was 5 m$M$, 1 m$M$ PMSF, and 20 m$M$ DTT were added to all solutions except for the case of NM III in which only PMSF was added.

[b] Digestion buffer consisted of 20 m$M$ Tris–HCl, pH 7.4, 0.05 m$M$ spermine, 0.125 m$M$ spermidine; 20 m$M$ KCl, 70 m$M$ NaCl, 10 m$M$ $MgCl_2$, 0.1% digitonin, 100 kIU/ml Trasylol, 0.1 m$M$ PMSF.

[c] Extraction buffer consisted of 5 m$M$ Hepes/NaOH pH 7.4, 0.25 m$M$ spermidine, 2 m$M$ EDTA/KOH pH 7.4, 2 m$M$ KCl, 0.1% digitonin, 25 m$M$ 3,5-diiodosalycylic acid lithium salt.

structure and a loss of refringency, especially after Triton X-100 extraction, in such a way that the final nuclear matrices appear as faint structures that show a reticulate organization after red ponceau staining (Fig. 1B). The assessment of DNA elimination of the residual structures by DAPI fluorescence confirmed that approximately 95% of DNA is extracted from nuclei by this method (Moreno Díaz de la Espina *et al.,* 1991). The nuclear matrices from onion cells obtained under these conditions do have a tripartite organization similar to those described in animal cells, with a lamina, a residual nucleolus and an internal network (Figs. 1D–G), although the internal components of the plant matrices appear to be more massive than those in animal cells, and to resist the digestion with RNase, even when freshly prepared and prior high salt extraction is used (Cerezuela and Moreno Díaz de la Espina, 1990).

The corresponding onion nuclear matrices display a complex polypeptide composition in 1-D SDS–PAGE with many bands between 12 and

90 kDa. These bands are also present in isolated nuclei and in the residual nuclear structures from the intermediate steps of the extraction (Fig. 2C; Moreno Díaz de la Espina *et al.*, 1991). Their major polypeptide bands show approximately the same range of molecular weight ($M_r$) as those

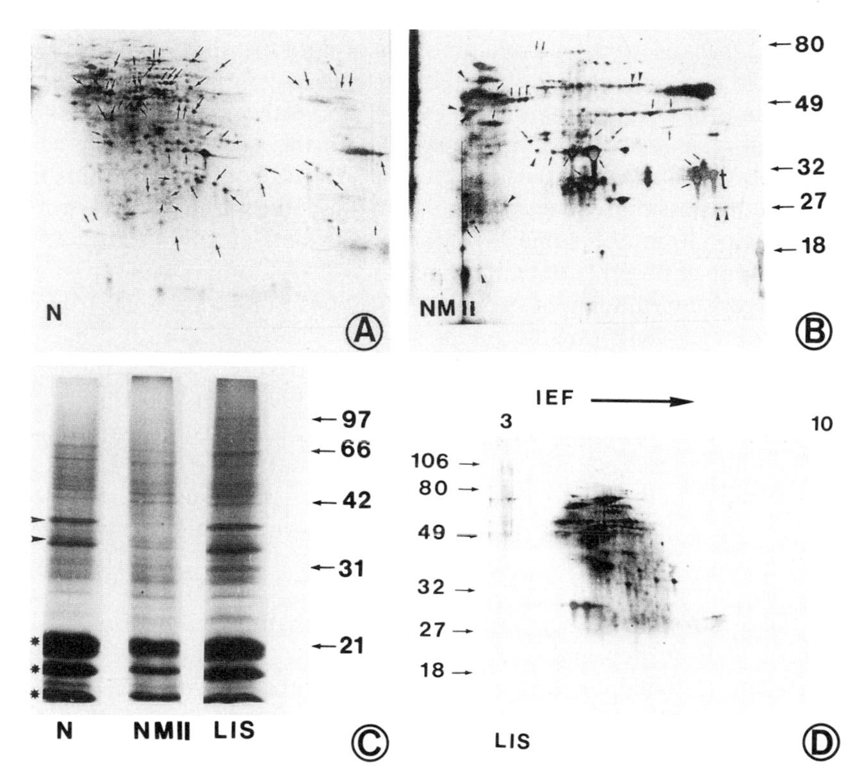


FIG. 2 1-D and 2-D polypeptide patterns of isolated nuclei and nuclear matrices obtained by either high ionic strength extraction (NM II) or the LIS method. Nuclear matrices were prepared from isolated nuclei as described in Table II (procedures II and V, respectively). Protein solubilization and gel running were as described in Minguez and Moreno Díaz de la Espina (1993a). (A) Total nuclear proteins from a nuclear fraction similar to those shown in Fig 1A and 1C were separated in 2-D gels according to O'Farrell (1975), after silver staining. Small arrows point to proteins also present in nucleolar fractions. (B) 2-D polypeptide pattern of a nuclear matrix fraction prepared by high ionic strength procedure II (see Table II), analogous to those shown in Fig. 1B and 1E. (C) 1-D polypeptide patterns of proteins corresponding to the same fractions shown in Fig. A, B, and D of this plate, separated in 10% PAGE after Coomassie blue staining. In the nuclear fraction (N), H1 variants (arrowheads) and core histones (asterisks) are the most prominent bands. High ionic strength extraction (NMII) is more efficient than LIS extraction in the elimination of histones, although some low $M_r$ bands comigrating with core histones are present on the fractions. (D) 2-D polypeptide pattern of a LIS fraction after silver staining.

from vertebrates, insects (Berezney, 1984; Verheijen *et al.,* 1988), and lower eukaryotes (Waitz and Loidl, 1988), and some of them have been identified by immunological methods, like lamins, topoisomerase II, fibrillarin (see Sections V,A,B,C). One conflicting result in the experiments was the apparent presence of detectable amounts of core histones (but not $H_1$) after extensive DNase digestion and 2 *M* NaCl extraction in the onion matrix, in spite of the very low amount of DNA detected in them by 4,6-diamidino-2-phenylindole (DAPI) fluorescence. However, this result is not completely surprising, because reproducible amounts of core histones have also been reported in nuclear matrices of vertebrates and lower eukaryotes with very low amounts of DNA present (Berezney, 1984; Waitz and Loidl, 1988). An artifactual reassociation of core histones with the abundant highly acidic proteins of the nuclear matrix detected in 2-D gels could explain this fact, although this possibility remains unclear. While some authors have reported specific association of histone $H_3$ with nuclear matrix proteins after cross-linking of intact nuclei (Grebanier and Pogo, 1979), others failed to detect high salt-resistant binding of exogenous histones to isolated nuclear matrices (Berezney, 1984). Low $M_r$ bands similar to those of core histones are also common in plant nuclear matrices from carrot cells obtained by similar procedures (Beven *et al.,* 1991; Masuda *et al.,* 1993). The possibility that these bands also include other proteins different from histones also has to be taken into consideration. In fact, in onion matrices, RNase I, which is very difficult to eliminate from the preparations after the digestion step, appears as a contamination of the final matrix, and comigrates with core histones at 17 kDa (data not shown here) as occurs in HeLa cells (Belgrader *et al.,* 1991).

In the electron microscope, the nuclear matrix obtained by the classical protocol of Berezney and Coffey (NM I) shows the three well-defined ultrastructural domains described in animal cells: the pore–lamina complex, nucleolar matrix (NuM), and internal matrix network, which we will analyze in detail later. In animal cells, the internal components of the matrix appear to be relatively labile, and by digesting with RNase A, varying the order of the extraction steps, or the extent of disulfide crosslinking, the resultant nuclear matrices show a wide range of morphologies and compositions (Kaufmann *et al.,* 1981; Belgrader *et al.,* 1991; Stuurman *et al.,* 1992b). For these reasons a stabilization step is needed in most protocols to prepare whole nuclear matrices from animal cells. Otherwise the internal network of the matrix dissociates during the extraction of chromatin. The molecular mechanisms responsible for stabilization are poorly understood. On the contrary, in onion cells the internal network of the matrix is very stable, even after extensive RNase digestion as shown in the first experiments of isolation (Barthelemy and Moreno Díaz de la Espina, 1984; Cerezuela and Moreno Díaz de la Espina, 1990; Moreno Díaz de la Espina *et al.,* 1991) and later confirmed in other plant species (Avramanova and Bennetzen, 1993). There still is a controversy in relation to the factors responsible for the structural integrity of the nuclear matrix. Disulfide bonds and RNA have been proposed to induce the internal matrix structure (Kaufmann *et al.,* 1981; Bouvier *et al.,* 1984; Stuurman *et al.,* 1992a), although in other cases no effect of RNase digestion was detected on the ultrastructure of the matrices (van Eekelen and van Venrooij, 1981).

In plant cells, the nuclear matrices obtained by different experimental procedures indicate a higher stability of the internal matrix components than in the case of animal matrices. In most cases the maintenance of the internal components is not dependent on RNase A digestion (Stolyarov, 1984; Ghosh and Dey, 1986; Cerezuela and Moreno Díaz de la Espina, 1990; Moreno Díaz de la Espina *et al.,* 1991; Frederick *et al.,* 1992; Minguez and Moreno Díaz de la Espina, 1993a), or only the nucleolar matrix appears affected by RNase (Masuda *et al.,* 1993). In other cases the stability to RNase of the internal matrix depends on the starting tissue rather than on the isolation procedure (Krachmarov *et al.,* 1991). Although some authors have reported a solubilization of the internal matrix by RNase (Galcheva-Gargova *et al.,* 1988), the putative lamina fraction appears highly contaminated by components of the internal matrix in other cases (McNulty and Saunders, 1992).

In addition, plant matrices obtained by the LIS method do not need a heat stabilization to maintain their integrity (Hall *et al.,* 1991; Breyne *et al.,* 1992; Avramanova and Bennetzen, 1993), supporting the hypothesis of the stability of the nuclear matrix in plants (Minguez and Moreno Díaz de la Espina, 1993a). These results suggested a different organization of the internal nucleoskeleton in plants. To test that, we analyzed the effects of various parameters of the isolation procedure in the composition and organization of the final nuclear matrix of the onion (see Table II).

To test the effects of exogenous RNase digestion on the stabilization of the onion nuclear matrix and avoid interferences with the high salt extraction step, we used digested NM with RNase A together with DNase I, under reducing conditions, and prior to the high salt extraction steps (see Table II). The resulting nuclear matrices (NM II) have a well-organized internal matrix with a prominent nucleolus and an internal matrix network including interchromatin granules (Fig. 1E), indicating that its structural integrity is not dependent on a kind of RNA accessible to exogenous RNase digestion (Cerezuela and Moreno Díaz de la Espina, 1990).

Protein analysis of this fraction by 1-D and 2-D PAGE demonstrates that this procedure is effective in solubilizing proteins from the nucleus, which presents a very different pattern of spots in 2-D gels (Figs. 2A and 2B). $H_1$ variants are efficiently removed from the fractions, but low-molecular-weight bands are still abundant in it (Fig. 2C). RNase A (migrating at 17 kDa) binds to the matrix during extraction and probably accounts for a considerable amount of protein in this band (not shown here), as reported in HeLa nuclear matrix (Belgrader *et al.,* 1991).

To investigate the role of intermolecular disulfide bond formation on the stabilization of the plant nuclear matrix, we oxidized nuclear matrices (NM III) isolated from nuclei incubated with the sulfhydryl oxidizing agent sodium tetrathionate (NaTT) (see Table II), which were later submitted to both DNase I and RNase A digestion and eventually high salt extraction. The corresponding matrices also have a tripartite structure (Fig. 1F). The most evident effects of the stabilization are observed in the nucleolar matrix, which presents a dense compact organization compared with the nucleolar matrices produced by other procedures. The electrophoretic profiles of these matrices are not identical to those of NM I and NM II, but rather show additional bands, especially in the high-molecular-weight regions of the gels, suggesting that the stabilization of NaTT produces more than just a general increase of proteins (Cerezuela and Moreno Díaz de la Espina, 1990). This fact correlates with the enhancement of the nucleolar matrix observed by electron microscopy (Fig. 1F). Further extraction of the oxidized matrices (NM III) with 2 *M* NaCl in the presence of the sulfhydryl reducing agent dithiothreitol (DTT) removes some bands from the gels, especially those in the high-molecular-weight range (Cerezuela and Moreno Díaz de la Espina, 1990). The nuclear matrices, and especially the nucleolar matrices inside them, appear much more extracted by electron microscope (Fig. 1G).

These results confirmed that the onion nuclear matrix has a tripartite structure with a complex polypeptide pattern whatever extraction procedure is used. The lamina is very stable to any of the stabilizing or destabilizing agents tested in these experiments. The internal matrix ultrastructure appears to be slightly sensitive to RNase A, while stabilization by NaTT has a marked effect on the nucleolus.

The combination of the destabilizing effects of DTT, RNase A, and 2 *M* NaCl extraction does not remove the internal matrix components in onion cells, whatever the sequence of extractions used (see NM I and NM II, Figs. 1D and 1E, respectively), as reported in nuclear matrices from carrot suspension cultures (Frederick *et al.,* 1992) and *Zea mays* roots (Krachmarov *et al.,* 1991), but in contrast with that reported for *Phaseolus* nuclear matrices (Galcheva-Gargova *et al.,* 1988). That indicates that the structural organization of the plant matrix is not exclusively dependent on RNA and disulfide bonds existing *in situ.* The nature of its components will be discussed in Section V,B.

## B. Experimental Procedures Based on Physiological Ionic Strength Extraction

The suspicion that high salt extraction could produce sliding of the attachment points of DNA loops, precipitation of the transcription complexes into the matrix, or rearrangement of their components, led to the development of a milder extraction procedure based on low salt and detergent extraction (LIS) to eliminate histones before endonuclease digestion (Mirkovitch *et al.,* 1984). In animal cells, LIS extraction requires a stabilization of nuclei by heat or divalent cations. This poses another problem, since thermal stabilization is known to induce unspecific protein association to the matrix (McConnell *et al.,* 1987). LIS extraction also results in the loss of transcribed sequences of native genes from the matrix (Mirkovitch *et al.,* 1984). Thus, both types of procedures have their advantages and disadvantages and the choice should depend on the kind of experiments to be performed in each case.

Plant nuclear matrices prepared by the LIS method do not require a stabilization step, as occurs with those of animal origin. Plant matrices prepared with or without thermal or LIS 2+ stabilization present the same protein patterns and DNA binding abilities (Breyne *et al.,* 1992; Avramova and Bennetzen, 1993). Thus apparently this extraction procedure is not as harsh for plant nuclei as it appears to be for nuclei of other systems. The molecular basis for the stability of the plant matrices is not yet understood.

Nonstabilized LIS nuclear matrices from onion cells have significantly higher amounts of histones than the matrices prepared by high salt extraction procedures (Fig. 2C). Not only have they a higher proportion of core histones than NM II, but also they present significant amounts of both H1 variants which are never found associated with any of the high ionic strength extracted matrices. This is a typical result in other plant (Hall *et al.,* 1991; Breyne *et al.,* 1992) and animal (Mirkovitch *et al.,* 1984) systems. Two-dimensional PAGE separation reveals that the protein composition of matrices prepared by high and low salt extraction procedures is different (Figs. 2B and 2D).

In summary, the protein composition of the nuclear matrix depends on the conditions of isolation procedures, and depending on that, different components would be removed from the basic matrix, producing different patterns of polypeptides in 2-D gels. As long as the nuclear matrix remains defined in an operational way, and we do not understand its basic organization, it is difficult to discriminate between proteins forming the essential basic nucleoskeleton, if it exists, and those forming the associated multimeric complexes that perform matrix-associated activities, and are associated to the nucleoskeleton with different affinities. But the protein pattern of the nuclear matrix should also be tissue-dependent, as different types of cells would present different protein complexes associated with the matrix.

## III. Structural Domains of the Plant Nuclear Matrix and 3-D Organization of Its Components

Although the protein compositions of the onion nuclear matrices prepared by different extraction procedures present some variations (see Section II), their basic structural organizations in thin sections are constant and comprise three main subcomponents: the lamina, the nucleolar matrix, and the internal matrix (Fig. 1D–G), which can be correlated with some of the structures of the unextracted nuclei (see Fig. 5A,B and Section VI).

The role of the nuclear matrix in coordinating nuclear function at the molecular level is now accepted (Jackson, 1991; Hoffman, 1993). The existence of a nuclear architecture acting at the supramolecular level is still controversial. Some authors argue that the intranuclear skeleton observed by electron microscope is artifactually created by the harsh procedures used to prepare the nuclear matrices. To clear this point, we reconstructed onion nuclear matrices prepared by different procedures (see Table II) from serial sections in 3-D (Figs. 3; 4, color plate 2). The first conclusion of our work was that at least at this level of resolution, there is little random aggregation or redistribution of the intranuclear domains in spite of the high salt extraction, as judged by their ultrastructural definition and maintenance of their territories within the matrix, which in the same way resembles results obtained in animal cells with the same approach (Bouteille *et al.*, 1983; Bouvier *et al.*, 1984) and by resinless electron microscopy (EM) of nuclear matrices prepared by milder procedures (Jackson and Cook, 1988; Nickerson *et al.*, 1990). Thus the matrices prepared in the most destabilizing conditions, that is digested with RNase and extracted with high ionic strength in reducing conditions (NM II), present a tridimensionally well-organized internal structure (Fig. 4). The analysis of reconstructed serial

FIG. 3 A–H. Eight consecutive sections from a serial through the same isolated onion nuclear matrix prepared by procedure II (see Table II). Then the pellets were fixed in 2.5% glutaraldehyde and 2% osmium tetroxide, processed as expressed in Moreno Díaz de la Espina *et al.* (1991), and embedded in Epon resin. Serial sections from the pellets were made with a diamond knife, mounted on single-hold grids coated with plastic supports, and stained with uranyl acetate and lead citrate. Several matrices were conventionally photographed at a constant magnification (×6000). Lamina (lam) nucleolar matrix (Num), internal matrix network (im). The stereological reconstruction of this structure is shown in Fig. 4. Bars = 1 $\mu$m.

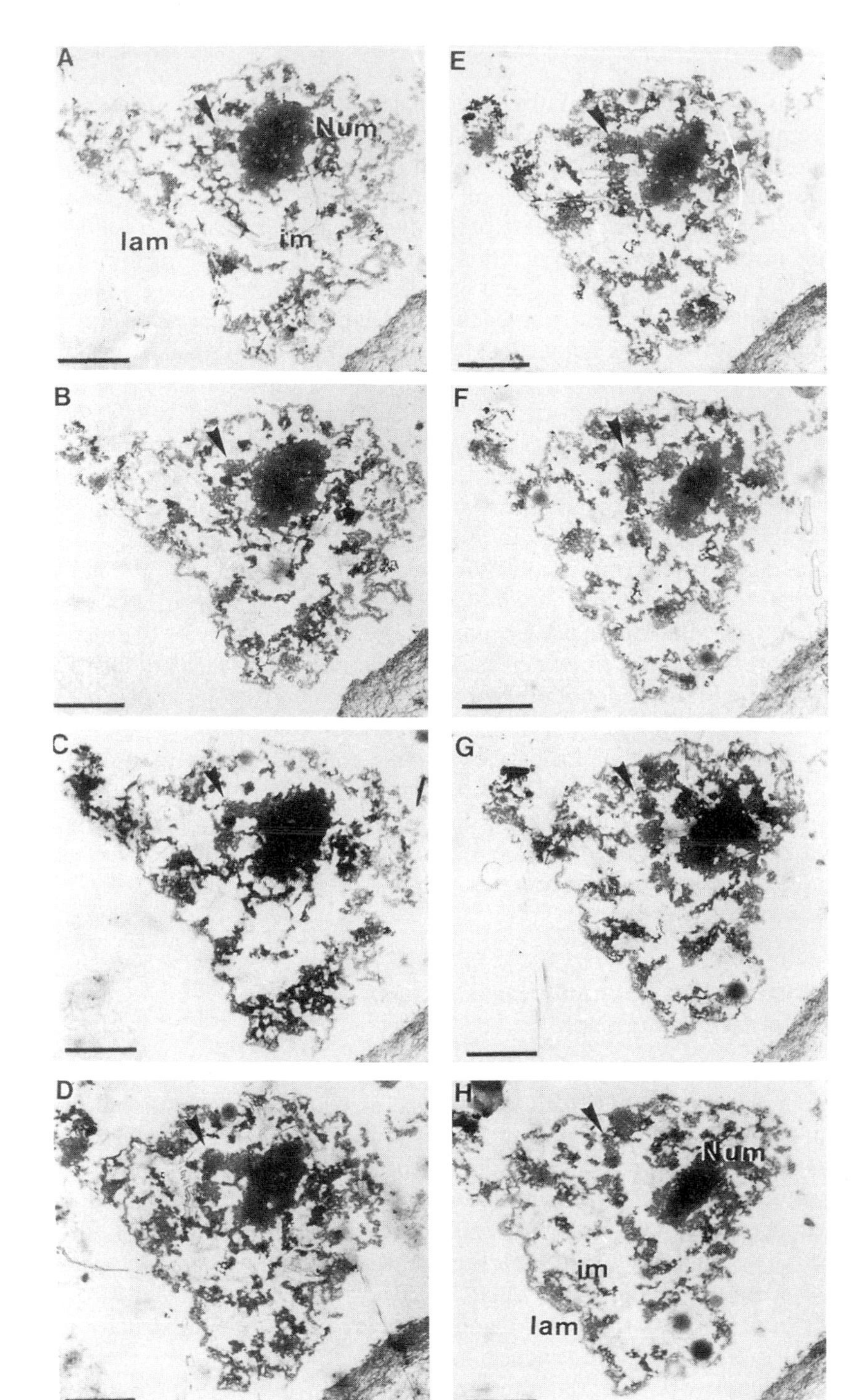

A
Num
lam
im
B
C
D
E
F
G
H
Num
im
lam

sections reveals that the matrices retain a highly ordered organization, with three main architectural domains each with a different structural organization: the lamina, nucleolar matrix, and internal matrix.

Stereological reconstructions of whole nuclear matrices prove that the lamina constitutes the boundary of the nuclear matrix, forming a continuous layer around it, that is physically associated with elements of the internal matrix (Figs. 3, 4). Whole nuclear matrices usually reveal a single nucleolar matrix, or two separated from each other. Nucleolar matrices do not show direct contact with the lamina, as reported for HeLa cell nucleolar matrices (Bouteille *et al.,* 1983). This is probably a consequence of the different architectural organization of nucleoli in animal and plant cells (Risueño and Medina, 1986). After stereological reconstruction the residual fibrillar centers appear as discrete entities immersed in the residual fibrillar nucleolar matrix without any apparent association with each other (not shown here) as is the case in the intact nucleolus (Medina *et al.,* 1983).

The reconstructed internal matrix forms a complex fibrogranular network attached to the lamina and extending through the nuclear interior coming into contact with the nucleolar matrix. The cross sections of the elements of the internal matrix show that they are composed of a mesh of fibrils with associated granules identified by immunocytochemistry as interchromatin granules (IGs) (see Section V,B).

In conclusion, elements of the residual nuclear matrix of onion cells retain a highly ordered architectural organization after elimination of DNA, RNA, and soluble proteins from the nucleus, supporting the idea that the plant nucleus has its own nucleoskeleton of insoluble protein fibers with a complex internal organization (see Fig. 15).

## IV. Interaction of the Nuclear Matrix with Plant Cytoskeleton

In animal cells, the cytoskeleton forms a structural and functional bridge from the cell periphery to the nuclear interior. It is composed of a structural interacting network of intermediate filaments, actin microfilaments, and microtubules with their corresponding sets of associated proteins that modulate their polymerization and functionality. In addition to their structural roles, cytoskeletal networks interact with cell signaling pathways and are involved in intracellular particulate transport. Each component of the cytoskeleton appears to play a distinct dynamic function in regulating cell structure, transport, and functioning.

The IF network extends from the cell membrane, where it contacts with desmosomes and tight junctions and binds to protein components of the cell membrane such as ankyrin and spectrin, to the nuclear envelope (Pienta *et al.,* 1991). The cytoplasmic IF network interacts directly with the nucleoskeleton or nuclear matrix apparently by polymerization of vimentin with nuclear lamins (Georgatos and Blobel, 1987; Foisner *et al.,* 1991). Microtubules act as tracks for intracellular particulate movement. Actin microfilaments are responsible for the control of cell shape, structure, and mobility, but they are also implicated in intracellular transport. Within the nucleus, the lamina, an IF network, constitutes the connecting structure between the nucleoskeleton and cytoskeleton. Although the existence of an intranuclear skeletal network is still in debate, the new preparative techniques have allowed the visualization of this network, which in some cases presents a similar morphology to that of IFs (Penman *et al.,* 1982; Jackson and Cook, 1988; He *et al.,* 1990; Wang and Traub, 1991). In plant cells, the basic cytoskeleton components appear to be similar to their animal counterparts although the knowledge of the cytoskeletal systems is much less advanced (Lloyd, 1989; Staiger and Lloyd, 1991; Lambert, 1993; Shibaoka and Nagai, 1994) (see Table III).

Immunolabeling with specific antibodies has allowed the identification of the major cytoskeletal arrays in higher plants composed of microtubules (MTs) and intermingling actin filaments (F-A): the cortical array, the radial cytoplasmic array, the preprophase band (PPB), the mitotic spindle, and the phragmoplast (Clayton, 1985; Lambert, 1993) (see Table III), which undergo dramatic changes in distribution correlating with the different phases of the cell cycle (Staiger and Lloyd, 1991). Recently, progress has been made in the identification of proteins involved in both cytoskeletal systems, as are some microtubule-associated proteins (MAPs), tubulin, actin-binding proteins, and myosin, although further characterization is needed (Staiger and Lloyd, 1991; Shibaoka and Nagai, 1994).

The most controversial cytoskeletal system in plants is that of IFs. Although IF-related proteins have been detected for a long time in plants (Shaw *et al.,* 1991), a cytoskeletal network of 10-nm filaments was not detected until recently in whole mount preparations of extracted cells (C. Yang *et al.,* 1992). On the contrary, the plant nucleoskeleton not only displays a well-organized lamina in which plant lamins with conserved epitopes have been characterized (Li and Roux, 1992; McNulty and Saunders, 1992; Masuda *et al.,* 1993; Minguez and Moreno Díaz de la Espina, 1993a), but also an intranuclear network of IF-type proteins have been suggested (Frederick *et al.,* 1992; Minguez and Moreno Díaz de la Espina, 1993a).

Many ultrastructural and immunological data confirm the importance of the nuclear surface as microtubule and F-actin organizing centers in higher plants (Staiger and Lloyd, 1991; Lambert, 1993). The evidence that isolated nuclei and nuclear particles from tobacco cells are able to nucleate microtu-

TABLE III

Main Components of the Plant Cytoskeleton and Nucleoskeleton

| | Structures | Type of proteins | Reference |
|---|---|---|---|
| Cytoskeleton | | | |
| Microtubules | Cortical microtubules | Tubulin | Goodbody *et al.* (1991) |
| | Preprophase band | MAP-like protein | Staiger and Lloyd (1991) |
| | Radial microtubules | | Lambert (1993) |
| | Mitotic microtubules | | Schellenbaum *et al.* (1993) |
| | Phragmoplast | | |
| Microfilaments | Cortical microfilaments | Actin | Clayton (1985) |
| | Radial microfilaments | Myosin heavy chains | Lloyd (1988) |
| | Preprophase band | Actin-binding-like proteins ABP | Staiger and Lloyd (1991) |
| | Phragmoplast microfilaments | (profilin, spectrin) | Shibaoka and Nagai (1994) |
| Intermediate filaments | Network of 10-nm filaments | Keratins (acidic and basic) | Shaw *et al.* (1991) |
| | | | Staiger and Lloyd (1991) |
| | | | C. Yang *et al.* (1992) |
| Nucleoskeleton | | | |
| | Lamina | Lamins (A and B) | Shaw *et al.* (1991) |
| | | | Li and Roux (1992) |
| | | | McNulty and Saunders (1992) |
| | | | Minguez and Moreno Díaz de la Espina (1993a) |
| | | | Masuda *et al.* (1993) |
| | Internal network | Intermediate filaments | Frederick *et al.* (1992) |
| | | | Minguez and Moreno Díaz de la Espina (1993a) |
| | | Topoisomerase II | (see this chapter) |
| | | Anti-P105 nuclear antigen | |
| | Nucleolar matrix | Fibrillarin | Minguez and Moreno Díaz de la Espina (1993b) |
| | | Nucleolin | |

bules (Mizuno, 1993) further supports the idea of close interactions between the nucleoskeleton and cytoskeleton in higher plants. Studies on the junctions between the cytoskeleton and nuclear matrix in higher plants are delayed mainly by methodological problems to prepare nuclear matrices *in situ* because of the plant cell wall. However, the delay is not only due to technical reasons, but also to the fact that except for the case of microtubules, the study of the plant cytomatrix (including both cytoskeleton and nucleoskeleton) has attracted few researchers in the past.

Preliminary studies of ultrastructural organization by transmission electron microscopy (TEM) of ultrathin sections of small blocks of undifferentiated meristem cells of onion roots submitted to the same extraction procedure used to prepare isolated nuclear matrices (see NM I in Table II) confirm structural relationships between the cytoskeleton and nucleoskeleton in these cells (Fig. 5B). The extraction procedure applied *in situ* is not as harsh as when applied to isolated nuclei, at least at the ultrastructural level, and especially in the case of the internal matrix which displays a complex structure in sections with abundant fibrils and granules, similar to that of the interchromatin region network of intact nuclei (Fig. 5A).

The lamina appears well defined constituting the boundaries of the nuclear matrix *in situ.* The cytoplasm appears to be extracted and devoid of organelles, such as membranous structures of the endoplasmic reticulum, Golgi apparatus, and plasma membrane. A cytoskeletal network is visible radiating from the lamina to the cell wall (Fig. 5B), composed of polymorphic fibers, which appear to be made of thin, less-stained 10- to 15-nm filaments and associated denser fibrils and granules. In favorable sections, these filaments appear anchored at the nuclear lamina (Fig. 5C). The resolution of the transmission electron microscope does not allow us to discriminate the fine structure of these filaments. Observation of embedment-free sections would probably help to elucidate the 3-D organization and ultrastructure of the cytoskeletal network that interacts with the nuclear matrix. In plant cells, both actin filaments and microtubules are involved in nuclear anchorage to the cell wall. But IF-related antigens have also been detected in these filament strands (Goodbody *et al.,* 1991; Staiger and Lloyd, 1991; Shibaoka and Nagai, 1994). The high concentration of DNase I used during the preparation of matrices would favor F-actin depolymerization, which would be unable to polymerize due to the additional high salt concentration (Hitchcock *et al.,* 1976). Further experiments using specific markers for each of the three cytoskeletal systems (i.e., antibodies against actin, tubulin, or the IFA epitope) and immunogold labeling are necessary to identify the cytoskeletal arrays which interact with the nuclear matrix and proteins involved in this process. Preliminary ultrastructural results strongly suggest a physical continuity from the cell wall to the nuclear

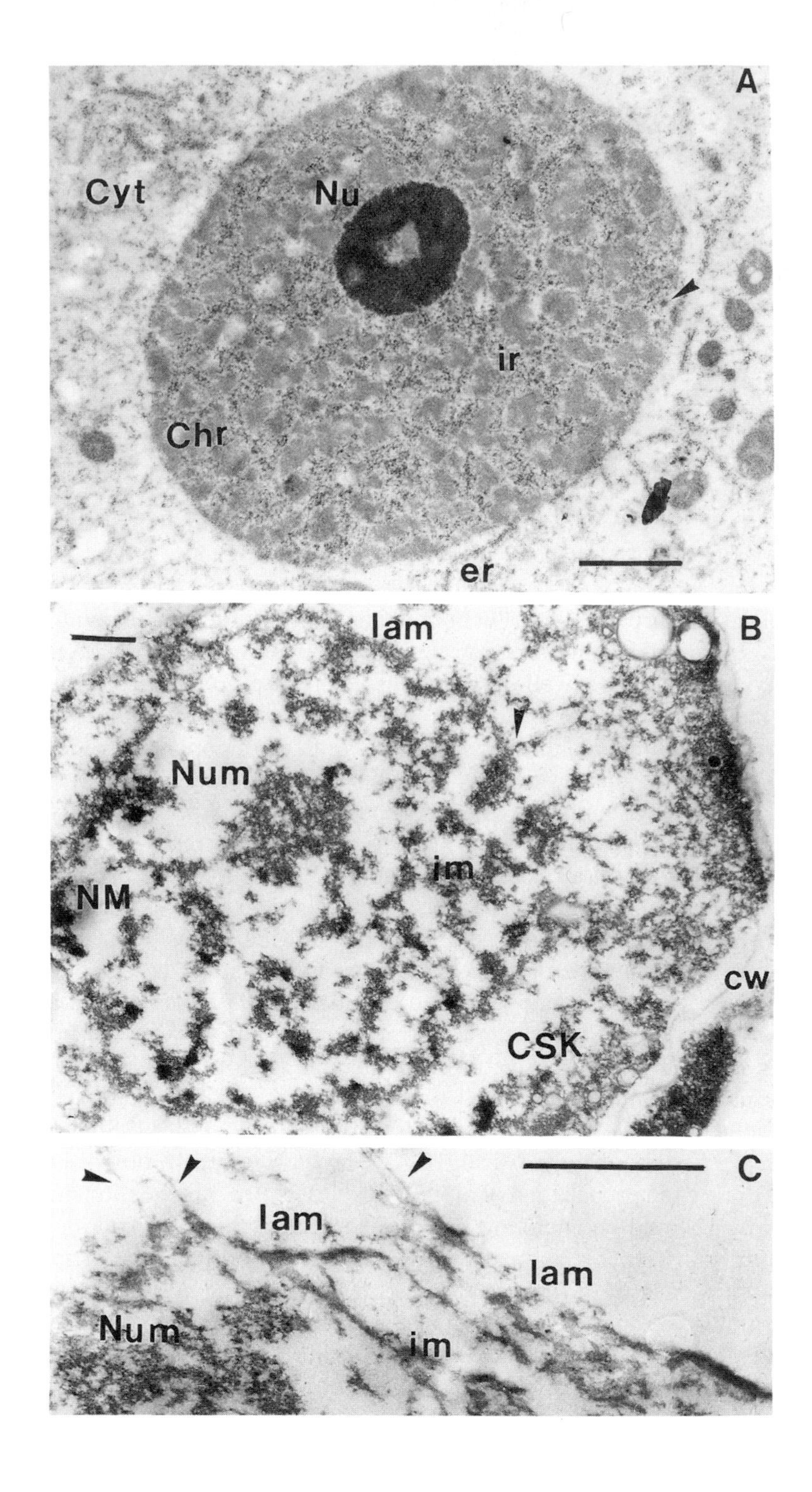
A
Cyt
Nu
ir
Chr
er
B
lam
Num
im
NM
cw
CSK
C
lam
lam
Num
im

lamina in plant cells provided by some undetermined fibrillar elements of the cytoskeleton (Fig. 6).

## V. Subfractionation of the Plant Nuclear Matrix

### A. The Lamina

The nuclear lamina is a scaffolding structure that corresponds to residual elements of the nuclear envelope, constituting a universal component of eukaryotic nuclei lying at the nucleoplasmic surface of the inner nuclear envelope (Dessev, 1992). It is basically made up of a network of orthogonal lamin filaments with which the residual pore complexes are associated (Aebi *et al.,* 1986; Dessev, 1992; Nigg, 1992), which makes contact with the cytoskeleton of IFs (Carmo Fonseca *et al.,* 1987). The principal structural components of the lamina are lamins, which constitute type V of IF proteins (McKeon, 1991; Nigg, 1992; Moir and Goldman, 1993). Lamins appear to be ancestral IF proteins (Weber *et al.,* 1989; Doring and Stick, 1990). They are ubiquitously distributed in eukaryotes, from dinoflagellates (Minguez *et al.,* 1994a,b) and yeast (Georgatos *et al.,* 1989) to higher plants (Minguez

---

FIG. 5 (A) Thin section of a whole onion meristematic root cell fixed and stained according to Bernhard's EDTA preferential staining for ribonucleoproteins (Bernhard, 1969), which reveals the nucleolus densely contrasted (Nu), and the nonnucleolar RNP network (ir) formed by granules and fibrils in the inter- and perichromatin regions of the nucleus. Membranes are not contrasted after this staining; thus the nuclear envelope is not evident although pore complexes appear contrasted (arrowhead). Within the cytoplasm (Cyt), the endoplasmic reticulum (er) connected to the nuclear envelope is recognized by the associated ribosomes. Mitochondria and plastids are also evident, but the elements of the cytoskeleton are not observed in thin sections. (B and C) Cells of the meristematic population of the onion root extracted *in situ* to prepare nuclear matrices. The last 2-mm regions of the tips of the roots containing this population were cut in very small pieces under a binocular microscope and submitted to sequential extraction with 25 m*M* Tris–HCl buffer containing in each case: (1) % Triton X-100, (2) 100 $\mu$g/ml DNase I and 100 $\mu$g/ml RNase A, (3) 0.025 m*M* $MgCl_2$, and (4) 2 *M* NaCl. After this extraction the nuclear matrix (NM) shows a similar structure and morphology to those prepared from isolated nuclei, with the three typical domains: the nucleolar matrix (Num), the lamina (lam), and the internal matrix (im). Within the cytoplasm, the extraction produces an elimination of membranes revealing a cytoskeletal network of fibrillogranular organization (CSK), which connects the lamina (arrowhead) with the cell wall (cw). In favorable sections and at higher magnification, a fibrillar network of 10–15 nm pervading the cytoskeletal network is observed. These fibers anchor in the lamina (arrowheads in C) and also in the cell wall (not shown in the figure), and in most cases appear masked by denser fibrils and some granules as shown in B. Bars = 1 $\mu$m.

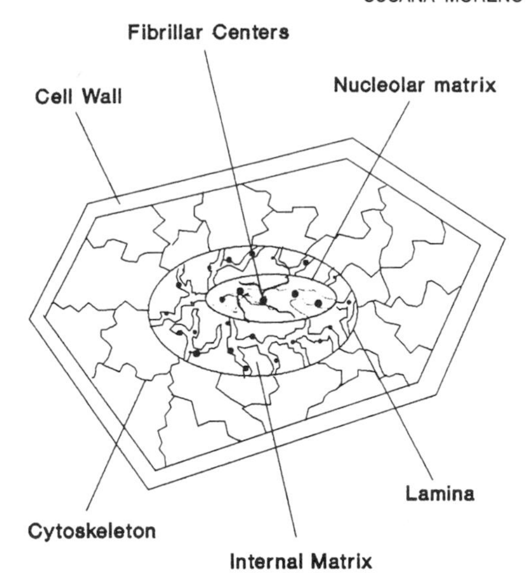


FIG. 6 Schematic representation of the interconnections between the different skeletal components in a plant cell. The structural elements of the plant cytoskeleton including microtubules, microfilaments, and the less-characterized intermediate filaments are represented as a unique network connecting the cell wall and the nuclear periphery. The cytoskeleton interacts with the cell wall through the skeleton of the plasma membrane and probably with the proteins of the extracellular matrix. The nuclear periphery is also an anchoring site for the elements of the cytoskeleton through the lamina, which constitutes the skeleton of the nuclear membrane and is made of lamins, the best-characterized IF-type proteins in plants, and further connects with the elements of the internal network of the nucleoskeleton, which are very stable in plants and also contain IF proteins.

and Moreno Díaz de la Espina, 1993a and vertebrates (Krohne and Benavente, 1986). Lamins present three well-defined separate domains: the $\alpha$-helical rod domain, typical of IF proteins that form coiled-coil dimers, is essential for higher order assembly; the nuclear localization signal that governs nuclear location; and the RAS-like carboxy-terminal CAAX domain responsible for membrane targeting (Nigg, 1989; McKeon, 1991). Lamins have been classified in two different subtypes A and B according to their antigenic determinants, sequence analysis, specific expression, post-translational modifications, and solubility during cell cycle (Nigg, 1989, 1992; McKeon, 1991; Moir and Goldman, 1993). The B lamins are constitutively expressed, while A lamins are expressed primarily in differentiated cells. There are several B-type lamins, transcribed by different genes. Usually cells contain one to five different lamins that can form homopolymers or heteropolymers. The assembly and disassembly of lamins appear to be controlled by phosphorylation and dephosphorylation (Nigg, 1989, 1992;

McKeon, 1991). Although lamins have been traditionally considered peripheral constituents of the nucleus, recent results demonstrate that these proteins are highly dynamic and also localize deep inside the nucleus (Goldman *et al.,* 1992; Bridger *et al.,* 1992; Minguez and Moreno Díaz de la Espina, 1993a).

Although the nuclear lamina is not easily observed in intact plant cells, it is a well-organized constant component of the nuclear matrices isolated from different species of plant cells such as the onion (Cerezuela and Moreno Díaz de la Espina, 1990; Moreno Díaz de la Espina *et al.,* 1991; Minguez and Moreno Díaz de la Espina, 1993a), carrot (Frederick *et al.,* 1992; Masuda *et al.,* 1993), pea (Li and Roux, 1992; McNulty and Saunders, 1992), maize (Krachmarov *et al.,* 1991), and beans (Galcheva-Gargova *et al.,* 1988) in a way that the nuclear lamina can be considered an ubiquitous structural component of the nuclei of higher plants. The lamina was the first IF-type network observed by EM in higher plants ( Moreno Díaz de la Espina *et al.,* 1991; Minguez and Moreno Díaz de la Espina, 1993a), before a cytoskeletal IF framework was demonstrated in whole-mount preparations of carrot cells (C. H. Yang *et al.,* 1992). Later, the immunological localization of lamins in this structure and the biochemical characterization of plant lamins confirmed that both the animal and plant laminae were homologous structures.

### 1. Fractionation of the Plant Lamina

In contrast to purified lamina fractions obtained from animal systems, it is difficult to separate the lamina from residual nuclear components in plants. Plant nuclear matrix fractions of different species prepared by procedures including both an RNase digestion step and strong reducing conditions present abundant internal components of both internal and nucleolar matrices (Figs. 1D–1G) (Cerezuela and Moreno Díaz de la Espina, 1990; Krachmarov *et al.,* 1991; Moreno Díaz de la Espina *et al.,* 1991; Frederick *et al.,* 1992; Masuda *et al.,* 1993; Minguez and Moreno Díaz de la Espina, 1993a). The only known exception is the matrix from inactive maize dry embryos, in contrast to that of the transcriptionally active root tissue (Krachmarov *et al.,* 1991), and that is probably related to the virtually inactive nuclear metabolism in this system, as occurs in adult chicken erythrocytes (Lafond and Woodcock, 1983). For this reason only lamina-enriched fractions, most frequently obtained by further urea extraction and which present variable amounts of internal components visible by EM, have been used for biochemical and functional studies of the plant lamina (Li and Roux, 1992; McNulty and Saunders, 1992; Minguez and Moreno Díaz de la Espina, 1993a). Accordingly, plant matrices prepared by the low ionic strength LIS method do not require a heat stabilization step (Avramanova and Bennetzen, 1993).

### 2. Ultrastructural Organization of the Plant Lamina

Onion nuclear matrices have a well-developed lamina with an intricate fibrillar structure with associated pore complexes (Fig. 7F) (Moreno Díaz de la Espina *et al.*, 1991; Minguez and Moreno Díaz de la Espina, 1993a), homologous to those found in other plant (Krachmarov *et al.*, 1991; Frederick *et al.*, 1992) and vertebrate (Krohne and Benavente, 1986) species. It forms a thin continuous filamentous layer at the periphery of the nuclear matrix, as demonstrated by 3-D reconstruction of thin sections of whole nuclear matrices, only interrupted by associated residual pore complexes (Fig. 4).

The ultrastructure of the onion lamina is insensitive to RNase digestion, disulfide stabilization, or strong reducing conditions (Figs. 1D–1F) forming in all cases a thin filamentous layer in thin sections (Fig. 7F). The plant lamina is basically a higher insoluble polymeric structure of IF proteins, whose primary organization does not depend on these kinds of interactions. We have never observed granular elements in the lamina, as reported in carrot matrices by Masuda *et al.* (1993), not even after NaTT stabilization which produces the stabilization of protein components on the internal and nucleolar matrices (Fig. 1F). The only granules observed near the lamina clearly correspond to those of the internal matrix network attached to the lamina, which clearly belong to a different nuclear matrix domain (Figs. 7F, 9B, and 9C; see also Section V,B).

The onion lamina constitutes the attachment site for fibers of the cytoskeleton (Fig. 5C), demonstrating that this structure is not only involved in maintaining nuclear integrity and functionality, but also in nuclear positioning and probably signal trafficking between nucleus and cytoplasm. The investigation of relationships between the cytoskeleton and nuclear matrix in plant cells would need further research using not only whole-mount preparations and resinless sectioning to investigate the spatial relationships between the two skeletal systems but also video and confocal microscopy, combined with immunolabeling with antibodies specific for their protein components, to clarify the interactions between the two systems. We should consider the pivotal role played by the cell nucleus in the organization of the cytoskeletal components along the cytoskeletal cycle in higher plants, mainly as a MT organizing center and anchoring site for actin and IFs (Staiger and Lloyd, 1991).

### 3. Protein Composition of the Lamina

The stability of internal matrix components in plants hampers the preparation of purified lamina fractions for biochemical and protein analysis pur-

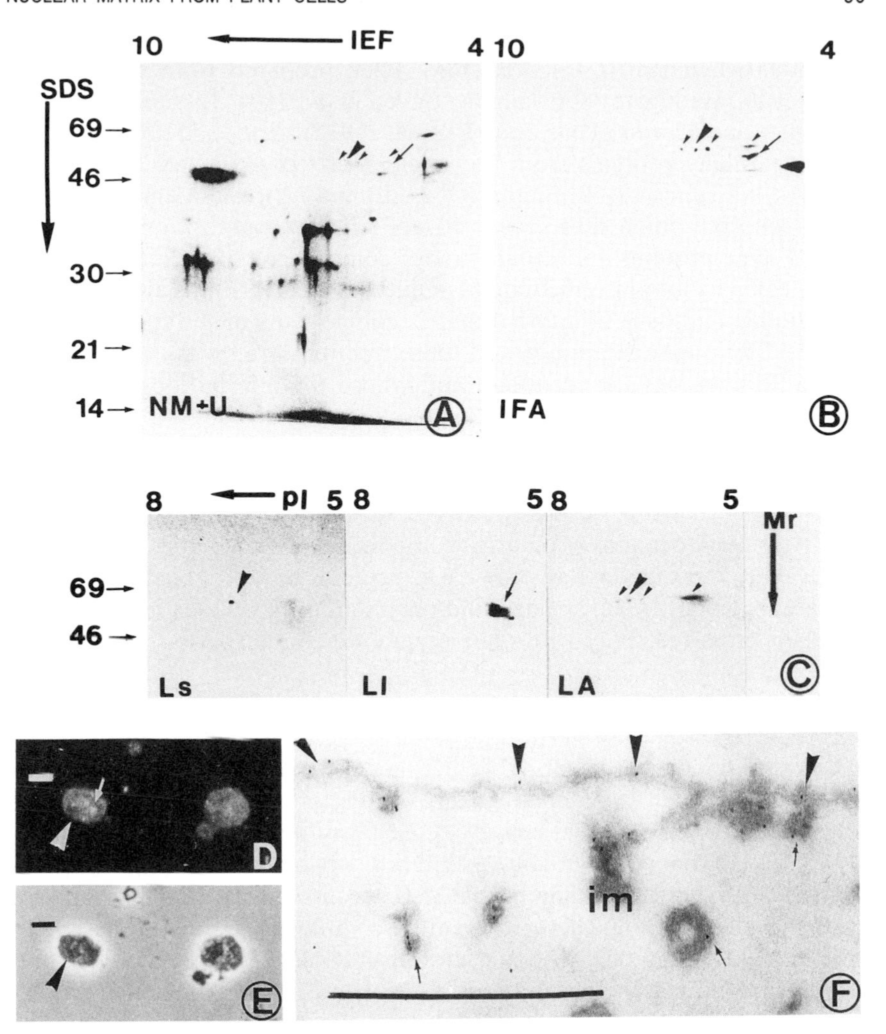


FIG. 7 Immunological detection of lamins in the onion nuclear matrix. Nuclear matrices prepared by method I as described in Table II were further extracted with 4 *M* urea to concentrate lamins, their proteins separated in a 2-D PAGE system, and either stained with silver or transferred to nitrocellulose membranes and incubated with different antibodies against IF proteins and lamins. (A) Total matrix proteins stained with silver. Arrows point to protein spots recognized by the antibodies in blots, which are minor components of the fraction. (B) IFA incubation. (C) Incubation with antilamins (see Table IV): Ls, chicken antilamin serum; LI, anti-*Xenopus* $L_I$; LA, antichicken lamin A. (D)–(F) Topological localization of lamins after incubation with the antichicken lamin serum. (D) After immunofluorescence staining the matrices show a spotlike decoration of their peripheries (arrows) that do not correspond with the nucleolar area (arrowhead). (E) Phase contrast of the same field. (F) Electron micrograph showing the immunolabeling of the lamina (arrowheads) and internal matrix (small arrows). When not indicated, Bars = 1 $\mu$m. D and E bars = 10 $\mu$m.

poses. Taking advantage of the stability of IF filaments in urea solutions, lamin-enriched matrix fractions have been prepared from several plant species to investigate their lamins (Beven *et al.,* 1991; Li and Roux, 1992; Minguez and Moreno Díaz de la Espina, 1993a). Although urea extraction removes many proteins from the nuclear matrix (Minguez and Moreno Díaz de la Espina, 1993a; compare Figs. 2B and 7A), lamin-enriched matrix fractions from onion cells show a heterogeneous composition in 2-D gels, and IF-type proteins are actually minor components of these preparations as revealed by anti-intermediate filaments antibody (IFA) staining, indicating the high amount of internal matrix components of the fractions (Figs. 7A,B). The major components of these fractions are spots in the range of 25 to 50 kDa, not yet identified and whose topological localizations are also not determined.

Although the presence of a well-developed lamina in plant nuclear matrices and the discovery that lamins are ancestral members of the IF family of proteins that are widely conserved in eukaryotes (Doring and Stick, 1990) pointed to lamins as putative components of plant matrices; in spite of several attempts made, until now these proteins have not been successfully isolated from any plant species, and plant lamins have been identified only by their cross-reactivity with their vertebrate counterparts.

## 4. Immunological Identification of Lamins in Plants

Early immunological results of detection of IF and lamin antigens in plant matrices were uncertain, and proteins with different molecular mass values were detected in different species (Galcheva-Gargova *et al.,* 1988; Beven *et al.,* 1991). That was probably caused not only by the lack of monospecific antibodies against plant lamins, but also because the set of nuclear IF-type proteins in plants could actually be more complex than previously thought. But the use of monoclonal antibodies on 2-D blots of lamina–matrix fractions has permitted confirmation of the conservation of lamins in plant cells. The use of several monoclonal antibodies from different sources not only provides strong evidence of the presence of lamins in the matrix, but also minimizes fortuitous cross-reaction with unrelated proteins. Contamination of the lamina–matrix fraction with cytoplasmic IF antigens is discounted by microscopic observations, and also by using antibodies against cytoplasmic IF reacting with plants like AFB, MAC 322 (Minguez and Moreno Díaz de la Espina, 1993a) (Table IV).

The IFA antibody recognizing a conserved epitope of IF, essential for the assembly of a normal IF network (Hatzfeld and Weber, 1992) and found in all lamins so far studied, reveals a set of proteins with relatively close $M_r$ and p*I* values in the onion matrix (Fig. 7B). Specific antibodies reacting with either A- or B-type vertebrate lamins allow discrimination

TABLE IV

Cross-reactivity of Antibodies Against Lamins and IF with the Onion Nuclear Matrix Proteins[a]

| Antibody | Type | Onion NM 1-D | Onion NM 2-D | Homologous system |
|---|---|---|---|---|
| B type lamins | | | | |
| Antichicken $B_2$ | M | 65 (+) | nd | $B_2$ lamin |
| Anti-*Xenopus* $L_I$ | M | 65 (+) | 65 kDa p*I* 5.75 (+) | $L_I$ lamin |
| Anti-*Xenopus* $L_{II}$ | M | 64 (+) | nd | $L_{II}$ lamin |
| A type lamins | | | | |
| Antichicken lamina | P | 65 (+++) | 65 p*I* 6.8 (+++) | 71 and 67 kDa |
| Antichicken A | M | 65 (+) | 65 p*I* 6.7 (+++) | A lamin |
| | | | 65 p*I* 5.65 (++) | |
| I filaments | | | | |
| IFA | M | 65 (+++) | 65 p*I* 7.1 (++) | Intermediate |
| | | 64 (+++) | 65 p*I* 6.8 (+++) | filaments |
| | | | 65 p*I* 6.5 (+++) | |
| | | | 64 p*I* 5.75 (+++) | |
| | | 58 (+) | 58 p*I* 5 (+++) | |
| MAC 322 | M | (−) | nd | 42, 50, 55 kDa |
| AFB | M | (−) | nd | 38, 62 kDa |
| Antivimentin | M | (−) | nd | Vimentin |

[a] P, polyclonal; M, monoclonal, nd, not determined. Intensity of reaction: strong (+++); medium (++); weak (+); no reaction (−).

between both types of lamins in onion matrices, on the basis of their cross-reactivity and their p*I* values and spot shapes. Anti-B-lamin antibodies recognize a polypeptide of about 65 kDa with the most acidic isoelectric point which migrates as a single spot in nonequilibrium pH-gradient electrophoresis (Fig. 7C; Minguez and Moreno Díaz de la Espina, 1993a). Antilamin-A antibodies, on the contrary, recognize several spots with different intensities, showing microheterogeneity, probably corresponding to modifications of peptides and with more basic p*I* values, except for one case in which the p*I* is similar to that shown by B-lamin (Fig. 7C). All these characteristics are typical of A-type lamins.

Our results on onion lamins are in agreement with those obtained by cross-reaction with IFA antibodies in nuclear matrices from carrot (Frederick *et al.,* 1992) and pea (Li and Roux, 1992), and with antilamin-B antibodies in pea (McNulty and Saunders, 1992). Based on these results, we think that not only B-lamins which are considered to be constitutive and correspond to the most ancestral lamins (Stick, 1992), but also A-lamins are conserved components of the meristematic plant nuclear matrix. Nevertheless, additional sequencing experiments will be necessary to clarify the molecular structure of plant lamins and classify them precisely as components of the already defined lamin subtypes. Although information about the functionality of plant lamins is scarce, recently a casein kinase II which phosphorylates a lamin-like protein (Li and Roux, 1992) and also an NTPase, probably derived from lamin precursors by proteolysis, biochemically and immunologically similar to the 46-kDa NTPase from rat liver nuclear matrix and involved in regulation of RNA export from the nucleus (Tong *et al.,* 1993), have been reported in the lamina–matrix fraction of pea. These results suggest a similar functionality for vertebrate and plant lamins.

## 5. Topological Distribution of Lamins

The pattern of lamin distribution in onion nuclear matrices after immunofluorescence staining shows not only a discontinuous peripheral localization that could reflect the distribution of lamins in the nuclear periphery, but also staining of foci deep inside the matrix (Fig. 7D; Minguez and Moreno Díaz de la Espina, 1993a). Immunogold labeling of thin sections of nuclear matrix confirms an intranuclear distribution of lamins in addition to the peripheral labeling of the lamina (Figs. 7F, 8). These

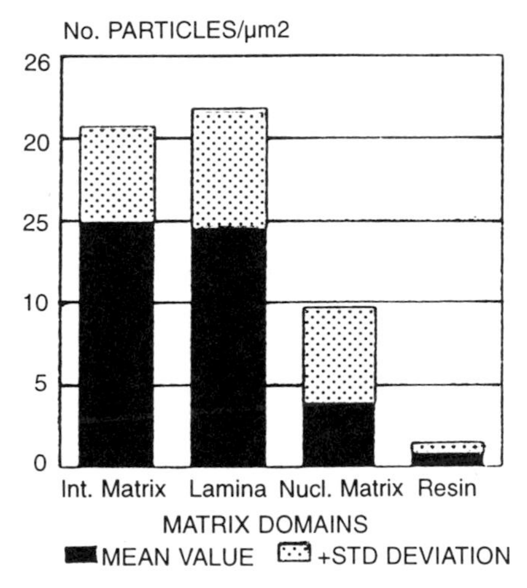


FIG. 8 Quantitative estimation of the particle density in the nuclear matrix after immunogold labeling with the chicken antilamin serum. The distribution of labeling was quantified by using 30 different micrographs from randomly chosen matrices taken from three different incubations and from two different grids from each experiment. Particle densities in the three main matrix components, and in areas containing only resin, were calculated using a semiautomatic procedure in a computer with a digitizer tablet. (Reproduced from Minguez and Moreno Díaz de la Espina, 1993a, with permission.)

results, and also the specificity of the labeling in 2-D blots, suggest a distribution of onion lamins (or some antigenically related proteins) not only in the lamina but also in the internal matrix components, in contrast with the classical peripheral distribution described for vertebrate lamins (Krohne and Benavente, 1986). However, the distribution of IF-related antigens in plant nuclear matrices (Frederick *et al.,* 1992; McNulty and Saunders, 1992) supports the idea of an intranuclear skeleton of IFs suggested by Jackson and Cook (1988). These results suggest that the set of IF-type proteins in the plant nuclear matrix are more complex than we think, including other proteins that could share antigenic determinants with lamins. But they could also reflect the physiological association of lamins with the intranuclear matrix as reported in cultured animal cells (Bridger *et al.,* 1993; Moir and Goldman, 1993).

### B. The Internal Networks of the Matrix

The internal ribonucleoprotein (RNP) network is the most controversial domain of the nuclear matrix. The contradictory results of different isolation procedures lead some authors to claim that the internal matrix is sensitive to RNase digestion and that a real internal network could not exist *in vivo.* Thus the internal network observed after the extraction would be an artifactual reassociation of RNP particles (Verheijen *et al.,* 1988; de Jong *et al.,* 1990). But it can also be argued that harsh methods of extraction, which are far from physiological conditions, destroy organized internal functional structure *in vivo.* The use of milder protocols would allow the quantitative recovery of transcription and replication activities associated with the residual nuclear structure (Cook, 1989). Also a well-organized internal nucleoskeleton has been visualized under different isolation conditions (Jackson and Cook, 1988; Wang and Traub, 1991; Nickerson *et al.,* 1992).

Considering all the results obtained by different groups working on the structure and composition of the nuclear matrix, not only is the protocol used critical for the observation of an internal network, but it also depends on the cell type and activity involved. The influence of RNase, divalent cations, molarity of extraction buffer, high ionic strength, extent of disulphyde cross-linking, use of different salts or detergents as extracting agents, use of nucleases or restriction enzymes, and even the sequence of various preparation steps or the stepwise increase in ionic strength (He *et al.,* 1990) affect the composition and organization of the matrix, and have been studied exhaustively. Nevertheless, there is no definitive answer to the controversy of the organization of the internal matrix, as is the case of the lamina.

Several lines of evidence demonstrate that most of the nuclear functions are associated with the matrix. There are permanent as well as dynamic DNA attachment sites on the matrix (Berezney, 1991). Also, replication and transcription factories are associated with the nucleoskeleton (Jackson and Cook, 1988; Hozák *et al.,* 1993b), and it appears plausible that RNA is permanently associated with the nuclear matrix from the time of synthesis to its transport into the cytoplasmic compartment. That means that RNA synthesis, processing, and transport all occur as a solid-state process in association with the matrix (Cook, 1989). This evidence suggests that the coordination of all these processes is unlikely to occur without an underlying nucleoskeletal array, the nature of which is still unknown.

### 1. Ultrastructural Organization and Stability of the Internal Matrix

In onion cells, the internal matrix is stable with the extraction procedures used and forms a ramified fibrogranular nucleoskeleton extending from the lamina to nucleolar matrices, as demonstrated by 3-D reconstruction of thin sections of whole nuclear matrices (Figs. 3, 4). The ultrastructural organization of the onion internal matrix is different from that of the lamina and nucleolar matrix, which never show associated granules, but similar to that described for the internal matrix of other plant (Frederick *et al.,* 1992) and vertebrate (Berezney, 1984) systems. It consists mainly of ~10-nm lightly stained fibers and 25- to 30-nm densely stained granules, which are associated to form a continuous network from the peripheral lamina to the nucleolar matrix. The degree of aggregation of fibers of the network varies according to the extraction procedure used. This network is apparently homologous to a part of the RNP network observed in intact onion nuclei after preferential staining of RNPs, which corresponds to the RNA "tracks" or domains of RNA synthesis, processing, and transport within the eukaryotic nucleus (Carter *et al.,* 1993; Spector, 1993a), as discussed later (compare Figs. 5A and 9A with 1D–G, 5B and 9B–9C).

The persistence of this network after exhaustive salt extraction and enzyme digestion indicates that it constitutes an insoluble component of the plant nucleoskeleton as described in other systems (Berezney, 1980; Kryzowska-Gruca *et al.,* 1983; Zborek *et al.,* 1990). The distribution of granules in the internal fibrillar network of the matrix is not homogeneous. In fact, some patches of the internal matrix do not have associated granules, although they show physical continuity with the fibrogranular ones (Fig. 9D). Thus, the internal matrix network in plants would contain different structural domains, with differences in their protein composition, as demonstrated by immunolabeling (see next paragraph), that could be supporting different functions.

In contrast to the results reported on animal systems, the internal matrix is a very resistent structure in plants. RNase digestion before or after high salt extraction does not disorganize the internal matrix network, and both the 10-nm fibers and 25-nm granules are apparently unaffected (Figs. 1D–1F). Although stabilization by disulfide bond formation increases the amount of materials associated with the internal matrix, further reduction of them during high salt extraction does not disorganize the internal network (Fig. 1F,G). Even after 4 *M* urea extraction, which solubilizes a lot of proteins (Figs. 2B and 2D), the matrices appear much more extracted in the electron microscope, but still granules and fibrils organized in an internal network are present (data not shown). The stability of the internal matrix in plants is supported by the fact that LIS matrices do not require a stabilization step in plants as in animal systems (Avramanova and Bennetzen, 1993; data not shown). All of these results support the idea that the organization of the internal matrix in plants is complex and probably depends on more than a single factor. As we will see later, IF probably are involved in its stabilization although interactions between RNA and proteins and between proteins cannot be discarded either.

## 2. Immunolocalization of Proteins and Characterization of Different Domains on the Internal Network of the Matrix

The difficulties in fractionating the internal network of the matrix in plants made the analysis of proteins in this matrix domain difficult. Although the NuM can be fractionated and its proteins identified on the gels, the impossibility to separate clean laminae, which would allow a further identification of their proteins, made the accurate adscription of protein spots to these two matrix domains uncertain.

Many of the protein spots migrating in the range of HnRNP and SmRNP proteins in the NM (Figs 2B and 2D), but not in the NuM (Fig. 12H) fractions, are most probably components of the internal matrix. The adscription of proteins in the range of about 60 kDa is more difficult, taking into account that the lamina has abundant components in this region of the gel. Immunolocalization with specific antibodies to different protein components of the internal matrix allowed us to identify the proteins forming the network, and also to define structural and functional domains according to the preferential distribution of these proteins (Table V). For these purposes we used antibodies against proteins considered to be structural components of the matrix (IF, topoisomerase II), enzymatic activities associated to the matrix (polymerase II), or protein components of the RNA-processing machinery.

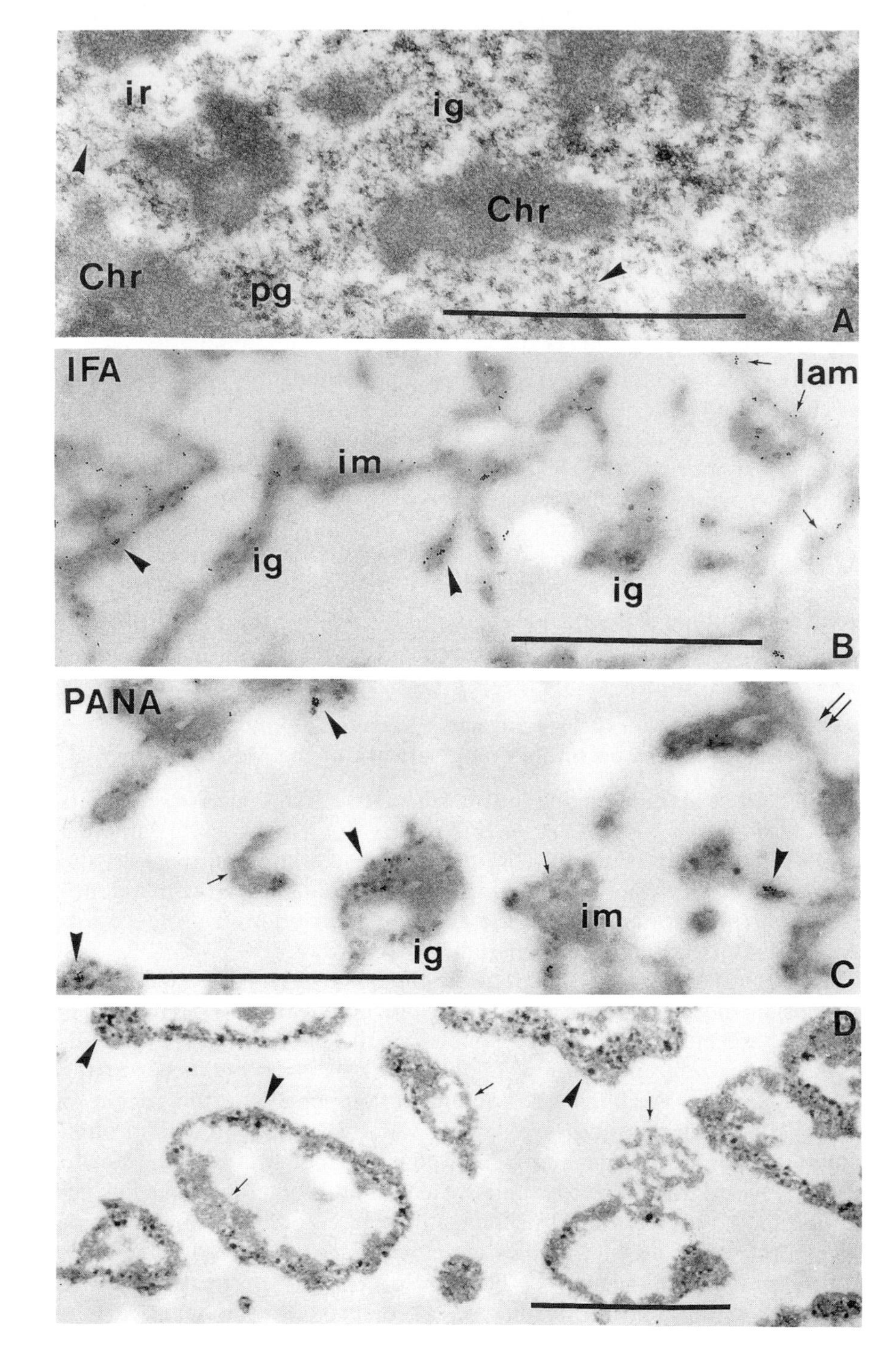
ir
ig
Chr
Chr
pg
A
IFA
lam
im
ig
ig
B
PANA
ig
im
C
D

Topoisomerase II is one of the main components of the NM involved in the attachment of chromatin loops by preferential binding to MAR sequences (Earnshaw *et al.,* 1985; Gasser *et al.,* 1989; Laemmli *et al.,* 1992). Our results, using an antibody against a synthetic peptide of the human nucleoplasmic isoform of the enzyme, demonstrate the existence of topoisomerase II in the nonstabilized onion NM with a similar Mr value to that reported for topoisomerase II of plant species, and immunologically related to the mammalian enzyme (Fig. 11A, Table V). The pattern of distribution of the fluorescence labeling, suggests a preferential distribution of the insoluble enzyme on the internal network of the matrix (Fig. 11C). In the electron microscope the more resolutive immunogold labeling localizes the topo II mainly on the fibrillar network of the internal matrix, lamina, and also in the fibrillar nucleolar matrices, suggesting that the anchoring of MAR sequences would localize in these domains (Figs. 11F and 10B).

Polymerase II, a component of the extranucleolar transcription machinery, appears to be associated with the internal matrix network (data not shown here), in agreement with the results of Jackson *et al.* (1993) demonstrating the attachment of transcription foci to a nucleoskeleton. Immunogold labeling is apparently associated with the fibrillar internal matrix (Table V). Antibodies against Sm protein components of SnRNPs that would bind to insoluble splicing foci apparently label both the fibrils and granules

FIG. 9 Composition and ultrastructural organization of the internal network of the plant nuclear matrix. (A) Thin section of a nonextracted meristematic cell nuclei *in situ,* fixed and stained according to the Bernhards's EDTA protocol which preferentially stains ribonucleoproteins (Bernhard, 1969), in order to enhance the visualization of the RNP network of the nucleus. The 20- to 25-nm interchromatin granules (ig) appear dispersed within the interchromatin regions (ir), while the 30- to 40-nm perichromatin granules (pg) appear, either isolated or clustered, at the borders of the bleached chromatin masses (Chr). There is also a network of fibers not preferentially contrasted pervading the interchromatin regions, to which the granules are associated (arrowheads). (B), (C) and (D) Thin sections of isolated nuclear matrices prepared by procedure II, incubated with antibodies against components of the internal matrix network (B,C), or with bismuth oxynitrate staining, preferential for interchromatin granules (Medina *et al.,* 1989). (B) The IFA antibody (Pruss *et al.*) reacting with all types of IF proteins that recognize several proteins in the nuclear matrix (see Fig. 7B), labels the lamina (lam) containing the lamins (arrows) and also the fibrillar network of the internal matrix (im) (arrowheads), but not the granules (ig). (C) The antiproliferation associated nuclear antigen antibody (PANA) (Clevenger and Epstein, 1984), recognizing a protein component of interchromatin granules, reacts with the clusters of interchromatin granules (ig) of the internal matrix (im), but not with the fibrillar zones devoid of granules (small arrows). The lamina (double arrows) and the nucleolar matrix do not show labeling. (D) The bismuth oxynitrate staining preferentially contrasts the interchromatin granules (arrowheads), confirming that they form a part of the internal matrix in plants. The fibrillar zones of the internal matrix do not bind busmuth (small arrows). Bars = 1 $\mu$m.

TABLE V
Protein Components of the Internal Frameworks of the Onion Nuclear Matrix Detected by Immunoblotting[a]

| Antibody | Type | Protein | Topological localization | Homologous system |
|---|---|---|---|---|
| Anti-topo II | P | 215 kDa<br>72 kDa | Internal matrix network (+++)<br>Lamina (++)<br>Nucleolar matrix (+) | 170 kDa |
| Anti-pol II | M | nd | Internal matrix network | 175 kDa |
| Anti-P105 NA | M | (−) | Granules of the internal matrix | 105 kDa |
| Y 12 | M | 29 kDa | Granules and fibrils of the internal matrix | Sm proteins (B) |
| JIM 63 | M | 100 kDa | Internal matrix network (+++)<br>Lamina (++)<br>Nucleolar matrix (+) | 92 kDa (carrot) |
| IFA | M | 65, 64, 58 kDa | Internal matrix network | IF |
| Antifibrillarin | | | | |
| S4 | P | 37 kDa | Nucleolar matrix | 34 kDa (human) |
| 72B9 | M | 37 kDa | Nucleolar matrix | |
| P2G3 | M | 37 kDa | Nucleolar matrix | 34 kDa (*Physarum*) |
| Antinucleolin | P | 58 kDa | Nucleolar matrix | 100 kDa (mouse) |

[a] M, monoclonal, P, polyclonal. Intensity of reaction: strong (+++); medium (++); weak (+); no reaction (−).

of the internal matrix network (Table V). These results give evidence that the plant NM provides structural support for the anchoring of DNA loops, and the transcription and splicing complexes, as demonstrated in animal cells (Jackson *et al.,* 1993).

The nature of the skeletal structures providing this support is uncertain. Some authors have demonstrated the existence of core filaments (Nickerson *et al.,* 1990, 1992) or IF-like arrays (Jackson and Cook, 1988; Wang and Traub, 1991) that would be good candidates for these functions. In plant NMs the IFA detects an intranuclear distribution of IF epitopes localized in the fibrils of the internal matrix (Table V, Fig. 9B). Additionally, one of the main protein spots of the NM recognized by the antibody in blots does not cross-react with lamins and presents an $M_r$ value similar to that described for an keratin-like IF protein of the nuclear matrix of rat liver cells (Aligué *et al.*, 1990). Jim 63, an antibody recognizing a high $M_r$ protein related to IF of the carrot NM (Beven *et al.*, 1991), also reacts to the internal filaments of the onion NM (Table V).

The detection of IF antigens in the internal plant nuclear matrix and the high stability of this structure to high salt and low urea extraction suggest that IF-type antigens could constitute a part of the pervading nucleoskeleton of plant cells, with which not only DNA loops, but also replication transcription and splicing factories forming heterogenous complexes could associate by different mechanisms. In animal cells, NuMA, a high-molecular-weight protein structurally related to IF, has been proven to be a component of the core filaments of the internal nucleoskeleton (C. Yang *et al.*, 1992) and to provide a bridge between the nucleoskeleton and the RNA processing machinery (Zeng *et al.*, 1994). A pervading nucleoskeleton of F-actin does not appear likely because DNase I depolymerizes F-actin and would destroy the internal architectural organization of the matrix. Unfortunately, an organized internal nucleoskeletal array of IF cannot be detected by TEM as occurs with the lamina. Resinless sections and protocols with stepwise increases of high salt extraction (He *et al.*, 1990) probably would reveal these structures and allow efficient fractionation that permits separation and biochemical analysis of the nucleoskeletal components and also defines which proteins are components of the nucleoskeletal arrays and which form

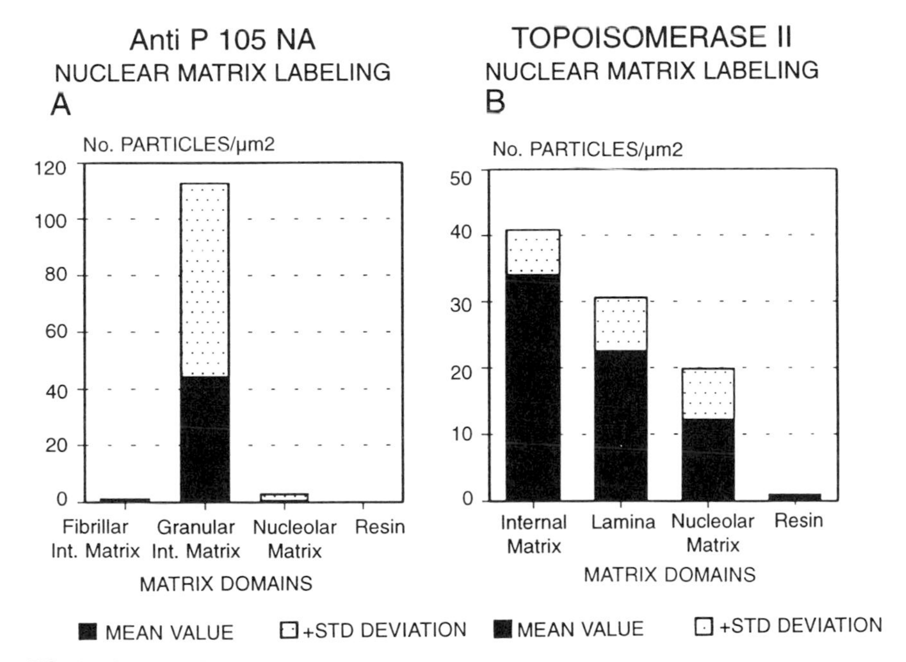


FIG. 10 Quantitative estimations of the particle densities in the nuclear matrix after immunogold labeling with the antiproliferating associated nuclear antigen antibody (A) and topoisomerase II (B). The distribution of labeling was quantified as described in Fig. 8.

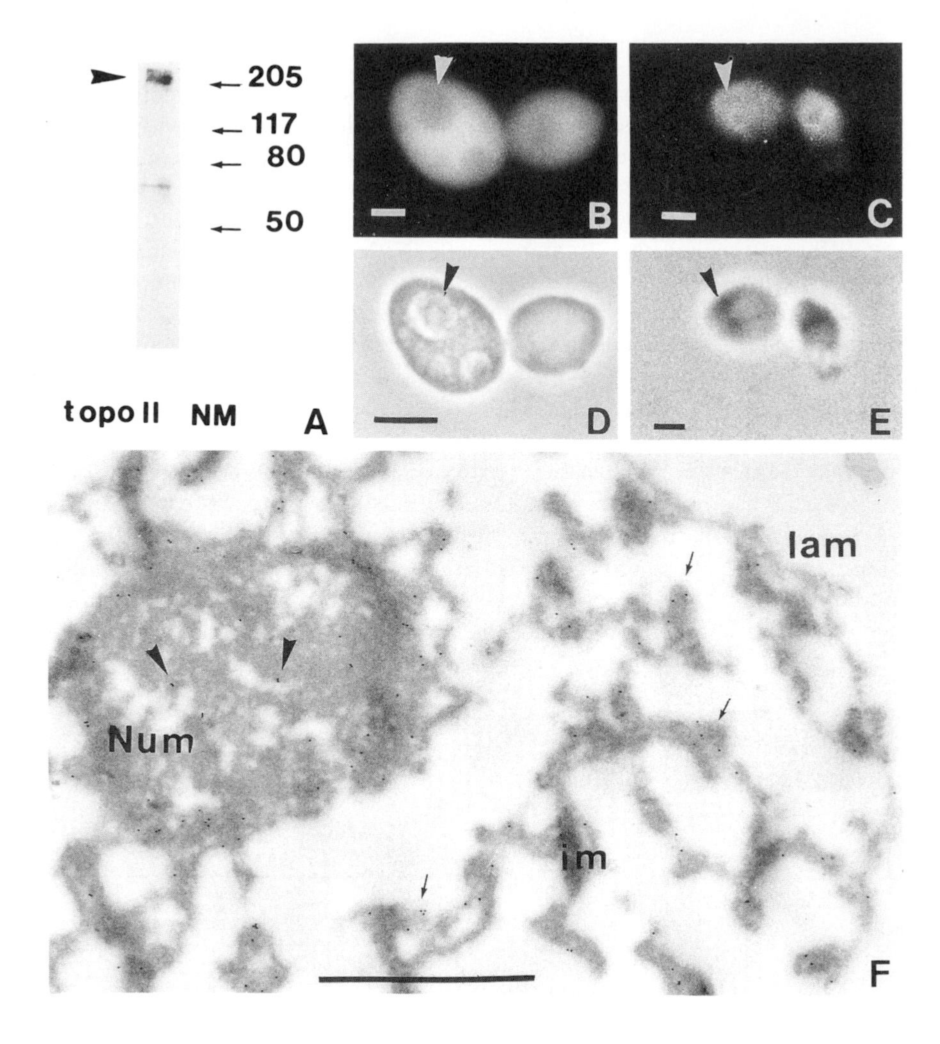


FIG. 11 Immunological detection of topoisomerase II in the nuclear matrix of onion cells. Nuclear matrices prepared by method II (see Table II), were either processed for PAGE, their proteins separated in 8% gels, transferred to nitrocellulose sheets, and incubated with a monoclonal antibody against human topoisomerase II or fixed for immunolocalization either in 0.3% paraformaldehyde, spread on polylysine-coated slides, washed in cold 1:1 ethanol:acetone, and incubated with the antibody for immunofluorescence or in 4% paraformaldehyde and embedded in LR white acrylic resin for immunogold labeling. (A) Immunoblotting with the anti-human topo II reveals a band at 215 kDa in the onion matrix. (B) Immunoflourescence staining of isolated nuclei reveals a nuclear distribution of the enzyme, while the nucleolar areas display a very low fluorescence (arrowhead). (D) Phase contrast of the same field. One of the nucleoli is marked with an arrowhead. (C) Immunofluorescence staining of the isolated nuclear matrices. The labeling is associated with the internal matrices, while the nucleolar matrices show a lower staining (arrowhead). (E) Phase contrast of the same field. (F) Electron micrograph of a portion of a nuclear matrix after incubation with the anti-topo II and gold-conjugated second antibody. The gold particles accumulate mainly in the internal matrix (im) (small arrows), while the nucleolar matrix (Num) and the lamina (lam) display a much lower labeling (arrowheads). Bar in (F) = 1 $\mu$m. Bars B–E = 10 $\mu$m.

a part of the multimeric complexes and have a functionality in replication, transcription, splicing, and transport.

Our data allow us to conclude that the internal matrix in plants is organized into two different structural domains. One is made up of ~10-nm fibers and contains structural proteins (IF-type antigens), but also topoisomerase II and polymerase II, implying that replication and transcription complexes are associated there. The second is fibrogranular in structure, contains epitopes of the snRNPs and interchromatin granules, and as I will discuss next, contains the intranuclear network of interchromatin granules and is probably involved in post-transcriptional RNA metabolism, providing support for the so-called RNA tracks (Spector, 1993b). In spite of the apparent structural homogeneity observed by TEM within each domain, most probably their composition and functionality vary between local sites in each domain. The existence of two different networks in the NM have also been suggested in animal systems (Berezney, 1980; Pouchelet *et al.,* 1986).

### 3. Interchromatin Granules Are Components of the Internal Matrix

Interchromatin granules are relatively well-characterized subnuclear structures in animal nuclei that form clusters of 20- to 25-nm granules linked by thin fibers, organized in a network, which apparently corresponds with the speckled fluorescent staining pattern of the nucleus revealed at the light microscope by antibodies against splicing factors and *in situ* hybridization with probes to snRNPs (Spector, 1993a,b). They show little or no associated RNA synthesis under physiological conditions (Puvion-Dutilleul, 1983), contain highly phosphorylated proteins (Wassef, 1979), and form a part of the isolated nuclear matrix (Zborek *et al.,* 1990). The interchromatin granules are related to the recruiting mechanism that regulates the storing and shuttling of splicing factors within the nucleus (Jiménez-García and Spector, 1993).

Plant nuclei lack the large interchromatin granule clusters typical of animal nuclei, although they present similar 20- to 25-nm granules interconnected by thin fibers forming a continuous network along the interchromatin spaces (Fig. 9A). On the basis of their fine structure, topological distribution, cytochemical characteristics, high phosphorylated protein content, and the presence of common epitopes, these ubiquitous structures have been identified with the interchromatin granules of animal systems (Medina *et al.,* 1989; Moreno Díaz de la Espina *et al.,* 1993). The evolutionary conservation of these structures in plant cells would suggest that in spite of the differences in the topological distribution of interchromatin granules, the nuclear mechanism of mRNA splicing has been widely conserved in eukaryotes.

Interchromatin-like granules are constant components of the internal matrix domain of the isolated onion nuclear matrix, as expressed above (Figs. 9B–9D), independent of the extraction procedure employed for its preparation (Figs. 1D–1G). Three-dimensional reconstruction of whole NMs demonstrate that the groups of IGs form a part of a continuous network extending between the lamina and the nucleolar matrix (Figs. 3, 4), similar to their distribution in intact nuclei (Figs. 5A, 9A). This network appears composed of two different domains when viewed by electron microscope, only one of which contains granules suggesting the heterogeneity of the internal matrix. Matrix interchromatin-like granules present more heterogeneous diameters than IG *in situ,* probably as a consequence of the harsh extraction protocols that sometimes produce aggregation of two or more granules (Fig. 9D), even in isolated nonextracted nuclei (Fig. 1C). But their high affinity for bismuth salts, which preferentially stain interchromatin granules (Fig. 9D; Medina *et al.,* 1989; Moreno Díaz de la Espina *et al.,* 1993), and also the identification of epitopes of interchromatin granules on them (Table V; Fig. 9C), allow the unequivocal identification of these structures.

The most convincing data on identifying these structures as interchromatin granules come from immunogold-labeling experiments with antibodies against proteins associated with interchromatin granules, which allow the highest resolution available. This method allowed us to detect the presence of P-105 (proliferation-associated nuclear antigen), a protein component of interchromatin granules (Clevenger and Epstein, 1984) (Figs 9C and 10A), as well as some Sm proteins of the snRNP particles (data not shown) (Table V) on the interchromatin granules forming part of the internal matrix, although not all the granules in a section are labeled. The low efficiency of this method, which only detects exposed epitopes on the surface of the sections, does not allow us to conclude whether all the interchromatin granules are homogeneous in composition or contain different proteins and snRNAs. Our results demonstrate that the interchromatin granule network is a differentiated domain of the internal matrix in plant cells, which provides physical support for the organization of the multimeric complexes involved in mRNA splicing.

### C. The Nucleolar Matrix

The nucleolus is the ribosome-producing factory of the cell. It has a complex organization with many different components: ribosomal DNA (rDNA), ribosomal RNA (rRNA), and also small nucleolar RNA (snoRNA), and many different kinds of proteins (Risueño and Medina, 1986; Scheer and

Benavente, 1990; Hozák *et al.,* 1994; Scheer and Weisenberger, 1994). All these components are synthesized either in the nucleus or the cytoplasm and need to be highly organized to interact efficiently and produce functional peribosomal particles. Up to now the structures or molecules responsible for maintaining the nucleolar structure and functionality have not been defined. In almost all higher eukaryotes the nucleolus has a similar pattern of ultrastructural and biochemical features and also of assembly and disassembly during the mitotic cycle (Scheer *et al.,* 1993; Scheer and Weisenberger, 1994), demonstrating that the mechanisms that regulate this essential function for the eukaryotic cell have been very well conserved in evolution. Nevertheless, the biochemical and structural basis for the maintenance of the nucleolar organization and function are not yet well understood, and the existence of a nucleolar matrix pervading the nucleolar structure, which could provide the anchoring sites for rDNA transcription and processing machineries, is still controversial.

In the most simplified case, the nucleolus may be self-assembled around the synthesized rRNA transcripts by a cascade of molecular interactions. rRNA could bind processing enzymes and eventually interact with 5sRNA and ribosomal proteins to produce the ribosomal subunits, and the nucleolus would be formed by accumulation of all these compounds. Although true self-assembly should be tested *in vitro* using purified components, *in vitro* reassociation of soluble-nucleolar extracts from *Amoeba* proved to be able to form nucleolus-like structures (Trimbur and Walsh, 1993) and would support this idea. On the other hand, experiments with transformants of rDNA genes suggest that the ability to organize a nucleolus from these materials is an intrinsic property of the rDNA genes, and that even a single rRNA gene is able to organize a nucleolus (Karpen *et al.,* 1988). Probably rDNA-associated proteins, nascent pre-rRNAs, and snoRNAs bound to the genes could be involved in the organization of the nucleolus architecture.

However, the experiments with a mutant yeast strain defective in RNA polymerase I suggest a structural role for RNA polymerase I in maintaining nucleolar structure (Oakes *et al.,* 1993). These and also experiments with 5,6-dichloro-(β-D-ribofuranosyl)-benzimidazole (DRB)-treated cells suggest that a conventional nucleolar structure is not a prerequisite for rRNA synthesis and processing (Scheer *et al.,* 1984) and vice versa. Thus, in certain cases, a nucleolar morphology exists in the absence of rRNA synthesis as occurs in resting tissues (Risueño and Moreno Díaz de la Espina, 1979) or even in the absence of rDNA genes as is the case of the anucleolate mutant of *Xenopus laevis* (Hay and Gurdon, 1967). In these cases, the organized nucleolar structure probably reflects an underlying organization independent of rRNA transcription, processing, and genes.

The formation of prenucleolar bodies *in vivo* (Moreno Díaz de la Espina and Risueño, 1976; Risueño *et al.,* 1976) or *in vitro* during reconstitution of nuclear extracts (Bell *et al.*, 1992), and of nucleolus-like particles from soluble nucleolar extracts (Trimbur and Walsh, 1993) support the idea that specific protein–protein interactions provide the bases for nucleolar morphology in the absence of rRNA synthesis. But the proteins responsible for these interactions are not known. According to the consensus, the nucleolar matrix is "the residual structure left after extraction procedures used to reveal the nuclear matrix" (Jordan, 1984). Studies on the underlying nucleolar matrix of eukaryotes are relatively scarce (Todorov and Hadjiolov, 1979; Franke *et al.*, 1981; Krohne *et al.*, 1982; Shiomi *et al.*, 1986; Olson and Thompson, 1983; Olson *et al.*, 1986), although an association of rDNA spacer sequences to the nucleolar matrix has been reported (Bolla *et al.*, 1985; Stephanova *et al.*, 1993). The existence of a real nucleolar skeleton is still controversial (Labhart *et al.*, 1984).

### 1. The Nucleolar Domain of the Nucleolar Matrix

Most of the studies undertaken so far on the nuclear matrix of plant cells have revealed the existence of an organized nucleolar matrix inside it that roughly retains the size and shape of the nucleolus (Figs. 1B, and 1D–1G) (Barthellemy and Moreno Díaz de la Espina, 1984; Stolyarov, 1984; Ghosh and Dey, 1986; Moreno Díaz de la Espina *et al.*, 1991; Frederick *et al.*, 1992). However, the fine structural organization of the nucleolar matrix apparently depends on the extraction procedure used (Figs. 1D–1G) (Cerezuela and Moreno Díaz de la Espina, 1990; Masuda *et al.*, 1993).

In onion cells, the nucleolar matrices survive RNase digestion prior to high salt extraction and do not depend on stabilization by disulfide bonds during extraction (Figs. 1D–1G) (Cerezuela and Moreno Díaz de la Espina, 1990), although the nucleolar domain presents a denser structure in oxidized matrices (Fig. 1F). The stability of the plant nucleolar matrix differs from results reported for other animal systems (Bourgeois *et al.*, 1987). They present a complex fibrillar structure in which the dense fibrillar and granular components are no longer evident. Moreover, the fibrillar centers are constant components of them (Fig. 1D), suggesting that these enigmatic structures which have been related to rDNA transcription (Thiry, 1992) or anchoring (Hozák *et al.*, 1994) could play a structural role on nucleolar functioning.

### 2. Preparation of Isolated Nucleolar Matrices from Isolated Nucleoli

Investigation of the protein components of the nucleolar matrix inside the nuclear matrix are hampered by the presence of abundant proteins from the lamina and internal matrix in this fraction. To avoid this problem nucleolar matrices can be prepared from isolated nucleoli; but to minimize the effects of the extraction procedures of nuclei and nucleoli on the final nucleolar matrices, special care should be taken with the following points in the case of plant cells. Nuclei must be isolated in a medium without divalent cations to favor chromatin dispersion, and that is especially important in reticulate plant species with a high amount of condensed chromatin like the onion. Nucleoli are fractionated by sonication, but right after isolation they must be resuspended in a medium with an optimal concentration of divalent cations to avoid their disorganization. This procedure produces clean nucleolar fractions, free of condensed chromatin as observed by light and electron microscopy, relatively stable when kept under appropriate conditions, and which maintain their structural domains and do not appear to be highly contaminated by nuclear proteins as shown in 2-D gels (Figs. 2A and 12G) (Figs. 12A and 12B and Figs. 1C and 12D), (Martin *et al.,* 1992a; Minguez and Moreno Díaz de la Espina, 1993b; 1995). For all these reasons the nucleolar fractions are the choice starting materials for the preparation of isolated nucleolar matrices (Table VI).

Successive extractions of this fraction with DNase I, RNase A, and 2 *M* NaCl, under reducing conditions, does not result in a disorganization of the residual structures and produces isolated nucleolar matrices that are smaller than the starting nucleoli (Fig. 12C), and appear well-organized in

TABLE VI

Schematic Representation of the Procedures Used to Prepare Nucleolar Matrices from Onion Meristematic Root Cells[a]

|  |  |
| --- | --- |
| Meristematic root tissue | |
| Freshly purified nuclei[a] | |
| | Sonication in $MgCl_2$ buffer[b] |
| Purified nucleoli[c] | |
| | 100 $\mu$g/ml DNase I<br>100 $\mu$g/ml RNase A<br>+<br>2 *M* NaCl |
| Nucleolar matrices | |

[a] Prepared in isolation medium without $MgCl_2$.

[b] Before sonication a concentration of 5 m*M* $MgCl_2$ was restored in the isolation medium.

[c] The corresponding solutions were prepared in 10 m*M* Tris–HCl, pH 7.4, containing 5 m*M* $MgCl_2$, 1 m*M* PMSF, and 20 m*M* DTT.

thin sections observed by TEM (Fig. 12F), suggesting that the stability of the isolated nucleoli and nucleolar matrices prepared there do not exclusively depend on other components of the nuclear matrix. These results differ from others in rat liver nucleolar matrices reporting the disappearance of the structured nucleolar matrix in sections when using buffers without divalent cations, but not in spread preparations (Stephanova and Valkov, 1991). The stability of the isolated nucleolar matrix in onion cells would be in relation to the already known stability of the internal components of the matrix in plant cells (Cerezuela and Moreno Díaz de la Espina, 1990; Avramova and Bennetzen, 1993; Minguez and Moreno Díaz de la Espina, 1993b), and considered a general characteristic of plant nucleolar matrices. But also the presence of divalent cations in the extraction media could contribute to stabilize the preexisting structure.

### 3. Ultrastructural Organization of the Isolated Nucleolar Matrix

The integrity and ultrastructure of the isolated nucleoli strongly depend on the isolation procedure used. As observed by EM, the isolated onion nucleoli obtained after mild sonication maintain their integrity and a distribution of their three main components similar to that of the *in situ* nucleoli. The pale fibrillar centers are surrounded by the dense fibrillar component which in turn is immersed in the peripheral loose granular component (Fig. 12D), in the same way that isolated nucleoli do (Olson *et al.,* 1986; Stephanova *et al.,* 1993), while a harsh sonication results in the loss of a typical granular component (Martin *et al.,* 1992a). The isolated nucleolar matrix obtained by simultaneous digestion with DNase I and RNase A and 2 *M* NaCl extraction under reducing conditions preserves its integrity and presents a similar organization to those inside the nuclear matrices (compare Fig. 12F and Figs. 1D–1G). It differs from the intact nucleolus mainly in the lack of preribosomal particles and a distinct granular component, but also in the organization of the dense fibrillar threads in both structures. Fibrillar centers are the only distinct nucleolar components present in both structures. As stated above the stability of the isolated onion nucleolar matrix could be caused by divalent cations, or be considered a characteristic of the plant nucleolus. Unfortunately there are no other similar studies performed on plant systems to allow a comparison with our data.

Until now, the structures or molecules responsible for maintaining the nucleolar structure and functionality have not been defined. They may involve several structural proteins that could provide the anchoring sites for rDNA transcription and processing machineries in the nuclear matrix (Dickinson *et al.,* 1990; Hozák *et al.,* 1993a). Experiments with antimetabolites have demonstrated that the association of two nucleolar components (fibrillar centers and the surrounding portion of the dense fibrillar component) is able to sustain efficient rRNA synthesis (Scheer and Benavente, 1990). The presence of fibrillar centers and the tightly bound surrounding portion of densely packed fibrillar threads in the nucleolar matrix and in mild loosening preparations of the whole plant (Figs 12E and 12F) and HeLa (Hozák *et al.*, 1992, 1994) cells, as well as the immunological detection of some components of the transcription complexes in these structures *in situ* like RNA polymerase I, polymerase 1 transcription factor UBF, and topoisomerase I (Scheer and Benevente, 1990; Rodrigo *et al.*, 1992), suggest that the fibrillar centers could be the anchoring sites of the multimeric transcription complexes of the nucleolus, while the more peripheral portions of the nucleolar matrix could serve for other functions as, for example, skeletal support for preribosomal particle attachment (Miller and Beatty, 1969; Franke *et al.*, 1981). Further experiments of immunodetection of some protein components of the nucleolar transcription and processing complexes on the nucleolar matrix (some of which will be referred to in Section V,C,5 of this section) will clear up this point.

### 4. Nucleolar Matrix Proteins

The differences in protein composition between the isolate onion nucleoli and residual nucleolar matrices are obvious even in 1-D gels (Fig. 13A), which reveal an enrichment in bands of the same $M_r$ range as those reported for rat and mouse nucleolar matrix (Todorov and Hadjiolov, 1979; Shiomi *et al.*, 1986; Olson *et al.*, 1986). The presence of core histones in the isolated nucleolar matrices after extensive DNase digestion and high salt extraction is in conflict with the very low amounts of DNA present in them, but trace amounts of core histones have also been detected in the isolated nucleolar matrices of rat liver (Stephanova and Valkov, 1991) and Novikoff hepatoma cells (Olson *et al.*, 1986). Artifactual reassociation of core histones with the abundant highly acidic proteins of the nucleolar matrix, as exposed for the nuclear matrix in Section II, could explain this fact, although nonhistone proteins of high mobility could also account for these bands. Two-dimensional separation of their proteins reveals a reproducible polypeptide pattern on the isolated nucleolar matrix that is different from those of the nucleolar and nuclear matrix fractions, respectively (Figs. 12G, 12H, 2B, and 2D), indicating that there is not an unrepresentative retention of proteins in these structures. This methodology also reveals that many of the apparently common bands of nucleoli and nucleolar matrices on 1-D gels really do correspond to several polypeptides with different p*I* values, which are present at different relative concentrations in both fractions.

Up to 127 different polypeptide spots are observed in nucleolar matrix fractions after silver staining against 180 on the isolated nucleoli, with a

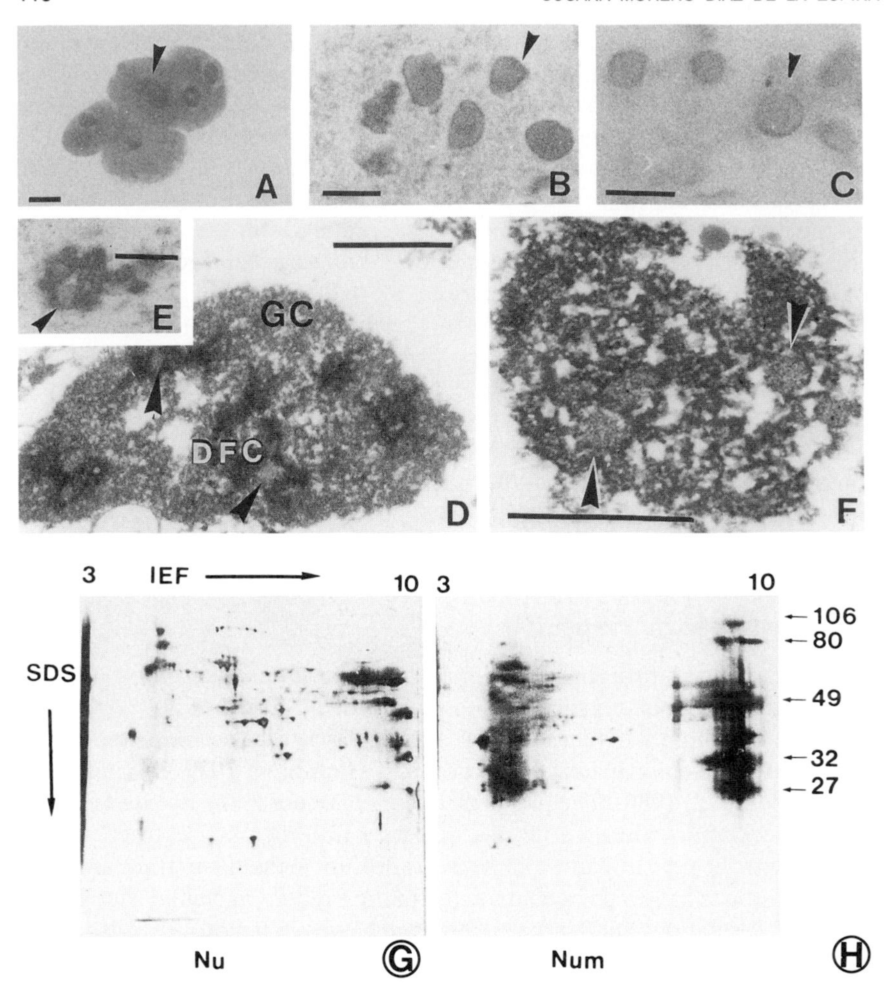


FIG. 12 Light and electron micrographs as well as 2-D polypeptide patterns of isolated nuclei (A), nucleoli (B, D, G), and nucleolar matrices (C, E, H). Isolation and purification of nuclei were performed as described in Moreno Díaz de la Espina *et al.* (1991), but omitted the $MgCl_2$ in the isolation medium to avoid nuclear stabilization. Purified nuclei were resuspended in a medium with 5m*M* $MgCl_2$ and submitted to sonication waves of 15 sec with pauses inbetween. The number of sonication waves in each experiment was determined according to the nuclear disruption and nucleolar integrity after each wave, as controlled under a phase-contrast microscope. Nucleoli were purified through a cushion of 1.5 *M* sucrose, digested with 200 $\mu$g/ml DNase I and 100 $\mu$g/ml freshly prepared RNase A, and extracted with 2 *M* NaCl to obtain the isolated nucleolar matrices. For light microscopy the fractions were stained with methyl green pyronine. For electron microscopy, pellets of nucleoli and nucleolar matrices were processed as described in Fig. 1. Partial decondensation of nucleolar structures *in situ* was performed using hypotonic formaldehyde in the presence of nonionic detergents as described in Vázquez Nin *et al.* (1992). For 2-D separation of proteins, proteins from both fractions were solubilized and gels run as described in Minguez and Moreno Díaz de la Espina (1993a).

wide range of $M_r$ and p*I* values. This complex composition suggests that the nucleolar matrix corresponds to a complex multifunctional structure rather than to a simple nucleolar skeleton whose protein composition could vary according to the extraction procedure. The fraction is enriched in acidic proteins, most of which are not a majority in the whole nucleoli; this fact has also been reported in the mouse L-cell nucleolar matrix (Shiomi *et al.*, 1986). The nucleolar matrices share 23 protein spots with the isolated nucleolus (Figs. 12G and 12H), but only 18 with the whole nuclear matrix (Figs. 12H, and 2B) confirming that the nucleolar matrix proteins are a subset of those of the nuclear matrix, some of which have been identified by immunoblotting (see next paragraph) (Minguez and Moreno Díaz de la Espina, 1993b). In fact, the residual nucleolar proteins should account for a high proportion of the nuclear matrix fraction, considering that the residual nucleolus is a massive structure compared to the lamina and internal matrix.

### 5. Immunological Characterization of Nucleolar Matrix Proteins

Incubation with antibodies against well-known nucleolar proteins fibrillarin and nucleolin allowed us to characterize the protein components of the residual nucleolar matrix. Fibrillarin is a highly conserved nucleolar protein that shares several biochemical characteristics between different species, as a molecular mass from 34 to 38 kDa, a basic isoelectric point (p*I* 8.5–10), and a high content of glycine and $N^7,N^8$-dimethylarginine. It associates with $U_3$, $U_8$, and $U_{13}$ small nucleolar RNAs, which form a part of the ribonucleoprotein particle involved in pre-rRNA processing. Fibrillarin appears to be involved in the initial processing step, that is the endonucleolytic removal of the 5′ externally transcribed spacer (Fournier and Maxwell, 1993). Immunolocalization has revealed that fibrillarin not only occurs at the 5′ end of spread rRNA transcripts and in the dense fibrillar component of the intact nucleolus (Scheer and

---

(A) Isolated nuclei, (B) isolated nucleoli, and (C) isolated nucleolar matrices, after methyl green pyronine staining. The arrowheads point to nucleoli. (D) Electron micrograph of an isolated nucleolus. The three typical nucleolar components remain after the isolation: fibrillar centers (arrowheads), dense fibrillar components (DFC), and granular component (GC). (E) Electron micrograph of a portion of a nucleolus after "mild loosening" *in situ* showing the association between fibrillar centers (arrowhead) and the surrounding fibrillar component under these conditions. (F) Electron micrograph of an isolated nucleolar matrix, displaying the fibrillar centers (arrowheads) and the fibrillar nucleolar matrix. (G) 2-D polypeptide patterns of isolated nucleoli (Nu) after silver staining. (H) 2-D polypeptide patterns of isolated nucleolar matrices after silver staining. When not indicated, bars = 1 μm. Bars A, B, and C = 5 μm. Bar E = .25 μm.

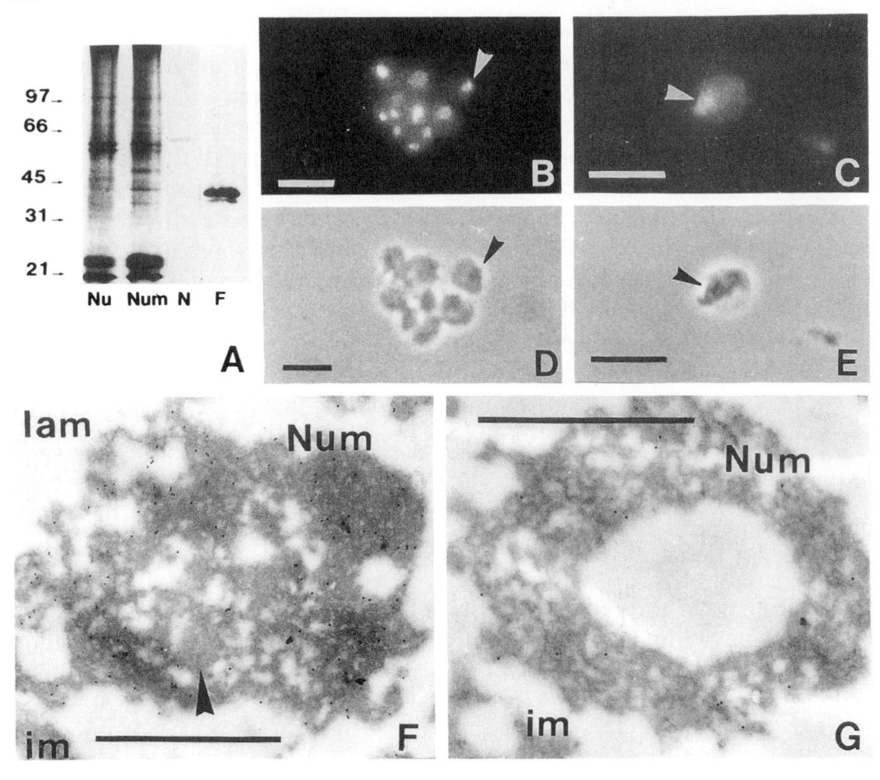


FIG. 13 Immunolocalization of fibrillarin and nucleolin in the isolated nucleolar matrices of the onion, by immunoblotting (A), immunofluorescence (B, C), and immunogold labeling (F, G). (A) 1-D PAGE on 10% gels of the total proteins of the nucleolar (Nu) and nucleolar matrix (Num) fractions after Coomassie blue staining. Immunoblotting of the nucleolar matrix proteins with the antinucleolin antibody revealed a band at 58 kDa (N). The antifibrillarin serum S4 recognizes a broad band at 37 kDa and also an additional minor band of slightly greater mobility, which reveals the microheterogeneity of fibrillarin in plants. (B) Immunofluorescence staining with the antifibrillarin sera S4 shows a clear staining of the nucleolar matrices inside the nuclear matrices (arrowhead). (D) Phase contrast of the same field. (C) Immunofluorescence staining with a polyclonal antimouse nucleolin produces a weaker staining of the nucleolar matrix (arrowhead). (E) Phase contrast of the same field. (F) Electron micrograph of a portion of a nuclear matrix after immunolabeling with the antifibrillarin S4 sera, analogous to those shown in (B). A high density of labeling accumulates in the nucleolar matrix (Num) except for the fibrillar centers (arrowheads). The lamina (lam) and internal matrix (im) show a very low labeling. (G) Electron micrograph of a portion of a nuclear matrix after immunogold labeling with an antinucleolin sera as in (C). The density of labeling on the nucleolar matrix (Num) is low compared with that of fibrillarin. When not indicated, bars B–E = 10 $\mu$m; bars F–G = 1 $\mu$m.

Benavente, 1990), but also in nonactively transcribing nucleolar structures such as the mitotic prenucleolar bodies (data not shown, Azum-Gélade *et al.,* 1994).

Three different antifibrillarin antibodies against human and *Physarum* fibrillarins (Table V) confirmed the presence of this protein in the isolated nucleolar matrix of the onion (Figs. 13A, 13B, and 13F). Residual fibrillarin appears to be relatively abundant and presents an $M_r$ of 37 kDa with an additional minor band of slightly greater mobility, which corresponds to the microheterogeneity shown by fibrillarin in plants (Minguez and Moreno Díaz de la Espina, 1993b, 1995; Guiltinan *et al.,* 1988). Immunofluorescent staining with the same antibodies reveals that the protein is associated with the nucleolar matrix, either isolated or inside the whole nuclear matrix, but not with the lamina or internal matrix (Fig. 13B). Immunogold labeling confirms the specificity of the fibrillarin distribution in the nuclear matrix and also allows an analysis of its distribution within the nucleolar matrix. Fibrillarin labeling is high and appears distributed over the nucleolar matrix except for the fibrillar centers (Fig. 13F). Because the experimental procedure used to obtain the matrices using low temperatures and reducing conditions excludes a stabilization of fibrillarin in the matrix by heat or disulfides (Gasser *et al.,* 1989; Martelli *et al.,* 1991), we conclude that a fraction of this protein is found in an insoluble form in the nucleolar matrix confirming previous immunocytochemical results in HeLa cells *in situ* (Ochs and Smetana, 1991). Optical serial section images of nuclear matrices incubated with antifibrillarin antibodies obtained by confocal laser microscopy after deblurring to eliminate the out-of-focus fluorescence show that the fibrillarin labeling lies in discrete domains within the nucleolar matrix (compare Figs. 14 and 13B), which in view of their number, topological distribution, and reactivity with this antibody, do not correspond to fibrillar centers (Fig. 14, color plate 3).

Because fibrillarin is a part of the multi-snoRNP complex or "processome" that achieves rRNA processing in the nucleolus (Fournier and Maxwell, 1993), our data would support the hypothesis that the nucleolar matrix could contain insoluble preassembled snoRNP complexes as occurs with the transcription factories and the snRNP complexes in the nuclear matrix (Hozák *et al.,* 1993b) and that they would be anchored in the dense fibrillar threads of the nucleolar skeleton. A further role of the insoluble fibrillarin in maintaining the structural integrity of the nucleolar matrix cannot be discarded either.

Nucleolin is an ubiquitous nucleolar protein in eukaryotic cells. It is a multifunctional protein that appears to be involved in the regulation of transcription by RNA polymerase I, and also associates with early rRNA precursors. It has three different domains apparently involved in different functions—an acidic amino-terminal region with several consensus phosphorylation sites, a central region with four RNA binding domains, and a glycine-rich carboxy terminus (Caizergues-Ferrer *et al.,* 1989). Plant nucleolin has a lower $M_r$ than its vertebrate counterparts (Martin *et al.,* 1992a; Minguez and Moreno Díaz de la Espina, 1993b), but its solubility distribution and phosphorylation patterns are similar to those of mammalian nucleolin (Medina *et al.,* 1993). Although a considerable portion of nucleolin is recovered in the soluble fraction after low-ionic-strength washing of nucleoli (Cerdido *et al.,* 1993), a small fraction of this protein resists not only low and high salt extractions but also DNase I and RNase A digestions, which have been reported to release nucleolin from Novikoff hepatoma nuclei (Olson and Thompson, 1983).

Two different antinucleolin antibodies as well as its affinity for bismuth oxynitrate identify the 58-kDa protein of the onion nuclear and nucleolar matrix fractions with nucleolin (Fig. 13A) (Minguez and Moreno Díaz de la Espina, 1993b). The microscopic estimations of the purity of final fractions as well as the specific localization of nucleolin in the nucleolar matrix by immunolabeling and bismuth staining excludes possible contamination of the residual fraction with unextracted nucleoli, confirming that the protein is actually a component of the residual nucleolar structure. Immunofluorescence and immunogold-labeling intensities of nucleolar matrices with antinucleolin sera are much lower than in the case of antifibrillarin, and that is also the case with the amount of proteins detected in the corresponding bands of immunoblottings of nucleolar matrix fractions (Fig. 13A–G).

Thus nucleolin proves to be a minor component of the nucleolar matrix in the onion. Nucleolin had previously been detected in the nuclear matrix of CHO cells after heat shock, which produces an overall increase in the proteins associated with the matrix (Caizergues-Ferrer *et al.,* 1984) but this is the first identification of nucleolin in the nucleolar matrix under nonstabilizing conditions. Although further experiments are necessary to ascertain the functions of the residual nucleolin in the nucleolar matrix, its relatively low concentration in the nucleolar matrix suggests that nucleolin is not a component of a nucleolar core scaffold. The ability of nucleolin to bind to DNA and histones and to control chromatin conformation (Suzuki *et al.,* 1993) makes it a good candidate to form a part of the protein complexes that anchor the rDNA loops to the nucleolar matrix (Dickinson *et al.,* 1990). Recently, Dickinson and Kohwi-Shigematsu (1995) confirmed that nucleolin is not only a nucleolar matrix component but also an MAR-binding protein in human K-562 cells. But also, its ability to bind RNA could contribute to its presence in the multimeric complexes containing the nucleolar processing machinery. Further characterization of the nucleolar matrix fraction by immunoblotting and of the topological distribution of its proteins will clear up many pending questions about the composition, organization, and functionality of the plant nucleolar matrix.

## VI. Correspondence of the Nuclear Matrix Components with the Functional and Structural Domains of Nuclei *in Situ*

Recent studies have demonstrated that nuclei are highly ordered and that nuclear order is related to function (de Jong *et al.,* 1990; Jackson, 1991; Spector, 1993a,b). The existence of structural and functional domains within the nucleus was already suggested in the 1960s on the basis of cytochemical staining, enzymatic digestion, and [$^{3}$H]uridine incorporation of the nuclear structures (Bernhard, 1969; Monneron and Bernhard, 1969). Later ultrastructural studies combined with high-resolution radioautography, immunogold labeling, and *in situ* hybridization confirmed the existence of discrete subnuclear domains for replication, RNA synthesis, processing, and transport (Puvion-Dutilleul, 1983), but the molecular details of composition and function of these domains are not completely well defined. Recent works confirm that nuclear function is driven by nuclear architecture at two fundamental levels. Formation of chromatin loops by means of specific recognition of MAR sequences and location of these loops to specific nuclear compartments allow specific expression of discrete groups of genes. But then the nuclear architecture should also provide necessary factors to ensure that the properly located genes are expressed, acting as a nucleation site for promoters, activating factors, and polymerases, as well as for the component of the splicing machinery (Gasser *et al.,* 1989). In that way, the "nuclear structure" would coordinate domain formation, assembly of competent complexes, and transcription, but would also support spatially ordered RNA processing, storing, and transport. However, the influence of structural organization does not stop at the nucleus, and the unified nucleoskeleton and cytoskeleton provide coordinated functioning in the whole cell (Penman *et al.,* 1982).

The nuclear matrix or nucleoskeleton constitutes the complex structure that provides the nuclear structural organization. Until now this domain has been defined in an operational way, and its composition and organization vary with the extraction procedure, making its interpretation rather confusing. Further investigations combining biochemical, immunological, genetical, and ultrastructural approaches would clarify in the near future the organization and functioning of the nucleoskeleton, as occurred with the cytoskeleton during past decades. Although studies on nuclear organization and functioning in plant cells are far less abundant than those in animal cells (Jordan *et al.*, 1980; Moreno Díaz de la Espina *et al.*, 1982a,b, 1992; Medina *et al.*, 1989; Vázquez-Nin *et al.*, 1992; Martin *et al.*, 1992b), available data support the view of a close similarity in the organization of nuclei in multicellular eukaryotes, in spite of minor differences that correlate with peculiarities of plant cells. The most obvious is the extensive chromatin reticulum which corresponds to the large amount of total DNA per nucleus typical of plant species, and the topological distribution of the interchromatin granule network (Medina *et al.*, 1989; Moreno Díaz de la Espina *et al.*, 1992, 1993) described above.

Four different ultrastructural domains can be distinguished in the plant nucleus—the nuclear envelope, chromatin, nucleolus, and nucleoplasm, which have their counterparts on the nuclear matrix (Figs. 5A and 15; Table VII). The nuclear envelope is also involved in transport between the nucleus and cytoplasm through the pore complexes. The chromatin domain formed by a reticulum of inactive condensed chromatin with decondensed active chromatin fibers is at its periphery. The nucleolus constitutes the nuclear domain in which rDNA genes are transcribed and rRNAs processed to form the preribosomal particles. Within the nucleolus, other structural subdomains correlated with synthesis (dense fibrillar components and fibrillar centers) and processing (outer dense fibrillar and granular components) of preribosomal precursors have also been described (Medina *et al.*, 1983, 1990; Risueño *et al.*, 1982). Finally, the nucleoplasm is the domain involved in RNA polymerase II synthesis, as well as splicing, storing, and transport of the hnRNA, providing the recruitment sites for the necessary factors. This domain forms an intricate network, delimited by the nuclear envelope and the edges of the nucleolar body and condensed chromatin masses (Figs. 5A and 9A). It is comprised of different ribonucleoprotein fibrils and granules localized at the periphery of the condensed chromatin or deep inside the interchromatin regions that contain the transcription and splicing machineries and appear associated to a nonribonucleoprotein fibrillar network (Medina *et al.*, 1989; Moreno Díaz de la Espina *et al.*, 1992; Chamberland and Lafontaine, 1993). This domain corresponds to the nuclear RNA tracks described in animal cells (Carter, *et al.*, 1993). In onion cells, the structures forming the nuclear matrix are well defined, have regular reproducible ultrastructural and cytochemical characteristics, and can be detected by EM in matrices prepared by different procedures, under various conditions used. These facts allow us to conclude that although the isolation procedures could stabilize a preexisting intranuclear nucleoskeleton, this structure would really be present in the intact nuclei (Cerezuela and Moreno Díaz de la Espina, 1990; Moreno Díaz de la Espina *et al.*, 1991).

Comparison of the ultrastructural organization of nuclei before and after extraction and the identification of residual proteins in the matrix by immunolabeling or cytochemical staining allows us to establish a correlation

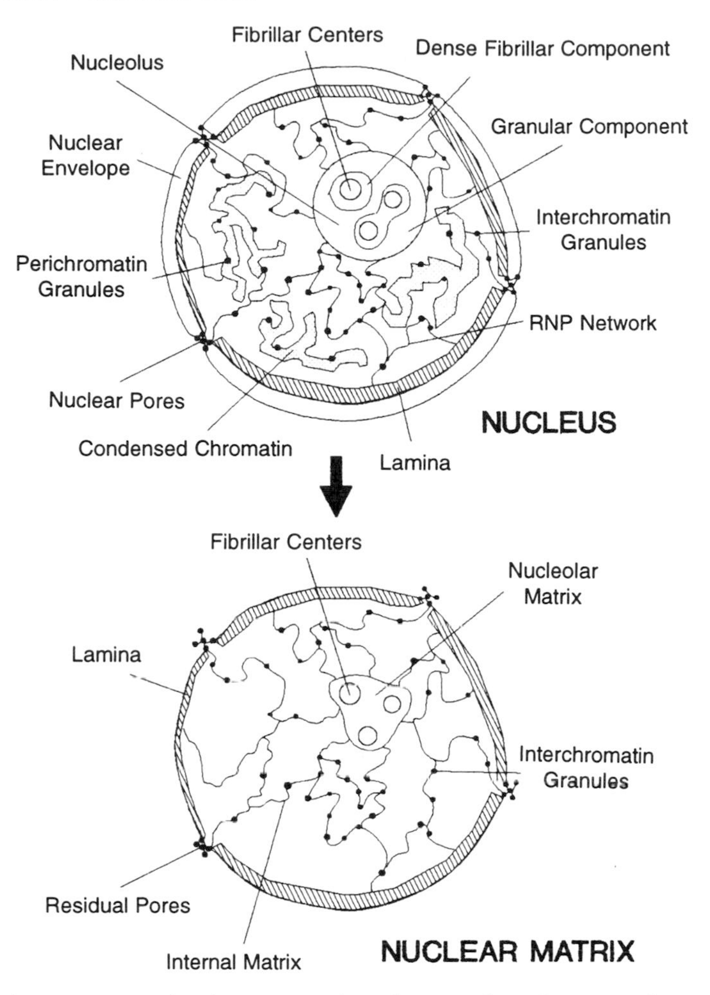


FIG. 15 Diagram representing the correspondence between the main structural components of the plant nucleus and nuclear matrix. The nuclear envelope constitutes the boundaries of the nucleus, which includes the nuclear membrane, nuclear pores, and the lamina. The chromatin is found as masses of condensed chromatin fibers. The nucleolus presents its three typical components: fibrillar centers, surrounded by the dense fibrillar component, and the granular component containing the preribosomal particles. The nonnucleolar ribonucleoproteins are represented by perichromatin granules found at the borders of the chromatin masses and interchromatin granules associated with an intranuclear fibrillar network, but not forming clusters. Perichromatin and interchromatin fibrils are not represented here. The sequential extraction that produces the nuclear matrix (represented by the black arrow) removes the nuclear membrane, the chromatin, and some of the ribonucleoproteins (perichromatin granules and preribosomal particles) from the nucleus. The nuclear matrix is composed of the lamina–pores complex, the fibrillar network of the internal matrix with associated interchromatin granules, and the fibrillar centers and the fibrillar network forming the nucleolar matrix.

TABLE VII

Correspondence of the Plant Nuclear Matrix Domains with the Functional Domains of the Nuclei *in Situ*

| Nuclear domains | Structural components | Nuclear matrix frameworks |
|---|---|---|
| Nuclear envelope | Membranous elements | |
| | Lamina | Lamina–pore fraction |
| | Pore complexes | |
| Chromatin | Condensed chromatin masses | — |
| | Decondensed chromatin | |
| Nucleoplasm | RNP network | Internal RNP framework |
| | Perichromatin granules | RNP fibrils (10–15 nm) |
| | Perichromatin fibrils | IG (20–25 nm) |
| | Interchromatin fibrils (IF) | |
| | Interchromatin granules (IG) | |
| | Non-RNP network | |
| Nucleolus | Fibrillar centers | Fibrillar centers |
| | Dense fibrillar component | Fibrillar nucleolar matrix |
| | Granular component | |

between some domains of the nuclei and corresponding structural components of the nuclear matrix (Table VII, Fig. 15). Thus the lamina would provide structural support to the nuclear envelope constituting the boundary of the nucleoskeleton, which connects it with the cytoskeleton, and would also provide physical support for nuclear transport through the pore complexes (Figs. 5, 6). The residual framework of the nucleolus would be formed by two distinct elements: the fibrillar centers (Fig. 12F), probably serving as anchoring sites for rDNA loops but also as recruitment sites for activating factors and polymerases, as they contain RNA polymerase I, topo I, and UBF transcription factor (Scheer and Weisenberg, 1994), and the fibrillar nucleolar matrix that will provide support for the processing and transcription complexes and probably attachment of preribosomal particles. The internal matrix presents two interconnected networks. One of them corresponds to the nuclear network of interchromatin granules, contains proteins of the splicing machinery (Table V), and probably provides physical support for extranucleolar RNA synthesis, splicing, and transport. The other with an exclusively fibrillar structure has not yet a defined correlate on the *in situ* nuclei. The presence of topoisomerase II associated to their fibers, with a similar distribution to that reported for MAR sequences *in situ* (Fig. 11F) (Ludérus *et al.,* 1992), could relate it to the nucleoskeleton involved in the organization of chromatin domains containing the multimeric complexes of replication origins and transcriptional enhancers. In this sense this fibrillar network would correlate with the pervading matrix associated to chromatin *in situ.* But the presence of IF-type epitopes in this network (Fig. 9B) would relate to the nucleoskeletal array of core filaments described in vertebrates (Jackson and Cook, 1988). Further experiments are needed to determine if these two networks are integrated in a single domain.

## VII. MAR Sequences in Plants

Eukaryotic chromatin is organized within the nucleus into topologically constrained domains that provide the basis for differential gene expression (Gasser *et al.,* 1989). MAR or scaffold-attached region (SAR) sequences constitute the specific DNA sequences that mediate attachment of chromatin loops to the nuclear matrix (Mirkovitch *et al.,* 1984; Ludérus *et al.,* 1992). They are usually 300–1000 bp long with a high AT content, contain specific recognizable DNA motifs present in a modular array, and build up MAR subdomains that interact cooperatively with the scaffold (Laemmli *et al.,* 1992). These motifs include A boxes and T boxes, the ATATTT sequence, and the consensus sequence of the minor groove of *Drosophila* topoisomerase II (Ludérus *et al.,* 1992).

MARs have been identified in many eukaryotes such as *Drosophila* (Mirkovitch *et al.,* 1984), vertebrates, yeast (Gasser *et al.,* 1989), plants (Breyne *et al.,* 1994; Table I), and also in the lower eukaryotic dinoflagellates, which do not have a nucleosomal organization in their chromatin (Minguez *et al.,* 1994a,b). Thus MARs appear to be highly conserved during evolution. Animal MARs can bind to plant and lower eukaryote matrices (Breyne *et al.,* 1992; Minguez *et al.,* 1994b) and vice versa (Mielke *et al.,* 1990; Dietz *et al.,* 1994), indicating that the mechanisms of higher order organization of chromatin and the functional arrangement of the genome are basically the same in all the eukaryotes. Nevertheless, different MARs do not appear to present extensive sequence homology and it is possible that the nuclear matrix recognizes some structural features of the MARs rather than just exclusively specific sequences (Laemmli *et al.,* 1992).

Although little information is available about the nuclear structures or proteins that bind MARs, several MAR-binding matrix components have been identified in the internal matrix: the attachment region-binding protein (ARBP) (von Kries *et al.,* 1991) and the topoisomerase II (Adachi *et al.,* 1989) and lamin-B, which bind MARs in a specific cooperative way (Ludérus *et al.,* 1992). Several classes of MARs have been proposed according to their functions. One class binds constitutively and with high affinity to the matrix, bordering chromatin loops, and interacts with abundant proteins of the matrix. Another class is near the regulatory elements of genes and binds in a transcription-related way and with less affinity to cell–type-specific nuclear matrix proteins. A third class comaps with replication origins (Gasser, 1991; Razin and Vanetzky, 1992).

Analyses of plant MARs have been relatively abundant in the 1990s, compared with other kinds of studies performed on plant nuclear matrices (see Table I). Plant MARs have been identified in several monocot and dicot plant species, mostly in matrices prepared by the low-ionic-strength LIS procedure (see Table I). In 1991, Hall *et al.* identified the MARs flanking three root-specific tobacco genes in isolated tobacco matrices that specifically bound a yeast MAR, and Slatter *et al.* (1991) identified a MAR downstream from a single-copy plastocyanine gene in pea, with homologies with topoisomerase II binding sites, A and T boxes, and replication origins. Later, other plant MARs were identified in tobacco (Breyne *et al.,* 1992) and the maize Adh 1 gene (Avramanova and Bennetzen, 1993; Stoilov *et al.,* 1992; Paul and Ferl, 1993). Recently, a MAR was detected near a T–DNA integration site in *Petunia,* which is active in mammalian cells (Dietz *et al.,* 1994). Plant MARs share many structural and functional characteristics with those described in insects, vertebrates, and yeast. Although with different binding abilities, plant matrices are able to bind yeast (Hall *et al.,* 1991), animal (Avramanova and Bennetzen, 1993), and insect (Minguez *et al.,* experiments in progress) MARs, and vice versa (Dietz *et al.,* 1994). This suggests that the higher-order organization of the genome is strongly conserved in Eukaryotes. The majority of plant MARs so far identified are located in nontranscribed regions either upstream or downstream from genes (Slatter *et al.,* 1991; Avramanova and Bennetzen, 1993), in agreement with the results in nonplant systems (Gasser *et al.,* 1989), although location within introns has been reported (Breyne *et al.,* 1994). They usually map to A + T sequences, A boxes, and T boxes as do animal and yeast MARs (Gasser *et al.,* 1989), although homology to the topoisomerase II consensus cleavage site is less frequent (Slatter *et al.,* 1991).

Some reported plant MARs comap or are close to the regulatory elements upstream from the genes (Hall *et al.,* 1991; Avramanova and Bennetzen, 1993; Paul and Ferl, 1993). The colocalization of MARs with transcription regulatory elements suggests that they may be involved in facilitating the interaction of DNA sequences with the *trans*-acting and transcription factors that would be assembled on the nuclear matrix (Hoffman, 1993). Apart from their effects on transcription, plant MARs are able to protect an integrated transgene from the influences of the surrounding chromatin, thereby conferring a position-independent gene expression to stably integrated genes (Breyne *et al.,* 1992; Dietz *et al.,* 1994). In conclusion, in spite of the very few studies yet performed, the available data support the view that the structural and functional higher order organization of DNA is basically identical in both animal and plant cells.

## VIII. Concluding Remarks

In plants, as in other eukaryotes, the accumulated evidence demonstrates that nuclei function as a highly ordered ensemble of integrated domains (Spector, 1993a; Fig. 15). The nuclear matrix or nucleoskeleton constitutes the complex structure responsible for the nuclear order, which provides physical support for the anchoring of loop domains and also for the multimeric complexes involved in replication, transcription, processing, and transport of RNA. Studies on the plant nuclear matrix ultrastructure and protein composition are advanced enough at present to allow the conclusion that the plant nuclear matrix is a complex heterogeneous structure in protein composition, organized in well-defined domains, and homologous to that of vertebrates. Although the definition of intranuclear matrix frameworks would need a combined ultrastructural, biochemical, and genetical approach, immunological identification of some plant nuclear matrix components such as lamins, topoisomerase II, some components of the transcription and splicing machineries, internal IF proteins, and also specific nucleolar proteins like fibrillarin and nucleolin, which associate to specific matrix domains, allow the establishment of a model of organization for the plant nuclear matrix very similar to that of other eukaryotes with components of the transcription, processing, and DNA-anchoring complexes associated with a very stable nucleoskeleton. Nevertheless, more insights are needed on the characterization of the protein components of the different matrix domains.

The immunological detection of lamins A and B in the plant lamina, with conserved values of molecular mass and p*I*, has been an important contribution to the understanding of the plant cytoskeletal system, in which IF had not been described as such (Shaw *et al.*, 1991), as lamins were the first members of the plant IF family to be described and topologically localized integrated into an organized network. These results reveal that higher plants in which an organized network of cytoplasmic IF has not yet been well documented have a well-organized lamina containing lamins, reinforcing the idea that lamins are old members of the IF family of proteins, apparently better conserved in plants than their cytoplasmic counterparts (Döring and Stick, 1990). Nevertheless, additional sequencing experiments of plant lamins are needed to characterize their molecular structures and classify them more precisely into lamin subtypes. Information about the functionality of plant lamins is very scarce, although casein kinase II phosphorylating lamins and also an NTPase have been reported in the lamina–matrix fraction of pea. Further investigation on the functionality and metablism of plant lamins as well as on their assembly and disassembly cycles is needed.

The differential distribution of the components of the internal matrix obtained by immunolabeling suggest the existence of at least two structural and functional subdomains highly interrelated within this compartment. One is fibrogranular in structure, contains proteins of the snRNPs and interchromatin granules, and is probably involved in RNA metabolism providing support for the nuclear RNA tracks (Carter *et al.,* 1993; Spector, 1993b). The other is fibrillar, contains structural (IF type) proteins, but also topoisomerase II and polymerase II, relating it with nucleoskeletal filaments, but also with DNA attachment, replication, and transcription sites. The close interactions observed between the two subdomains would reflect the highly dynamic functioning of the nuclear matrix. The molecular characterization of the proteins or structures that provide stability to the internal matrix in plants is open to investigation. The detection of IF antigens in the fibrillar network of the matrix and their resistance to solubilization would suggest an internal nucleoskeleton of IF-related core filaments, although other kinds of interactions cannot be excluded either. Analysis of whole-mount plant matrices combined with resinless microscopy and stepwise extraction protocols would contribute to define the macromolecular organization of the internal nucleoskeleton in plants. The possibility of preparing isolated nucleolar matrices from isolated nucleoli under good condition for biochemical and microscopical purposes has allowed a detailed characterization of the ultrastructural organization and protein composition of this matrix domain in plants. At the ultrastructural level the most relevant result is the discovery that fibrillar centers, whose functionality is still in controversy (Scheer *et al.,* 1993), are components of the nucleolar matrix. The protein pattern of the nucleolar matrix has been already outlined, and some of its proteins identified by immunological approaches as is the case of fibrillarin.

The analyses of the relationships between the nuclear matrix and the cytoskeleton in plants are very rudimentary. Although the physical continuity between the lamina and a yet unidentified cytoskeletal fibrillar array has been confirmed by TEM of matrices prepared *in situ,* further experiments using specific probes for each of the three cytoskeletal systems would be necessary to determine the cytoskeletal arrays interacting with the nuclear matrix, as well as the proteins involved in these processes. The study of the mechanisms that control the integrated functioning of the nucleus and cytoplasm is very important in plant cells that are acentriolar and in which the nucleus is the main microtubule organizing center and actin nucleation site (Staiger and Lloyd, 1991; Shibaoka and Nagai, 1994). The analysis of the functions played by the nucleoskeleton in the organization and functioning of the plant cytoskeleton is a very promising field of research in plant cell biology.

Analysis of plant MARs is relatively well advanced. MARs flanking several genes have been described in both monocot and dicot species, which present sequence homology with MARs from insects and vertebrates and affinity for animal matrices. Plant MARs not only interact with the transcription regulatory elements, but also confer position-independent gene expression to stably integrated genes (Dietz *et al.*, 1994). These data confirm that the structural and functional higher order organization of DNA is basically identical in animal and plant cells. But up to the present there is no information about proteins that bind MARs in plants. The increasing current interest in plant molecular and cell biology will probably promote further investigations combining biochemical, immunological, genetical, and ultrastructural approaches that should clarify in the near future many of the unsolved questions about the organization and functioning of the plant nucleoskeleton.

## Acknowledgments

I am indebted to the members of my lab who collaborated in the work on the onion nuclear matrix exposed in the text and most especially to my former and present students Dr. I. Barthelemy, Dr. M. A. Cerezuela, and Dr. A. Minguez for their kind permission to include unpublished results. I also thank the many individuals who kindly provided us with antibodies for our experiments: Dr. Reimer Stick (Gottingen, Germany) for the antilamins; Dr. D. Fairbairn (Norwich, United Kingdom) for the AFB, MAC 322, and IFA; Dr. R. Ochs (La Jolla, California) for the $S_4$ and 72 B antifibrillarin antibodies; Dr. M. Christensen (Kansas City, Missouri) for the P2G3 antifibrillarin; Dr. T. Martin (Chicago, Illinois) for Y12; and Dr. Nicole Gas (Toulouse, France) and Dr. R. Petrishyn (Syracuse, New York) for the antinucleolin antibodies. I am also grateful to Dr. R. E. Fernández Gómez for critical reading of the manuscript, Dr. F. J. Medina for cooperation with the statistical representation of data, Mrs. N. Fontúrbel for excellent photographic work and for typing the manuscript, Mrs. M. Carnota for expert technical assistance, Mr. R. Bermejo for computerized artwork, and Mrs. B. L. Walker for revising the English style of the text. The work reported from my lab has been supported by CICYT/Spain, Grants PB880037 and PB91-0124.

## References

Adachi, Y., Käs, E., and Laemmli, U. K. (1989). Preferential, cooperative binding of DNA topoisomerase II to scaffold-associated regions. *EMBO J.* **8,** 3997–4006.

Aebi, U., Cohn, J., Buhle, L., and Gerace, L. (1986). The nuclear lamina is a meshwork of intermediate-type filaments. *Nature* **323,** 560–564.

Aligué, R., Bastos, R., Serratosa, J., Enrich, C., James, P., Pujades, C., and Bachs, O. (1990). Increase in a 55-kDa keratin-like protein in the nuclear matrix of rat liver cells during proliferative activation. *Exp. Cell Res.* **186,** 346–353.

Avramanova, Z., and Bennetzen, J. L. (1993). Isolation of matrices from maize leaf nuclei: Identification of a matrix-binding site adjacent to the Adh1 gene. *Plant Mol. Biol.* **22,** 1135–1143.

Azum-Gélade, M. C., Noaillac-Depeyre, J., Caizergues-Ferrer, M., and Gas, N. (1994). Cell cycle redistribution of U3 snRNA and fibrillarin. *J. Cell Sci.* **107,** 463–475.

Barthelemy, I., and Moreno Díaz de la Espina, S. (1984). Preliminary studies on the plant nuclear matrix ultrastructure. *Cienc. Biol., Mol. Cell. Biol.* **9,** 138–139.

Belgrader, P., Siegel, A. J., and Berezney, R. (1991). A comprehensive study on the isolation and characterization of the HeLa S3 nuclear matrix. *J. Cell Sci.* **98,** 281–291.

Bell, P., Dabauvalle, M. C., and Scheer, U. (1992). In vitro assembly of prenucleolar bodies in *Xenopus* egg extract. *J. Cell Biol.* **118,** 1297–1304.

Berezney, R. (1980). Fractionation of the nuclear matrix. *J. Cell Biol.* **85,** 641–650.

Berezney, R. (1984). Organization and functions of the nuclear matrix. *In* "Chromosomal Nonhistone Proteins" (L. S. Hnilica, ed.), Vol. 4, pp. 119–180. CRC Press, Boca Raton, FL.

Berezney, R. (1991). The nuclear matrix: A heuristic model for investigating genomic organization and function in the cell nucleus. *J. Cell Biochem.* **47,** 109–123.

Bernhard, W. (1969). A new staining procedure for electron microscopical cytology. *J. Ultrastruct. Res.* **27,** 250–265.

Beven, A., Guan, Y., Peart, J., Cooper, C., and Shaw, P. (1991). Monoclonal antibodies to plant nuclear matrix reveal intermediate filament-related components within the nucleus. *J. Cell Sci.* **98,** 293–302.

Bolla, R. I., Braaten, D. C., Shiomi, Y., Hebert, M. B., and Schlessinger, D. (1985). Localization of specific rDNA spacer sequences to the mouse L-cell nucleolar matrix. *Mol. Cell. Biol.* **5,** 1287–1294.

Bourgeois, C. A., Bouvier, D., Seve, A. P., and Hubert, J. (1987). Evidence for the existence of a nucleolar skeleton attached to the pore complex-lamina in human fibroblasts. *Chromosoma* **95,** 315–323.

Bouteille, M., Bouvier, D., and Seve, A. P. (1983). Heterogeneity and territorial organization of the nuclear matrix and related structures. *Int. Rev. Cytol.* **83,** 135–181.

Bouvier, D., Hubert, J., Seve, A. P., Bouteille, M., and Moens, P. B. (1984). Three-dimensional approaches to the residual structure of histone-depleted HeLa cell nuclei. *J. Ultrastruct. Res.* **87,** 112–123.

Breyne, P., van Montagu, M., Depicker, A., and Gheysen, G. (1992). Characterization of a plant scaffold attachment region in a DNA fragment that normalizes transgene expression in tobacco. *Plant Cell* **4,** 463–471.

Breyne, P., Van Montagu, M., and Gheysen, G. (1994). The role of scaffold attachment regions in the structural and functional organization of plant chromatin. *Transgenic Res.* **3,** 195–202.

Bridger, J. M., Kill, I. R., O'Farrell, M., and Hutchison, C. J. (1993). Internal lamin structures within $G_1$ nuclei of human dermal fibroblasts. *J. Cell Sci.* **104,** 297–306.

Caizergues-Ferrer, M., Dousseau, F., Gas, N., Bouche, G., Stevens, B., and Amalric, F. (1984). Induction of new proteins in the nuclear matrix of CHO cells by a heat shock: Detection of a specific set in the nucleolar matrix. *Biochem. Biophys. Res. Commun.* **118,** 444–450.

Caizergues-Ferrer, M., Mariottini, P., Curie, C., Lapeyre, B., Gas, N., Amalric, F., and Amaldi, F. (1989). Nucleolin from *Xenopus laevis:* cDNA cloning and expression during development. *Genes Dev.* **3,** 324–333.

Carmo-Fonseca, M., Cidadao, A. J., and David-Ferreira, J. F. (1987). Filamentous cross-bridges link intermediate filaments to the nuclear port complexes. *Eur. J. Cell Biol.* **45,** 282–290.

Carter, K. C., Bowman, D., Carrington, W., Fogarty, K., McNeil, J. A., Fay, F. S., and Lawrence, J. B. (1993). A three-dimensional view of precursor messenger RNA metabolism within the mammalian nucleus. *Science* **259,** 1330–1335.

Cerdido, A., Caizergues-Ferrer, M., and Medina, F. J. (1993). Characterization of a set of major proteins extracted from plant cell nuclei in low ionic strength conditions. *Proc. Eur. Workshop Cell Nucleus, 13th,* Budapest, p. 15.

Cerezuela, M. A., and Moreno Díaz de la Espina, S. (1990). Plant nuclear matrix. Effects of different extraction procedures on its structural organization and chemical composition. *In*

"Nuclear Structure and Function" (J. R. Harris and I. B. Zbarsky, eds.), pp. 317–322. Plenum, New York.
Chamberland, H., and Lafontaine, J. G. (1993). Localization of snRNP antigens in nucleolus-associated bodies: Study of plant interphase nuclei by confocal and electron microscopy. *Chromosoma* **102,** 220–226.
Clayton, L. (1985). The cytoskeleton and the plant cell cycle. *Semin. Ser. Soc. Exp. Biol.* **26,** 114–131.
Clevenger, C. V., and Epstein, A. L. (1984). Identification of a nuclear protein component of interchromatin granules using a monoclonal antibody and immunogold electron microscopy. *Exp. Cell. Res.* **151,** 194–207.
Cook, P. R. (1989). The nucleoskeleton and the topology of transcription. *Eur. J. Biochem.* **185,** 487–501.
Corben, E., Butcher, G., Hutchings, A., Wells, B., and Roberts, K. (1989). A nucleolar matrix protein from carrot cells identified by a monoclonal antibody. *Eur. J. Cell Biol.* **50,** 353–359.
de Jong, L., van Driel, R., Stuurman, N., Meijne, A. M. L., and van Renswoude, J. (1990). Principles of nuclear organization. *Cell Biol. Int. Rep.* **14,** 1051–1074.
Dessev, G. N. (1992). Nuclear envelope structure. *Curr. Opin. Cell Biol.* **4,** 430–435.
Dickinson, L. A., and Kohwi-Shigematsu, T. (1995). Nucleolin is a matrix attachment region DNA-binding protein that specifically recognizes a region with high base-unpairing potencial. *Mol. Cell. Biol.* **15,** 456–465.
Dickinson, P., Cook, P. R., and Jackson, D. A. (1990). Active RNA polymerase I is fixed within the nucleus of HeLa cells. *EMBO J.* **9,** 2207–2214.
Dietz, A., Kay, V., Schlake, T., Landsmann, J., and Bode, J. (1994). A plant scaffold attached region detected close to a T-DNA integration site is active in mammalian cells. *Nucleic Acids Res.* **22,** 2744–2751.
Döring, V., and Stick, R. (1990). Gene structure of nuclear lamin LIII of *Xenopus laevis;* a model for the evolution of IF proteins from a lamin-like ancestor. *EMBO J.* **9,** 4073–4081.
Earnshaw, W. C., Halligan, B., Cooke, C. A., Heck, M. M. S., and Liu, L. F. (1985). Topoisomerase II is a structural component of mitotic chromosome scaffolds. *J. Cell Biol.* **100,** 1706–1715.
Foisner, R., Traub, P., and Wiche, G. (1991). Protein kinase A- and protein kinase C- regulated interaction of plectin with lamin B and vimentin. *Proc. Natl. Acad. Sci. U.S.A.* **88,** 3812–3816.
Fournier, M. J., and Maxwell, E. S. (1993). The nucleolar snRNAs: Catching up with the spliceosomal snRNAs. *Trends Biochem. Sci.* **18,** 131–135.
Franke, W. W., Kleinschmidt, J. A., Spring, H., Krohne, G., Gründ, C., Trendelenburg, M. F., Stoehr, M., and Scheer, U. (1981). A nucleolar skeleton of protein filaments demonstrated in amplified nucleoli of *Xenopus laevis. J. Cell Biol.* **90,** 289–299.
Frederick, S. E., Mangan, M. E., Carey, J. B., and Gruber, P. J. (1992). Intermediate filaments antigens of 60 and 65 kDa in the nuclear matrix of plants: Their detection and localization. *Exp. Cell Res.* **199,** 213–222.
Galcheva-Gargova, Z. I., Marinova, E. I., and Koleva, S. T. (1988). Isolation of nuclear shells from plant cells. *Plant Cell Environ.* **11,** 819–825.
Gasser, S. M. (1991). Replication regions, factors and attachment sites. *Curr. Opin. Cell Biol.* **3,** 407–413.
Gasser, S. M., Amati, B. B., Cardenas, M. E., and Hofmann, J. F.-X. (1989). Studies on scaffold attachment sites and their relation to genome function. *Int. Rev. Cytol.* **119,** 57–96.
Georgatos, S. D., and Blobel, G. (1987). Lamin B constitutes an intermediate filament attachment site at the nuclear envelope. *J. Cell Biol.* **105,** 117–125.
Georgatos, S. D., Maroulakou, I., and Blobel, G. (1989). Lamin A, lamin B, and lamin B receptor analogues in yeast. *J. Cell Biol.* **108,** 2069–2082.
Getzenberg, R. H., Pienta, K. J., Ward, W. S., and Coffey, D. S. (1991). Nuclear structure and the three-dimensional organization of DNA. *J. Cell. Biochem.* **47,** 289–299.

Ghosh, S., and Dey, R. (1986). Nuclear matrix network in *Allium cepa. Chromosoma* **93,** 429–434.

Goldman, A. E., Moir, R. D., Montag-Lowy, M., Stewart, M., and Goldman, R. D. (1992). Pathway of incorporation of microinjected lamin A into the nuclear envelope. *J. Cell Biol.* **119,** 725–735.

Goodbody, K. C., Venverloo, C. J., and Lloyd, C. W. (1991). Laser microsurgery demonstrates that cytoplasmic strands anchoring the nucleus across the vacuole of premitotic plant cells are under tension. Implications for division plane alignment. *Development* (*Cambridge, UK*) **113,** 931–939.

Grebanier, A. E., and Pogo, A. O. (1979). Cross-linking of proteins in nuclei and DNA-depleted nuclei from friend erythroleukemia cells. *Cell* (*Cambridge, Mass.*) **18,** 1091–1099.

Guiltinan, M. J., Schelling, M. E., Ehtesham, N. Z., Thomas, J. C., and Christensen, M. A. (1988). The nucleolar RNA-binding protein B-36 is highly conserved among plants. *Eur. J. Cell Biol.* **46,** 547–553.

Hall, G., Jr., Allen, G. C., Loer, D. S., Thompson, W. F., and Spiker, S. (1991). Nuclear scaffolds and scaffold-attachment regions in higher plants. *Proc. Natl. Acad. Sci. U.S.A.* **88,** 9320–9324.

Hatzfeld, M., and Weber, K. (1992). A synthetic peptide representing the consensus sequence motif at the carboxy-terminal end of the rod domain inhibits the intermediate filament assembly and disassembles performed filaments. *J. Cell Biol.* **116,** 157–166.

Hay, E. D., and Gurdon, J. B. (1967). Fine structure of the nucleolus in normal and mutant *Xenopus* embryos. *J. Cell Sci.* **2,** 151–162.

He, D., Nickerson, J. A., and Penman, S. (1990). Core filaments of the nuclear matrix. *J. Cell Biol.* **110,** 569–580.

Hitchcock, S. E., Carlsson, L., and Lindberg, U. (1976). Depolymerization of F-actin by deoxyribonuclease I. *Cell* (*Cambridge, Mass.*) **7,** 531–542.

Hoffman, M. (1993). The cell's nucleus shapes up. *Science* **259,** 1257–1259.

Hozák, P., Géraud, G., and Hernandez-Verdun-D. (1992). Revealing nucleolar architecture by low ionic strength treatment. *Exp. Cell Res.* **203,** 128–133.

Hozák, P., Hassan, A. B., Jackson, D. A., and Cook, P. R. (1993a). Visualization of Replication factories attached to a nucleoskeleton. *Cell* (*Cambridge, Mass.*) **73,** 361–373.

Hozák, P., Schöfer, C., Sylvester, J., and Wachtler, F. (1993b). A study on nucleolar DNA: Isolation of DNA from fibrillar components and ultrastructural localization of different DNA probes. *J. Cell Sci.* **104,** 1199–1205.

Hozák, P., Cook, P. R., Schöfer, C., Mosgöller, W., and Wachtler, F. (1994). Site of transcription of ribosomal RNA and intranucleolar structure in HeLa cells. *J. Cell Sci.* **107,** 639–648.

Ivanchenko, M., and Avramanova, Z. (1992). Interaction of MAR sequences with nuclear matrix proteins. *J. Cell. Biochem.* **50,** 190–200.

Ivanchenko, M., Tasheva, B., Stoilov, L., Christova, R., and Zlatanova, J. (1993). Characterization of some nuclear matrix proteins in maize. *Plant Sci.* **91,** 35–43.

Jackson, D. A. (1991). Structure-function relationship in eukaryotic nuclei. *BioEssays* **13,** 1–10.

Jackson, D. A., and Cook, P. R. (1988). Visualization of a filamentous nucleoskeleton with a 23 nm axial repeat. *EMBO J.* **7,** 3667–3677.

Jackson, D. A., Yuan, J., and Cook, P. R. (1988). A gentle method for preparing cyto- and nucleoskeletons and associated chromatin. *J. Cell Sci.* **90,** 365–378.

Jackson, D. A., Hassan, A. B., Errington, R. J., and Cook, P. R. (1993). Visualization of focal sites of transcription within human nuclei. *EMBO J.* **12,** 1059–1065.

Jiménez-Garcia, L. F., and Spector, D. L. (1993). In vivo evidence that transcription and splicing are coordinated by a recruiting mechanism. *Cell* (*Cambridge, Mass.*) **73,** 47–59.

Jordan, E. G. (1984). Nucleolar nomenclature. *J. Cell Sci.* **67,** 217–220.

Jordan, E. G., Timmis, J. N., and Trewavas, A. J. (1980). The plant nucleus. *In* "Biochemistry of plants. The Plant Cell" (N. E. Tolbert ed.), Vol. 1, pp. 489–588. Academic Press. New York and London.

Karpen, G. H., Schaefer, J. E., and Laird, C. D. (1988). A Drosophila rRNA gene located in euchromatin is active in transcription and nucleolus formation. *Genes Dev.* **2,** 1745–1763.

Kaufmann, S. H., Coffey, D. S., and Shaper, J. H. (1981). Considerations in the isolation of rat liver nuclear matrix, nuclear envelope, and pore complex lamina. *Exp. Cell Res.* **132,** 105–123.

Krachmarov, C., Stoilov, L., and Zlatanova, J. (1991). Nuclear matrices from transcriptionally active and inactive plant cells. *Plant Sci.* **76,** 35–41.

Krohne, G., and Benevente, R. (1986). The nuclear lamins. A multigene family in evolution and differentiation. *Exp. Cell Res.* **162,** 1–10.

Krohne, G., Stick, R., Kleinschmidt, J. A., Moll, R., Franke, W. W., and Hausen, P. (1982). Immunological localization of a major karyoskeletal protein in nucleoli of oocytes and somatic cells of *Xenopus laevis. J. Cell Biol.* **94,** 749–754.

Krzyzowska-Gruca, S., Zborek, A., and Gruca, S. (1983). Distribution of interchromatin granules in nuclear matrices obtained from nuclei exhibiting different degree of chromatin condensation. *Cell Tissue Res.* **231,** 427–437.

Labhart, P., Banz, E., Ness, P. J., Parish, R. W., and Koller, T. (1984). A structural concept for nucleoli of *Dictyostelium discoideum* deduced from dissociation studies. *Chromosoma* **89,** 111–120.

Laemmli, U. K., Käs, E., Poljak, L., and Adachi, Y. (1992). Scaffold-associated regions: cis-acting determinants of chromatin structural loops and functional domains. *Curr. Opin. Genet. Dev.* **2,** 275–285.

Lafond, R. E., and Woodcock, C. L. F. (1983). Status of nuclear matrix in mature embryonic chick erythrocyte nuclei. *Exp. Cell Res.* **147,** 31–39.

Lambert, A. M. (1993). Microtubule-organizing centers in higher plants. *Curr. Opin. Cell Biol.* **5,** 116–122.

Li, H., and Roux, S. J. (1992). Casein kinase II protein kinase is bound to lamina-matrix and phosphorylates lamin-like protein in isolated pea nuclei. *Proc. Natl. Acad. Sci. U.S.A.* **89,** 8434–8438.

Lloyd, C. W. (1988). Actin in plants. *J. Cell Sci.* **90,** 185–192.

Lloyd, C. W. (1989). The plant cytoskeleton. *Curr. Opin. Cell Biol.* **1,** 30–35.

Ludérus, M. E. E., de Graaf, A., Mattia, E., den Blaauwen, J. L., Grande, M. A., de Jong, L., and van Driel, R. (1992). Binding of matrix attachment regions to lamin $B_1$. *Cell (Cambridge, Mass.)* **70,** 949–959.

Martelli, A. M., Falcieri, E., Gobbi, P., Manzoli, L., Gilmour, R. S., and Cocco, L. (1991). Heat-induced stabilization of the nuclear matrix: A morphological and biological analysis in murine erythroleukemia cells. *Exp. Cell Res.* **196,** 216–225.

Martín, M., García-Fernandez, L. F., Moreno Díaz de la Espina, S., Noaillac-Depeyre, J., Gas, N., and Medina, F. J. (1992a). Identification and localization of a nucleolin homologue in onion nucleoli. *Exp. Cell Res.* **199,** 74–84.

Martín, M., Moreno Díaz de la Espina, S., Jiménez-García, L. F., Fernández-Gómez, M. E., and Medina, J. F. (1992b). Further investigations on the functional role of two nuclear bodies in onion cells. *Protoplasma* **167,** 175–182.

Masuda, K., Takahashi, S., Nomura, K., Arimoto, M., and Inoue, M. (1993). Residual structure and constituent proteins of the peripheral framework of the cell nucleus in somatic embryos from *Daucus carota* L. *Planta* **191,** 532–540.

McConnell, M., Whalen, A. M., Smith, D. E., and Fisher, P. A. (1987). Heat shock-induced changes in structural stability of proteinaceus karyoskeletal elements in vitro and morphological effects in situ. *J. Cell Biol.* **105,** 1087–1098.

McKeon, F. (1991). Nuclear lamin proteins: Domains required for nuclear targeting, assembly, and cell-cycle-regulated dynamics. *Curr. Opin. Cell Biol.* **3,** 82–86.

McNulty, A. K., and Saunders, M. J. (1992). Purification and immunological detection of pea nuclear intermediate filaments: Evidence for plant nuclear lamins. *J. Cell Sci.* **103,** 407–414.

Medina, F. J., Risueño, M. C., and Moreno Díaz de la Espina, S. (1983). 3-D reconstruction and morphometry of fibrillar centres in plant cells in relation to nucleolar activity. *Biol. Cell* **48,** 31–38.

Medina, F. J., Martin, M., and Moreno Díaz de la Espina, S. (1990). Implications for the function-structure relationship in the nucleolus after immunolocalization of DNA in onion cells. *In* "Nuclear Structure and Function" (J. R. Harris, and I. B. Zbarsky, eds.), pp. 231–235. Plenum, New York.

Medina, F. J., Cerdido, A., and Caizergues-Ferrer, M. (1993). Insights into the characterization of the plant nucleolin homologue. *Proc. Eur. Workshop Cell Nucleus, 13th,* Budapest, p. 66.

Medina, M. A., Moreno Díaz de la Espina, S., Martín, M., and Fernandez-Gómez, M. E. (1989). Interchromatin granules in plant nuclei. *Biol. Cell* **67,** 331–339.

Mielke, C., Kohwi, Y., Kohwi-Shigematsu, T., and Bode, J. (1990). Hierarchical binding of DNA fragments derived from scaffold-attached regions: Correlation of properties in vitro and function in vivo. *Biochemistry* **29,** 7475–7485.

Miller, O. R., and Beatty, B. R. (1969). Nucleolar structure and function. *In* "Handbook of Molecular Cytology" (A. Lima de Faria, ed.), pp. 605–619. North-Holland Publ., Amsterdam.

Mínguez, A., and Moreno Díaz de la Espina, S. (1993a). Immunological characterization of lamins in the nuclear matrix of onion cells. *J. Cel Sci.* **106,** 431–439.

Mínguez, A., and Moreno Díaz de la Espina, S. (1993b). Investigation of the protein components of the nucleolar matrix of onion cells by immunoblotting, immunogoldlabelling and ultracytochemistry. *Proc. Interamerican Congress on Elution Microscopy, 2nd,* Cancún, Mejico, pp. 233–234.

Mínguez, A., and Moreno Díaz de la Espina, S. (1995). The surface molecular organization of the plant nucleoles. Evidence for an underlying nucleolar matrix comprising fibrillar centers and discute processing domains. *Chrom. Res.* (submitted).

Mínguez, A., Franca, S., and Moreno Díaz de la Espina, S. (1994a). Dinoflagelates have a eukariotic nuclear matrix with lamin-like proteins and topoisomerase II. *J. Cell Sci.* **107,** 2861–2873

Mínguez, A., Franca, S., and Moreno Díaz de la Espina, S. (1994b). Characterization of some DNA binding proteins by immunological methods in the nuclear matrix of Dinoflagelates. *Cell Biol. Int. Rev.* **18,** 419.

Mirkovitch, J., Mirault, M. E., and Laemmli, U. K. (1984). Organization of the higher order chromatin loop: Specific DNA attachment sites on nuclear scaffold. *Cell* (*Cambridge, Mass.*) **39,** 223–232.

Mizuno, K. (1993). Microtubule-nucleation sites on nuclei of higher plant cells. *Protoplasma* **173,** 77–85.

Moir, R. D., and Goldman, R. D. (1993). Lamin dynamics. *Curr. Opin. Cell Biol.* **5,** 408–411.

Monneron, A., and Bernhard, W. (1969). Fine structural organization of the interphase nucleus in some mammalian cells. *J. Ultrastruct. Res.* **27,** 266–288.

Moreno Díaz de la Espina, S., and Risueño, M. C. (1976). Effect of $\alpha$-amanitine on the nucleolus of meristematic cells of Allium cepa in interphase and mitosis: An ultrastructural analysis. *Cytobiologie* **12,** 175–188.

Moreno Díaz de la Espina, S., Risueño, M. C., and Medina, F. J. (1982a). Ultrastructural, cytochemical and autoradiographic characterization of coiled bodies in the plant cell nucleus. *Biol. Cell.* **44,** 229–238.

Moreno Díaz de la Espina, S., Sanchez-Pina, M. A., and Risueño, M. C. (1982b). Localization of acid phosphatasic activity, phosphate ions and inorganic cations in plant nuclear coiled bodies. *Cell Biol. Int. Rep.* **6,** 601–607.

Moreno Díaz de la Espina, S., Barthelemy, I., and Cerezuela, M. A. (1991). Isolation and ultrastructural characterization of the residual nuclear matrix in a plant cell system. *Chromosoma* **100,** 110–117.

Moreno Díaz de la Espina, S., Mínguez A., Vázquez-Nin, G. H., and Echeverría, O. M. (1992). Fine structural organization of a non-reticulate plant cell nucleus. *Chromosoma* **101,** 311–321.

Moreno Díaz de la Espina, S., Medina, A., Mínguez, A., and Fernández-Gómez, M. E. (1993). Detection by bismuth staining of highly phosphorylated nucleo-proteins in plants. Determination of its specificity by X-ray microanalysis, SDS-PAGE and immunological analysis. *Biol. Cell.* **77,** 297–306.

Nickerson, J. A., He, D., Fey, E. G., and Penman, S. (1990). The nuclear matrix. *In* "The Eukaryotic Cell Nucleus: Molecular Biochemistry and Macromolecular assemblies" (P. R. Strauss and Wilson, eds.), Vol. 2, pp. 763–782. Telford Press, Nd.

Nickerson, J. A., Krockmalnic, G., Wan, K. M., Turner, C. D., and Penman, S. (1992). A normally masked nuclear matrix antigen that appears at mitosis on cytoskeleton filaments adjoining chromosomes, centrioles, and midbodies. *J. Cell Biol.* **116,** 977–987.

Nigg, E. A. (1989). The nuclear envelope. *Curr. Opin. Cell Biol.* **1,** 435–440.

Nigg, E. A. (1992). Assembly-disassembly of the nuclear lamina. *Curr. Opin. Cell Biol.* **4,** 105–109.

Oakes, M., Nogi, Y., Clark, M. W., and Nomura, M. (1993). Structural alterations of the nucleolus in mutants of *Saccharomyces cerevisiae* deffective in RNA polymerase I. *Mol. Cell. Biol.* **13,** 2441–2455.

Ochs, R. L., and Smetana, K. (1991). Detection of fibrillarin in nucleolar remnants and the nucleolar matrix. *Exp. Cell Res.* **197,** 183–190.

Ohyama, T., Iwaikawa, Y., Takegami, M., and Ito M. (1987). Isolation of residual nuclear structures from microsporocytes of *Trillium kamtschaticum. Cell Struct. Funct.* **12,** 433–442.

Olson, M. O. J., and Thompson, B. A. (1983). Distribution of proteins among chromatin components of nucleoli. *Biochemistry* **22,** 3187–3193.

Olson, M. O. J., Wallace, M. O., Herrera, A. H., Marshall-Carlson, L., and Hung, R. C. (1986). Preribosomal ribonucleoprotein particles are a major component of a nucleolar matrix fraction. *Biochemistry* **25,** 484–491.

Paul, A. L., and Ferl, R. J. (1993). Osmium tetroxide footprinting of a scaffold attachment region in the maize Adh1 promoter. *Plant Mol. Biol.* **22,** 1145–1151.

Penman, S., Fulton, A., Capco, D., Ben Ze'ev, A., Wittelsberger, S., and Tse, C. F. (1982). Cytoplasmic and nuclear architecture in cells and tissue: Form, functions, and mode of assembly. *Cold Spring Harbor Symp. Quant. Biol.* **46,** 1013–1028.

Peters, K. E., and Comings, D. E. (1980). Two-dimensional gel electrophoresis of rat liver nuclear washes, nuclear matrix, and hnRNA proteins. *J. Cell Biol.* **86,** 135–155.

Pienta, K. J., Getzenberg, R. H., and Coffey, D. S. (1991). Cell structure and DNA organization. *CRC Crit. Rev. Eukaryotic Gene Express.* **1,** 355–385.

Pouchelet, M., Anteunis, A., and Gansmuller, A. (1986). Correspondence of two nuclear networks observed in situ with the nuclear matrix. *Biol. Cell.* **56,** 107–112.

Puvion-Dutilleul, F. (1983). Morphology of transcription at cellular and molecular levels. *Int. Rev. Cytol.* **84,** 57–100.

Razin, S. V., and Vanetzky, Y. S. (1992). Domain organization of eukaryotic genome. *Cell Biol. Int. Rev.* **16,** 697–708.

Risueño, M. C., and Medina, F. J. (1986). The nucleolar structure in plant cells. *Cell Biol. Rev.* **7.**

Risueño, M. C., and Moreno Díaz de la Espina, S. (1979). Ultrastructural and cytochemical study of the quiescent root meristematic cell nucleus. *J. Submicrosc. Cytol.* **11,** 85–98.

Risueño, M. C., Moreno Díaz de la Espina, S., Fernández-Gómez, M. E., and Giménez-Martin, G. (1976). Ultrastructural study of nucleolar material during plant mitosis in the presence of inhibitors of RNA synthesis. *J. Microsc. Biol. Cell.* **26,** 5–18.

Risueño, M. C., Medina, F. J., and Moreno Díaz de la Espina, S. (1982). Nucleolar fibrillar centres in plant meristematic cells: Ultrastructure, cytochemistry, and autoradiography. *J. Cell Sci.* **58,** 313–329.

Rodrigo, R. M., Rendón, M. C., Torreblanca, J., García-Herdugo, G., and Moreno, F. J. (1992). Characterization and immunolocalization of RNA polymerase I transcription factor

UBF with anti-NOR serum in protozoa, higher plant and vertebrate cells. *J. Cell Sci.* **103,** 1053–1063.

Rzepecki, R., Szmidzinsky, R., Bode, J., and Szopa, J. (1992). The 65 kDa protein affected endonuclease tightly associated with plant nuclear matrix. *J. Plant Physiol.* **139,** 284–288.

Scheer, U., and Benavente, R. (1990). Functional and dynamic aspects of the mammalian nucleolus. *BioEssays* **12,** 14–22.

Scheer, U., and Weisenberger, D. (1984). The nucleolus. *Curr. Opin. Cell Biol.* **6,** 354–359.

Scheer, U., Hügle, B., Hazan, R., and Rose, K. M. (1984). Drug-induced dispersal of transcribed rRNA genes and transcriptional products: Immunolocalization and silver staining of different nucleolar components in rat cells treated with 5,6-Dichloro-β-D-ribofuranosylbenzimidazole. *J. Cell Biol.* **99,** 672–679.

Scheer, U., Thiry, M., and Goessens, G. (1993). Function and assembly of the nucleolus. *Trends Cell Biol.* **3,** 236–241.

Schellenbaum, P., Vantard, M., Peter, C., Fellous, A., and Lambert, A. M. (1993). Co-assembly properties of higher plant microtubule-associated proteins with purified brain and plant tubulins. *Plant J.* **3,** 253–260.

Schöffl, F., Schröder, Klierm, M., and Rieping, M. (1993). An SAR sequence containing 395 bp DNA fragments mediates enhanced, gene-dosage-correlated expression of a chimaeric heat shock gene in transgenic tobacco plants. *Transgenic Res.* **2,** 93–100.

Shaw, P. J., Fairbairn, D. J., and Lloyd, C. W. (1991). Cytoplasmic and nuclear intermediate filament antigens in higher plant cells. *In* "The Cytoskeletal Basis of Plant Growth and Form" (C. W. Lloyd, ed.). Academic Press, San Diego, CA.

Shibaoka, H., and Nagai, R. (1994). The plant cytoskeleton. *Curr. Opin. Cell Biol.* **6,** 10–15.

Shiomi, Y., Powers, J., Bolla, R. I., van Nguyen, T., and Schlessinger, D. (1986). Proteins and RNA in mouse L cell corenucleoli and nucleolar matrix. *Biochemistry* **25,** 5745–5751.

Slatter, R. E., Duprée, P., and Gray, J. C. (1991). A scaffold-associated DNA region is located downstream of the pea plastocyanin gene. *Plant Cell* **3,** 1239–1250.

Spector, D. L. (1993a). Macromolecular domains within the cell nucleus. *Annu. Rev. Cell Biol.* **9,** 265–315.

Spector, D. L. (1993b). Nuclear organization of pre-mRNA processing. *Curr. Opin. Cell Biol.* **5,** 442–448.

Staiger, C. J., and Lloyd, C. W. (1991). The plant cytoskeleton. *Curr. Opin. Cell Biol.* **3,** 33–42.

Stephanova, E., and Valkov, N. (1991). Rat liver nucleoli: Biochemical characterization of the residual skeletal structure. *C. R. Acad. Bulg. Sci.* **44,** 63–66.

Stephanova, E., Stancheva, R., and Avramanova, Z. (1993). Binding of sequences from the 5′- and 3′-nontranscribed spacer of the rat rDNA locus to the nucleolar matrix. *Chromosoma* **102,** 287–295.

Stick, R. (1992). The gene structure of *Xenopus* nuclear lamin A: A model for the evolution of A-type from B-type lamins by exon shuffling. *Chromosoma* **101,** 566–574.

Stoilov, L. M., Mirkova, V., Zlatanova, J., and Djondjurov, L. (1992). Matrix associated DNA from maize is enriched in repetitive sequences. *Plant Cell Rep.* **11,** 355–358.

Stolyarov, S. D. (1984). Isolation and characterization of the nuclear matrix of the onion *Allium cepa. Cytologia* **26,** 874–878.

Stuurman, N., de Jong, L., and van Driel, R. (1992a). Nuclear frameworks: Concepts and operational definitions. *Cell Biol. Int. Rep.* **16,** 837–852.

Stuurman, N., Floore, A., Colen, A., de Jong, L., and van Driel, R. (1992b). Stabilization of the nuclear matrix by disulfide bridges: Identification of matrix polypeptides that form disulfides. *Exp. Cell Res.* **200,** 285–294.

Suzuki, T., Suzuki, N., and Hosoya, T. (1993). Limited proteolysis of rat liver nucleolin by endogenous proteases: Effects of polyamines and histones. *Biochem. J.* **289,** 109–115.

Thiry, M. (1992). New data concerning the functional organization of the mammalian cell nucleolus: Detection of RNA and rRNA by in situ molecular immunocytochemistry. *Nucleic Acids Res.* **20,** 6195–6200.

Todorov, I. T., and Hadjiolov, A. A. (1979). A comparison of nuclear and nucleolar matrix proteins from rat liver. *Cell Biol. Int. Rev.* **3,** 753–757.

Tong, C. G., Dauwalder, M., Clawson, G. A., Hatem, C. L., and Roux, S. J. (1993). The major nucleoside triphosphatase in pea (*Pisum sativum* L.) nuclei and in rat liver nuclei share common epitopes also present in nuclear lamins. *Plant Physiol.* **101,** 1005–1011.

Trimbur, G. M., and Walsh, C. J. (1993). Nucleolus-like morphology produced during the in vitro reassociation of nucleolar components. *J. Cell Biol.* **122,** 753–766.

van Eekelen, C. A. G., and van Venrooij, W. (1981). HnRNA and its attachment to a nuclear protein matrix. *J. Cell Biol.* **88,** 554–563.

Vázquez-Nin, G. H., Echeverría, O. M., Mínguez, A., Moreno Díaz de la Espina, S., Fakan, S., and Martin, T. E. (1992). Ribonucleoprotein components of root meristematic cell nuclei of the tomato characterized by application of mild loosening and immunocytochemistry. *Exp. Cell Res.* **200,** 431–438.

Verheijen, R., van Venrooij, W., and Ramaekers, F. (1988). The nuclear matrix: Structure and composition. *J. Cell Sci.* **90,** 11–36.

von Kries, J. P., Buhrmester, H., and Strätling, W. H. (1991). A matrix scaffold attachment region binding protein: Identification, purification and mode of binding. *Cell* (*Cambridge, Mass.*) **64,** 123–135.

Waitz, W., and Loidl, P. (1988). In situ preparation of the nuclear matrix of *Physarum polycephalum:* Ultrastructural and biochemical analysis of different matrix isolation procedures. *J. Cell Sci.* **90,** 621–628.

Wang, X., and Traub, P. (1991). Resinless section immunogold electron microscopy of karyocytoskeletal frameworks of eukaryotic cells cultured in vitro. *J. Cell Sci.* **98,** 107–122.

Wassef, M. (1979). A cytochemical study of interchromatin granules. *J. Ultrastruct. Res.* **69,** 121–133.

Weber, K., Plessmann, U., and Ulrich, W. (1989). Cytoplasmic intermediate filament proteins of invertebrates are closer to nuclear lamins than are vertebrate intermediate filament proteins: sequence characterization of two muscle proteins of a nematode. *EMBO J.* **8,** 3321–3327.

Yang, C., Xing, L., and Zhay, Z. (1992). Intermediate filaments in higher plant cells and their assembly in a cell-free system. *Protoplasma* **171,** 44–54.

Yang, C. H., Lambie, E. J., and Snyder, M. (1992). NuMA: An unusually coiled-coil related protein in the mammalian nucleus. *J. Cell Biol.* **116,** 1305–1317.

Zborek, A., Krzyzowska-Gruca, S., and Gruca, S. (1990). Nuclear fraction enriched in interchromatin granules. *Cell Biol. Int. Rep.* **14,** 79–88.

Zeng, C., He, D., Berget, S. M., and Brinkley, B. R. (1994). Nuclear-mitotic apparatus protein: A structural protein interface between the nucleoskeleton and RNA splicing. *Proc. Natl. Acad. Sci. U.S.A.* **91,** 1505–1509.

# The Dynamic Properties and Possible Functions of Nuclear Lamins

Robert D. Moir,[1] Timothy P. Spann,[1] and Robert D. Goldman
Department of Cell and Molecular Biology, Northwestern University Medical School, Chicago, Illinois 60611

The nuclear lamins are thought to form a thin fibrous layer called the nuclear lamina, underlying the inner nuclear envelope membrane. In this review, we summarize data on the dynamic properties of nuclear lamins during the cell cycle and during development. We discuss the implications of dynamics for lamin functions. The lamins may be involved in DNA replication, chromatin organization, differentiation, nuclear structural support, and nuclear envelope reassembly. Emphasis is placed on recent data that indicate that the lamina, contrary to previous views, is not a static structure. For example, the lamins form nucleoplasmic foci, distinct from the peripheral lamina, which vary in their patterns of distribution as well as their composition in a cell cycle-dependent manner. During the S phase, these foci colocalize with chromatin and sites of DNA replication. At other points during the cell cycle, they may represent sites of lamin post-translation processing that take place prior to incorporation into the lamina. Secondary modifications of the lamins such as isoprenylation and phosphorylation are involved in the regulation of the dynamic properties and the assembly of lamins. In addition, a number of lamin-associated proteins have been recently identified and these are described along with their potential functions.

**KEY WORDS:** Nuclear lamins, Chromatin, DNA replication, Phosphorylation, Isoprenylation, Intermediate filaments, Nuclear lamina.

## I. Introduction

Nuclear processes such as DNA replication, transcription, and RNA processing appear to be carefully regulated in time and space. It now appears that there is an underlying structural component to the nucleus, termed

[1] The first two authors contributed equally to this work.

the nuclear matrix, that helps to coordinate each of these processes possibly by organizing interphase chromatin. The original definition of the matrix was a biochemical one that referred to those proteins and structures remaining after nuclei are extracted to remove most of the protein, DNA and RNA, and according to some definitions, the nuclear envelope membranes. The nuclear lamins are a prominent group of proteins within the nuclear matrix fraction. *In vivo,* the lamins appear to assemble to form the nuclear lamina, a thin fibrous structure immediately underlying the inner nuclear membrane of most eukaryotic cell nuclei. The lamins were the first component of the matrix to be described in detail and remain the most well characterized. The lamina probably provides structural support for the nucleus, helping to maintain nuclear shape. The lamina may also be involved in the organization of chromatin and consequently may influence DNA replication and transcription during interphase. In addition, it may have a critical role in nuclear envelope reassembly following cell division.

In this review, we survey the information available on nuclear lamins, emphasizing their dynamic properties. We define dynamics as the changes in the state of assembly and composition of lamins during the cell cycle and during development. The first part of this review describes data on the regulation of the dynamics of nuclear lamins. The second describes the functional implications of these dynamic properties. There have been numerous, shorter reviews published recently that focus on different aspects of the properties of lamins (McKeon, 1991; Nigg, 1992a,b; Paddy *et al.,* 1992; Moir and Goldman, 1993; Georgatos *et al.,* 1994; Gerace and Foisner, 1994; Heins and Aebi, 1994; Hutchison *et al.,* 1994; Lourim and Krohne, 1994).

## II. Molecular Aspects of Nuclear Lamins

### A. Identification of Nuclear Lamins

A lamina-like structure was first described in thin sections of the nucleus of *Ameoba proteus* cells where there is a particularly distinctive electron-dense layer underlying the nuclear envelope (Pappas, 1956; Mercer, 1959). The term fibrous lamina was first used by Gray and Guillery (1963) and Coggeshall and Fawcett (1964) to describe the structure seen in the nuclei of invertebrate cells. An equivalent structure was later described in mammalian nuclei (Fawcett, 1966; Patrizi and Poger, 1967). Subsequent biochemical fractionation of nuclei demonstrated the existence of a highly insoluble structure termed the pore complex–lamina fraction (Aaronson and Blobel, 1975; Dwyer and Blobel, 1976) and the principal proteins of the lamina, termed nuclear lamins A, B, and C, were identified (Gerace *et al.,* 1978; Krohne *et al.,* 1978; Gerace and Blobel, 1980). Biochemical, immunological, and structural characterization of the lamins led to the suggestion that they were related to cytoplasmic intermediate filament (IF) proteins (Zackroff *et al.,* 1984; Goldman *et al.,* 1986; Aebi *et al.,* 1986) and this was confirmed by cloning of the first lamin cDNAs (McKeon *et al.,* 1986; Fisher *et al.,* 1986).

## B. Primary Structure: Relationship to Cytoplasmic Intermediate Filaments

The cDNAs for the human lamins A and C were the first to be cloned (Fisher *et al.,* 1986; McKeon *et al.,* 1986). Analysis of these sequences as well as those obtained for lamin B confirmed that nuclear lamins are closely related to the cytoplasmic IF proteins and therefore the lamins are classified as Type V IFs (Parry and Steinert, 1992; Fig. 1). As in cytoplasmic IFs, the lamin sequences have a central rod domain of approximately 350 amino acids with the heptad repeat indicative of an $\alpha$-helical protein structure (Steinert and Roop, 1988). This domain begins and ends with highly conserved sequences found in almost all IF proteins (Steinert and Roop, 1988). Unlike the cytoplasmic IFs, the lamin rod sequences are not interrupted by breaks (linker regions) in the heptad motif. In addition, lamin rod domains have six additional heptads (42 amino acids; Fig. 1) in the central rod domain when compared to cytoplasmic IF proteins of vertebrates (Fisher *et al.,* 1986; McKeon *et al.,* 1986). Several invertebrate cytoplasmic IF sequences have been cloned, and these sequences have an additional 42-amino acid insertions in the rod (Weber *et al.,* 1989a, 1991; Dodemont *et al.,* 1990; Way *et al.,* 1992). Based on this, lamins are considered to be the progenitor of all IFs and cytoplasmic IFs (Weber *et al.,* 1989a; Stick, 1992).

The rod domain is flanked by N- and C-terminal non-$\alpha$-helical domains. The N-terminal domain is relatively short (30–40 amino acids) and contains a phosphorylation site for p34cdc2, the mitotic kinase. The C-terminal domain is longer (210–300 amino acids) and contains a number of sequence motifs involved in lamin dynamics and assembly that are not found in cytoplasmic IFs. A series of basic residues immediately following the end of the rod domain defines a nuclear localization signal that is required for transport from the cytoplasm to the nucleus (Loewinger and McKeon, 1988; Frangioni and Neel, 1993). In addition, A and B lamins end in the sequence CaaX, where C is cysteine, a is an aliphatic residue, and X is any residue (Marshall, 1993). The cysteine residue in this motif is a target for farnesylation and carboxymethylation (Fig. 2 for summary of the farnesylation and associated reactions on lamins). The same sequence is also found at the

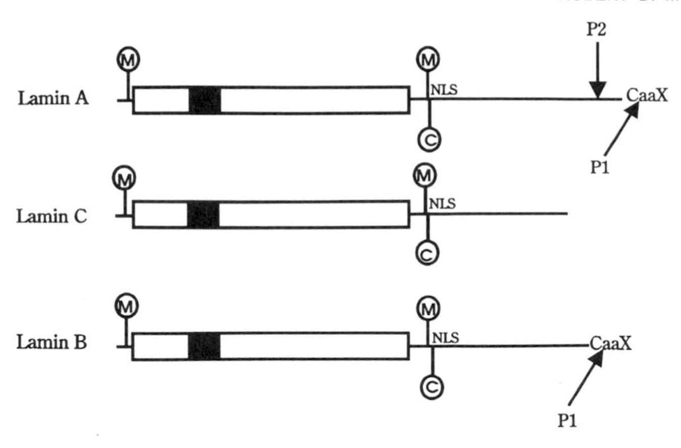


FIG. 1 Molecular schematic of nuclear lamins. A schematic of the primary structure of nuclear lamins from mammalian cells. Lamins A and C are the A-type somatic lamins. There are two B-type somatic lamins that have very similar sequences shown by a single diagram in this figure. Lamins A and C are identical in sequence except that lamin C has a unique 6 amino acid extension at its C-terminus, while lamin A has a 90 amino acid extension. The open rectangles indicate the central rod domain with the $\alpha$-helical heptad motif. The filled area within the rectangle shows the location of the 42-amino-acid insertion in the rod domain that is present in lamin but not cytoplasmic IFs of vertebrates. The N- and C-terminal nonhelical domains are shown by lines extending from the rectangle. The relatively short N-terminal domain contains a phosphorylation site for the p34cdc2 mitotic kinase, as indicated by a circle with an M. The C-terminus contains several sequence motifs. The nuclear localization signal (NLS) is a motif of 6–7 basic amino acids required for transport of the protein from the cytoplasm to the nucleus. Immediately preceding the NLS are the known *in vivo* phosphorylation sites for p34cdc2 (indicated by M) and protein kinase C (indicated by C). There are actually several closely spaced phosphorylation sites for p34cdc2 and protein kinase C in this region (see text). Lamins A and B end in the motif CaaX, where C is a cysteine residue, a is an aliphatic residue, and X is any residue. This motif undergoes a series of modifications, including farnesylation, proteolytic cleavage (P1), and carboxymethylation. The details of these modifications are shown in Fig. 2. Only lamin A undergoes a second proteolytic cleavage (P2) after the first set of modifications at the CaaX motif. This second cleavage occurs 14 residues from the farnesylated cysteine.

C-terminus of the *ras* family of proteins and appears to be required for *ras* binding to the plasma membrane (see Marshall, 1993, for discussion of *ras* isoprenylation). In the case of lamins, this sequence appears to be required for transport from the nucleoplasm to the nuclear envelope (Holtz *et al.*, 1989; Krohne *et al.*, 1989; Kitten and Nigg, 1991; Lutz *et al.*, 1992). The role of this sequence in lamin dynamics is described more fully in Section III,C,3. In addition, the C-terminus carries a number of phosphorylation sites for protein kinase C and p34cdc2 kinase. These sites are clustered near the nuclear localization sequence (Fig. 1). In a recent study, the sequence domains unique to lamin IFs were inserted into a neural IF

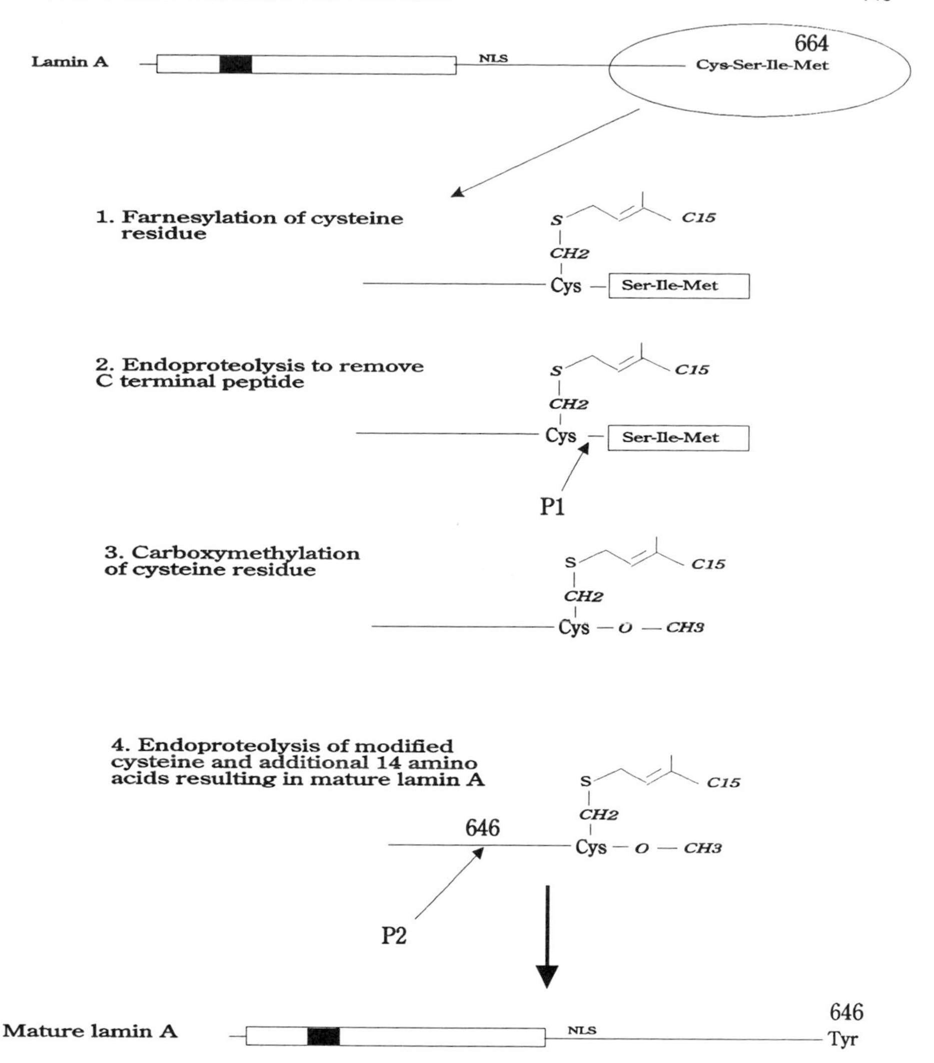


FIG. 2 The sequence of modifications undergone by lamin A at the C-terminal CaaX sequence. The B-type lamins undergo the same series of modifications with the exception of the second proteolytic cleavage (P2). The C-terminal cysteine is farnesylated (1) and the three terminal amino acids removed by proteolysis (2). The cysteine is then carboxymethylated (3). A C-terminal peptide containing 15 amino acids, including the modified cysteine, is removed by proteolysis (4), resulting in mature lamin A.

(Type IV) cDNA and expressed in cells using transient transfection assays (Monteiro *et al.,* 1994). The results suggest that all the lamin-specific domains are essential for the proper regulation of their dynamic properties.

## C. Expression of the Lamin Multigene Family

### 1. Vertebrate Lamins

Over 25 lamin sequences from *Caenorhabditis elegans* to *Homo sapiens* are now present in DNA databases and the lamin gene family of a given species is becoming increasingly complex as more members are identified. For example, six different lamin cDNAs have been cloned in mouse species (Hoger *et al.*, 1988, 1990a,b; Riedel and Werner, 1989; Furukawa and Hotta, 1993; Nakajima and Sado, 1993; Furukawa *et al.*, 1994). These cDNA sequences are grouped into two families, A- or B-type, depending on sequence and expression patterns. In addition, there are several lamins identified on the basis of immunological evidence that have not been cloned (Kaufmann, 1989).

A lamin B isotype is found in all tissues examined to date. Three B-type lamins, B1, B2, and B3, have been identified in mouse species (Hoger *et al.*, 1988, 1990a,b; Weber *et al.*, 1991; Furukawa and Hotta, 1993). B1 and B2 are somatic lamins that are closely related in sequence, but are products of different genes. There has not been a detailed examination of the differences in expression between B1 and B2 during development. B3 is a sperm cell-specific lamin that appears to be a splicing variant of lamin B2 (Furukawa and Hotta, 1993).

There are three different A-type lamins, apparently all transcribed from the same gene (Lin and Worman, 1993; Furukawa *et al.*, 1994). Lamins A and C are somatic lamins that are apparently expressed in approximately equal amounts (but see Martelli *et al.*, 1992) and are identical in sequence except that lamin C has a unique 6 amino acid extension at its C-terminus, while lamin A has a 90-amino acid extension. Lamin C lacks the CaaX target sequence for isoprenylation. These C-terminal differences are the result of the splicing of different C-terminal exons for lamin A and C mRNAs (Lin and Worman, 1993). In addition a sperm-specific lamin, termed lamin C2, has been recently described (Furukawa *et al.*, 1994). This latter lamin is similar to lamin C and is also apparently transcribed from the same gene.

The A-type lamins are expressed exclusively in differentiating tissues. When sections of developing mouse embryos are stained with lamin antibodies, lamin A/C staining does not appear until 8–11 days postfertilization, initially in developing muscle (Stewart and Burke, 1987; Rober *et al.*, 1989, 1990a,b; but see Coonen *et al.*, 1993, for differing results). The induction of lamin A/C expression seems to coincide with major changes in tissue structure, but occurs well after cells have committed to form a particular tissue. In the most striking example of this expression pattern, lamin A/C expression does not appear in the developing brain until 10 days after birth.

Hemopoietic cells as well as intestinal epithelial cells do not appear to express lamins A/C even in the adult (Rober *et al.,* 1989, 1990a,b). The correlation of the induction of lamins A/C with tissue differentiation has led to the suggestion that these lamins are involved in the maintenance of the differentiated phenotype.

The correlation between lamin A/C induction and differentiation is also seen in a number of tumor cell lines that have been used to examine lamin function and dynamics as discussed in detail in Section IV,A (Hass *et al.,* 1990; Peter and Nigg, 1991; Collard *et al.,* 1992; Horton *et al.*, 1992). In the undifferentiated state these cell lines express lamin B, but do not express lamins A/C or express them at lower levels. The cells can be induced to differentiate with various agents and this results in an increase in the levels of mRNA and protein for lamins A/C (Lanoix *et al.,* 1992; Mattia *et al.,* 1992), although it is not clear if this is due to an increase in transcription (Mattia *et al.,* 1992) or an increase in mRNA stability (Lanoix *et al.,* 1992). These cell lines also raise the possibility that nuclear lamins may be useful as markers for cancer diagnosis. Low lamin A/C expression may correlate with certain cancer subtypes (Kaufmann, 1992; Broers *et al.,* 1993; Broers and Ramaekers, 1994; Muller *et al.,* 1994).

In other vertebrates a similar pattern of gene expression is seen. Human lamins A, C, B1, and B2 have been cloned (Fisher *et al.,* 1986; McKeon *et al.,* 1986; Pollard *et al.,* 1990; Biamonti *et al.,* 1992). Lamin A, B1, and B2 cDNAs have been cloned from chicken, but a lamin C-like protein has not yet been identified (Peter *et al.,* 1989; Vorburger *et al.,* 1989). In chickens, lamin A is expressed only in differentiating cells, and interestingly there appear to be differences between lamin B1 and B2 expression. Lamin B2 levels appear to be relatively constant while B1 expression is high in embryonic tissues but drops in adult tissues (Lehner *et al.,* 1986). Finally, in the amphibian *Xenopus laevis,* there are five different lamins; lamin A, LI, LII, LIII, and LIV (Krohne *et al.,* 1984; Benavente *et al.,* 1985; Stick and Hausen, 1985; Wolin *et al.,* 1987; Lourim and Krohne, 1993). Lamins I–IV are also termed B1–B4, but this numbering system does not reflect equivalence with the corresponding mammalian B lamins (Stick, 1992). LI and LII are the primary B-type somatic lamins while LIII is found in few somatic tissues. LIII is the major lamin expressed in oocytes although LII is also present (Lourim and Krohne, 1993). LIV is a male germ line-specific lamin that has not yet been cloned.

A number of genomic clones for vertebrate lamins have been characterized. Analysis of these sequences shows that intron positions are remarkably well conserved among A- and B-type lamins of different species (Stick, 1992). Since B-type lamins are expressed in all tissues, it has been argued that they are the progenitor lamin sequence and a model for evolution of

A-type lamins from B-type lamins by exon shuffling has been proposed (Stick, 1992).

### 2. Invertebrate Lamins

The characterization of nuclear lamins expressed in invertebrate cells is less complete than that in vertebrates. Biochemical and immunological analysis indicates a single lamin in the mollusk *Spisula solidissima* that has not yet been cloned (Maul *et al.,* 1984; Dessev and Goldman, 1990). A B-type lamin has been cloned from *Drosophila melanogaster* and its expression in tissues has been described (Gruenbaum *et al.,* 1988). A second *Drosophila* cDNA with similarities to vertebrate lamin C, lacking an isoprenylation signal, has been cloned, although its expression pattern is not clear (Bossie and Sanders, 1993; Riemer and Weber, 1994). Finally, a single lamin cDNA has been cloned from *Caenorhabditis elegans* (Riemer *et al.,* 1993). Since these studies are preliminary, it seems likely that there are lamin sequences remaining to be identified in invertebrate species.

### 3. Unicellular Organisms and Plants

Lamins have not been convincingly identified in the widely studied organisms *Saccharomyces cerevisiae, Schizosaccharomyces pombe,* or *Dictyostelium discoideum.* Some immunological evidence exists for the presence of lamins in *S. cerevisiae* (Georgatos *et al.,* 1989) although electron microscopic studies of yeast nuclei have been contradictory on the presence of IF-like proteins in the nuclear envelope (Allen and Douglas, 1989; Cardenas *et al.,* 1990). When human lamin B1 is expressed in *S. cerevisiae* (Smith and Blobel, 1994), or when chicken lamin B2 is expressed in *S. pombe* (Enoch *et al.,* 1991), the expressed protein appears to be properly targeted to the nuclear envelope, although it is not clear if this protein is actually incorporated into a lamina-like structure. Lamin cDNAs have not been cloned in plants, but there is good biochemical and immunological evidence for lamin-like proteins (Frederick *et al.,* 1992; McNulty and Saunders, 1992).

## III. Structure and Dynamic Properties of Nuclear Lamins

The lamina is an assembly of up to five different lamin proteins, but its precise structure is yet to be determined. However, there is evidence accumulating on how it might be assembled. For example, there is substantial evidence that post-translational modifications such as phosphorylation and isoprenylation modulate this process. Furthermore, assembly *in vivo* seems to involve interactions with other nuclear structures such as chromatin and lamin-associated proteins (LAPs). Recently, nucleoplasmic lamin foci have been described by several groups (Goldman *et al.*, 1992; Lutz *et al.*, 1992; Bridger *et al.*, 1993; Mancini *et al.*, 1994; Moir *et al.*, 1994). These lamin foci appear to be distinct from the lamin assembly at the nuclear periphery and the number of these foci varies in a cell cycle-dependent manner (Goldman *et al.*, 1992; Bridger *et al.*, 1993; Moir *et al.*, 1994). Furthermore, some of these foci appear to be associated with chromatin (Bridger *et al.*, 1993; Moir *et al.*, 1994). These nucleoplasmic structures may represent sites of lamin assembly or modification prior to incorporation into the peripheral lamina and they also seem to be part of previously unrecognized structures that may have additional roles in nuclear organization and function.

## A. Structure of the Nuclear Lamina

There have been only a few attempts to directly visualize and discern the structure of the nuclear lamina. The germinal vesicle of the *Xenopus laevis* oocyte is one system that has been exploited for the study of nuclear structure. If preparations of nuclear envelopes of the germinal vesicle are examined in the electron microscope, then the presumptive lamina appears as an array of 10-nm-diameter filaments in close association with the nuclear membrane (Aebi *et al.*, 1986; Akey, 1989; Stewart *et al.*, 1991; Goldberg and Allen, 1992). These arrays occasionally show remarkable tetragonal order, implying that the lamina fibers are organized, perhaps through association with membrane proteins or other components of the nuclear matrix. The lamina in these cells is composed primarily of one lamin, B3, as well as small amounts of B2 (Lourim and Krohne, 1993), but it still remains to be demonstrated conclusively that the filaments seen in the electron microscope are composed of lamins. This is a particularly important point because the structure of the oocyte nucleus is much more complex than originally thought. For example, a second highly ordered structure, termed the nuclear envelope lattice (NEL), has also been described recently in amphibian oocytes (Goldberg and Allen, 1992). The NEL appears to lie at the level of the nucleoplasmic face of the nuclear pore and is thought to be distinct from the lamina, but it still may interconnect with it (Goldberg and Allen, 1992). The precise relationship between the lamina and the nuclear envelope lattice remains to be established and it must also be established that the lattice is present in nuclei of other cell types.

Paddy *et al.* (1990) used fluorescence microscopy to obtain three-dimensional reconstructions of both HeLa cell and *Drosophila* nuclei stained with lamin antibodies. The reconstructions suggest that the lamins form aggregates that appear as fibrillar networks. However, the fibers are not straight and are not organized in regular arrays. Furthermore, there appear to be large areas of the nuclear periphery that are not covered by the lamina when studied using these light microscopic methods. More recent electron microscopic analyses (Belmont *et al.,* 1993) suggest that these gaps may be areas where the lamina is thinner than at other points. These thin areas of the lamina may have been missed in previous studies because of the use of imaging techniques that exclude low levels of fluorescence signal in the light microscope (Belmont *et al.,* 1993). Other work using immunogold electron microscopy also suggests that the lamina forms a continuous layer around the inner nuclear envelope (Hoger *et al.,* 1991a). The structural basis for the differences in lamina thickness is not clear, but the thicker areas frequently appear to be located adjacent to chromatin (Belmont *et al.*, 1993).

## B. Physical Properties of the Lamins

Although the detailed structure of the lamina is still not clear, a substantial amount of information has accumulated on the properties of the isolated lamins. The nuclear lamina possesses the same overall solubility properties as cytoplasmic IFs and other nuclear matrix proteins. An insoluble pore complex–lamina fraction is prepared by treating isolated nuclei with RNase, DNase, nonionic detergents, and high concentrations of NaCl (Aaronson and Blobel, 1975; Dwyer and Blobel, 1976; Aebi *et al.,* 1986). Although numerous other proteins, including other putative nuclear matrix proteins, are present in the insoluble fraction, the predominant proteins are the nuclear lamins as well as contaminating cytoplasmic IF proteins. In fact, resistance to extraction is used as a biochemical assay to demonstrate that lamins have been incorporated into a polymer. The lamins can only be solubilized by extraction with high salt/high pH buffers or with chaotropes such as urea (Gerace *et al.,* 1984; Zackroff *et al.,* 1984; Aebi *et al.,* 1986). Like cytoplasmic IF proteins, the lamins are almost completely insoluble under physiological conditions. However, the solubility properties of the lamins are not identical with cytoplasmic IFs (see Moir *et al.,* 1991, for details of lamin solubility *in vitro* and differences relative to other IF proteins).

The ability to express the lamins in *E. coli* has greatly facilitated examination of lamin interactions and their assembly properties *in vitro.* As with other IFs (and other $\alpha$-helical proteins), the first step of nuclear lamin assembly *in vitro* is the formation of a dimer of two protein chains (see Stewart, 1993, and Heins and Aebi, 1994, for detailed discussion of IF assembly). The formation of the dimer is driven by the formation of a coiled-coil between $\alpha$-helical regions of the subunit. Electron microscopy suggests the two chains in the dimer are parallel and unstaggered with respect to each other. The lamin dimers associate *in vitro* in a distinctive head-to-tail arrangement (Heitlinger *et al.,* 1991, 1992; Peter *et al.,* 1991). It has not been established if this interaction is significant *in vivo,* but cross-linking experiments using isolated nuclei do suggest that the dimer is the major cross-linked product *in vivo* (Dessev *et al.,* 1990). The head-to-tail interaction of dimers has not been observed with cytoplasmic IFs and therefore may represent a difference in the assembly of nuclear and cytoplasmic IF proteins.

There appear to be other differences in the assembly of lamins and cytoplasmic IFs. Cytoplasmic IFs, either purified from tissues or made by expression in *E. coli,* readily assemble *in vitro* to form 10-nm-diameter filaments, similar in appearance to those seen in cells (Chou *et al.,* 1990). However, purified lamins assemble *in vitro* into paracrystals, rather than filaments (Zackroff *et al.,* 1984; Aebi *et al.,* 1986; Goldman *et al.,* 1986; Parry *et al.,* 1987; Moir *et al.,* 1990, 1991; Heitlinger *et al.,* 1991, 1992). Such large, ordered aggregates have not been observed *in vivo,* and therefore, the assembly of the lamin proteins must be precisely regulated *in vivo.* Despite this, the paracrystalline structures formed *in vitro* can provide information about lamin–lamin interactions. For example, lamins A and C form precisely the same paracrystals with the same repeat unit (24-nm) whereas lamin B1 and B2 have different repeats (Parry *et al.,* 1987; Moir *et al.,* 1990, 1991; Heitlinger *et al.,* 1991, 1992). This implies that the protein–protein interactions governing the two types of lamins may be different. In addition, the deletion of N- and C-terminal domains does not prevent the formation of paracrystals. However, the solubility properties of these molecules are altered, suggesting that the N- and C-terminal domains play a role in assembly (Moir *et al.,* 1991; Heitlinger *et al.,* 1992). Surprisingly, there has been little work carried out on the *in vitro* interactions among the different lamin isotypes, although some evidence suggests that the N-terminus of an A-type lamin binds to the C-terminus of lamin B (Georgatos *et al.,* 1988).

## C. Assembly and Disassembly *in Vivo*

### 1. Nucleoplasmic Lamin Foci

Despite their relatively stable properties *in vitro*, the lamins are dynamic throughout the cell cycle. The most dramatic example of this occurs at mitosis when the lamina is disassembled and then reassembled as daughter cell nuclei form. Phosphorylation appears to be the main regulatory mechanism involved in this process. The lamina also possesses dynamic properties during interphase. For example, as the nucleus grows during interphase, the lamina must also grow and rearrange itself to accommodate changes in nuclear shape and size. In addition, there may be other specific rearrangements of the lamina during the cell cycle involved in facilitating other nuclear processes.

As nuclei reform at the completion of mitosis, daughter cells reutilize at least some of the lamin protein available from the previous cell cycle (Gerace *et al.,* 1984). Synthesis of lamins A and C seems to take place throughout the cell cycle (Gerace *et al.,* 1984), while there is evidence that lamin B may be synthesized specifically during S phase (Foisy and Bibor-Hardy, 1988). Interestingly, biochemical extractions of interphase cells suggest that there is not a significant pool of unincorporated lamins (Gerace and Blobel, 1980), indicating that newly synthesized lamins must be rapidly integrated into the lamin polymer. Therefore, the lamina must be able to incorporate lamin subunits throughout interphase and not just during nuclear reassembly following mitosis. The incorporation of the newly synthesized subunits may be responsible for the growth of the lamina during the cell cycle and, in fact, this may be directly linked to the growth of the nucleus itself (Swanson *et al.,* 1991).

When cytoplasmic IFs, in a soluble form, are injected into cells and examined at very early time points postinjection, the injected protein first accumulates as cytoplasmic foci, then becomes incorporated into the endogenous IF network (Vikström *et al.,* 1989; Miller *et al.,* 1991). Similar observations have been made with injection of nuclear lamins (Goldman *et al.,* 1992). For example, when lamin A, made by expression in *E. coli,* is injected into the cytoplasm of 3T3 cells, the injected protein rapidly accumulates in nucleoplasmic foci that are not part of the peripheral lamina (Goldman *et al.,* 1992). This protein eventually becomes incorporated into the lamina over a time period of several hours. Similar foci are seen in uninjected cells stained with lamin antibodies, and therefore, the foci of injected protein are not thought to be accumulations of excess protein, but instead seem to be part of normal nucleoplasmic structures (Goldman *et al.,* 1992; Bridger *et al.,* 1993; Moir *et al.,* 1994; Sasseville and Raymond, 1995; see Fig. 3). Furthermore, they do not appear to be invaginations of the nuclear membrane as suggested by several observations. The lamin foci do not colocalize either with membrane-specific dyes (Bridger *et al.,* 1993) or with a nuclear pore antibody (Moir *et al.,* 1994; but see Belmont *et al.,* 1993). These data suggest that the lamin foci are not part of the nuclear membrane/nuclear envelope.

Although it has not been clearly established, several observations suggest functions for nucleoplasmic lamin foci. The number of foci is greatest in early $G_1$ and in mid-S phase (Goldman *et al.,* 1992; Bridger *et al.,* 1993; Moir *et al.,* 1994). In $G_1$, nucleoplasmic foci of lamins A and C are most

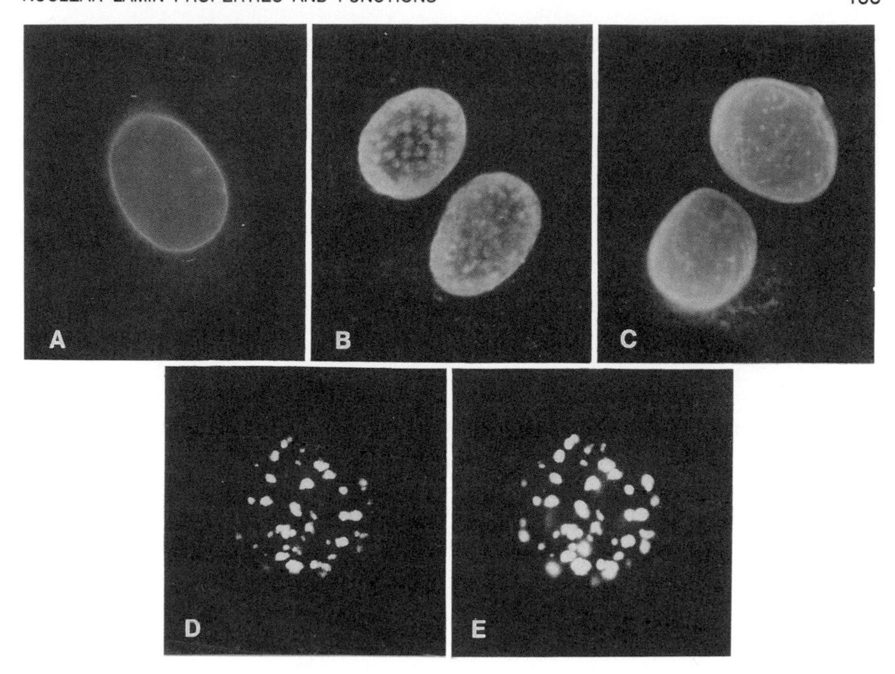


FIG. 3 Examples of nucleoplasmic lamin foci. The top series (A, B, C) are immunofluorescence micrographs of nuclei stained with a polyclonal antibody that reacts with all lamins. A represents the common view of the lamina, showing a distinctive peripheral rim staining. B and C show the nucleoplasmic lamin foci seen in addition to the rim staining (see Goldman *et al.*, 1992). D and E are confocal micrographs of a section of a nucleus stained with antibromodeoxyuridine (D), a DNA replication marker and antilamin B (E). The nucleoplasmic foci detected with both antibodies coalign (see Moir *et al.*, 1994).

obvious and appear to be associated with heterochromatin (Bridger *et al.*, 1993). These foci may be involved in the reorganization of chromatin at this stage of the cell cycle (Goldman *et al.*, 1992; Bridger *et al.*, 1993). Similarly, foci staining with a lamin B antibody are associated with replicating DNA at a specific stage of S phase, and lamin B, therefore, is implicated in chromatin organization during S phase (Moir *et al.*, 1994; see Fig. 3).

Several observations indicate that foci may also represent sites of assembly and post-translational modification of newly synthesized lamins. For example, when prelamin A cDNA (see Fig. 2) is transfected into cells treated with lovastatin (mevinolin), an inhibitor of isoprenylation, all of the expressed protein accumulates in foci similar to those seen with lamin antibodies (Lutz *et al.*, 1992). These structures disappear when the lovastatin is removed, suggesting that isoprenylation occurs within foci. This result also implies that isoprenylation and associated reactions are required for incorporation into the lamina. Foci are also observed when cells are treated with lovastatin and then are stained with an antibody specific for prelamin A (Sasseville and Raymond, 1995), although in this case, the lovastatin does not completely block lamin A incorporation into the peripheral lamina. Taken together, these results suggest that newly synthesized lamins accumulate in foci prior to assembly into the lamina and these lamins are associated with chromatin. The subsequent assembly of lamins into the peripheral lamina may drive reorganization of chromatin (Goldman *et al.,* 1992; Bridger *et al.,* 1993; Moir *et al.,* 1994).

### 2. The Role of Phosphorylation in Nuclear Lamin Dynamics

***a. Phosphorylation in Mitotic Cells*** The nuclear lamina is rapidly disassembled during mitosis and then reassembled in daughter cells. The disassembly process is correlated with an increase in the phosphorylation state of the lamins while reassembly is correlated with dephosphorylation (Gerace and Blobel, 1980; Ottaviano and Gerace, 1985). The increase in phosphorylation at mitosis is approximately five times that at interphase and apparently results in the addition of two phosphates per lamin molecule (Ottaviano and Gerace, 1985; Dessev and Goldman, 1988; Dessev *et al.,* 1991). As a consequence of phosphorylation and disassembly, the lamins become "soluble" and are more easily extracted than at any time during interphase (Gerace and Blobel, 1980). Furthermore, the majority of disassembled lamin B apparently remains associated with nuclear membrane-derived vesicles in mitotic cells, while lamins A/C do not (Gerace and Blobel, 1980; Chaudhary and Courvalin, 1993).

An essential kinase activity that promotes the transition from interphase ($G_2$) to mitosis is the p34cdc2 kinase complexed to cyclin B (Murray and Hunt, 1993). This kinase appears to be the major kinase responsible for lamina disassembly. For example, when a purified $p34^{cdc2}$/cyclin complex from HeLa cells is added to detergent-extracted nuclei of the surf clam (*Spisula solidissima*) oocyte, the lamina is disassembled and the nuclei break down (Dessev *et al.,* 1991). Similarly, if p34cdc2 purified from starfish embryos is added to nuclei of cultured chick embryo cells, the lamina is depolymerized, but the nucleus does not disassemble (Peter *et al.,* 1990). This implies that p34cdc2 is sufficient for lamin depolymerization, but other factors are required for the breakdown of the entire nuclear envelope. The same conclusion is reached when rat liver nuclei are incubated with a *Xenopus laevis* oocyte extract containing active p34cdc2 (Newport and Spann, 1987). Therefore, disassembly of the nuclear lamina appears to be separable from other mitotic events in these experimental systems.

Analysis of lamin sequences reveals the SPTR motif, thought to be the primary consensus recognition sequence for p34cdc2. Two such motifs immediately flank the rod domain; one in the N-terminal nonhelical domain and the second in the at C-terminal domain immediately after the end of the rod domain. In fact, phosphorylation sites for protein kinase C (PKC) also occur at serine or threonine residues immediately flanking the rod, suggesting this region is critical for modulating the dynamic properties of the lamins.

The evidence for direct phosphorylation of lamins by p34cdc2 comes from comparisons of lamins labeled with $^{32}P$ by p34cdc2 *in vitro,* with lamins isolated from mitotic cells grown in medium with radioactive phosphate. In several cases, the tryptic peptide maps of lamins labeled *in vitro* are very similar to those of lamins isolated from mitotic cells (Peter *et al.,* 1990; Dessev *et al.,* 1991; Eggert *et al.,* 1991, 1993; Luscher *et al.,* 1991). The amino acids labeled *in vitro* have been identified by microsequencing of peptides or by using lamins with mutated serine/threonine residues. The results of these experiments using several different lamins as substrates indicate that there are two major phosphorylation sites, one serine located in the N-terminus and one serine in the C-terminus (Peter *et al.,* 1990; Eggert *et al.,* 1991, 1993). Several serine/threonine residues close to the latter phosphorylation site are also labeled, but to a much lesser extent. The C-terminal sites are also labeled *in vivo* as shown by microsequencing, but because the N-terminus of lamins is blocked to Edman degradation, this site has not yet been sequenced (Eggert *et al.,* 1993). In one case, a synthetic peptide containing only the N-terminal serine was efficiently labeled *in vitro* by p34cdc2, and this peptide comigrated with a tryptic peptide seen in maps of chicken lamin B2, strongly suggesting this site is labeled *in vivo* (Peter *et al.,* 1990). It is not clear at the moment if both sites are labeled exclusively in mitotic cells. Ward and Kirschner (1990) detected phosphorylation at both N- and C-terminal sites when human lamin C was incubated in *Xenopus* mitotic extracts. However, the N-terminal site was also phosphorylated in interphase extracts in which p34cdc2 was inactivated. Other results have also suggested that p34cdc2 sites are labeled in interphase cells (Eggert *et al.,* 1993). Since the p34cdc2/cyclin B complex is exclusively mitotic, this may indicate that other p34cdc2/cyclin complexes act on lamins at other phases of the cell cycle. It is also possible that there are other kinases with specificities similar to p34cdc2. For example, it has been shown that mitogen-activated kinases (MAP kinases) phosphorylate lamin B2 at the same sites *in vitro* as p34cdc2 (Peter *et al.,* 1992).

Assays conducted both *in vivo* and *in vitro* suggest that both of the two major p34cdc2 sites are involved in lamin assembly/disassembly reactions. Chicken lamin B2 does not assemble into head-to-tail lamin dimers *in vitro* following phosphorylation of the N-terminal p34cdc2 site alone (Peter *et al.,* 1990; in these experiments the C-terminal site was mutated to prevent its phosphorylation). This head-to-tail interaction is unperturbed in experiments in which phosphorylation occurs at the C-terminal site. Phosphorylation at this site may modulate later steps of assembly. It should be emphasized that it remains to be demonstrated that the head-to-tail assemblies of lamin dimers reflect interactions seen *in vivo.* There is good *in vivo* evidence for a role of these p34cdc2 sites in lamin assembly. Human lamin A cDNAs with mutations at p34cdc2 sites have been transfected into CHO (Heald and McKeon, 1990) or chicken cells (Haas and Jost, 1993). The expressed proteins produce dominant negative phenotypes in mitosis. The protein is correctly assembled into the endogenous lamina during interphase, but apparently cannot be disassembled in mitosis, resulting in aberrant mitotic phenotypes. Mutation of the N-terminal serine residue has a more dramatic effect than the C-terminal serine, but cells tranfected with the double mutant construct exihibit the highest percentage of aberrant mitotic phenotypes. Interestingly, a large number of cells expressing the mutant protein (30%) escape the mitotic block (Heald and McKeon, 1990).

It should be pointed out that Type III cytoplasmic IFs are also phosphorylated by p34cdc2 at mitosis. For example, vimentin is phosphorylated by p34cdc2 at an N-terminal residue that induces disassembly *in vitro* (Chou *et al.,* 1990) and *in vivo* (Y. H. Chou, P. Opal, and R. Goldman, personal communication). There is a second mitotic kinase that phosphorylates vimentin at a C-terminal site (Chou *et al.,* 1989).

There are also other mitotic lamin kinases besides those in the p34cdc2 family. For example, the phosphopeptide maps of lamins from mitotic cells show $^{32}$P-labeled peptides in addition to those labeled *in vitro* by p34cdc2 (Luscher *et al.,* 1991; Goss *et al.,* 1994). These kinases may act in concert with p34cdc2, or in place of it in some cells. The beta II isoform of PKC may be one candidate for another mitotic lamin kinase (Fields *et al.,* 1988; Goss *et al.,* 1994). In a human leukemic cell line, PKC appears to phosphorylate human lamin B1 at one site in mitosis. This serine residue lies in the C-terminus, again very close to the end of the rod domain. Surprisingly, there does not appear to be a p34cdc2-specific mitotic phosphorylation site in these cells. These cells proliferate in response to stimuli such as phorbol esters and this may account for a different pattern of lamin phosphorylation. In addition, beta II PKC is transported to the nucleus immediately prior to mitosis in these cells where it appears to become associated with the nuclear envelope (Murray *et al.,* 1994). Other isoforms of PKC are less efficient with respect to phosphorylating lamin B1, suggesting that there may be a specific activator of beta II PKC in the nuclear envelope.

***b. Phosphorylation in Interphase Cells*** When lamins metabolically labeled with $^{32}$P are isolated from interphase cells and subjected to phosphopeptide analysis, there are a number of labeled peptides (Dessev *et al.*, 1988; Eggert *et al.*, 1991, 1993; Luscher *et al.*, 1991). This indicates an appreciable level of phosphorylation during interphase. Phosphorylation in interphase may modulate lamin dynamics in a number of ways. It may facilitate the incorporation of new lamin subunits during interphase of the cell cycle. It may also be required for the remodeling of the lamina required for the changes in chromatin structure that appear to take place during the cell cycle.

Both protein kinases C and A phosphorylate lamins *in vitro* and probably *in vivo* (Peter *et al.*, 1990; Eggert *et al.*, 1991, 1993; Hennekes *et al.*, 1993). Hennekes *et al.* (1993) mapped several serine residues in chicken lamin B2 that were phosphorylated by PKC both *in vivo* and *in vitro*. These serines, phosphorylated when cells are treated with phorbol esters, are closely spaced near the end of the rod domain adjacent to the p34cdc2 site. Protein phosphorylated *in vitro* by PKC still assembles into head-to-tail dimers. However, when the protein is used in nuclear transport assays in permeabilized cells, the phosphorylated protein is transported less efficiently into the nucleus relative to the unphosphorylated protein (Hennekes *et al.*, 1993). In addition, if a human lamin A cDNA is mutated at one of the PKC sites and transfected into chicken cells, then the expressed protein is not transported into the nucleus (Haas and Jost, 1993). Both of these results argue for a role of PKC in the nuclear transport of lamins, but they differ significantly from each other. In one case phosphorylation inhibits transport (Hennekes *et al.*, 1993) and in the other, the loss of a phosphorylation site prevents transport (Haas and Jost, 1993). PKC also labels lamin B during mitosis, at least in some cells (Goss *et al.*, 1994; see above). Thus, this kinase appears to have a particularly complicated role in lamin dynamics that remains to be more precisely delineated.

Protein kinase A (PKA; cAMP-dependent kinase) also has been implicated in the regulation of lamin assembly in interphase. PKA can phosphorylate lamins *in vitro* (Peter *et al.*, 1990; Eggert *et al.*, 1991), and it has been implicated in regulating dynamics *in vivo* (Peter *et al.*, 1990: Lamb *et al.*, 1991; Molloy and Little, 1992). For example, if the mitotic kinase p34cdc2 is injected into cells, the lamina is not disassembled (Lamb *et al.*, 1991). However, if an inhibitor of PKA is included in the injection solution, then disassembly occurs. Furthermore, microinjection of the inhibitor alone results in lamin dephosphorylation. This argues that both inhibition of PKA and the subsequent lamin dephosphorylation by unknown phosphatases are required for the action of p34cdc2. Similar conclusions are reached when nuclear envelope ghosts prepared from rat liver are incubated with a *Xenopus laevis* extract containing active p34cdc2 (Molloy and Little, 1992). If an inhibitor of PKA is included in the extract, then the release of lamins from the ghosts by p34cdc2 phosphorylation is maximal. PKA phosphorylation may "protect" the lamins from p34cdc2 phosphorylation by altering the structure or accessibility of the p34cdc2 site.

In summary, there is evidence for the action of several well-known kinases on the various lamins both during interphase and mitosis. However, their precise roles in regulating lamin dynamics is not clear. They may alter the interactions between the lamins in the lamina polymer or modulate some other aspect of lamin physiology such as transport into the nucleus or interactions with other proteins. In addition, there are almost certainly other lamin kinases to be identified. For example, Dessev *et al.* (1988) identified a 52-kDa kinase closely associated with the lamina fraction in EAT cells. This kinase phosphorylated lamins A, B, and C *in vitro*. There is also evidence for S6 kinase phosphorylation of lamin C during mitosis (Ward and Kirschner, 1990).

### 3. Role of Isoprenylation in Lamin Dynamics

Nuclear lamins A and B belong to a group of proteins that are isoprenylated and methylated at a C-terminal cysteine residue. These modifications take place when the C-terminus ends in the motif CaaX, where C is cysteine, a is an aliphatic residue, and X is any residue (see Fig. 2). Lamin C as well as cytoplasmic IF proteins do not end in this motif and are not isoprenylated. Isoprenes are carbon chains of alternating double and single bonds. If the X residue in CaaX is serine, alanine, or methionine, then a farnesyl moiety (a C15 chain) is linked to the cysteine. If the X is leucine, then the cysteine is geranylgeranylated (a C20 chain). All isoprenylated lamins are farnesylated. Other farnesylated proteins are p21 *ras*, a yeast pheromone, transducin, and rhodopsin (Sinensky and Lutz, 1992). The sequence of modifications has been determined for *ras* and by analogy, the same events occur for the lamins (see Fig. 2 for summary of farnesylation and associated reactions on lamins). In the first step, the isoprenyl group is linked to the cysteine through a thioether linkage. The C-terminal residues (aaX) are then removed by proteolysis and the cysteine is carboxymethylated (Marshall, 1993).

A number of papers have indicated that isoprenylation is required for targeting of lamins A and B to the nuclear envelope. Transfection with lamin cDNAs coding for mutated C-terminal cysteines results in the accumulation of the expressed protein in the nucleoplasm rather than becoming incorporated into the lamina (Holtz *et al.,* 1989; Kitten and Nigg, 1991; Lutz *et al.,* 1992). Lamin C lacks the modified cysteine and might be expected to behave as the mutated protein, and this appears to be the case (Horton *et al.,* 1992). When lamin C is transfected into cells, the expressed protein remains in the nucleoplasm while expressed lamin A is incorporated rapidly into the lamina (Horton *et al.,* 1992). Apparently the expressed lamin C can only be incorporated into the lamina when it is reassembles after mitosis. However, it remains to be demonstrated that this is the case *in vivo* since lamin A and C are made throughout the cell cycle in equal amounts (Gerace *et al.*, 1984). Lamin C appears to be more easily extracted from interphase cells, which may indicate a different pattern of incorporation for this lamin (Dagenais *et al.*, 1990).

Similar results on the role of CaaX targeting are obtained when mutated *Xenopus laevis* lamins B1 or A proteins are injected into *Xenopus* oocytes (Krohne *et al.*, 1989). Wild-type protein apparently associates with the nuclear envelope, but protein with a mutated cysteine accumulates in the nucleoplasm and is extractable with detergents or high salt. Furthermore, if wild-type human lamin C is injected, it also accumulates in the nucleoplasm, but if a CaaX sequence is added, the protein is targeted to the envelope.

The prenyl moiety may insert into the lipid bilayer and "anchor" the lamins to the membrane. However, there is also evidence that the prenyl groups may mediate protein–protein interactions (Marshall, 1993). For example, farnesylated p21 *ras* interacts with a GDP releasing factor in the absence of membranes while the nonfarnesylated form does not. Thus, the mutation of the C-terminal cysteine in lamins may stop membrane association by preventing interaction with integral proteins (LAPs or lamin receptors) in the membrane (Firmbach-Kraft and Stick, 1993; Hennekes and Nigg, 1994). In some cases the expressed protein carrying the mutated cysteine appears to form nucleoplasmic foci (Holtz *et al.*, 1989; Lutz *et al.*, 1992). Nucleoplasmic foci are also observed if cells are transfected with wild-type protein and treated with lovastatin, an inhibitor of the isoprenyl synthesis pathway (Lutz *et al.*, 1992). When the cells are released from the block, the expressed protein becomes incorporated. These foci could represent sites where the lamin are modified before becoming assembled into the lamina (Goldman *et al.*, 1992). This also argues that the farnesylation and associated reactions occur in the nucleus (Lutz *et al.*, 1992).

Lamin A only undergoes additional processing after isoprenylation and methylation. The C-terminal 15 amino acids, including the farnesylated and methylated cysteine, are removed by proteolytic cleavage (Weber *et al.*, 1989b; see Fig. 2). The final processed form of lamin A is termed mature lamin A. Pulse/chase experiments suggest that this occurs rapidly after newly synthesized lamin A becomes resistant to detergent extraction and therefore presumably is assembled into the lamina (Gerace *et al.*, 1984). This final processing step is dependent on the previous steps: if farnesylation is inhibited by lovastatin then the endoproteolytic cleavage does not occur (Beck *et al.*, 1990). Furthermore, in transiently transfected cells, the CaaX cysteine mutants do not undergo this proteolytic cleavage (Hennekes and Nigg, 1994). If the mature form of lamin A is expressed by transfection, it does not assemble into the lamina at early timepoints post-transfection but is incorporated at later time points (Lutz *et al.,* 1992; Hennekes and Nigg, 1994). This suggests that it is behaving like lamin C in that the lack of a farnesylation signal appears to prevent immediate assembly into the lamina, and transit through the cell cycle is required for incorporation (Hennekes and Nigg, 1994).

The function of this endoproteolytic processing of lamin A is not clear. It has suggested that this modification may be involved in the different distributions of the lamina at mitosis (Hennekes and Nigg, 1994). In this scheme, farnesylation is required for targeting of lamin A and B to the nuclear membrane where the farnesyl group could either interact with the membrane itself or with an integral membrane protein (a lamin receptor; see Section IV,G). The subsequent proteolysis of lamin A would release the link to the membrane. Consequently, lamin A (and presumably C) appears to be "soluble" at the subsequent mitosis, unattached to another structure, while lamin B remains attached to nuclear membrane vesicles (Gerace and Blobel, 1980). The interaction of lamins A and C with chromatin may be an initial, critical step in nuclear envelope assembly (see Section IV,E) and the solubility of lamins A and C may facilitate the speed and sequence of assembly events (Hennekes and Nigg, 1994; see below).

## IV. Functions of Nuclear Lamins

As we have described above, the nuclear lamins have been studied in considerable depth at the biochemical and molecular levels, and yet their functions remain largely undetermined. In spite of this, numerous functions have been proposed for nuclear lamins, most of which are based either on ultrastructural observations or on temporal alterations in lamin composition that coincide with changes in nuclear function. Data from ultrastructural studies have been used to suggest that the nuclear lamins provide the infrastructure that mechanically supports the nuclear membrane and forms anchorage sites for chromatin. Furthermore, alterations in nuclear lamin composition have been proposed to be involved in the regulation of DNA replication and the types of chromatin reorganization that may modulate transcription. However, until quite recently there has been very little direct evidence to support lamin involvement in any of these proposed functions. Current studies using *in vitro* systems and studies of lamin dynamics *in vivo* have provided some data to support these ideas; in particular the possible involvement of lamins in DNA replication and in the reformation of the nuclear envelope at the completion of mitosis. In this section, evidence supporting the possible functions of nuclear lamins is presented.

### A. Changes in Lamin Expression Are Correlated with Changes in Nuclear Function

Studies of the differential expression of lamins by cells and tissues as they develop and differentiate have helped to shape theories regarding their function. As described above (Section II,B), all cell systems characterized to date express B-type lamins, but in the majority of cases A-type lamins are expressed only as cells differentiate (Stewart and Burke, 1987; Rober *et al.,* 1989). This developmental regulation was first observed in *Xenopus* embryos, where the changes in nuclear lamin composition coincide with significant changes in nuclear function (Krohne *et al.,* 1984; Benavente *et al.,* 1985; Stick and Hausen, 1985). In the oocyte and during early development up to the midblastula transition (MBT, 2000- to 4000-cell embryo), the major lamin is B3 (originally called lamin III). Interestingly, the majority of this lamin is not associated with nuclear envelope-derived vesicles at mitosis as is the case with most B-type lamins (Benavente *et al.,* 1985; Stick and Hausen, 1985). The cells comprising the preMBT embryo divide synchronously every 20–30 min, requiring that each cell replicate its entire genome in 10–15 min. During this period little transcription is taking place (Graham and Morgan, 1966; Satoh *et al.,* 1976; Gerhart, 1980).

At the MBT, another B-type lamin, lamin B1 (also called L I), is induced presumably from the initiation of translation of a previously masked maternal RNA (Benavente *et al.,* 1985; Stick and Hausen, 1985). Coincidental with the expression of lamin B1, the length of the cell cycle increases, the synchronous cleavages of the embryonic cells cease, and nuclei become transcriptionally active (Graham and Morgan, 1966; Satoh *et al.,* 1976; Gerhart, 1980; Newport and Kirschner, 1982a,b). Lamin B2 (also called L II) is induced during gastrulation (Benavente *et al.,* 1985; Stick and Hausen, 1985). The expression of this latter lamin takes place at another major developmental milestone, the organization of the embryo into three germ layers. Similarly, in the mouse, cells that do not express lamins A and C include nondifferentiated cells, dedifferentiated cancer cells, intestinal epithelial cells, and some cell types of the immune system. The precise expression patterns of the different lamins vary slightly, depending upon the system and the laboratory carrying out the experiments (see Section II,B). However, it is clear that lamin expression does vary throughout development The expression pattern seems to follow the general rule in which a B-type lamin is expressed first in all cell types, and subsequently lamins A and C are induced as cells differentiate (Stewart and Burke, 1987; Rober *et al.,* 1989, 1990a,b). The induction of A-type lamins or other specialized lamins seems to occur after the initiation of differentiation. This has led to the suggestion that changes in nuclear lamin composition leads to locking-in of the differentiated state, possibly by playing a role in the reorganization of chromatin (Rober *et al.,* 1989; Peter and Nigg, 1991). While the temporal correlation between an alteration in lamin composition and nuclear function is very provocative, experimental evidence has yet to clearly link these phenomena.

Attempts have been made to determine if the premature expression of A-type lamins would lead to a change in the pattern of gene expression (Peter and Nigg, 1991; Lourim and Lin, 1992). For example, P19 cells are an undifferentiated mouse cell line in which A-type lamins are not expressed until exposure to retinoic acid induces differentiation. In stably transfected cell lines, ectopically expressed lamin A was correctly targeted to the nuclear rim, but did not induce differentiation-specific genes (Peter and Nigg, 1991). These results were interpreted to mean that expression of lamin A alone cannot cause the differentiation of P19 cells. In contrast, high levels of expression of lamin A in a chicken muscle cell line has been linked to a transient increase in the expression of two, but not all of the muscle-specific genes normally induced during development (Lourim and Lin, 1992). In conclusion, both sets of experiments demonstrate that simply expressing lamin A is insufficient to induce a developmental program.

### B. Lamins as a Structural Support System for the Nuclear Membrane

Since the earliest ultrastructural studies described the nuclear lamina as an electron-dense structure underlying the nuclear membrane and adjacent to chromatin, it has been suggested that the nuclear lamins act as an interface linking the proximal face of the inner nuclear membrane to chromatin (Gray and Guillery, 1963; Coggeshall and Fawcett, 1964; see Section II,A). The localization of the lamins along with their resistance to harsh extraction conditions suggests that the lamins help to maintain the structural integrity of the nucleus through the formation of a relatively stable polymer adjacent to and interacting with the nuclear membrane (Aaronson and Blobel, 1975; Dwyer and Blobel, 1976; Gerace *et al.,* 1978; Gerace and Blobel, 1980). The fact that lamins are members of the IF family of proteins supports the idea that they form a polymer that provides structural integrity to the nuclear envelope (Zackroff *et al.,* 1984; Aebi *et al.,* 1986; Fisher *et al.,* 1986; Goldman *et al.,* 1986). Data from recent experiments carried out in *Xenopus* egg extracts lend further support to this model. The *Xenopus* egg is quite large and each egg has been estimated to contain the components required to make 2000–4000 nuclei (Newport and Kirschner, 1982b; Forbes *et al.,* 1983). When DNA (demembranated sperm nuclei or protein-free $\lambda$ DNA) is added to these extracts, nuclei form that contain a lamina and replicate DNA in a semiconservative manner (Lohka and Masui, 1983; Blow and

Laskey, 1986; Newport, 1987). When the major lamin of these extracts, lamin B3, is immunodepleted from these extracts, nuclei still appear to form (Newport *et al.,* 1990; Meier *et al.,* 1991). However, these nuclei are very fragile and are crushed during routine manipulations that have no effect on nuclei formed in intact extracts. These results indicate that lamin B3 and possibly molecules immunoprecipitating with lamin B3 form a structure that strengthens the nuclear envelope.

In a related report, an alteration of the lamin compositon of the nucleus has been found to cause obvious changes in nuclear morphology. When mammalian lamin B3, a spermatocyte-specific lamin, was expressed in somatic cells, the nucleus seemed to become hook shaped (Furukawa and Hotta, 1993). Presumably the entire nuclear envelope is rearranged to achieve this change in shape. It will be interesting to determine if this change in shape is caused by perturbing the endogenous lamina or through direct interactions with the nuclear envelope.

## C. Lamin Interactions with DNA and/or Chromatin

It has long been assumed that lamins bind to chromatin both during interphase and in different stages of mitosis. During interphase, these interactions are thought to influence both the organization and regulation of chromatin. At the end of mitosis, the early stages of nuclear envelope assembly may involve the binding of nuclear lamins to chromatin (see Section IV,E). Early observations of thin sections of nuclei show heterochromatin immediately abutting the inner nuclear membrane, implying direct contact with the lamina (Coggeshall and Fawcett, 1964). More recently, three-dimensional reconstructions of nuclei stained with lamin antibody and DNA dyes indicate that approximately 10–15% of interphase chromatin is close enough to be in direct contact with the lamina at the nuclear periphery (Paddy *et al.,* 1990). Other results using EM tomography suggest that the lamina varies in thickness. The thickest regions are most closely associated with chromatin (Belmont *et al.,* 1993). However, the structural basis for variations in lamina thickness is not clear. Most recently, it has been shown that lamin B can be present within the nucleus as nucleoplasmic foci which are associated with DNA (Moir *et al.*, 1994).

There are also several studies that report interactions between purified lamins and DNA/chromatin. Lamins can bind telemeric DNA sequences in filter-binding assays (Shoeman and Traub, 1990) and matrix-attached DNA regions (MARs) in gel overlay assays (Ludérus *et al.,* 1992, 1994). Furthermore, lamin-rich fractions extracted from nuclear matrices also bind MAR sequences in filter-binding assays (Ludérus *et al.,* 1992, 1994). These reports are interesting and may demonstrate a direct interaction between lamins and functionally important regions of DNA. However, the stoichiometric ratio of lamins to MAR DNA was high (approximately 500 : 1 calculated from Ludérus *et al.,* 1994) in these studies, and therefore the DNA-binding activity in these preparations could be due to other minor proteins present in the lamin fractions. Considering the numerous DNA-binding affinities that have been attributed to nuclear matrices, it is important that these intriguing results are confirmed with bacterially expressed or *in vitro* synthesized lamins.

Bacterially expressed and *in vitro* synthesized lamins have been reported to bind assembled chromatin *in vitro* (Hoger *et al.,* 1991b; Yuan *et al.,* 1991; Glass *et al.,* 1993). Hoger *et al.* (1991b) have been able to show the specificity of the binding of certain lamins and have been able to define a chromatin-binding region. Using plasmid DNA, they assembled minichromosomes and assessed the binding capacity of various lamins to this chromatin. *Xenopus* lamins A and B2 bound to these chromosomes, while *Xenopus* lamin B1 and human lamin C did not. Using this same assay with lamin fragments produced by site-directed mutagenesis, a DNA-binding domain in the non-$\alpha$-helical C-terminus was identified. Moreover, a specific peptide from the extreme C-terminus of lamin A competed for both the lamin B2- and lamin A-binding sites (Hoger *et al.,* 1991b). In contrast, Glass *et al.* (1993) found a second chromatin-binding site in the rod region of lamins A/C. This site was mapped using mitotic chromosomes as the binding substrate. Therefore, there may be two distinct chromatin-binding domains in A-type lamins, one in the non-$\alpha$-helical C-terminus and one in the $\alpha$-helical rod region. The component(s) of chromatin to which the A-type lamins bind has/have not been identified.

## D. Lamin B and DNA Replication

Immunodepletion and inhibition experiments in *Xenopus laevis* interphase extracts have provided the first direct evidence that lamins are involved in DNA replication (Newport *et al.,* 1990; Meier *et al.,* 1991; Jenkins *et al.,* 1993). Monoclonal antibodies directed against lamin B3 were used to immunodeplete these extracts. When demembranated sperm nuclei were added to immunodepleted extracts, the nuclei that formed were incompetent to replicate DNA, and a lamina was not detected using immunofluorescence assays. However, transport of molecules into these nuclei seemed normal, as these nuclei contained nuclear pores, grew larger with time, and were competent to import fluorescently tagged molecules linked to nuclear localization signal peptides (Newmeyer *et al.,* 1986; Newport *et al.,* 1990; Meier *et al.,* 1991; Jenkins *et al.,* 1993). The *in vitro* nuclei formed in the depleted extract retained the capacity to import PCNA, a DNA polymerase cofactor, although the imported protein was not resistant to detergent and salt extraction (Meier *et al.,* 1991; Jenkins *et al.,* 1993). In nuclei formed in extracts containing lamin B3, PCNA was resistant to extraction. This result supports a model in which lamin B3 is required to establish DNA replication foci or centers. However, the possibility remains that the immunodepletion also removes other nuclear components that are complexed with lamin B3 and are required for DNA replication.

Evidence from another system also supports the possible involvement of lamin B in DNA replication. In mid-S phase in several types of mammalian cells in culture, lamin B colocalizes with PCNA and sites of BrDU incorporation as evidenced by indirect immunofluorescence (Moir *et al.,* 1994). These sites are found within the nucleoplasm and are not restricted to the nuclear surface. Originally nucleoplasmic foci were detected following the microinjection of purified lamins into cells (Goldman *et al.,* 1992). In these studies, the injected lamins were initially located both in nucleoplasmic foci and subsequently at the nuclear periphery (Goldman *et al.,* 1992). Similar structures have been found in uninjected cells, and these foci have been correlated with the cell cycle. These foci do not appear to be invaginations of the nuclear envelope, since they do not colocalize with gp 210, a nuclear pore antigen (Moir *et al.,* 1994). In further support of the possible involvement of lamin B inDNA synthesis is the report that lamin B binds to MARs that have been proposed to serve as origins of replication (Ludérus *et al.,* 1992, 1994; also see Section IV,C). Although the association between lamin B and DNA replication is intriguing, much work is now required to determine the molecular mechanisms underlying this interaction. In this regard lamin B could bind directly to replication factors and/or origins of DNA replication and in so doing facilitate replication. It is also possible that lamin B could form a scaffold upon which replication factors bind either directly or through lamin-associated proteins to form a replication center.

## E. A Role for Lamins in the Formation of the Nuclear Envelope.

Evidence concerning the role of nuclear lamins in the formation of the nuclear envelope in daughter cells following the completion of mitosis is controversial (Lourim and Krohne, 1994). Initially, experiments using a mammalian cell-free system suggested that the addition of antibodies directed against lamins A/C or B slowed the process of membrane reformation, with the antibody directed against lamin A/C having a greater effect (Burke and Gerace, 1986). From these and other results that have shown that lamins A and C may be binding to chromosomes at the completion of mitosis (Glass and Gerace, 1990; Hoger *et al.,* 1991b), a model has emerged in which lamins A and C first bind to chromatin and subsequently lamin B associated with membrane vesicles binds to the lamin A/C-coated chromosomes (Burke and Gerace, 1986). In this model, it is thought that these vesicles then fuse to initiate the formation of the nuclear envelope. A key element in this scheme is the proposed interaction between the A- and B-type lamins at the surface of chromosomes.

This model does not explain how the nuclear envelope forms in cells that reportedly contain only B-type lamins. It is also inconsistent with the results obtained in *Xenopus* nuclear assembly extracts from which the vast majority of lamins were immunodepleted. In these extracts, nuclear envelopes still assembled around chromatin (Newport *et al.,* 1990; Meier *et al.,* 1991; Jenkins *et al.,* 1993). In one set of experiments (Newport *et al.,* 1990), no lamin B3 could be detected, and in the second only trace amounts could be detected by Western blot analyses (Jenkins *et al.,* 1993). At the time that these results were obtained, it was believed that lamin B3 was the only lamin present in the extracts, and therefore it was argued that the binding of nuclear membrane precursor vesicles to chromatin and subsequent nuclear reassembly occurred via a lamin-independent pathway. To address concerns that an additional lamin, B4, might be present in the chromatin added to the extract (the chromatin was isolated from sperm nuclei), the ability of the lamin-depleted extract to form nuclei around protein-free λ DNA was assayed (Newport *et al.,* 1990). Nuclei still formed under these conditions. However, these results are in contrast to those obtained by Dabauvalle *et al.* (1991), who have found that adding an antibody directed against both lamins B3 and B2 to *Xenopus* interphase extracts blocked nuclear membrane assembly around sperm chromatin. Likewise, addition of an antibody directed against the *Drosophila* lamin blocked nuclear formation in *Drosophila* interphase extracts (Ulitzur *et al.,* 1992).

These contradictory results may soon be resolved. A recent study shows that *Xenopus* egg extracts contain lamin B2, in addition to lamin B3 (Lourim and Krohne, 1993). Lamin B3 accounts for approximately 95% of the lamins found in the extract, and interestingly a portion (5–7%) of lamin B3 has been found to be associated with a membrane fraction even after extensive washes. This fraction of lamin B3 would not have been depleted by the Newport *et al.* (1990) protocol which removes lamin B3 only from the membrane-free cytosolic fraction. Furthermore, the monoclonal antibody used by Newport *et al.* (1990) and Meier *et al.* (1991) binds only to lamin B3, while the monoclonal antibody used by Dabauvalle *et al.* (1991) binds to both lamin B2 and B3 (see above). These results indicate that if the vast majority of lamin protein B3 is removed from the extract, then that extract remains competent to form nuclei but unable to support DNA replication (Newport *et al.,* 1990; Jenkins *et al.,* 1993).

Alternatively, if both lamins B3 and B2 are immunoabsorbed, then a nuclear membrane does not form around chromatin (Dabauvalle *et al.,*

1991). Taken together these results point to the involvement of lamins in the process of nuclear envelope formation and argue against a lamin-independent pathway for this process. However, a limitation of these experiments is that the lamins may be in a complex with other proteins so that immunodepletion of lamins may cause the loss of other proteins required for nuclear assembly. In order to determine more precisely the role of specific lamins in this process, purified lamins must be added back to lamin-depleted extracts. If such an addition rescues membrane formation in the immunodepletion extract then the lamins will have been shown to have a central role in assembly of the nuclear membrane.

### F. Lamin Interactions with Cytoskeletal IF Networks: Possible Roles in Signal Transduction or Targeting

Electron microscopic observations of thin sections of cells and extracted whole cells suggest that IFs interact with both nuclear pores and the nuclear envelope (Fey *et al.,* 1984; Jones *et al.,* 1985). Similar studies of IF-enriched cytoskeletal preparations of cells have revealed a detergent and high salt-resistant association between the cytoskeletal IF network and the karyoskeletal lamin network of cells (R. Goldman *et al.,* 1985). The interaction between these two networks could act to mechanically connect the nucleus and the cell membrane (Bissell *et al.,* 1982; R. Goldman *et al.,* 1985). *In vitro* binding assays have supported these observations to some extent and have led some researchers to propose that lamins interact directly with cytoplasmic IF networks, possibly by the IF passing through the nuclear pore and binding directly to the nuclear lamins (Georgatos and Blobel, 1987). If this indeed is the case, it would provide a mechanism by which nuclear lamins could be linked to the cell surface through the cytoskeletal IF system. The latter are known to associate with the plasma membrane and membrane-associated structures such as desmosomes and hemidesmosomes. Linkages between the nuclear and the cytoplamic IF networks could therefore provide structural pathways between the nucleus and the cell surface. Further support for cytoskeletal IF–lamin interactions comes from recent findings that during mitosis, vesicles associated with lamin B may bind to vimentin IFs (Maison *et al.,* 1993). Such an interaction could provide a mechanism for targeting these vesicles to specific locations in the cytoplasm.

### G. Lamin-Associated Proteins May Mediate Functions Associated with Nuclear Lamins

A number of proteins that bind to or associate with nuclear lamins have been identified (Senior and Gerace, 1988; Worman *et al.,* 1988; Powell and

Burke, 1990; Bailer *et al.,* 1991; Foisner and Gerace, 1993). Most of these appear to be inner nuclear envelope integral membrane proteins that may be acting as lamin "receptors"; perhaps involved in linking the envelope to the lamina. They may also play a role in establishing interactions with other nuclear components such as chromatin. These proteins, collectively termed lamin-associated proteins, probably play a critical role in many of the processes in which the lamins have been implicated.

The first such protein to be identified was a putative lamin B receptor (LBR), also termed p58 (Worman *et al.,* 1988). This protein was identified and subsequently purified by assaying lamin B binding to nuclear envelope fractions in solution and in solid-phase binding assays (Worman *et al.,* 1988). When p58 was cloned, sequence analysis suggested that there are eight potential transmembrane domains as well as a large, highly charged N-terminal domain (Worman *et al.,* 1990). The N-terminal domain binds both lamin B and DNA *in vitro,* suggesting that it probably extends into the nucleoplasm (Ye and Worman, 1994). This domain also contains a p34/cdc2 phosphorylation site (Courvalin *et al.,* 1992) as well as phosphorylation sites for protein kinase A which appear to modulate the binding affinity of p58 for lamin B (Appelbaum *et al.,* 1990). In addition to lamin B, LBR appears to interact with other nuclear proteins (Simos and Georgatos, 1992, 1994). For example, a kinase activity that phosphorylates the receptor as well as the lamins coimmunoprecitates with p58 (Simos and Georgatos, 1992). One of the other proteins that coprecipitates is related to a splicesome factor, suggesting that p58 and the nuclear envelope may play a role in anchoring these complexes (Simos and Georgatos, 1994).

A number of integral membrane proteins that may be related to LBR have been identified in invertebrate species. A LAP that binds the single *Drosophila* lamin has been cloned. It has only a single membrane domain and a large nucleoplasmic domain, although it apparently cross-reacts with p58 antibodies suggesting some similarity in structure (Padan *et al.,* 1990). Immunological evidence indicates a possible LBR homolog in *S. cerevisiae* (Georgatos *et al.,* 1989). Interestingly, if the human lamin B receptor is expressed in *S. cerevisiae,* then some of the expressed protein localizes appropriately to the inner nuclear membrane, with the remainder localizing to membrane stacks found adjacent to the nucleus or the plasma membrane (Smith and Blobel, 1994). Similarly, although a lamin homolog from *S. pombe* has not yet been identified, an integral membrane protein has been described that shows significant sequence homology to a portion of the LBR (Shimanuki *et al.,* 1992).

A number of reports have described other LAPs, some of which appear to preferentially bind to or interact with only one lamin subtype (Senior and Gerace, 1988; Powell and Burke, 1990; Bailer *et al.,* 1991; Foisner and Gerace, 1993). Four integral membrane proteins, LAPs 1A, 1B, 1C, and

2, were identified with monoclonal antibodies generated against rat liver nuclear envelope fractions (Senior and Gerace, 1988; Foisner and Gerace, 1993). Following differential extractions of nuclear preparations, LAPs 1A, 1B, and 2 were found to be associated with the lamina fraction, and *in vitro* binding studies suggest that LAPs 1A and 1B bind to both lamins A and B, while LAP 2 binds only to lamin B (Senior and Gerace, 1988; Foisner and Gerace, 1993). Interestingly, *in vitro* experiments indicate that LAP 2 also binds to mitotic chromosomes (Foisner and Gerace, 1993). Although LAP 2 and LBR have similar molecular weights and each binds to both lamin B and chromatin, they are thought to be different proteins (Foisner and Gerace, 1993). In contrast to LAPs 1A, 1B, and 2, LAP 1C is readily extracted from nuclear envelopes and it does not bind to lamins in *in vitro* binding assays (Foisner and Gerace, 1993). However, in transfection studies, LAP 1C is localized to the inner nuclear envelope when lamin A is expressed in cells that normally do not express A-type lamins, indicating that the localization of LAP 1C may be lamin dependent (Powell and Burke, 1990). Another LAP-like integral membrane protein, p54, has been isolated from chicken nuclei. Immunofluorescence and immunoblotting studies of detergent-extracted chicken nuclei and nuclear envelope fractions indicate that p54 is associated with the lamina. However, no direct interaction of p54 with lamins has yet been demonstrated (Bailer *et al.,* 1991). Fs(1)Ya, a potential LAP from *Drosophila*, is required to complete mitosis in the early embryo (Lin and Wolfner, 1991; Lopez *et al.,* 1994). In immunofluorescence assays, fs(1)Ya colocalized with lamins at the nuclear envelope (Lin and Wolfner, 1991; Lopez *et al.,* 1994). However, the cloned cDNA has not yet been assigned to a known gene family, and the protein has not been shown to bind to or interact with nuclear lamins.

The fate of LAPs during cell division is unclear. Since lamin B remains largely membrane associated at mitosis and LBR is an integral membrane protein, one might expect the two proteins to remain associated in vesicles throughout mitosis. Indeed, this seems to be the case in chicken cells where immunofluorescence studies suggest that LBR colocalizes with lamins B1 and B2 throughout mitosis (Meier and Georgatos, 1994). However, Chaudhary and Courvalin (1993) followed the fate of the LBR, as well as the lamins in HeLa cells through mitosis and did not find such colocalization. In these cells, the LBR appears to be found in membrane vesicles that are the first to reassemble on decondensing chromatin at anaphase, whereas vesicles containing lamin B do not appear to assemble on chromatin until telophase. Similarly, immunofluorescence studies show that during mitosis LAP2 and lamin B1 do not colocalize, indicating that they are not interacting during mitosis (Foisner and Gerace, 1993). As the nucleus reforms in daughter cells, LAP2 and LAP1, which colocalize throughout the cell cycle, bind to decondensing chromosomes prior to lamin B and lamins A/C

(Foisner and Gerace, 1993). These results have led to the speculation that LAPs 1 and 2 may target membrane vesicles to chromatin during nuclear assembly and may also mediate lamin repolymerization. Although the results from these different systems must be reconciled, it seems clear that at least some LAPs are involved in nuclear envelope reassembly. Ultimately in order to understand this process, it is important to precisely define the nature and temporal sequence of the interactions among the different LAPs, the various lamins, and decondensing chromatin during nuclear reassembly. Similarly, we suggest that there are other yet unidentified LAPs involved in other lamin-related functions.

## V. Concluding Remarks: Summary and Prospects

In this review, we have summarized data on the dynamic properties of the nuclear lamins and have discussed the implications of these properties for lamin function. One obvious example of the dynamic behavior occurs at cell division when the lamina is depolymerized during mitosis and then repolymerized as daughter cell nuclei form. However, even during interphase, the lamina is not an inert structure exclusively at the nuclear periphery. Instead, the lamins are dynamic throughout the cell cycle. Recent evidence for dynamic properties of the nuclear lamins comes from the identification of nucleoplasmic lamin foci. These foci appear to be cell cycle-dependent assemblies of lamins distinct from the peripheral lamina. The dynamics of nuclear lamins are also displayed in the post-translational modification of lamins and in the changing complement of lamin isotypes found in the nucleus as cells differentiate. Studies of these aspects of lamin dynamics have provided valuable information about the functions of lamins in the nucleus.

These potential functions of the nuclear lamins include involvement in DNA replication, nuclear membrane formation, structural support of the nuclear membrane, chromatin organization, cell differentiation, and signal transduction. In this review we have discussed the data upon which each of these proposed functions are based. In this section, we discuss each function and the extent to which it is supported by experimental evidence. In the last part of this section, we briefly discuss avenues of research that will further our understanding of the roles of lamins in the nucleus.

Lamin B seems to be involved in DNA replication. In recent experiments, nucleoplasmic foci of lamin B colocalize with markers for replicating DNA and PCNA, a DNA polymerase cofactor (Moir *et al.*, 1994). These results complement those obtained in *Xenopus* cell-free nuclear reconstitution extracts. In these latter experiments, depletion of extracts of lamin B3 blocked DNA replication in the nuclei that formed *in vitro* (Newport *et al.*, 1990; Meier *et al.*, 1991; Jenkins *et al.*, 1993). Interestingly, the localization of PCNA was perturbed, indicating that depletion of lamins inhibits the establishment of DNA replication centers (Meier *et al.*, 1991). Lamin B may be directly involved in DNA replication by binding to chromatin, or to DNA polymerase and its cofactors. Alternatively, lamin B could be indirectly involved by providing a scaffold upon which the components required for DNA replication assemble.

The role of lamins in the formation of a nuclear envelope around mitotic chromosomes at the end of mitosis is currently unclear, but this is unlikely to remain the case. The contradictory findings with regard to lamin depletion and nuclear membrane assembly in *Xenopus* extracts (Newport *et al.*, 1990; Dabauvalle *et al.*, 1991; Meier *et al.*, 1991; and see Section IV,B) appears to be have been partially resolved with the finding that the extracts contained lamin B2 in addition to lamin B3 (Lourim and Krohne, 1993). Therefore, experiments in which only B3 is depleted (Newport *et al.*, 1990; Meier *et al.*, 1991) can no longer be used to argue for a lamin-independent pathway of nuclear envelope formation. If experiments (which remain to be performed) involving the immunodepletion of both lamins B2 and B3 demonstrate that lamins are required for envelope reformation, then the results obtained from *Xenopus* systems would be in agreement with those obtained in mammalian and *Drosophila* cell-free systems (Burke and Gerace, 1986; Ulitzur *et al.*, 1992). In all of these cell-free systems, a portion of the lamins involved in reassembly is associated with membrane fractions (Burke and Gerace, 1986; Newport, 1987). These membrane fractions also contain a number of LAPs, other proposed chromatin receptors, and certainly other yet unidentified proteins and some of these proteins have also be implicated in nuclear envelope assembly (see Section IV,G; Lourim and Krohne, 1994). The various cell-free systems should eventually allow each of these components to be identified. It will then be possible to deplete or ablate each component and then add them back in order to define the sequence of interactions that are required for envelope reformation.

When lamin B3 is immunodepleted from *Xenopus* nuclear formation extracts, the nuclei that formed are more fragile than those formed in control extracts (Newport *et al.*, 1990). This result supports the often-proposed role of the nuclear lamina as a structural support system for the nuclear envelope. Studies that further examine lamin structure and dynamics should be informative with respect to how the lamina strengthens the envelope and maintains nuclear architecture. The ratio of nuclear surface area to cell volume is maintained as the cell grows during the cell cycle (Swanson *et al.*, 1991), and the lamina is presumably remodeled during this process. In addition, some recent evidence suggests that cytoplasmic IF may also modulate nuclear shape, possibly through their interactions with lamins and/or the nuclear envelope (Sarria *et al.,* 1994). This latter observation may also relate to a proposed involvement of lamins in signal transduction. This was originally suggested on the basis of structural studies of cytoskeletal preparations where the lamina and the nuclear matrix appear to be contiguous with the cytoplasmic intermediate filament network at the level of the nuclear pore complex (Fey *et al.,* 1984; Jones *et al.,* 1985; Goldman *et al.,* 1986). *In vitro* binding assays also indicate nuclear and cytoplasmic IFs interact (Georgatos and Blobel, 1987). It is tempting to speculate that cell signals could be transmitted mechanically through the cytoplasmic IF/nuclear matrix network to the genome, but experiments designed to test this are difficult and have yet to be carried out. Interestingly, recent studies of lamin B–IF interactions during mitosis have led to the proposal that lamin B may also be targeting membrane vesicles to specific subcellular locations within the mitotic cell (Maison *et al.,* 1993; Georgatos *et al.,* 1994).

Although lamins are a major component of the nuclear matrix, there is little evidence on how they interact with other matrix components. Undoubtedly, other protein–protein interactions with the lamina remain to be identified. For example, the retinoblastoma (Rb) gene product appears to bind lamins A/C *in vivo* and *in vitro.* Rb has also been shown to be matrix-associated (Mancini *et al.,* 1994). These types of interactions may help define the long-assumed structural and functional interdependence between the lamina and the rest of the nuclear matrix.

The suggestion that lamins are involved in cell differentiation comes from the observations of changes in the pattern of lamin isotype expression that are coincident with cell differentiation (see Section IV,A). Experimental results involving the premature expression of A-type lamins in undifferentiated cells have been inconclusive with respect to inducing differentiation or cell-type-specific gene expression (Peter and Nigg 1991; Lourim and Lin, 1992). However, the altering of the lamin composition of the nucleus may be part of an additive process, which after many steps results in the changes observed in a fully differentiated cell (Peter and Nigg, 1991). It may be necessary to assay several parameters of differentiation rather than just induction of gene expression to determine a role for the lamins in this process. It would, for example, be interesting to examine the effect of lamin A expression on some feature of chromatin organization and function. Unfortunately, the role of lamins in chromatin organization is poorly understood. While there is much *in vitro* and some *in vivo* evidence that lamins do interact with DNA and chromatin, there is little evidence that these interactions have any effect on chromatin organization. To address these questions, new approaches must be developed to determine if lamins are involved in chromatin organization and as a consequence of this involvement affect cell differentiation.

*In vitro* approaches have been somewhat successful in examining lamin function and dynamics, particularly in the cases of nuclear reformation and DNA replication. However, *in vivo* approaches, particularly genetic ones, must now be employed to complement the *in vitro* results. Systems that are genetically manipulatable have not been used extensively to examine lamin function. However, the identification of lamin genes in *C. elegans* and *D. melanogaster* will allow the application of genetic approaches to study lamin function *in vivo.* Gene knockout and temperature-sensitive mutants are constructed readily in these species, so that it should be possible to determine if lamins are essential for survival. In addition, transformation of mutated lamin genes into these species can be used to define functions for specific domains of the lamin molecule. The same approach can be used for the LAPs that have also been identified in *Drosophila* (Padan *et al.,* 1990; Lin and Wolfner, 1991). Furthermore, it is now important to initiate molecular genetic approaches in mammalian systems. For example, it should be possible to ablate the expression of one or more of the lamins using antisense methods and thereby to examine the effects on differentiation, chromatin organization, and other proposed nuclear lamin functions. Together with cell-free nuclear assembly systems and traditional biochemical purification and characterization methods, these approaches will lead to a clearer understanding of the structural and functional roles of lamins in the nucleus.

## References


Aaronson R. P., and Blobel, G. (1975). Isolation of nuclear pore complexes in association with a lamina. *Proc. Natl. Acad. Sci. U.S.A.* **72,** 1007–1011.

Aebi, U., Cohn, J., Buhle, L., and Gerace, L. (1986). The nuclear lamina is a meshwork of intermediate-type filaments. *Nature (London)* **323,** 560–564.

Akey, C. W. (1989). Interactions and structure of the nuclear pore complex revealed by cryo-electron microscopy. *J. Cell Biol.* **109,** 955–970.

Allen, J. L., and Douglas, M. G. (1989). Organization of the nuclear pore complex in *Saccharomyces cerevisiae. J Ultrastruct. Mol. Struct. Res.* **102,** 95–108.

Appelbaum, J., Blobel, G., and Georgatos, S. D. (1990). *In vivo* phosphorylation of the lamin B receptor. Binding of lamin B to its nuclear membrane receptor is affected by phosphorylation. *J Biol. Chem.* **265,** 4181–4184.

Bailer, S. M., Eppenberger, H. M., Griffiths, G., and Nigg, E. A. (1991). Characterization of A 54-kD protein of the inner nuclear membrane: Evidence for cell cycle-dependent interaction with the nuclear lamina. *J. Cell Biol.* **114,** 389–400.

Beck, L. A., Hosick, T. J., and Sinensky, M. (1990). Isoprenylation is required for the processing of the lamin A precursor. *J. Cell Biol.* **110,** 1489–1499.

Belmont, A. S., Zhai, Y., and Thilenius, A. (1993). Lamin B distribution and association with peripheral chromatin revealed by optical sectioning and electron microscopy tomography. *J. Cell Biol.* **123,** 1671–1685.

Benavente, R., Krohne, G., and Franke, W. W. (1985). Cell type-specific expression of nuclear lamina proteins during development of *Xenopus laevis*. *Cell (Cambridge, Mass.)* **41,** 177–190.

Biamonti, G., Perini, G., Weighardt, F., Riva, S., Giacca, M., Norio, P., Zentilin, L., Diviacco, S., Dimitrova, D., and Falaschi, A. (1992). A human DNA replication origin: Localization and transcriptional characterization. *Chromosoma* **102,** S24–S31.

Bissell, M. J., Hall, H. G., and Parry, G. (1982). How does the extracellular matrix direct gene expression? *J. Theor. Biol.* **99,** 31–68.

Blow, J. J., and Laskey, R. A. (1986). Intiation of DNA replication in nuclei and purified DNA by a cell-free extract of *Xenopus* eggs. *Cell (Cambridge, Mass.)* **47,** 577–587.

Bossie, C. A., and Sanders, M. M. (1993). A cDNA from *Drosophila melanogaster* encodes a lamin C-like intermediate filament protein. *J. Cell Sci.* **104,** 1263–1272.

Bridger, J. M., Kill, I. R., O'Farrell, M., and Hutchison, C. J. (1993). Internal lamin structures within G1 nuclei of human dermal fibroblasts. *J. Cell Sci.* **104,** 297–306.

Broers, J. L., and Ramaekers, F. C. (1994). Differentiation markers for lung-cancer subtypes. A comparison of their expression *in vitro* and *in vivo*. *Int. J. Cancer, Suppl.* **8,** 134–138.

Broers, J. L., Raymond, Y., Rot, M. K., Kuijpers, H., Wagenaar, S. S., and Ramaekers, F. C. (1993). Nuclear A-type lamins are differentially expressed in human lung cancer subtypes. *Am. J. Pathol.* **143,** 211–220.

Burke, B., and Gerace, L. (1986). A cell-free system to study reassembly of the nuclear envelope at the end of mitosis. *Cell (Cambridge, Mass.)* **44,** 639–652.

Cardenas, M. E., Laroche, T., and Gasser, S. M. (1990). The composition and morphology of yeast nuclear scaffolds. *J. Cell Sci.* **96,** 439–450.

Chaudhary, N., and Courvalin, J. C. (1993). Stepwise reassembly of the nuclear envelope at the end of mitosis. *J. Cell Biol.* **122,** 295–306.

Chou, Y. H., Rosevear, E., and Goldman, R. D. (1989). Phosphorylation and disassembly of intermediate filaments in mitotic cells. *Proc. Natl. Acad. Sci. U.S.A.* **86,** 1885–1889.

Chou, Y. H., Bischoff, J. R., Beach, D., and Goldman, R. D. (1990). Intermediate filament reorganization during mitosis is mediated by p34cdc2 phosphorylation of vimentin. *Cell (Cambridge, Mass.)* **62,** 1063–1071.

Coggeshall, R. E., and Fawcett, D. W. (1964). The fine structure of the central nervous system of the leech *Hirudo medicinalis*. *J. Neurophysiol.* **27,** 229–289.

Collard, J. F., Senecal, J. L., and Raymond, Y. (1992). Redistribution of nuclear lamin A is an early event associated with differentiation of human promyelocytic leukemia HL-60 cells. *J. Cell Sci.* **101,** 657–670.

Coonen, E., Dumoulin, J. C., and Ramaekers, F. C. (1993). Intermediate filament protein expression in early developmental stages of the mouse. A confocal scanning laser microscopy study of *in vitro* fertilized and *in vitro* cultured pre-implantation mouse embryos. *Histochemistry* **99,** 141–149.

Courvalin, J. C., Segil, N., Blobel, G., and Worman, H. J. (1992). The lamin B receptor of the inner nuclear membrane undergoes mitosis-specific phosphorylation and is a substrate for p34cdc2-type protein kinase. *J. Biol. Chem.* **267,** 19035–19038.

Dabauvalle, M. C., Loos, K., Merkert, H., and Scheer, U. (1991). Spontaneous assembly of pore complex-containing membranes ("annulate lamellae") in *Xenopus* egg extract in the absence of chromatin. *J. Cell Biol.* **112,** 1073–1082.

Dagenais, A., LeMyre, A., and Bibor-Hardy, V. (1990). Differential transport and integration into the nuclear lamina for lamins A, B, and C. *Biochem. Cell Biol.* **68,** 827–831.

Dessev, G., and Goldman, R. (1988). Meiotic breakdown of nuclear envelope in oocytes of *Spisula solidissima* involves phosphorylation and release of nuclear lamin. *Dev. Biol.* **130,** 543–550.

Dessev, G., and Goldman, R. (1990). The oocyte lamin persists as a single major component of the nuclear lamina during embryonic development of the surf clam. *Int. J. Dev. Biol.* **34,** 267–274.

Dessev, G., Iovcheva, C., Tasheva, B., and Goldman, R. (1988). Protein kinase activity associated with the nuclear lamina. *Proc. Natl. Acad. Sci. U.S.A.* **85,** 2994–2998.

Dessev, G. N., Iovcheva-Dessev, C., and Goldman, R. D. (1990). Lamin dimers. Presence in the nuclear lamina of surf clam oocytes and release during nuclear envelope breakdown. *J. Biol. Chem.* **265,** 12636–12641.

Dessev, G., Iovcheva-Dessev, C., Bischoff, J. R., Beach, D., and Goldman, R. (1991). A complex containing p34cdc2 and cyclin B phosphorylates the nuclear lamin and disassembles nuclei of clam oocytes *in vitro. J. Cell Biol.* **112,** 523–533.

Dodemont, H., Riemer, D., and Weber, K. (1990). Structure of an invertebrate gene encoding cytoplasmic intermediate filament (IF) proteins: Implications for the origin and the diversification of IF proteins. *EMBO J.* **9,** 4083–4094.

Dwyer, N., and and Blobel, G. (1976). A modified procedure for the isolation of a pore complex-lamina fraction from rat liver nuclei. *J. Cell Biol.* **70,** 581–591.

Eggert, M., Radomski, N., Tripier, D., Traub, P., and Jost, E. (1991). Identification of phosphorylation sites on murine nuclear lamin C by RP-HPLC and microsequencing. *FEBS Lett.* **292,** 205–209.

Eggert, M., Radomski, N., Linder, D., Tripier, D., Traub, P., and Jost, E. (1993). Identification of novel phosphorylation sites in murine A-type lamins. *Eur. J. Biochem.* **213,** 659–671.

Enoch, T., Peter, M., Nurse, P., and Nigg, E. A. (1991). p34cdc2 acts as a lamin kinase in fission yeast. *J. Cell Biol.* **112,** 797–807.

Fawcett, D. W. (1966). On the occurrence of a fibrous lamina on the inner aspect of the nuclear envelope in certain cells of vertebrates. *Am. J. Anat.* **119,** 129–146.

Fey, E. G., Wan, K. M., and Penman, S. (1984). Epithelial cytoskeletal framework and nuclear matrix-intermediate filament scaffold: three-dimensional organization and protein composition. *J. Cell Biol.* **98,** 1973–1984.

Fields, A. P., Pettit, G. R., and May, W. S. (1988). Phosphorylation of lamin B at the nuclear membrane by activated protein kinase C. *J. Biol. Chem.* **263,** 8253–8260.

Firmbach-Kraft, I., and Stick, R. (1993). The role of CaaX-dependent modifications in membrane association of *Xenopus* nuclear lamin B3 during meiosis and the fate of B3 in transfected mitotic cells. *J. Cell Biol.* **123,** 1661–1670.

Fisher, D. Z., Chaudhary, N., and Blobel, G. (1986). cDNA sequencing of nuclear lamins A and C reveals primary and secondary structural homology to intermediate filament proteins. *Proc. Natl. Acad. Sci. U.S.A.* **83,** 6450–6454.

Foisner, R., and Gerace, L. (1993). Integral membrane proteins of the nuclear envelope interact with lamins and chromosomes, and binding is modulated by mitotic phosphorylation. *Cell (Cambridge, Mass.)* **73,** 1267–1279.

Foisy, S., and Bibor-Hardy, V. (1988). Synthesis of nuclear lamins in BHK-21 cells synchronized with aphidicolin. *Biochem. Biophys. Res. Commun.* **156,** 205–210.

Forbes, D. J., Kirschner, M. W., and Newport, J. W. (1983). Spontaneous formation of nucleus-like structures around bacteriophage DNA microinjected into *Xenopus* eggs. *Cell (Cambridge, Mass.)* **34,** 13–23.

Frangioni, J. V., and Neel, B. G. (1993). Use of a general purpose mammalian expression vector for studying intracellular protein targeting: Identification of critical residues in the nuclear lamin A/C nuclear localization signal. *J. Cell Sci.* **105,** 481–488.

Frederick, S. E., Mangan, M. E., Carey, J. B., and Gruber, P. J. (1992). Intermediate filament antigens of 60 and 65kDa in the nuclear matrix of plants: Their detection and localization. *Exp. Cell Res.* **199,** 213–222.

Furukawa, K., and Hotta, Y. (1993). cDNA cloning of a germ cell specific lamin B3 from mouse spermatocytes and analysis of its function by ectopic expression in somatic cells. *EMBO J.* **12,** 97–106.

Furukawa, K., Inagaki, H., and Hotta, Y. (1994). Identification and cloning of an mRNA coding for a germ cell-specific A-type lamin in mice. *Exp. Cell Res.* **212,** 426–430.

Georgatos, S. D., and Blobel, G. (1987). Lamin B constitutes an intermediate filament attachment site at the nuclear envelope. *J. Cell Biol.* **105,** 117–125.

Georgatos, S. D., Stournaras, C., and Blobel, G. (1988). Heterotypic and homotypic associations between the nuclear lamins: Site-specificity and control by phosphorylation. *Proc. Natl. Acad. Sci. U.S.A.* **85,** 4325–4329.

Georgatos, S. D., Maroulakou, I., and Blobel, G. (1989). Lamin A, lamin B, and lamin B receptor analogues in yeast. *J. Cell Biol.* **108,** 2069–2082.

Georgatos, S. D., Meier, J., and Simos, G. (1994). Lamins and lamin-associated proteins. *Curr. Opin. Cell Biol.* **6,** 347–353.

Gerace, L., and Blobel, G. (1980). The nuclear envelope lamina is reversibly depolymerized during mitosis. *Cell* (*Cambridge, Mass.*) **19,** 277–287.

Gerace, L., and Foisner, R. (1994). Integral membrane proteins and dynamic organization of the nuclear envelope. *Trends Cell Biol.* **4,** 127–131.

Gerace, L., Blum, A., and Blobel, G. (1978). Immunocytochemical localization of the major polypeptides of the nuclear pore complex-lamina fraction. Interphase and mitotic distribution. *J. Cell Biol.* **79,** 546–566.

Gerace, L., Comeau, C., and Benson, M. (1984). Organization and modulation of nuclear lamina structure. *J. Cell Sci., Suppl.* **1,** 137–160.

Gerhart, J. G. (1980). Mechanisms regulating pattern formation in the amphibian egg and early embryo. *In* "Biological Regulation and Development" (R. F. Goldberger, ed.), Vol. 2, pp. 133–315. Plenum, New York.

Glass, C. A., Glass, J. R., Taniura, H., Hasel, K. W., Blevitt, J. M., and Gerace, L. (1993). The alpha-helical rod domain of human lamins A and C contains a chromatin binding site. *EMBO J.* **12,** 4413–4424.

Glass, J. R., and Gerace, L. (1990). Lamins A and C bind and assemble at the surface of mitotic chromosomes. *J. Cell Biol.* **111,** 1047–1057.

Goldberg, M. W., and Allen, T. D. (1992). High resolution scanning electron microscopy of the nuclear envelope: Demonstration of a new, regular, fibrous lattice attached to the baskets of the nucleoplasmic face of the nuclear pores. *J. Cell Biol.* **119,** 1429–1440.

Goldman, A. E., Maul, G., Steinert, P. M., Yang, H.-Y., and Goldman, R. D. (1986). Keratin-like proteins that coisolate with intermediate filaments of BHK-21 cells are nuclear lamins. *Proc. Natl. Acad. Sci. U.S.A.* **83,** 3839–3943.

Goldman, A. E., Moir, R. D., Montag-Lowy, M., Stewart, M., and Goldman, R. D. (1992). Pathway of incorporation of microinjected lamin A into the nuclear envelope. *J. Cell Biol.* **119,** 725–735.

Goldman, R., Goldman, A., Green, K., Jones, J., Lieska, N., and Yang, H.-Y. (1985). Intermediate filaments: Possible functions as cytoskeletal connecting links between the nucleus and cell surface. *Ann. N.Y. Acad. Sci.* **455,** 1–17.

Goss, V. L., Hocevar, B. A., Thompson, L. J., Stratton, C. A., Burns, D. J., and Fields, A. P. (1994). Identification of nuclear beta II protein kinase C as a mitotic lamin kinase. *J. Biol. Chem.* **269,** 19074–19080.

Graham, C. F., and Morgan, R. W. (1966). Changes in the cell cycle during early amphibian development. *Dev. Biol.* **14,** 439–460.

Gray, E. G., and Guillery, R. W. (1963). On nuclear structure in the ventral nerve cord of the leech Hirudo medicinalis. *Z. Zellforsch. Mikrosk. Anat.* **59,** 738–745.

Gruenbaum, Y., Landesman, Y., Drees, B., Bare, J. W., Saumweber, H., Paddy, M. R., Sedat, J. W., Smith, D. E., Benton, B. M., and Fisher, P. A. (1988). *Drosophila* nuclear lamin precursor Dm0 is translated from either of two developmentally regulated mRNA species apparently encoded by a single gene. *J. Cell Biol.* **106,** 585–596.

Haas, M., and Jost, E. (1993). Functional analysis of phosphorylation sites in human lamin A controlling lamin disassembly, nuclear transport and assembly. *Eur. J. Cell Biol.* **62,** 237–247.

Hass, R., Giese, G., Meyer, G., Hartmann, A., Dork, T., Kohler, L., Resch, K., Traub, P., and Goppelt-Strube, M. (1990). Differentiation and retrodifferentiation of U937 cells: Reversible induction and suppression of intermediate filament protein synthesis. *Eur. J. Cell Biol.* **51,** 265–271.

Heald, R., and McKeon, F. (1990). Mutations of phosphorylation sites in lamin A that prevent nuclear lamina disassembly in mitosis. *Cell (Cambridge, Mass.)* **61,** 579–589.

Heins, S., and Aebi, U. (1994). Making heads and tails of intermediate filament assembly, dynamics and networks. *Curr. Opin. Cell Biol.* **6,** 25–33.

Heitlinger, E., Peter, M., Haner, M., Lustig, A., Aebi, U., and Nigg, E. A. (1991). Expression of chicken lamin B2 in *Escherichia coli:* Characterization of its structure, assembly, and molecular interactions. *J. Cell Biol.* **113,** 485–495.

Heitlinger, E., Peter, M., Lustig, A., Villiger, W., Nigg, E. A., and Aebi, U. (1992). The role of the head and tail domain in lamin structure and assembly: Analysis of bacterially expressed chicken lamin A and truncated B2 lamins. *J. Struct. Biol.* **108,** 74–89.

Hennekes, H., and Nigg, E. A. (1994). The role of isoprenylation in membrane attachment of nuclear lamins. A single point mutation prevents proteolytic cleavage of the lamin A precursor and confers membrane binding properties. *J. Cell Sci.* **107,** 1019–1029.

Hennekes, H., Peter, M., Weber, K., and Nigg, E. A. (1993). Phosphorylation on protein kinase C sites inhibits nuclear import of lamin B2. *J. Cell Biol.* **120,** 1293–1304.

Hoger, T. H., Krohne, G., and Franke, W. W. (1988). Amino acid sequence and molecular characterization of murine lamin B as deduced from cDNA clones. *Eur. J. Cell Biol.* **47,** 283–290.

Hoger, T. H., Zatloukal, K., Waizenegger, I., and Krohne, G. (1990a). Characterization of a second highly conserved B-type lamin present in cells previously thought to contain only a single B-type lamin. *Chromosoma* **99,** 379–390.

Hoger, T. H., Zatloukal, K., Waizenegger, I., and Krohne, G. (1990b). Characterization of a second highly conserved B-type lamin present in cells previously thought to contain only a single B-type lamin. *Chromosoma* **100,** 67–69.

Hoger, T. H., Gründ, C., Franke, W. W., and Krohne, G. (1991a). Immunolocalization of lamins in the thick nuclear lamina of human synovial cells. *Eur. J. Cell Biol.* **54,** 150–156.

Hoger, T. H., Krohne, G., and Kleinschmidt, J. A. (1991b). Interaction of *Xenopus* lamins A and LII with chromatin in vitro mediated by a sequence element in the carboxyterminal domain. *Exp. Cell Res.* **197,** 280–289.

Holtz, D., Tanaka, R. A., Hartwig, J., and McKeon, F. (1989). The CaaX motif of lamin A functions in conjunction with the nuclear localization signal to target assembly to the nuclear envelope. *Cell (Cambridge, Mass.)* **59,** 969–977.

Horton, H., McMorrow, I., and Burke, B. (1992). Independent expression and assembly properties of heterologous lamins A and C in murine embryonal carcinomas. *Eur. J. Cell Biol.* **57,** 172–183.

Hutchinson, C. J., Bridger, J. M., Cox, L. S., and Kill, I. R. (1994). Weaving a pattern from disparate threads: Lamin function in nuclear assembly and DNA replication. *J. Cell Sci.* **107,** 3259–3269.

Jenkins, H., Holman, T., Lyon, C., Lane, B., Stick, R., and Hutchison, C. (1993). Nuclei that lack a lamina accumulate karyophilic proteins and assemble a nuclear matrix. *J. Cell Sci.* **106,** 275–285.

Jones, J. C. R., Goldman, A. E., Yang, H.-Y., and Goldman, R. D. (1985). The organizational fate of intermediate filament networks in two epithelial cell types during mitosis. *J. Cell Biol.* **100,** 93-102.

Kaufmann, S. H. (1989). Additional members of the rat liver lamin polypeptide family. Structural and immunological characterization. *J. Biol. Chem.* **264,** 13946–13955.

Kaufmann, S. H. (1992). Expression of nuclear envelope lamins A and C in human myeloid leukemias. *Cancer Res.* **52,** 2847–2853.

Kitten, G. T., and Nigg, E. A. (1991). The CaaX motif is required for isoprenylation, carboxyl methylation, and nuclear membrane association of lamin B2. *J. Cell Biol.* **113,** 13–23.

Krohne, G., Franke, W. W., and Scheer, U. (1978). The major polypeptides of the nuclear pore complex. *Exp. Cell Res.* **116,** 85-102.

Krohne, G., Debus, E., Osborn, M., Weber, K., and Franke, W. W. (1984). A monoclonal antibody against nuclear lamina proteins reveals cell type-specificity in *Xenopus laevis*. *Exp. Cell Res.* **150,** 47–59.

Krohne, G., Waizenegger, I., and Hoger, T. H. (1989). The conserved carboxy-terminal cysteine of nuclear lamins is essential for lamin association with the nuclear envelope. *J. Cell Biol.* **109,** 2003–2011.

Lamb, N. J., Cavadore, J. C., Labbé, J. C., Maurer, R. A., and Fernandez, A. (1991). Inhibition of cAMP-dependent protein kinase plays a key role in the induction of mitosis and nuclear envelope breakdown in mammalian cells. *EMBO J.* **10,** 1523–1533.

Lanoix, J., Skup, D., Collard, J. F., and Raymond, Y. (1992). Regulation of the expression of lamins A and C is post-transcriptional in P19 embryonal carcinoma cells. *Biochem. Biophys. Res. Commun.* **189,** 1639–1644.

Lehner, C. F., Kurer, V., Eppenberger, H. M., and Nigg, E. A. (1986). The nuclear lamin protein family in higher vertebrates. Identification of quantitatively minor lamin proteins by monoclonal antibodies. *J. Biol. Chem.* **261,** 13293–13301.

Lin, H. F., and Wolfner, M. F. (1991). The *Drosophila* maternal-effect gene fs(1)Ya encodes a cell cycle-dependent nuclear envelope component required for embryonic mitosis. *Cell (Cambridge, Mass.)* **64,** 49–62.

Lin, F., and Worman, H. J. (1993). Structural organization of the human gene encoding nuclear lamin A and nuclear lamin C. *J. Biol. Chem.* **268,** 16321–16326.

Loewinger, L., and McKeon, F. (1988). Mutations in the nuclear lamin proteins resulting in their aberrant assembly in the cytoplasm. *EMBO J.* **7,** 2301–2309.

Lohka, M. J., and Masui, Y. (1983). Formation *in vitro* of sperm pronuclei and mitotic chromosomes induced by amphibian ooplasmic components. *Science* **220,** 719–721.

Lopez, J. M., Song, K., Hirshfeld, A. B., Lin, H., and Wolfner, M. F. (1994). The *Drosophila* fs(1)Ya protein, which is needed for the first mitotic division, is in the nuclear lamina and in the envelopes of cleavage nuclei, pronuclei, and non-mitotic nuclei. *Dev. Biol.* **163,** 202–211.

Lourim, D., and Krohne, G. (1993). Membrane-associated lamins in *Xenopus* egg extracts: Identification of two vesicle populations. *J. Cell Biol.* **123,** 501–512.

Lourim, D., and Krohne, G. (1994). Lamin-dependent nuclear envelope reassembly following mitosis. *Trends Cell Biol.* **4,** 314–18.

Lourim, D., and Lin, J. J. (1992). Expression of wild-type and nuclear localization-deficient human lamin A in chick myogenic cells. *J. Cell Sci.* **103,** 863–874.

Ludérus, M. E. E., de Graaf, A., Mattia, E., den Blaauwen, J. L., Grande, M. A., de Jong, L., and van Driel, R. (1992). Binding of matrix attachment regions to lamin B1. *Cell (Cambridge, Mass.)* **70,** 949–959.

Ludérus, M. E. E., den Blaauwen, J. L., de Smit, O. J., Compton, D. A., and van Driel, R. (1994). Binding of matrix attachment regions to lamin polymers involves single-stranded regions and the minor groove. *Mol. Cell. Biol.* **14,** 6297–6305.

Luscher, B., Brizuela, L., Beach, D., and Eisenman, R. N. (1991). A role for the p34cdc2 kinase and phosphatases in the regulation of phosphorylation and disassembly of lamin B2 during the cell cycle. *EMBO J.* **10,** 865–875.

Lutz, R. J., Trujillo, M. A., Denham, K. S., Wenger, L., and Sinensky, M. (1992). Nucleoplasmic localization of prelamin A: Implications for prenylation-dependent lamin A assembly into the nuclear lamina. *Proc. Natl. Acad. Sci. U.S.A.* **89,** 3000–3004; erratum: p. 5699.

Maison, C., Horstmann, H., and Georgatos, S. D. (1993). Regulated docking of nuclear membrane vesicles to vimentin filaments during mitosis. *J. Cell Biol.* **123,** 1491–1505.

Mancini, M. A., Shan, B., Nickerson, J. A., Penman, S., and Lee, W. H. (1994). The retinoblastoma gene product is a cell cycle-dependent, nuclear matrix-associated protein. *Proc. Natl. Acad. Sci. U.S.A.* **91,** 418–422.

Marshall, C. J. (1993). Protein prenylation: A mediator of protein–protein interactions. *Nature* (*London*) **259,** 1865–1866.

Martelli, A. M., Billi, A. M., Gilmour, R. S., Manzoli, L., DiPrimo, R., and Cocco, L. (1992). Mouse and human hemopoietic cell lines of erythroid lineage express lamins A, B and C. *Biochem. Biophys. Res. Commun.* **185,** 271–276.

Mattia, E., Hoff, W. D., den Blaauwen, J., Meijne, A. M., Stuurman, N., and van Renswoude, J. (1992). Induction of nuclear lamins A/C during *in vitro*-induced differentiation of F9 and P19 embryonal carcinoma cells. *Exp. Cell Res.* **203,** 449–455.

Maul, G. G., Baglia, F. A., Newmeyer, D. D. and Ohlsson-Wilhelm, B. M. (1984). The major 67,000 moelcular weight protein of the clam oocyte nuclear envelope is lamin-like. *J. Cell Sci.* **67,** 69–85.

McKeon, F. (1991). Nuclear lamin proteins: Domains required for nuclear targeting, assembly, and cell-cycle-regulated dynamics. *Curr. Opin. Cell Biol.* **3,** 82–86.

McKeon, F., Kirschner, M., and Caput, D. (1986). Homologies in both primary and secondary structure between nuclear envelope and intermediate filament proteins. *Nature* (*London*) **319,** 439–468.

McNulty, A. K., and Saunders, M. J. (1992). Purification and immunological detection of pea nuclear intermediate filaments: Evidence for plant nuclear lamins. *J. Cell Sci.* **103,** 407–414.

Meier, J., and Georgatos, S. D. (1994). Type B lamins remain associated with the integral nuclear envelope protein p58 during mitosis: Implications for nuclear reassembly. *EMBO J.* **13,** 1888–1898.

Meier, J., Campbell, K. H., Ford, C. C., Stick, R., and Hutchison, C. J. (1991). The role of lamin L III in nuclear assembly and DNA replication, in cell-free extracts of *Xenopus* eggs. *J. Cell Sci.* **98,** 271–279.

Mercer E. H. (1959). An electron microscopic study of Amoeba proteus. *Proc. R. Soc. London, Ser. B* **150,** 216–232.

Miller, R. K., Vikström, K., and Goldman, R. D. (1991). Keratin incorporation into intermediate filament networks is a rapid process. *J. Cell Biol.* **113,** 843–855.

Moir, R. D., and Goldman, R. D. (1993). Lamin dynamics. *Curr. Opin. Cell Biol.* **5,** 408-411.

Moir, R. D., Quinlan, R. A., and Stewart, M. (1990). Expression and characterization of human lamin C. *FEBS Lett.* **268,** 301–305.

Moir, R. D., Donaldson, A. D., and Stewart, M. (1991). Expression in *Escherichia coli* of human lamins A and C: Influence of head and tail domains on assembly properties and paracrystal formation. *J. Cell Sci.* **99,** 363–372.

Moir, R. D., Montag-Lowy, M., and Goldman, R. D. (1994). Dynamic properties of nuclear lamins: Lamin B is associated with sites of DNA replication. *J. Cell Biol.* **125,** 1201–1212.

Molloy, S., and Little, M. (1992). p34cdc2 kinase-mediated release of lamins from nuclear ghosts is inhibited by cAMP-dependent protein kinase. *Exp. Cell Res.* **201,** 494–499.

Monteiro, M. J., Hicks, C., Gu, L., and Janicki, S. (1994). Determinants for intracellular sorting of cytoplasmic and nuclear intermediate filaments. *J. Cell Biol.* **127,** 1327–1343.

Muller, P. R., Meier, R., Hirt, A., Bodmer, J. J., Janic, D., Leibundgut, K., Luthy, A. R., and Wagner, H. P. (1994). Nuclear lamin expression reveals a surprisingly high growth fraction in childhood acute lymphoblastic leukemia cells. *Leukemia.* **8,** 940–945.

Murray, A. W., and Hunt, T. (1993). "The Cell Cycle." Freeman, New York.

Murray, N. R., Burns, D. J., and Fields, A. P. (1994). Presence of a beta II protein kinase C-selective nuclear membrane activation factor in human leukemia cells. *J. Biol. Chem.* **269,** 21385–21390.

Nakajima, N., and Sado, T. (1993). Nucleotide sequence of a mouse lamin A cDNA and its deduced amino acid sequence. *Biochim. Biophys. Acta.* **1171,** 311–314.

Newmeyer, D. D., Lucocq, J. M., Burglin, T. R., and DeRobertis, E. M. (1986). Assembly *in vitro* of nuclei active in nuclear protein transport: ATP is required for nucleoplasmin accumulation. *EMBO J.* **5,** 501–510.

Newport, J. (1987). Nuclear reconstitution *in vitro:* Stages of assembly around protein-free DNA. *Cell (Cambridge, Mass.)* **48,** 205–217.

Newport, J., and Kirschner, M. (1982a). A major developmental transition in early *Xenopus* embryos: I. Characterization and timing of cellular changes at the midblastula stage. *Cell (Cambridge, Mass.)* **30,** 675–686.

Newport, J. and Kirschner, M. (1982b). A major developmental transition in early *Xenopus* embryos: II. Control of the onset of transcription. *Cell (Cambridge, Mass.)* **30,** 687–696.

Newport, J., and Spann, T. (1987). Disassembly of the nucleus in mitotic extracts: Membrane vesicularization, lamin disassembly, and chromosome condensation are independent processes. *Cell (Cambridge, Mass.)* **48,** 219–230.

Newport, J. W., Wilson, K. L., and Dunphy, W. G. (1990). A lamin-independent pathway for nuclear envelope assembly. *J. Cell Biol.* **111,** 2247–2259.

Nigg, E. A. (1992a). Assembly–disassembly of the nuclear lamina. *Curr. Opin. Cell Biol.* **4,** 105–109.

Nigg, E. A. (1992b). Assembly and cell cycle dynamics of the nuclear lamina. *Semin. Cell Biol.* **3,** 245–253.

Ottaviano, Y., and Gerace, L. (1985). Phosphorylation of the nuclear lamins during interphase and mitosis. *J. Biol. Chem.* **260,** 624–632.

Padan, R., Nainudel-Rpszteyn, S., Goitein, R., Fainsod, A., and Gruenbaum, Y. (1990). Isolation and characterization of the *Drosophila* nuclear envelope otefin cDNA. *J. Biol. Chem.* **265,** 7808–7813.

Paddy, M. R., Belmont, A. S., Saumweber, H., Agard, D. A., and Sedat, J. W. (1990). Interphase nuclear envelope lamins form a discontinuous network that interacts with only a fraction of the chromatin in the nuclear periphery. *Cell (Cambridge, Mass.)* **62,** 89-106.

Paddy, M. R., Agard, D. A., and Sedat, J. W. (1992). An extended view of nuclear lamin structure, function, and dynamics. *Semin. Cell Biol.* **3,** 255–266.

Pappas, G. D. (1956). The fine structure of the nuclear envelope of *Amoeba proteus. J. Biophys. Biochem. Cytol.* **2,** 431–435.

Parry, D. A. D., and Steinert, P. M. (1992). Intermediate filament structure. *Curr. Opin. Cell Biol.* **4,** 94–98.

Parry, D. A. D., Conway, J. F., Goldman, A. E., Goldman, R. D., and Steinert P. M. (1987). Nuclear lamin proteins: Common structures for paracrystalline, filamentous and lattice forms. *Int. J. Biol. Macromol.* **9,** 137–145.

Patrizi, G., and Poger, M. (1967). The ultrastructure of the nuclear periphery. *J. Ultrastruct. Res.* **17,** 127–136.

Peter, M., and Nigg, E. A. (1991). Ectopic expression of an A-type lamin does not interfere with differentiation of lamin A-negative embryonal carcinoma cells. *J. Cell Sci.* **100,** 589–598.

Peter, M., Kitten, G. T., Lehner, C. F., Vorburger, K., Bailer, S. M., Maridor, G., and Nigg, E. A. (1989). Cloning and sequencing of cDNA clones encoding chicken lamins A and B1 and comparison of the primary structures of vertebrate A- and B-type lamins. *J Mol. Biol.* **208,** 393–404.

Peter, M., Nakagawa, J., Dorée, M., Labbé, J. C., and Nigg, E. A. (1990). *In vitro* disassembly of the nuclear lamina and M phase-specific phosphorylation of lamins by cdc2 kinase. *Cell (Cambridge, Mass.)* **61,** 591–602.

Peter, M., Heitlinger, E., Haner, M., Aebi, U., and Nigg, E. A. (1991). Disassembly of *in vitro* formed lamin head-to-tail polymers by CDC2 kinase. *EMBO J.* **10,** 1535–1544.

Peter, M., Sanghera, J. S., Pelech, S. L., and Nigg, E. A. (1992). Mitogen-activated protein kinases phosphorylate nuclear lamins and display sequence specificity overlapping that of mitotic protein kinase p34cdc2. *Eur. J. Biochem.* **205,** 287–294.

Pollard, K. M., Chan, E. K., Grant, B. J., Sullivan, K. F., Tan, E. M., and Glass, C. A. (1990). *In vitro* post-translational modification of lamin B cloned from a human T-cell line. *Mol. Cell. Biol.* **10,** 2164–2175.

Powell, L., and Burke, B. (1990). Internuclear exchange of an inner nuclear membrane protein (p55) in heterokaryons: *In vivo* evidence for the interaction of p55 with the nuclear lamina. *J. Cell Biol.* **111,** 2225–2234.

Riedel, W., and Werner, D. (1989). Nucleotide sequence of the full-length mouse lamin C cDNA and its deduced amino-acid sequence. *Biochim. Biophys. Acta* **1008,** 119–122.

Riemer, D., and Weber, K. (1994). The organization of the gene for *Drosophila* lamin C: Limited homology with vertebrate lamin genes and lack of homology versus the *Drosophila lamin* Dmo gene. *Eur. J. Cell Biol.* **63,** 299–306.

Riemer, D., Dodemont, H., and Weber, K. (1993). A nuclear lamin of the nematode *Caenorhabditis elegans* with unusual structural features; cDNA cloning and gene organization. *Eur. J. Cell Biol.* **62,** 214–223.

Rober, R. A., Weber, K., and Osborn, M. (1989). Differential timing of nuclear lamin A/C expression in the various organs of the mouse embryo and the young animal: A developmental study. *Development* (*Cambridge, U.K.*) **105,** 365–378.

Rober, R. A., Gieseler, R. K., Peters, J. H., Weber, K., and Osborn, M. (1990a). Induction of nuclear lamins A/C in macrophages in *in vitro* cultures of rat bone marrow precursor cells and human blood monocytes, and in macrophages elicited *in vivo* by thioglycollate stimulation. *Exp. Cell Res.* **190,** 185–194.

Rober, R. A., Sauter, H., Weber, K., and Osborn, M. (1990b). Cells of the cellular immune and hemopoietic system of the mouse lack lamins A/C: Distinction versus other somatic cells. *J. Cell Sci.* **95,** 587–598.

Sarria, A. J., Lieber, J. G., Nordeen, S. K., and Evans, R. M. (1994). The presence or absence of a vimentin-type intermediate filament network affects the shape of the nucleus in human SW-13 cells. *J. Cell Sci.* **107,** 1593–1607.

Sasseville, A. M.-J., and Raymond, Y. (1995). Lamin A precursor is localized to intranuclear foci. *J. Cell Sci.* **106,** 273–285.

Satoh, N., Kageyama, T., and Sirakami, K. I. (1976). Motility of dissociated embryonic cells in *Xenopus laevis*: Its significance to morphogenetic movements. *Dev., Growth Differ.* **18,** 55–67.

Senior, A., and Gerace, L. (1988). Integral membrane proteins specific to the inner nuclear membrane and associated with the nuclear lamina. *J. Cell Biol.* **107,** 2029–2036.

Shimanuki, M., Goebl, M., Yanagida, M., and Toda, T. (1992). Fission yeast sts1+ gene encodes a protein similar to the chicken lamin B receptor and is implicated in pleiotropic drug-sensitivity, divalent cation-sensitivity, and osmoregulation. *Mol. Biol. Cell.* **3,** 263–273.

Shoeman, R. L., and Traub, P. (1990). The *in vitro* DNA-binding properties of purified nuclear lamin proteins and vimentin. *J. Biol. Chem.* **265,** 9055–9061.

Simos, G., and Georgatos, S. D. (1992). The inner nuclear membrane protein p58 associates *in vivo* with a p58 kinase and the nuclear lamins. *EMBO J.* **11,** 4027–4036.

Simos, G., and Georgatos, S. D. (1994). The lamin B receptor-associated protein p34 shares sequence homology and antigenic determinants with the splicing factor 2-associated protein p32. *FEBS Lett.* **346,** 225–228.

Sinensky, M., and Lutz, R. J. (1992). The prenylation of proteins. *BioEssays* **14,** 25–31.

Smith, S., and Blobel, G. (1994). Colocalization of vertebrate lamin B and lamin B receptor (LBR) in nuclear envelopes and in LBR-induced membrane stacks of the yeast Saccharomyces cerevisiae. *Proc. Natl. Acad. Sci. U.S.A.* **91,** 10124–10128.

Steinert, P. M., and Roop, D. R. (1988). Molecular and cellular biology of intermediate filaments. *Annu. Rev. Biochem.* **57,** 593–625.

Stewart, C., and Burke, B. (1987). Teratocarcinoma stem cells and early mouse embryos contain only a single major lamin polypeptide closely resembling lamin B. *Cell* (*Cambridge, Mass.*) **51,** 383–392.

Stewart, M. (1993). Intermediate filament structure and assembly. *Curr. Opin. Cell Biol.* **5,** 3–11
Stewart, M., Whytock, S., and Moir, R. D. (1991). Nuclear envelope dynamics and nucleocytoplasmic transport. *J. Cell Sci., Suppl.* **14,** 79–82.
Stick, R. (1992). The gene structure of *Xenopus* nuclear lamin A: A model for the evolution of A-type from B-type lamins by exon shuffling. *Chromosoma* **101,** 566–574.
Stick, R., and Hausen, P. (1985). Changes in the nuclear lamina composition during early development of *Xenopus laevis. Cell (Cambridge, Mass.)* **41,** 191–200.
Swanson, J. A., Lee, M., and Knapp, P. E. (1991). Cellular dimensions affecting the nucleocytoplasmic volume ratio. *J. Cell Biol.* **115,** 941–948.
Ulitzur, N., Harel, A., Feinstein, N., and Gruenbaum, Y. (1992). Lamin activity is essential for nuclear envelope assembly in a *Drosophila* embryo cell-free extract. *J. Cell Biol.* **119,** 17–25.
Vikström, K. L., Borisy, G. G., and Goldman, R. D. (1989). Dynamic aspects of intermediate filament networks in BHK-21 cells. *Proc. Natl. Acad. Sci. U.S.A.* **86,** 549–543.
Vorburger, K., Kitten, G. T., and Nigg, E. A. (1989). Modification of nuclear lamin proteins by a mevalonic acid derivative occurs in reticulocyte lysates and requires the cysteine residue of the C-terminal CXXM motif. *EMBO J.* **8,** 4007–4013.
Ward, G. E., and Kirschner, M. W. (1990). Identification of cell cycle-regulated phosphorylation sites on nuclear lamin C. *Cell (Cambridge, Mass.)* **61,** 561–577.
Way, J., Hellmich, M. R., Jaffe, H., Szaro, B., Pant, H. C., Gainer, H., and Battey, J. (1992). A high-molecular-weight squid neurofilament protein contains a lamin-like rod domain and a tail domain with Lys-Ser-Pro repeats. *Proc. Natl. Acad. Sci. U.S.A.* **89,** 6963–6967.
Weber, K., Plessmann, U., and Ulrich, W. (1989a). Cytoplasmic intermediate filament proteins of invertebrates are closer to nuclear lamins than are vertebrate intermediate filament proteins: Sequence characterization of two muscle proteins of a nematode. *EMBO J.* **8,** 3221–3227.
Weber, K., Plessmann, U., and Traub, P. (1989b). Maturation of nuclear lamin A involves a specific carboxy-terminal trimming, which removes the polyisoprenylation site from the precursor: Implications for the structure of the nuclear lamina. *FEBS Lett.* **257,** 411–414.
Weber, K., Riemer, D., and Dodemont, H. (1991). Aspects of the evolution of the lamin/intermediate filament protein family: A current analysis of invertebrate intermediate filament proteins. *Biochem. Soc. Trans.* **19,** 1021–1023.
Wolin, S. L., Krohne, G., and Kirschner, M. W. (1987). A new lamin in *Xenopus* somatic tissues displays strong homology to human lamin A. *EMBO J.* **6,** 3809–3818.
Worman, H. J., Evans, C. D., and Blobel, G. (1990). The lamin B receptor of the nuclear envelope inner membrane: A polytopic protein with eight potential transmembrane domains. *J. Cell Biol.* **111,** 1535–1542.
Worman, H. J., Yuan, J., Blobel, G., and Georgatos, S. D. (1988). A lamin B receptor in the nuclear envelope. *Proc. Natl. Acad. Sci. U.S.A.* **85,** 8531–8534.
Ye, Q., and Worman, H. J. (1994). Primary structure analysis and lamin B and DNA binding of human LBR, an integral protein of the nuclear envelope inner membrane. *J. Biol. Chem.* **269,** 11306–11311.
Yuan, J., Simos, G., Blobel, G., and Georgatos, S. D. (1991). Binding of lamin A to polynucleosomes. *J. Biol. Chem.* **266,** 9211–9215.
Zackroff, R. V., Goldman, A. E., Jones, J. C., Steinert, P. M., and Goldman, R. D. (1984). Isolation and characterization of keratin-like proteins from cultured cells with fibroblastic morphology. *J. Cell Biol.* **98,** 1231–1237.

# Intracellular Structure and Nucleocytoplasmic Transport

Paul S. Agutter
Department of Biological Sciences, Napier University, Edinburgh, Scotland, EH14 1DJ, United Kingdom

---

Intracellular movement of any solute or particle accords with one of two general schemes: either it takes place predominantly in the solution phase or it occurs by dynamic interactions with solid-state structures. If nucleocytoplasmic exchanges of macromolecules and complexes are predominantly solution-phase processes, i.e., if the former ("diffusionist") perspective applies, then the only significant structures in nucleocytoplasmic transport are the pore complexes. However, if such exchanges accord with the latter ("solid-state") perspective, then the roles of the nucleoskeleton and cytoskeleton in nucleocytoplasmic transport are potentially, at least, as important as that of the pore complexes. The role of the nucleoskeleton in mRNA transport is more difficult to evaluate than that of the cytoskeleton because it is less well characterized, and current evidence does not exclude either perspective. However, the balance of evidence favors a solid-state scheme. It is argued that ribosomal subunits are also more likely to migrate by a solid-state rather than a diffusionist mechanism, though the opposite is true of proteins and tRNAs. Moreover, recent data on the effects of viral proteins on intranuclear RNA processing and migration accord with the solid-state perspective. In view of this balance of evidence, three possible solid-state mechanisms for nucleocytoplasmic mRNA transport are described and evaluated. The explanatory advantage of solid-state models is contrasted with the heuristic advantage of diffusion theory, but it is argued that diffusion theory itself, even aided by modern computational techniques and numerical and graphical approaches, cannot account for data describing the movements of materials within the cell. Therefore, the mechanisms envisaged in a diffusionist perspective cannot be confined to diffusion alone, but must include other processes such as bulk fluid flow.

**KEY WORDS:** Diffusionist perspective, Solid-state models, Nucleoskeleton, Pore complexes, Viral proteins.

---

## I. Introduction

Evidence for the role of the nuclear matrix (nucleoskeleton) in transcription, RNA processing, and RNA transport has been reviewed by other contributors to this volume (see chapters by Cook and Jackson, van Driel and de Jong, Stein *et al.,* and Berezney) and elsewhere (Agutter, 1988, 1991). The nucleoskeleton is held to be an RNA–protein network that forms the internal framework of the nucleus and serves as a workbench for many intranuclear activities. Specifically, it is claimed to organize transcription and pre-mRNA metabolism in topologically discrete centers and to provide pathways of mRNA transport to the nuclear periphery (Getzenberg *et al.,* 1991; van Driel *et al.,* 1991), and so by implication it has a role in the regulation of these processes (Pienta and Ward, 1994). These beliefs can be evaluated critically through the following three questions

1. Can the available evidence about RNA metabolism and transport be interpreted in terms of an alternative model that allocates no significant role to the nucleoskeleton?
2. If the answer to this first question is "no," how can the role of the nucleoskeleton be reconciled with the presumption that nucleocytoplasmic exchanges are largely or entirely functions of the pore complex? This presumption is implicit in most current research on pore-complex structure and function (for the status of this field, see chapters in this volume by Pante and Aebi and by Bastos and Burke).
3. Can a nucleoskeleton-dependent model of RNA transport, or a substantially similar model, be valid for transportants other than RNAs?

These questions pertain to the "solid-state" perspective on mRNA transport, which presumes that mRNA and its precursors are essentially nondiffusible in both nucleoplasm and cytoplasm (Agutter, 1988, 1991). The solid-state perspective implies that although events at the pore complex are interesting and important, they constitute only one step (which we have termed *translocation*) in the overall migration of messengers from their nuclear transcription and processing sites to their cytoplasmic translation and degradation sites. Intranuclear migration and *release* of the mRNA, and *cytoskeletal binding* and intracytoplasmic mRNA movement are potentially just as important as events at the pore complex for regulating the overall transport process. Thus, the solid-state perspective accords with the apparent modern consensus that the nucleoskeleton is necessary for RNA transport. The logical alternative, which I shall call the *diffusionist* perspective, presumes that RNA *in vivo* is in solution (diffusible) for enough of its sojourn in nucleus and cytoplasm for its movement through these compartments to be explained by diffusion. On the face of it, a diffusionist model implies that translocation through the pore complex is the only significant aspect of transport. Conversely, the presumption that the pore complex is uniquely important in transport seems to entail the diffusionist perspective.

Evidence for the solid-state perspective has been reviewed several times (e.g., Agutter, 1984, 1988; Schroder *et al.,* 1987a). However, precise characterization of the putative solid-state apparatus and its *modus operandi* has still not been achieved, not surprisingly in view of the inherent complexity of the analytical problems. As a result, "solid-state transport" remains a general idea, or perhaps ideology, rather than a critically testable or predictive model. This makes it both easy and difficult to defend: it is too imprecise to be especially vulnerable to contrary evidence, and too poorly articulated to answer criticisms convincingly. Therefore, a diffusionist model, if it were precisely articulated and could accommodate all the evidence, would make the solid-state concept scientifically redundant, at least for the time being. (In principle, a diffusionist model is easier to articulate than a solid-state model because the only cellular structures that it implicates in transport are the pore complexes, which are much better characterized than the rest of the putative solid-state transport machinery.) The solid-state concept could survive such a diffusionist critique only if it were able to make at least equally coherent sense of the data, i.e., if it were articulated as a predictive, testable model, perhaps generalizable to cell components other than mRNA. A sound diffusionist critique of solid-state transport would therefore enhance understanding, irrespective of its ultimate credibility. Hence the three questions formulated above, which imply a diffusionist critique.

Rather than addressing these questions in sequence (because discussions of them are bound to overlap substantially), I shall assess the principles of the rival theoretical perspectives and possible models in Section II of this chapter and bring the relevant evidence to bear on them in Section III. Recent studies on the effects of viral infections on RNA transport seem particularly likely to give new insights, and I shall reserve a separate section for them (Section IV). In Section V, I shall summarize what seem to me are the current best answers to the three questions in the light of theoretical considerations and evidence.

Nomenclature in the field remains inconsistent. For the purposes of this chapter, I shall use the terms *release, translocation,* and *cytoskeletal binding* as indicated above; i.e., to denote mRNA transport events within the nucleus, at the pore complex, and in the cytoplasm, respectively. I shall use *efflux* for the export of RNA from isolated nuclei and other cell fractions. I shall reserve the term *nuclear matrix* for isolated preparations of the nuclear substructure and denote the *in situ* intranuclear framework by *nucleoskeleton.* This usage has the merits of explicity distinguishing the *in vitro* from the *in vivo,* and of consistency with our own previous contributions to the literature, but it is far from universal, and the reader should note that other contributors to this volume will use the same words in different ways. The term *cytoskeleton,* at least, is presumably unambiguous.

## II. Intracellular Macromolecule Transport: Theoretical Considerations

### A. Explanatory and Heuristic Assessments of Diffusion Theory as an Account of Intracellular Transport

All intracellular transport processes are complicated by compartmentalization. Internal membrane systems and other structures have a high total surface area (Peters, 1986; Gershon *et al.,* 1983), and even those parts of the cytoplasm lacking visible structure are not monophasic (Wolosewick and Porter, 1976; Fulton, 1982). Therefore, intracellular transport of any solute or particle is likely to be influenced by adsorption and desorption events, specific binding, the relative dimensions of "free water" channels and intervening solid structures, and cytoplasmic streaming and other net fluid movements. However, there is a tacit consensus assumption that diffusion is the main mechanism of intracellular transport, albeit modified by the topology, orientation, and composition of the soid phase, the composition, concentration, and mobility of the fluid phase, and biological processes such as specific binding, active transport, and cytoplasmic streaming (Barrer, 1968).

This assumption is dubious. Classical diffusion theory comprises Fick's law and the Einstein–Smoluchowski model of Brownian motion, both of which give predictions that correspond with experimental data only under certain limiting conditions, to which the cell internum does not approximate. Brownian movement is inevitable in any fluid, but within the cell its effects are likely to be overwhelmed by the consequences of compartmentalization and net fluid flow. Even if it did contribute significantly to molecular movements within a more-or-less isotropic cellular compartment, the numbers of molecules involved wouild be too small for the statistical assumptions implicit in diffusion theory to be applicable. These reservations are hardly novel (Donnan, 1927; Halling, 1989) and they are not difficult to grasp; so what sustains the application of diffusion theory in cell biology? Broadly, it seems that although its explanatory value can be doubted, its heuristic value cannot. Classical diffusion theory is quantitative, intuitively simple, and free of *ad hoc* assumptions. These qualities have made it the background to much experimental work that has generated valuable knowledge about transport processes. There seems no prospect of any alternative theory of intracellular transport that can match diffusion theory in simple mathematical articulation or as a source of coherent interpretations of experimental data and novel testable hypotheses.

This judgment of diffusion theory needs explanation. A key issue is that the derivation of Fick's law from first principles (Carlslaw and Jaeger, 1959; Crank, 1975) depends on several assumptions, of which the following are salient. First, the time–intervals over which diffusion is assumed to occur must be several orders of magnitude greater than the duration of an average Brownian movement of a particle in the medium. Second, molecular bombardments of a particle by the solvent must be equally probable (i.e., equally frequent) from all directions. Third, the system must be unstirred, i.e., stationary; there must be no bulk directional solvent flow. Fourth, each particle (suspended particle or solute molecule) must migrate independently of all others. There must be no significant particle–particle interactions, or any particle–solvent interactions other than random elastic collisions. Unless the assumptions are made, Fick's law cannot be derived from first principles; the law is therefore invalid for systems that depart significantly from the assumed conditions.

The inside of the cell is undoubtedly such a system. Eukaryotic cytoplasm, for instance, can be considered homogenous only within very narrow boundaries that have short lifetimes because the macromolecular complexes that constitute them are often transient. The average size of such a compartment, to judge from the dimensions of the putative "microtrabecular lattice" (Wolosewick and Porter, 1976), is in the order of 10–20 nm, and protein migration rates through the cytoplasm imply mean lifetimes for structural protein–protein interactions in this "lattice" in order of 1–100 $\mu$sec (Peters, 1986). Therefore the homogeneous volume encompassing any individual particle and the time intervals between successive changes in the system are incompatible with the first two assumptions. Net solvent flow is commonplace, as indicated by cytoplasmic streaming, so the third assumption is invalid as well. Finally, "particles" within the cell are not independently mobile because of adsorption and specific binding to surfaces and association between macromolecules; this even applies to small solute molecules such as ATP (Horowitz and Miller, 1986). Moreover, Fick's implicit assumption of continuous resistance to diffusion in the medium (Fick, 1855) inevitably ignores electrostatic and Van der Waal's interactions between particle and solvent molecules. This becomes problematic when the particle is comparable in size to the solvent molecule and the distance over which migration takes place is in the same order as the particle diameter, as may often be the case in intracellular transport processes.

Because these implicit assumptions are invalid in the cell, no measured intracellular diffusivity (diffusion coefficient) can in principle be a constant; it must have different values over different time intervals and it must be a function of the concentration of the relevant particle or solute (Agutter *et al.,* 1994). Strictly, Fick's diffusion equation should be replaced by a nonlinear alternative, but the exact form of this would depend on the topology of the system (Crank, 1975). This creates two difficulties. First, since the topology is not usually known in sufficient detail and is in any case labile, the appropriate form of the equation is undecidable. Second, even if a suitable equation could be selected, its nonlinearity would preclude anything but a numerical or graphical solution, and the relevant parameters could not be measured accurately enough to allow the calculation of "diffusivities."

It seems clear that the case for applying Fick's law in biology and for purporting to measure "diffusivities" inside the cell is very weak. What about the molecular mechanism of diffusion (Brownian motion)? The mathematical model of Brownian motion was obtained by Einstein (1905) and von Smoluchowski (1906) from kinetic theory. Their derivations imply not only the same assumptions as the derivation of Fick's law (above), but also the assumptions that (a) the medium is homogeneous and contains large numbers of particles of identical mass, (b) the particle mass is not large, (c) the particles are rigid (all collision are perfectly elastic) and approximately spherical (the Stokes model for the viscous drag force is valid), (d) the basic principles of the kinetic theory of gases apply in the medium, (e) the viscosity of the medium does not vary substantially from one region to another. Clearly none of these conditions applies in the cell. Another difficultly in applying the Einstein–Smoluchowski model to the cell internum is that viscosity and thermodynamic temperature (both explicit terms in the equations) are bulk properties that have significance only when statistical assumptions can be made, and therefore their physical interpretation in the microcompartments of a heterogeneous system such as cytoplasm is unclear. To compound the difficulties, attempts to generalize the diffusion model to three dimensions are handicapped by the anisotropies of intracellular compartments and the labilities of relevant structural arrangements.

During the past two decades there has been revolutionary progress in the provision of computer-dependent numerical solutions to problems in non-Fickian diffusion, diffusion in heterogeneous media, and moving-boundary problems (see Crank, 1975, for an exposition of these terms). All these topics seem *prima facia* relevant to cell biology because of the phase heterogeneity, phase lability, and polymer absorption properties of the cell internum. Briefly, non-Fickian diffusion occurs in systems containing a polymer and a solution of penetrant molecules; the penetrant molecules interact with the polymer and change its surface properties, in turn changing the adsorption and migration of the penetrant species. Analytically, non-Fickian diffusion is especially problematic when the polymer relaxation rate is comparable with the diffusivity of the penetrant, and this might often be the case within the cell. Heterogeneous media are taken to be either particulate dispersions or laminates. Moving boundary problems relate to phase changes such as melting ice in contact with liquid water. Advances in these topics have many applications in physics and engineering, but their apparent relevance to the cell internum is chimeric because molecular movements are not random in the cell, not all the relevant parameters can be measured, and the uncertain topological and other characteristics of the system mean that there is no clear basis for constructing a mathematical model. Thus, although diffusion theory has advanced beyond its simple classical formulation, none of these developments has made it applicable to intracellular media, and it seems that no possible development can do so. This conclusion is implicit in recent biophysical literature that models movements of ions, molecules, and particles by combining the relevant approaches to the diffusion equation with treatments of obstruction by immobile structures, adsorption, specific binding and flow. The complexity of the mathematical results renders them practically inapplicable in many biological situations, other than the particular ones for which they were developed (Saxton, 1982; Bers and Peskoff, 1991; Eckstein and Begacem, 1991; Zwanzig and Szabo, 1991).

In summary, no possible solution of the diffusion equation (Crank, 1975) can properly be applied to intracellular transport phenomena, and the Einstein–Smoluchowski model does not provide a mechanistic explanation for such phenomena. This means that *in principle* diffusionist models of processes such as mRNA transport cannot be explanatorily adequate. On the other hand, the heuristic advantage of the diffusionist perspective is unquestionable, so long as experimental measurements are not interpreted simplistically, e.g., by calculation of diffusivities. Also, I have used the term diffusionist to label perspectives and models that presume non-solid-state transport, i.e., movement of the transportant within the solution phase. Therefore, rejection of diffusion theory as an explanatory tool need not entail rejection of diffusionism, so long as solution-phase transport mechanisms other than diffusion itself are assumed.

## B. The Solid-State Perspective: Assessment of Possible Models

Peters (1930) suggested that cells have an internal solid mosaic responsible for "action at a distance" and for targeting substances to specific intracellular locations. During the past two decades this view has been justified by the introduction of a solid-state concept of the organization of metabolism (Welch and Clegg, 1987) and a consensus opinion that the cell internum is heterogeneous, crowded, and characterized by a high surface-area solid phase (Wolosewick and Porter, 1976; Fulton, 1982; Gershon *et al.,* 1983; Luby-Phelps, 1994). The notion of solid-state mRNA transport developed as these ideas were emerging (Agutter, 1984).

This general perspective needs to be realized in terms of models that can serve as viable alternatives to diffusionism. Any solid-state model seems likely to have the following distinctive characteristics.

1. It implies that the **steady-state** level of any cytoplasmic messenger depends only on the relative rates of transcription and mRNA degradation. Diffusionist models would share this characteristic only if adsorption, degradation, and processing rates were insignificant compared to diffusion rates in every cellular subcompartment, which would be a very dubious assumption. Consistent with this inference, current views on the control of gene expression are that most control is exerted at the levels of transcription and mRNA degradation, and that regulation of the intervening steps (post-transcriptional processing, nucleocytoplasmic migration and translation) is of minor significance.
2. However, it allows **pre-steady-state** levels of a cytoplasmic messenger to depend on the individual rate constants of the multistep transport process. Intermediate steps in transport might therefore be important in switching the system from one steady state to another; any change in a slow rate constant at any stage in the transport process would alter the translation rate. Splicing and translocation through the pore complex might often be slow steps, so regulation of the splicing and mRNA translocation rates may affect the pre-steady state. The finding that individual pore complexes can be "open" or "closed" could therefore be significant for the control of transport during a switch between steady states (Dworetzky and Feldherr, 1988).
3. It implies the existence of large numbers of mRNA and pre-mRNA "location signal" receptors, because receptors must be distributed throughout the solid-state system rather than confined to the pore complexes. Obviously, this inference contrasts sharply with diffusionist predictions.

Maul (1982) suggested that macromolecule transport in nuclei might operate by analogy with either a railcar or a cable car; i.e., the transportants might move relative to a fixed fibril, or might be attached to a fibril which itself moves, e.g., by polar assembly/disassembly processes. As an alternative, solid-state transport can be seen as analogous to metabolic channeling (Srivastava and Bernhard, 1986; Welch and Clegg, 1987). These three suggestions can be developed into the following outlines of models.

1. *Direct transfer model:* the transport is transferred from the binding site on one receptor to the binding site on the next without any intervening

release into the aqueous phase. This is analogous to the behavior of metabolic complexes as described by Srivastava and Bernhard (1986). A model of this type does not require a stable continuous fibrillar system extending between nucleoplasm and cytoplasm. The links between successive receptors can be labile, so long as they dissociate more slowly than the particle migrates. Also, there is no implicit requirement for ATP. However, it is difficult to deduce critically testable predictions from a direct transfer model, so no really well-directed experiments can be generated. Perhaps cross-linking studies could be employed to identify candidate receptor molecules (van Eekelen and van Venrooij, 1981), but results from experiments of this kind are difficult to interpret, especially when the affinity of ligand for receptors is low. Also, it is not clear from a direct transfer model why mRNA transport should be vectorial, though this is evidently the case. On the other hand, some RNAs (and proteins including the ribonucleosome core protein A1 and the heat-shock protein hsp70) are known to shuttle between nucleus and cytoplasm (Goldfarb, 1991; Imamoto *et al.,* 1992; Pinol Roma and Dreyfuss, 1992; Shi and Thomas, 1992), and a direct-transfer solid-state model could be consistent with this phenomenon.

2. *Motor driven model:* the ligand is attached via a molecular motor to a cytoskeletal or other fibril. Transport is analogous to the migration of chromatids on the mitotic spindle or of neurotransmitter vesicles along a nerve axon. If this model is valid for any transcript then the motor is most likely to be myosin (Adams and Pollard, 1986; Warrick and Spudlich, 1987; Korn and Hammer, 1988). A continuous fibrillar system linking the internal spaces of nucleus and cytoplasm is implicit, and the vectorial character of mRNA transport is consistent with this model. The nucleocytoplasmic shuttling of some proteins and RNAs is therefore problematic, but no more so than the twin phenomena of forward and retrograde axonal transport that are attributed to a common fibrillar system (Maul, 1982). A motor-driven model makes clear predictions (transport requires ATP hydrolysis and a continuous fibrillar system; a homogenous population of receptors is distributed on fibrils in nucleus and cytoplasm) and it is therefore experimentally testable in principle; but in practice, evidence pertinent to this prediction would be difficult to interpret unequivocally.

3. *Assembly driven model:* the ligand is attached to a fibrillar subunit, and the attaching subunit is moved through the cell by net fibril assembly at the proximal site and disassembly at the distal site. This reflects ideas of microtubule and microfilament dynamics that were current in the early 1980s. If fibril assembly can proceed in any direction from a nucleation point, this model can reconcile the nucleocytoplasmic shuttling of some species with the generally vectorial nature of the transport of mRNA and of other macromolecules. According to this model, transport should be more efficient if the affinity of receptor for ligand is high. This contrasts

with the direct transfer model, which can operate efficiently only with low receptor–ligand affinities (and high receptor numbers). The model also predicts a positive correlation between the rate of a cell's mRNA transport and its rate of fibril turnover. This prediction should be testable, though again the experimental difficulties are considerable. There is presumably a requirement for ATP or an alternative energy source (only cells with low mRNA transport rates, and therefore low fibril turnover rates, could give the appearance of ATP-independence). However, a directional assembly-driven mechanism is unlikely in view of current understanding of cytoskeletal fibril dynamics. Microtubule assembly/disassembly, for example, shows chaotic fluctuations at both poles.

Even this brief outline suggests that the solid-state perspective can potentially be articulated in terms of models that are explanatorily superior to diffusionist models. However, although testable predictions can be deduced from at least some of these models, the relevant experiments would be very difficult to perform and interpret. It therefore seems that the explanatory advantage of solid-state over diffusionist models is compensated by a heuristic disadvantage.

### C. Generalizability of Diffusionist and Solid-State Models

Mobile entities in cells range from small solute molecules and ions to organelles. There is no question of explaining the intracellular transport of glucose molecules or potassium ions in solid-state terms; adsorption and exclusion processes notwithstanding, this is a matter of movement in the solution phase. Conversely, there is no question of explaining migrations of mitochondria in diffusionist terms; this is a matter of cytoskeletal dynamics (Alberts *et al.,* 1988). However, there is room for debate about transport mechanisms for entities between these extremes of size. In principle, macromolecules and complexes such as ribonucleoproteins could migrate by either diffusionist or solid-state processes. The uncertainty is acute in relation to nucleocytoplasmic transport. Before the evidence is brought to bear on this problem (Section III), some further general considerations will be helpful.

Molecular traffic across the nuclear envelope must be precisely choreographed because constant communication between nucleus and cytoplasm is a precondition for the survival of the cell. Modulation of this traffic might play a role in regulating cellular activities; nucleocytoplasmic transport processes change qualitatively and quantitatively during development, aging, and carcinogenesis (Muller *et al.,* 1985; Agutter, 1988; Pienta and Ward, 1994). Although characterization of the location signals and knowledge of the functional organization of the pore complex are prerequisites for elucidating nucleocytoplasmic transport mechanisms (Dingwall and Laskey, 1986; Feldherr, 1992), they will not account for all our knowledge of nucleocytoplasmic transport. Intracompartmental binding, not translocation across the nuclear envelope, determines the nucleocytoplasmic distribution ratios of many macromolecules (Paine 1984, 1993). Potentially, therefore, both signal–receptor interaction at the pore complex and nuclear and cytoplasmic binding kinetics are relevant to intracellular macromolecule distributions. If a negligible fraction of the transport substrate is "in solution," i.e., most or all of the material remains bound to intracellular structures, a solid-state transport process is implicit. If less of the transportant is bound, a diffusionist process is more likely.

The inferences from this are as follows. To decide which **perspective** better accommodates nucleocytoplasmic transport data, two issues must be resolved: how many of the data can be explained by translocation (events at the pore complex); and how much of the transportant can be shown to be mobile in nucleus and cytoplasm. If these judgments can be made, it is then necessary to decide which **model** best accommodates the data. Here, critiques from the opposed perspective, and the criterion of experimental testability, will be useful guides.

## III. Nucleocytoplasmic Transport: Perspectives, Models, and Evidence

### A. Messenger RNA Transport: Evidence for the Solid-State Perspective and a Diffusionist Critique

Important lines of evidence supporting the solid-state perspective for nucleocytoplasmic mRNA transport include the following.

1. Normal nucleocytoplasmic RNA distributions survive puncturing of the oocyte nuclear envelope *in situ* (Feldherr, 1980), suggesting that much of the RNA is structure-bound in both nucleus and cytoplasm.

2. Transcription is initiated on DNA sequences associated with the nucleoskeleton (Jackson and Cook, 1985), indicating that nascent transcripts are structure-bound.

3. Most nuclear RNA copurifies with nuclear matrix preparations (Long *et al.,* 1979; Berezney, 1980; van Eekelen and van Venrooij, 1981) and is not exchangeable *in vitro* with soluble RNA (Agutter and Birchall, 1979; Gruss *et al.,* 1979), suggesting that the bulk of the nuclear RNA is bound rather than free. Moreover, some transcripts are distributed along definite

curved tracks, several micrometers long, leading through the intranuclear space to the nuclear envelope. This evidence is difficult to reconcile with a diffusionist perspective. Examples include c-*fos* transcripts in mouse fibroblasts (Huang and Spector, 1991) and Epstein–Barr virus transcripts (Lawrence *et al.*, 1989). These tracks are apparent in both the *in situ* nucleus and the chromatin-depleted matrix, indicating that RNA association with isolated nuclear matrix preparations is not a simple artifact. Also, intron-containing RNA microinjected into nuclei has been found to associate with the "speckle" structures in which splicing components seem to be concentrated, suggesting processing sites that are maintained as topologically discrete entities (Wang *et al.*, 1991).

4. Isolated nuclei show normal nuclear RNA restriction *in vitro* as long as swelling and RNase and proteinase activities are inhibited. Aqueous isolation procedures are known to damage nuclei so that any unbound constituents will be more or less completely lost (Paine, 1984), so these findings strongly support the concept of solid-state intranuclear attachments. If any more than a trace of the nuclear RNA were soluble, nuclei isolated in aqueous media could not serve as experimental models for RNA transport (Agutter, 1994).

5. Most (though not all) translationally active polysomes are cytoskeleton-linked after cultured cells are extracted with detergents and dilute buffers (Cervera *et al.*, 1981; van Venrooij *et al.*, 1981; Jones and Kirkpatrick, 1987; Hesketh and Pryme, 1988). These findings corroborate the idea that most messengers are structure-bound in the cytoplasmic as well as the nuclear compartment. The relevance of cytoskeletal binding of mRNAs to translation and to protein sorting among cytoplasmic subcompartments has been reviewed by Singer (1992) and will be discussed further below.

6. Nuclear RNA returns to the nucleus after mitosis, but exogenous RNA injected into the cytoplasmic compartment remains localized and does not redistribute (Rao and Prescott, 1967). This further implies that cytoplasmic messengers do not behave as if they were freely diffusible so RNAs injected into a cell cannot be assumed to behave physiologically (Agutter, 1994).

7. In *Drosophila* embryos, pair-rule segmentation gene transcripts are found in definite cytoplasmic locations, indicting that they have been translocated through a few specific pore complexes and have subsequently been immobilized (Davis and Ish-Horowicz, 1991). This apparent polarity of mRNA transport is difficult to reconcile with the existence of soluble, diffusible messengers and is consistent with the "gene gating" hypothesis (Blobel, 1985).

8. In all known cases, the affinity of the receptor for the signal is in the order of $10^7$ *M*. This value is strikingly low. It implies that if translocation is to be efficient, the number of receptors must be large, by analogy with

extracellular matrix receptors, which typically have affinities two to three orders of magnitude lower than (say) plasma-membrane hormone receptors but are much more abundant. (It is noteworthy that the interaction of receptors with hormones is clearly a diffusionist process, while cell migration along the extracellular matrix is patently a solid-state one.) This in turn suggests that the receptors might not be confined to the pore complexes, which presumably could house only limiting numbers of specific receptors (Garcia-Bustos *et al.,* 1991). Indeed there is evidence for widespread intracellular distributions of both protein (Yamasaki *et al.,* 1989, Breeuwar and Goldfarb, 1990; Stochaj and Silver, 1992) and mRNA (Scroder *et al.,* 1988) receptors, perhaps involving the cytoskeletal and nucleoskeletal fibers involved in solid-state transport.

However, a paper by Zachar *et al.* (1993) throws doubt on the interpretation of some of this evidence. These workers studied the large, relatively uncluttered polytene nuclei of *Drosophila* salivary glands *in situ,* avoiding the possibilities of artifacts from extraction or fractionation procedures and of confusion of the microscopic evidence by multiple chromosomes and transcription sites. They used fluorescence microscopy to study the intranuclear migration of transcripts of a chimeric gene (SgSLAC1), one intron of which is cotranscriptionally spliced and another very slowly spliced (a prerequisite for sufficiently prolonged nuclear retention for practicable microscopy). Weblike fluorescent arrays were observed throughout the nucleus, suggesting that the gene products are conveyed unselectively to all pore complexes rather than specifically to a few. Moreover, evidence was presented that the web was defined not by the presence of nucleoskeletal fibrils, but as zones of exclusion from the nucleoli and chromosome axes; the "web" system was therefore called an *extrachromosomal network.* The slow splicing step occurred within this network, after detachment of the pre-mRNA from the transcription zone. Rates of migration to the nuclear periphery were calculated to be consistent with diffusion, implying that immature messenger precursors are retained in the nucleus by the permeability limitation of the pore complexes. Zachar *et al.* (1993) suggested that their conclusions might be valid for metazoan nuclei in general. This paper throws doubt on most of the above lines of evidence for the solid-state perspective.

1. The pathways in the extrachromosomal network are narrow and sparsely distributed. Therefore, if the nuclear envelope were cut open, the rate of egress of the RNA would not be significantly accelerated; diffusion from the endogenous tracks to the incision site would be very slow for the most part.
2. It is not doubted that transcription occurs at the nucleoskeleton, but this is irrelevant to the presumed role of the nucleoskeleton in mRNA processing and transport.

3. The appearance of defined tracks for some transcripts could be explained as extended "primary zones," i.e., long loops of DNA bearing active transcription sites. They need not be interpreted as sites of processing and transport. Also, artifactual reassociation of RNA with the chromatin or with nucleoskeletal fibrils could occur during chromatin depletion, generating the appearance of tracks. The association of microinjected RNA with the speckle structures in the nucleus can be interpreted as entry by molecular exclusion into the extrachromosomal network, rather than binding to the nucleoskeleton, particularly since the occurrence of cotranscriptional splicing ensures the association of some splicing components with primary zones.

4. The capacity of isolated nuclei to simulate physiological transport was not addressed. Because of the time courses involved, the explanation given in (1), above, cannot be applied here. However, there is no evidence that *Drosophila* polytene nuclei do simulate physiological transport *in vitro;* mRNA efflux studies have been conducted on nuclei from very different sources.

5,6. Association of mRNA with cytoskeletal elements has no bearing on its presumed association with the nucleoskeleton, except as an analogy (see below).

7. In the case of those few messengers for which export to the cytoplasm is demonstrably anisotropic, Zachar *et al.* (1993) allow the possibility of a separate, specific mechanism that might involve the nucleoskeleton; but these cases are relatively rare.

8. Low receptor affinity is less problematic when the concentration of transportant is high. The concentration of mRNA within the confined space of the extrachromosomal network would ensure this, so there is no need to infer from the kinetic data that signal receptors are too numerous to be confined to the pore complexes.

In short, except for special cases (point 7), the only substantial line of evidence for a nucleoskeletal role in mRNA transport that seems to survive this diffusionist critique is the physiological validity of RNA efflux from isolated nuclei. However, there are difficulties with the critique itself. For example, the extrachromosomal network represents about 10% of the nuclear volume, implying a 10-fold concentration of splicing components within the pathways. At such high concentrations, snRNPs and other splicing components are unlikely to remain freely soluble; they will probably aggregate or associate with insoluble elements in the nucleus. This casts doubt on the authors' assumption that these components have diffusional access to active genes. Moreover, they assume that diffusivities calculated from the rates of movement of microinjected dextrans and other inert particles (d'Angelo, 1946; Peters, 1986) are simply interpretable and are applicable to RNP movements. This is invalid (see Section II,A); in any case, to judge from the micrographs, the diameters of the channels do not greatly exceed the radius of gyration of RNP complexes, which precludes the possibility of diffusion (Luby-Phelps, 1994). Also, these workers assume that solid-state transport must occur by a motor-driven mechanism (Section II,B) and, moreover, that this operates at the highest known rate for cytoskeletal processes (in the order of 1 $\mu m\ sec^{-1}$). This value need not be valid even for a motor-driven model, let alone either of the alternative models outlined in Section II,B. Therefore, the argument excluding solid-state transport of the transcripts to the nuclear periphery is not necessarily sound.

However, the data and the arguments in Zachar *et al.* (1993) emphasize that the controversy cannot be resolved without the development of well-focused models for solid-state transport, and although diffusion in the strict sense of the term cannot account for intracellular movement, particularly of molecules and complexes with large radii of gyration, the possibility remains that some transported species can be carried in solution phase by bulk flow, as envisaged e.g., in the fluid dynamic model of Remenyik and Kellermeyer (1978). A well-conducted test of bulk fluid flow through the putative extrachromosomal matrix would therefore corroborate the essence of the Zachar *et al.* (1993) viewpoint.

Since the 1970s, a serious obstacle to the generation of testable models of nucleoskeletal function has been the poorly characterized state of the structure. Recent advances (see Section III,D) have partly but not completely demolished this obstacle, so more convincing tests of the solid-state position may soon be possible. Meanwhile, the diffusionist critique has been reinforced by the observation that nascent RNA from the transcription of some individual genes has been located on several hundred, rather than just a few, intranuclear domains (Jackson *et al.*, 1993). This casts doubt on the specific association of active genes with discrete points on the nucleoskeleton and *a fortiori* on the solid-state view of transport as it is currently accepted. Rosbash and Singer (1993) have interpreted these data in a way similar to Zachar *et al.* (1993).

## B. Association between mRNA and the Cytoskeleton

The preceding discussion leaves open the question of how much intranuclear pre-mRNA is bound and how much is diffusible *in situ*. Because the cytoskeleton is better characterized than the nucleoskeleton, the parallel question might be more readily answered for the cytoplasmic compartment. Conclusions about the extent of association of mRNA with the cytoskeleton might, by analogy, be used to generate hypotheses about the situation in the nucleus.

The earliest studies indicating cytoskeletal binding of all translationally active mRNAs were performed on cultured cells depleted of membranes and soluble components (Cervera *et al.,* 1981; van Venrooij *et al.,* 1981), but most of the recent evidence has been obtained from intact cells. Specific messengers have been located by *in situ* hybridization at their translation sites. For instance, actin mRNA is concentrated in several cell types in regions where actin is synthesized or polymerized (Lawrence and Singer, 1986; Cheng and Bjerknes, 1989; Hoock *et al.,* 1991), and anchoring of the messenger to these sites depends on the RNA itself, not on a nascent polypeptide chain (Sundell and Singer, 1990, 1991). This contrasts with the signal recognition complex in the endoplasmic reticulum, where the linkage occurs via the nascent secreted polypeptide (Walter *et al.,* 1984). Messengers for microtubule-associated proteins are anchored in neuronal dendrites (Garner *et al.,* 1988), while messengers for other proteins in the neuron, including those for tubulin, actin, and GAP 3, are restricted to the cell bodies (Kleiman *et al.,* 1990). Only as neurites mature into neurons does this sorting occur; moreover, messengers are imported only into the dendrites, not the axons. Myosin mRNA is located near the muscle sarcomeres (Pomeroy *et al.,* 1991). Data such as these are now widely reported, and they imply specific partitioning of messengers within different cytoplasmic regions. This inference more easily reconciled with the solid-state than with the diffusionist perspective.

Current evidence suggests that most of the cytoplasmic mRNA is bound to microfilaments. Cytochalasin releases messengers from isolated cytoskeletal preparations, actin-depolymerizing agents randomize messenger distributions in intact cultured cells (Sundell and Singer, 1990, 1991), and the *Dyctostelium* actin-binding protein ABP-50 is identical with the elongation factor EF1 (Yang *et al.,* 1990). This last observation is striking in view of the apparent dependence of translation on cytoskeletal binding (Cervera *et al.,* 1981). These experiments seem to imply that attachment to the actin cytoskeleton is a precondition for mRNA translation (Cervera *et al.,* 1981; van Venrooij *et al.,* 1981; Vedeler *et al.,* 1991), as long as the actin-containing protein gel of the rough endoplasmic reticulum membrane can be counted as part of this cytoskeleton (Cervera *et al.,* 1981; van Venrooij *et al.,* 1981); but these last results may be difficult to interpret (Hesketh and Pryme, 1988; Agutter, 1991).

In some cell types, mRNAs may be microtubule-associated for at least some of their cytoplasmic lifespans (Yisraeli *et al.,* 1990), or in some cases bound to intermediate filaments (Jeffery, 1989). However, since colchicine does not affect polysome anchoring, translationally active messengers are presumably not associated with microtubules or intermediate filaments

(Lenk *et al.*, 1977; Vedeler *et al.*, 1991). The possibility that only translationally inactive messengers are bound to intermediate filaments is supported by immunohistochemical and other studies on the distributions of untranslated messenger particles (Scherrer, 1990). These findings may indicate that polysome redistribution between active and inactive pools, e.g., in response to extracellular signals (Moon *et al.*, 1983; Vedeler *et al.*, 1990), occurs by messenger migration from one binding site to another. Redistribution of messengers to nonactin cytoplasm anchoring sites could be important in controlling both translation (Hesketh and Pryme, 1988) and degradation rates (Zambetti *et al.*, 1990).

All these data seem to support a solid-state rather than a diffusionist interpretation. However, in their review of cytoskeleton–mRNA association, Hesketh and Pryme (1988) concluded that only 70–80% of the polysomal messenger population was firmly anchored in most cell types. The other 20–30% was readily solubilized. This conclusion can be reconciled with the solid-state interpretation by hypotheses such as: the "free" polysomes contain the older mRNAs, as suggested by the increase in the size of this free pool when the protein synthesis rate is increased (Moon *et al.*, 1983); or they represent mRNA in the process of migration from one type of binding site to another; or they indicate low affinity of receptor for transportant. However, the existence of a free pool reopens diffusionist possibilities. For instance, if the actin cytoskeleton restricts polysome movement not by specific binding but by molecular exclusion into restricted aqueous channels, then the data on the effects of F-actin disruption and the role of ABP-50 are readily explicable. The different partitioning of active and inactive messengers within the cytoplasm might be attributed to the structural and physical (e.g., size and surface charge) characteristics of the two pools, though this hypothesis would need more precise formulation. Only the specific association of some active polysomes with defined cytoplasmic regions seems genuinely irreconcilable with the diffusionist perspective.

On balance, however, the current evidence seems more in line with a solid-state than a diffusionist interpretation. Its only difficulty is the free pool, and potentially testable hypotheses can be advanced to explain this (see above). In contrast, there seems no convincing way to reconcile the subcompartmentalization data with the diffusionist perspective. Moreover, if the solid-state view of cytoplasmic mRNA distribution is justified, a number of interesting questions are raised. For instance, to what extent does the location of a messenger in a particular cytoplasmic region depend on the polarity of translocation? The gene gating hypothesis of Blobel (1985) and the pair-rule transcript localization described by Davis and Ish-Horowicz (1991) imply some such correlation. Further to this question: how does mRNA leaving the pore complex become associated with the solid-state machinery of the cytoplasm, and by what (presumably) actin-associated proteins is the mRNA anchored at its translation site? Attempts to address these questions might *inter alia* help to clarify the significance of the free polysome pool. In contrast, the diffusionist perspective does not seem to lead to any further significant inquiries other than the possible relationships between the physical properties of different RNP classes and their association with actin or nonactin regions. In short, this is one case where the solid-state perspective has a heuristic advantage over the diffusionist.

### C. Association between Pre-mRNA and the Nucleoskeleton

This discussion justifies the provisional assumption that translationally active messengers are associated with the actin cytoskeleton. Actin also seems to be among the main components of the nucleoskeleton (Schindler and Jiang, 1987), so there is *prima facie* support for the hypothesis that pre-mRNA is associated with an actin nucleoskeleton. Using a method (Comerford *et al.,* 1986) that yields nuclear matrix preparations ultrastructurally indistinguishable from the nucleoskeleton visualized *in situ* (Brasch, 1982; Fey *et al.,* 1984), i.e., a dense reticular system of fine (approx. 8–15 nm diameter) fibrils, Schroder *et al.* (1987b,c) were able to show that incompletely spliced ovalbumin premessengers from hen oviduct nuclei were selectively solubilized by cytochalasin B. On the face of it, this result supports the view that intranuclear migration of the RNA depends on actin fibrils and myosin motors (Berrios and Fisher, 1986; Schindler and Jiang, 1986). In contrast, mature ovalbumin mRNA was liberated by DNA topoisomerase II activators, and its splicing intermediates were not; DNA topoisomerase II is considered another nucleoskeletal component (Berrios *et al.,* 1986). The findings of Schroder *et al.* (1987b,c) have interesting relationships with other data. First, the matrix preparation used in these studies is mechanically delicate and cannot be isolated from nuclei that have been swollen or subjected to proteolysis or RNases. These are precisely the insults that prevent normal RNA restriction in isolated nuclei used for efflux studies (Agutter, 1988, 1994). Second, isolated nuclei incubated in the presence of cytochalasin B liberate most of their RNA, including immature messenger precursors. Third, cytochalasin B treatment liberates the Hn-RNP C-group proteins, which have been implicated in RNA–matrix binding by ultraviolet cross-linking studies (van Eekelen and van Venrooij, 1981).

Despite these indications that pre-mRNA is associated with the nucleoskeleton, the contrary evidence discussed in Section III,A insists, minimally, that nucleoskeletal attachments must differ among RNA species. Better characterization of the nucleoskeleton is a prerequisite for accomodating these differences within a unified solid-state model. This improved characterization is now being achieved. General findings such as the increased protein mass and HnRNP association of the nuclear matrix after heat shock (Wachsbereger and Cos, 1993) are of little value for the purpose, but identification of specific nucleoskeletal components such as the high $M_r$ component H1B2 Ag (Nickerson and Penman, 1992) and the 68,000 molecular mass phosphoprotein B2 (Chew *et al.,* 1992) is potentially more helpful. Two specific component, PI1 and PI2, seem to be essential for normal genomic expression, which makes them particularly interesting in relation to RNA metabolism (Prather and Schatten, 1992). The specific associations that have been reported between actin and snRNPs and the splicing factor SC-35 (Carter *et al.,* 1993; Sahlas *et al.,* 1993) may contribute significantly to answering the diffusionist critique (Zachar *et al.,* 1993). These associations between actin and the splicing components become more intimate in differentiating cells and support "a nuclear model in which there is a specific topological arrangement of noncontiguous centers involved in precursor mRNA metabolism from which RNA transport towards the nuclear envelope radiates" (Carter *et al.,* 1993). Most interesting of all, perhaps, is the apparent role of the nuclear mitotic apparatus protein in providing a functional interface between the nuclear skeleton and the spliceosome (Lydersen and Pettijohn, 1980; Zeng *et al.,* 1994) and the localization of EGF receptor messenger splicing intermediates in the nucleoskeletal region immediately surrounding the nucleolus (Sibon *et al.,* 1993). This last observation is the first clear indication of functional compartmentalization of the nucleoskeleton in respect of the metabolism of specific mRNAs, and like the analogous phenomenon in the cytoskeleton (Section III,B), it is very difficult to reconcile with the diffusionist perspective.

However, the nucleoskeleton remains less fully characterized than the actin cytoskeleton, so the diffusionist arguments advanced for cytoplasmic mRNA distributions (Section III,B) have greater force when applied to the nucleus. In the cytoskeleton, characterization includes not only the identification of core fibril components and the appropriate immunomicroscopic demonstrations of fibrillar structures *in situ,* but also the identification of conditions for reversible assembly/disassembly and insights into the molecular basis of monomer binding. Characterization of the nucleoskeleton has not reached this point, and this is crucial for the interpretation of evidence about specific associations with RNAs and splicing components. For example, a diffusionist might argue the following. Copurification with the nuclear matrix can be dismissed as an artifact; colocalization with the nucleoskeleton might be fortuitous and functionally insignificant; demonstration that a component binds to actin is weak evidence because so many things bind to actin; and the nucleoskeleton might fulfill a molecular exclusion rather than a workbench role. If the actin nucleoskeleton and DNA topoisomerase II-dependent structures act as exclusion zones, then disruption of either will increase the relative volume of the putative extrachromosomal network and render the pre-mRNAs (or mature mRNAs) more readily mobile.

Such interpretations could accommodate many of the data reviewed in this section. The only serious surviving difficulty for the diffusionist perspective seems, once again, to lie in the evidence for subcompartmentalization: why should pre-mRNA (after detachment from the transcription sites) and mature nuclear mRNA be mobilized by different agents, and why should the EGF receptor messenger be processed only in the region surrounding the nucleolus? Overall, though, the argument is that the solid-state transport is weaker for pre-mRNA in the nucleus than for mRNA transport in the cytoplasm, so the diffusionist position is correspondingly stronger, provided that (a) the diffusionist mechanism is understood in terms of e.g., bulk fluid flow through the extrachromosomal network rather than diffusion per se (Sections II,A and III,A), (b) it is accepted that the elements of the nucleoskeleton, i.e., the intranuclear actin and DNA topoiomerase II-dependent structures within the nucleus act as exclusion zones contributing to the definition of the extrachromosomal network. These provisos do not fundamentally change the position argued by Zachar *et al.* (1993). Therefore, the question of the nucleoskeleton's role in intranuclear RNA transport (release) remains open. It may function positively, as the framework for migration; or negatively, by defining restricted aqueous compartments within which soluble ribonucleoproteins migrate. This question is highly pertinent to the overall theme of this volume and central to the present chapter, and further elucidation is desirable.

### D. Explanation of Nucleocytoplasmic Transport Data by Reference to Translocation Alone

In Sections III,A–C it has become apparent that current information about intracompartmental binding fails to give definitive support to either major perspective on mRNA transport, particularly in regard to the nucleus, though the balance of evidence seems to favor the solid-state perspective. The other pertinent line of inquiry into the general perspectives (Section II,C) concerns the explanatory value of "translocation-only" models of transport. Most knowledge of the mRNA translocation mechanism has come from studies of efflux from isolated nuclei and resealed nuclear envelope vesicles. These studeis show that in most cell types, translocation involves ATP hydrolysis by a nuclear envelope-associated NTPase, and receptors in the envelope for signals that reside in the mature messenger molecule. One such receptor, a protein of $M_r$ 110,000 (P110), binds poly(A) with an affinity (around $10^7$ $M$) that is changed by phosphorylation by endogenous protein kinases. Poly(A)$^+$ mRNA binding to P110 precedes NTPase-dependent migration across the nuclear envelope, making mRNA translocation a two-step process (McDonald and Agutter, 1980; Bernd *et al.,* 1982) analogous to nuclear protein import (Newmeyer and Forbes, 1988; Richardson *et al.,* 1988). The interpretability of these data depends on the physiological validity of efflux studies, which has been established for some types of nuclei under certain experimental conditions (Agutter, 1988). Messenger transport also involves the pore-complex nucleoporin gp62 because the process is blocked by agents that interact with this component (Baglia and Maul, 1982; Davis and Blobel, 1986). There is now *in situ* evidence not only supporting the ATP requirement in translocation (Dargemont and Kuhn, 1992) and the necessity for a mature 3′ end on the transportant (Eckner *et al.,* 1991), but also indicating the requirement for 5′ cap-bindings proteins (Hamm and Mattaj, 1990; Hamm *et al.,* 1990).

Given that the efflux data are valid, translocation is better understood than any other aspect of mRNA transport (Agutter, 1994), though the topology of the translocation apparatus (the relative arrangements of e.g., the NTPase, P110, and the relevant nucleoporins such as gp62) remains obscure. Correspondingly, there are kinetic models for translocation (Schroder *et al.,* 1986), but kinetic models for solid-state transport as a whole have not been articulated. The kinetic complexity of at least some translocation processes is well attested (Michaud and Goldfarb, 1991). Nevertheless, it seems that simple kinetic models of translocation, in which enough of the parameters have experimentally determined values for the equations to be explicitly soluble, can account quite satisfactorily for some experimental findings (Prochnow *et al.,* 1994). However, they have limitations.

Analysis of such models shows that under steady-state conditions the concentrations of poly(A)$^+$ mRNA required for a half-maximal translocation velocity is equal to the dissociation constant of the mRNA–P110 complex, which is consistent with experimental findings (McDonald and Agutter, 1980). A difficulty with the argument is the referent of "concentration" when solid-state perspective is assumed. Provided that translocation is rate-limiting in transport, this term might be taken to signify the number of messenger molecules transferred per unit time from the nucleoskeleton to the pore complex. According to the model, the translocation system amplifies nuclear concentration differences for low and intermediate abundance messengers, so that a 100-fold increase in nuclear concentration of a low abundance poly(A)$^+$ mRNA, from $10^{-8}$ to $10^{-6}M$ elicits a 1000 (if P110 is phosphorylated) or 3000-fold (if P110 is unphosphorylated) increase in cytoplasmic concentration. This is broadly consistent with the known abundance differences of mRNAs and mRNA precursors in the two cellular compartments (Sippel *et al.,* 1977). However, the amplification effect does not occur for high-abundance messengers, implying that regulation of the transport of these molecules is not exerted at the translocation stage.

In other respects, the model is less successful. Since extracellular effectors including insulin and epidermal growth factor can alter the extent of phosphorylation of P110 (Schroder *et al.,* 1990), the simple kinetic model should be able to predict the influences of these effectors on cytoplasmic mRNA levels. However, it does not explain their selectivity; the model implies that they should increase (insulin) or suppress (EGF) the amplification effect for all messengers except high-abundance ones, and this does not accord with the evidence. The biological relevance of these effects is controversial (how is the insulin internalized *in vivo* so that it can interact with nuclear envelope components?) but might be explained by the hormone-induced migration of protein kinase C to the nuclear envelope (Divecha *et al.,* 1991) and the consequent phosphorylation of P110. This explanation does not evade the specificity issue. This point supports the contention of Riedel and Fasold (1992) that the translocation machinery is not selective in the messengers it conveys from nucleus to cytoplasm. More obviously, translocation-only models do not account for anisotropic transport phenomena or for cytoplasmic compartmentalization of messengers (see earlier). In Section IV, their inadequacy for explaining the effects of viral proteins on mRNA transport will also be demonstrated.

If the available data cannot be accommodated within a translocation-only model, then perhaps it follows, as suggested in Section I, that the general diffusionist perspective is inadequate to explain current information. Certainly, taking a diffusionist rather than a solid-state standpoint does not overcome the difficulties of explaining the insulin and EGF effects or cytoplasmic compartmentalization. Moreover, if it is assumed that mRNA is presented to the nuclear face of the pore complex by way of a "diffusion channel" in the extrachromosomal network rather than a solid-state fibril, then the effective concentration of low or intermediate abundance class messengers will be around 10× higher than determined by Sippel *et al.* (1977) and others. This will make the kinetic model of Schroder *et al.* (1994) less, not more, adequate; it will cease to explain the nucleocytoplasmic class abundance differences in terms of a simple amplification effect.

The argument in this section favors the solid-state over the diffusionist perspective. Specifically, it provides at least an indirect justification for the solid-state view of mRNA presentation to the pore complex. This completes the general evaluation of perspectives by the criteria proposed in Section II,C.

### E. Evaluation of the Three Models for Solid-State Transport

The evidence reviewed in Sections III,A–III,D implies overall that although the solid-state perspective remains open to diffusionist critique, and in particular that the solid-state character of intranuclear RNA transport remains uncertain, the diffusionist perspective is inadequate because it cannot account for cytoplasmic mRNA distributions and because translocation-only models of transport have serious limitations. This necessitates a fuller evaluation of the three models of solid-state transport that were outlined and assessed theoretically in Section II,B.

Some evidence is more easily reconciled with a direct transfer model than with the alternatives; e.g., the extrachromosomal network distributions of some transcripts (Zachar *et al.,* 1993), and the fact that up to 30% of cytoplasmic messengers are not apparently bound to the cytoskeleton (Hesketh and Pryme, 1988), could be interpreted as the results of low-affinity transportant–receptor binding and of labile interactions between adjacent receptors. The low affinity of the P110–mRNA complex (see Section III,D) supports this interpretation. However, the facts that some transcripts form defined intranuclear tracks (Huang and Spector, 1991; Xing and Lawrence, 1991) and that most cytoplasmic messengers are apparently bound to the cytoskeleton (Hesketh and Pryme, 1988) require modification of the direct-transfer model: interreceptor linkages must be stable, and transportant–receptor affinities must be relatively high, for some transcripts but not others. Therefore, either different sets of receptors are required for different transcripts, which may be uneconomical; or at least some receptors are structurally modified when some transcripts but not others are bound to them. The latter alternative is plausible in principle, but the mechanism is obscure. These difficulties make the inherent untestability of direct-transfer models (Section II,B) especially problematic.

More of the available evidence can be recruited in support of the motor-driven model. First, there is evidence for the involvement of myosin (and actin) at least at the translocation stage of mRNA transport (Berrios and Fisher, 1986; Schindler and Jiang, 1986), and of actin at both the release and cytoskeletal binding stages (see Sections III,C and III,D). Second, although the NTPase has been identified as a nuclear envelope component and implicated in translocation (Agutter, 1984, 1988), it has also been reported on the nucleoskeleton (Clawson *et al.,* 1984), and the possibility that it also occurs in the cytoskeleton cannot be excluded (Singer, 1992). This is consistent with the requirement of this model for ATP hydrolysis throughout the solid-state system. Third, there is evidence for a continuous nucleocytoplasmic fibrillar system, extending through the pore complexes, such as this model requires. High-voltage transmission electron micrographs

(Fey *et al.*, 1984) exemplify this evidence, although such pictures are difficult to interpret unambiguously. Furthermore, although many details of pore-complex organization remain to be elucidated, there is considerable evidence for fibrillar connections between an individual pore complex and (1) the nucleoskeleton, (2) the cytoskeleton, and (3) other pore complexes, and for the role of longitudinal fibrils within the pore complex itself in the movement of transportants (Agutter and Prochnow, 1994). The "basket" of intranuclear fibrils attached to the pore complex may be particularly significant for the presentation of translocatable messengers.

Fourth, P110 is found in association with cytoplasmic actin fibrils as well as the nuclear envelope (Schroder *et al.*, 1988) and probably occurs within the nucleus as well (Schweiger and Kostka, 1984). This suggests that P110 may be involved in mRNA–microfilament attachments. Since the mobility of P110–mRNA complexes in the translocation apparatus depends on ATP hydrolysis, this possibility is consistent with the suggestion (Davis and Ish-Horowicz, 1991) that mRNA migration along the cytoskeleton is ATP-dependent. This in turn is consistent with the requirement in a motor-driven model for receptors throughout the solid-state system.

The model also predicts the cytoskeletal location of translationally active mRNAs, the defined tracks formed by some intranuclear transcripts, and the vectorial character of mRNA transport. However, it has some difficulties. In some cells, the transport of at least some messengers appears to be ATP-independent (Schumm *et al.*, 1977); the known properties of the nucleoside triphosphatase are difficult to reconcile with those of myosin ATPases; transportant–receptor affinities would be expected to be high rather than low, as found experimentally; and even if cells have the kind of continuous fibrillar system required by the model, the fibrils are different in kind at different cellular locations. For instance, nuclear actin does not form fibrils like those in the cytoplasm; and fibrils that have been reported to link the pore complexes to the cytoskeleton and to the internum of the nucleus are morphologically different from any other known intracellular structure (Fey *et al.*, 1984). Therefore, the model may need to be modified; different regions of the cell contain different fibrils, each with their own characteristic receptors. Finally, as Zachar *et al.* (1993) observed, the model is difficult to reconcile with the weblike distributions of some intranuclear transcripts. It is also unclear how it can accomodate the free polysomal pool in the cytoplasmic compartment (see Section III,C).

The assembly–disassembly model could be consistent with both reported types of intranuclear transcript distribution (definite tracks and transnuclear webs), provided some linkages in the system are labile. Detachment of mRNAs from fibrils at the disassembly pole could account for the free fraction of cytoplasmic polysomes. (The model predicts that the relative size of this "soluble" fraction should be greater if the fibril disassembly rate is greater, implying that the rate of mRNA transport should correlate with the size of the soluble cytoplasmic fraction.) However, it was argued earlier that only cells with low mRNA transport rates, and therefore low fibril turnover rates, could give the appearance of ATP independence; but ATP independence seems to be incident on carcinogenesis, which entails higher, not lower, mRNA transport rates than are found in the untransformed cell (Schumm *et al.,* 1977). Also, not all transported macromolecules have the same requirement for ATP hydrolysis, and the model is not obviously consistent with this finding. Translocation of poly(A)$^-$ mRNAs such as histone messengers seems in some systems to have a lower ATP dependence than that of poly(A)$^+$ mRNAs such as globin and albumin messengers (Schroder *et al.,* 1989; however, see Prochnow *et al.,* 1994).

In summary, each of the three models considered has significant advantages, and significant disadvantages, in the light of current evidence. The fact that none of the models is wholly satisfactory does not necessarily oppose the solid-state perspective; it might simply indicate that further, less simplistic models are needed, combining the more satisfactory features of the three considered here. It certainly indicates that more data are required for more precise model articulation.

### F. Application of the Solid-State Perspective to Other Transportants

If mRNA transport is a solid-state process, does the same apply to the transport of other macromolecules (Section II,B)? The following arguments can be advanced for a solid-state view of nucleocytoplasmic protein transport.

1. Many proteins in the cytoplasm are nondiffusible (Paine, 1984; Peters, 1986). Those that are not immobile show very limited "diffusivities".
2. Many nuclear proteins are also nondiffusive (Feldherr, 1980).
3. The nucleocytoplasmic ratios of some karyophilic proteins cannot be accounted for by active transport at the nuclear pore complex (Paine, 1993).
4. Intracellular proteins can cross the diameter of a HeLa cell in 2–3 sec, whereas the time taken for them to diffuse a similar distance in dilute buffer would be an order of magnitude greater. Given the calculated diffusivities of many cytoplasmic proteins, the time taken to cross the cell by a process such as diffusion would be in the order of 20–30 min Luby-Phelps, 1994).
5. The phosphoserine-rich nuclear location signal-binding protein Nopp140, which shuttles between nucleolus and cytoplasm, does so on tracks connecting the structural fibrils of the nucleolus with the pore com-

plexes (Meier and Blobel, 1992). These tracks have been visualized by immunoelectron microscopy.

6. Some karyophilic proteins microinjected into the cytoplasm seem, from immunofluorescence evidence, to align themselves with fibrils in transit to the nucleus. Micrographs illustrate the point; but interpretation of this evidence is difficult.

7. Nuclear location signals in proteins have affinities for their receptors that are similar (about $10^7$ $M$) to those of mRNA. The same argument applies: large numbers of low-affinity receptors distributed across the space that the ligand must cross might provide a transport mechanism analogous to that provided for motile cells by the extracellular matrix. Identification of receptors for nuclear location signals in proteins has proven difficult because (i) their affinities are in the same order as nonspecific binding affinities and (ii) in many studies it has been assumed that they are restricted to the nuclear envelope, so samples containing small receptor populations have been studied. Nevertheless, it is now apparent that there are several receptors with different but overlapping specificities for protein location signals (Adam *et al.,* 1989; Benditt *et al.,* 1989; Lee and Malesi, 1989; Yamasaki *et al.,* 1989). $M_r$ values of 140, 100, 70–76, 67–70, 55–60 $\times$ $10^3$ have been reported. A common feature of these molecules may be an acidic oligopeptide sequence, because antibodies specific for the sequence DDDED seem to block all nucleus-targeted protein receptors (Imamoto-Sonobe *et al.,* 1990). Present knowledge of protein import receptors has been reviewed by Yamasaki and Lanford (1992).

The possibility that protein transport from cytoplasm to nucleus might be a solid-state process was also implied by Ambron *et al.* (1992); but data such as the foregoing might be interpreted in a diffusionist way. If by molecular exclusion the cytoskeleton concentrates soluble protein within restricted aqueous compartments, and also adsorbs some of them (with low affinity), then soluble proteins will move very slowly in the cytoplasm and will seem to align themselves with defined pathways. In other words, the cytoskeleton might work analogously to the putative extrachromosomal network (Zachar *et al.,* 1993). If this is so, the low affinities of location signals for pore-complex receptors become less problematic, because the actual concentration of protein in the microenvironment of the pore complex will be many times greater than the concentration calculated on the presumption that it has diffusional access to all parts of the cytoplasm. Similar arguments apply to the nucleus, where a soluble protein would presumably be confined to the extrachromosomal matrix, which *inter alia* would restrict its access to an artificial incision in the nuclear envelope. This diffusionist model, which seems to accord as well as any solid-state alternative with the review by Paine (1993), might also allow exclusion and adsorption mechanisms to account for higher nucleocytoplasmic concentration ratios than could be explained by translocation (active transport) alone.

Protein import to the nucleus seems to be regulated by soluble factors. Different factors are required for binding to the nuclear envelope and for ATP-dependent translocation (Newmeyer and Forbes, 1990; Moore and Blobel, 1992). Some of these have been isolated and characterized and their possible relationship to factors involved in protein uptake to mitochondria and endoplasmic reticulum has been considered (Goldfarb, 1992). However, it seems likely that the transport-facilitating factors so far identified represent only a small subset of the whole (Sterne-Marr *et al.,* 1992). Their existence does not elucidate the general character of protein transport (diffusionist or solid-state), but the fact that they themselves are proteins suggests that at least some proteins can move through the cell within the solution phase. Soluble factors also appear to regulate the transport of endogenous mRNAs. One example is the regulatory protein RCC1 which controls the cell-cycle-regulatory protein Ran/TC4 (Coutavas *et al.,* 1993). Cytoplasmic proteins associated with the signal recognition complex of the endoplasmic reticulum might be *trans*-acting factors, regulating the export of the mRNA for the complex from the nucleus (He *et al.,* 1994).

Interestingly, tRNA translocation has no requirement for ATP at all, either *in situ* or *in vitro* (Prochnow *et al.,* 1994); the *in vitro* model system used resealed nuclear envelope ghosts in which the RNA was entrapped; in contrast, mRNAs could not be exported from the ghosts without ATP (Prochnow *et al.,* 1994). Therefore, if tRNA transport is a solid-state process, the mechanism must accord more closely with the direct-transfer model (above) than with either of the proposed alternatives. One argument for a solid-state view of tRNA transport is its apparently vectorial character combined with the relative nuclear and cytoplasmic abundances of tRNAs, but as in the case of proteins some combination of exclusion and adsorption mechanisms might provide a satisfactory diffusionist explanation. On the other hand, the LOS1 gene, which controls tRNA splicing, encodes a protein that is apparently a nucleoskeleton component, suggesting a solid-state mechanism at least for pre-tRNA metabolism (Shen *et al.,* 1993).

Ribosome transport is more likely to be a solid-state process. A specific isoform of topoisomerase II is located in the nucleolus and may have a particular role in ribosomal processing (Zini *et al.,* 1992). The nucleolar structural protein B23 undergoes hormonally regulated phosphorylation when increased protein biosynthesis, and therefore increased ribosome assembly, is induced (Tawfic *et al.,* 1993). Export of ribosomal subunits from *Xenopus* oocyte nuclei seems to be ATP dependent (Bataille *et al.,* 1990), implying some solid-state mechanism (e.g., motor-driven) other than direct transfer. Certainly the rate of efflux of ribosomal subunits from resealed nuclear envelope vesicles, in the absence of nucleoskeletal structures, is very slow and does not seem to be affected by the presence of ATP (I. Hassell and H. Fasold, personal communication). Given the abundances of ribosomes in the cytoplasm, the apparently vectorial character of their transport would be difficult to explain from a diffusionist perspective, and this makes solid-state models worth considering.

## IV. Messenger Transport in Virally Infected Cells

Some recent *in situ* studies of virus-infected cells (Krug, 1993) have provided a further challenge to understanding and may elucidate the central problem of this chapter, viz., the role of the nucleoskeleton in mRNA transport. For instance, the adenovirus E1B–E4 complex seems to immobilize host transcripts while allowing viral messengers to migrate normally through the nucleus (Leppard and Shenk, 1989). This result might be explained in terms of a soluble factor that is required for nuclear RNA mobility and is sequestered by the E1B–E4 complex. Although this hypothesis could be reconciled to either major perspective, it does at least suggest an experimental tool by which pertinent information can be extracted. In HIV-1- and influenza virus-infected cells, the nuclei export both spliced and unspliced viral messengers that are translated to give different viral protein products. The mechanisms of action of such proteins as the HIV-1 product Rev and the influenza virus NS1 have been much studied. The export of HIV-1 mRNA from the infected nucleus is promoted by Rev, which interacts directly with the translocation apparatus (Malim *et al.,* 1989; Pfeifer *et al.,* 1991). It is also possible that HIV-1 transcripts contain sequences that inhibit both splicing and the export of host RNAs, and that Rev modulates these actions (Krug, 1993). NS1 selectively inhibits the transport of poly(A)$^+$ mRNAs (Krug, 1993).

The first step in the expression of the HIV-1 genome is the production of short, fully spliced 2-kb RNAs encoding the *trans*-acting regulatory proteins Tat, Rev, and Nef. Rev, a 19-kDa nuclear phosphoprotein located mainly in the nucleoli of HIV-1 expressing cells (Cullen *et al.,* 1988; Felber *et al.,* 1989; Cochrane *et al.,* 1990a), allows the synthesis of the full-length 9-kb viral RNA and a singly spliced 4-kb RNA; together, these RNAs encode the other viral proteins (Rosen *et al.,* 1988; Kim *et al.,* 1989). Rev acts by binding to a *cis*-acting RNA sequence, the Rev-responsive element (RRE), apparently stabilizing a complex secondary structure (Daly *et al.,* 1989; Malim *et al.,* 1989; Cochrane *et al.,* 1990b; Daefler *et al.,* 1990; Heaphy *et al.,* 1991; Kjems *et al.,* 1991). It also promotes the export of RRE-containing RNAs to the cytoplasm (Hadzopoulou-Cladaras *et al.,* 1989). This effect of Rev has been attributed by some workers to the inhibition of spliceosome assembly of HIV-1 transcripts, or alternatively to promotion of spliceosome disassembly (Chang and Sharp, 1989; Lu *et al.*, 1990), but since Rev can act on mRNAs that lack functional splice sites (Rosen *et al.*, 1988; Emerman *et al.*, 1989) and the RRE has been shown not to be absolutely required for its action (Venkatesan *et al.*, 1992), other workers have suggested that Rev effects the nuclear export of both spliced and unspliced HIV-1 message by direct interaction with some part of the transport apparatus (Hadzopoulou-Cladaras *et al.*, 1989; Malim *et al.*, 1989; Cullen and Malim, 1991, Rosen, 1991). More specifically, Pfeifer *et al.* (1991) obtained evidence for an effect on translocation. They demonstrated that Rev binds to P110 in the nuclear envelope, and in consequence both the NTPase and the export of poly(A)$^+$ mRNA were strongly inhibited. Concomitantly, the export of RRE-containing viral *env* RNA was stimulated. Independently, Clawson *et al.* (1991) showed that Rev inhibits the NTPase in nuclear matrix preparations and that isolated RRE reverses the inhibition. This study implicated Rev, via the nucleoskeleton-associated NTPase, in splicing and release. However, the debate about whether Rev acts primarily on splicing or on translocation continues. In this context, the possible presence of P110 in the nucleoskeleton (Schweiger and Kostka, 1984) may be interesting.

Extension of the simple kinetic model of translocation to the effects of Rev reveals that under steady-state conditions the ratio of spliced mRNA (2- and 4-kb HIV-1 mRNA) to unspliced mRNA (9-kb HIV-1 mRNA) in the cytoplasm is equal to the ratio of the limiting rate constant of splicing to the limiting rate constant of translocation of unspliced mRNA. Also, the ratio of spliced to unspliced mRNA in whole cells is also equal to this same ratio, as long as both species are degraded at similar rates in the cytoplasm. Therefore, if the limiting translocation rate constant is much greater than the limiting splicing rate constant, most of the cytoplasmic HIV-1 mRNA will be unspliced; the converse is also true. If the splicing rate constant is the same in two different cell systems, then the relative amounts of spliced RNA in the cytoplasms of the two systems will be determined by the relative translocation rates.

These inferences are intuitively reasonable, but the steady-state assumption is dubious in cells in which expression of the viral genome is being initiated. Under pre-steady-state conditions, assuming only a constant rate of transcription of the relevant HIV-1 gene, it is shown (Schroder *et al.*, 1994) that the relative rate of cytoplasmic accumulation of unspliced to spliced messenger can be increased by (i) more rapid transcription, (ii) faster degradation of spliced messenger in either compartment, and (iii) more obviously, quicker transport of the unspliced messenger. It is decreased by (i) faster splicing and (ii) quicker transport of the spliced messenger. In principle, therefore, Rev could modulate the cyotplasmic unspliced/spliced messenger ratios by changing any one of these parameters, and its modulating effect would be more marked if it changed two or more of them simultaneously. Thus, a kinetic model based entirely on translocation implies that Rev is likely to affect both translocation and splicing.

This inference that Rev has a dual action may imply that it recognizes two or more protein ligands. If one of these is P110, as the evidence (Clawson *et al.*, 1991; Pfeifer *et al.*, 1991) suggests, then the simplest interpretation is that the other is also P110; in which case, P110 is involved in splicing. Rev also binds in the nucleolus (Cullen *et al.*, 1988; Felber *et al.*, 1989; Cochrane *et al.*, 1990a), which could imply that P110, or a protein with strong homology to P110, is also to be found in the nucleolus. Prochnow *et al.* (1994) suggest ways in which the predictions of this model can be tested quantitatively.

If HIV-1 mRNA transport acts by a solid-state mechanism, then the motor-driven model accords most simply with the data. It is possible that P110 migrates along the solid-state framework with the mRNA attached. This would account for the presumed dual action of Rev, and also for the observation that Rev down-regulates the poly(A) binding sites in the nuclear envelope (Pfeifer *et al.*, 1991). It is notable that the complex formed between Rev and RRE-containing RNA generates nucleoprotein filaments that could act as a virus-specific solid-state transport framework (Heaphy *et al.*, 1991); Tat seems to be essential for forming this structure (Muller *et al.*, 1990). Rev might therefore be a valuable tool for elucidating the intranuclear migration/release stage of mRNA transport as well as the translocation stage; in particular, it could throw new light on the structural relationship between splicing and intranuclear migration (Zeitlin *et al.*, 1987). The distribution of P110 and related proteins *in situ* certainly deserves further investigation.

The influenza virus protein NS1 is structurally analogous to Rev; it has an RNA binding domain with a high affinity for poly(A), and an effector protein domain. It might therefore be expected to affect both splicing/release and translocation, as proposed for Rev. However, NS1 apparently inhibits the export of spliced poly(A)$^+$ mRNAS from the host nucleus without any concomitant effect on splicing(Krug, 1993). Moreover, cells transfected with influenza virus genes can export unspliced viral messengers from their nuclei without requiring NS1 or any other viral protein, though in these cells only about 25% of the viral transcripts are exported, in contrast to the situation in virus-infected cells (Alonso-Caplan *et al.*, 1992).

Because NS1 seems to bind to poly(A) rather than to a specific sequence such as RRE, its action might be explained by simple competition for poly(A) with P110 in the nuclear envelope, i.e. in terms of a translocation-only model. This is certainly not the case for the adenovirus E1B–E4 complex (Sarnow *et al.*, 1984), which in late infection ensures that only viral rather than host mRNA appears in the cytoplasm (Beltz and Flint, 1979; Pilder *et al.,* 1986). The E1B–E4 complex is mainly located at the large viral replication–transcription sites in the nucleus and if it is absent (e.g., if either component is defective), then viral mRNA is not moved efficiently from these sites to the nuclear periphery (Leppard and Shenk, 1989; Ornelles and Shenk, 1991). In this case, therefore, a viral protein complex appears to be acting specifically on intranuclear movement, i.e., on the release stage of mRNA transport alone and not on translocation. Krug (1993) suggests that a host factor required for intranuclear RNA migration (another of the soluble factors mentioned in Section II,F) might be sequestered by the E1B–E4 complex, making it unavailable to participate in the movement of host messengers. If Krug's view is correct, therefore, influenza virus transcripts must be able to recruit the postulated intranuclear migration factor from the E1B–E4 complex, although host transcripts cannot do so. Interestingly, when adenovirus-infected cells are superinfected with influenza virus, the influenza virus mRNA evades the adenovirus block (Katze *et al.,* 1984). Kinetic modeling of this hypothesis suggests that the rate of host mRNA transport to the cytoplasm is linearly related to the availability of the proposed soluble factor (Schroder *et al.,* 1994). In contrast, the influenza viral RNA must be able to reach the target site without the participation of this factor. Until the soluble factor is identified, however, this model is untestable.

These studies indicate both the limitations of translocation-only models in accommodating the viral protein data, and the current difficulty of elaborating kinetic models for other parts of the mRNA transport process such as the release stage.

In summary, kinetic models for translocation can cross-relate data from *in vitro* and *in situ* studies of mRNA transport. Also, they provide at least formal explanations for some bodies of data such as nucleocytoplasmic RNA ratios, and they lead to some testable predictions, e.g., in repect of the actions of the HIV-1-associated Rev protein. However, they fail to account for all the known effects of viral proteins on mRNA transport, and since the translocation mechanism is fairly well understood, this failure suggests that models addressing the translocation step alone have limited explanatory capacity. Indeed, there is direct evidence for the effects of viral proteins on intranuclear migration and splicing of pre-mRNA. The growing literature on the control of viral messenger transport is making the need for less restricted, more broadly based models of mRNA transport increasingly urgent.

## V. Conclusions and Future Prospects

The first question raised at the start of this chapter concerned the capacity of models that allot no significant role to the nucleoskeleton to account for known RNA transport data. A review of the evidence has failed to give definitive support to the hypothesis that the nucleoskeleton plays a positive part in this process, i.e., to substantiate the solid-state view that intranuclear RNA migration/release takes place along the fibrils of the system. However, the contrary diffusionist view was not corroborated either, and even if the mechanism turns out to be diffusionist rather than solid-state, it seems that (a) the actual process involved is highly unlikely to be diffusion per se, (b) the nucleoskeleton retains at least a negative role in molecular exclusion. Analogies with the cytoskeleton and the possible cellular distribution and roles of the poly(A) receptor P110, are among the lines of evidence that accord better with the solid-state view.

The second question concerned the compatibility between the presumption that transport consists exclusively of translocation through the pore complex, and the possible role of the nucleoskeleton in transport. Our increasing understanding of translocation shows that it is a complex process, interesting not least because of its potential role in switching between steady-state nucleocytoplasmic messenger distributions. However, it seems clear that translocation-only models cannot account for important parts of our knowledge about nucleocytoplasmic mRNA transport and distribution. Some of the difficulties of such models can be answered in diffusionist terms; for example, the low affinities of signal receptors in the pore complex could be reconciled with molecular exclusion effects that markedly increase the local concentrations of potential transportants. However, these difficulties have alternative solid-state interpretations that are at least as likely *a priori,* and there are some issues, notably the cytoplasmic subcompartmentalization of messengers, that are very difficult to reconcile with diffusionist interpretations.

The third question, concerning the generalizability of the solid-state transport concept to other macromolecules and complexes, remains open. With regard to proteins, the situation is comparable to pre-mRNA migration in the nucleus; most of the evidence can be interpreted equally well in diffusionist or solid-state terms. However, the existence of soluble regulatory factors which are themselves proteins puts the balance of argument in favor of the diffusionist perspective. Transport of tRNAs is more likely to involve a diffusionist than a solid-state mechanism; the converse is true for ribosomal subunits. In no case, however, is the situation clear.

So far as mRNA transport is concerned, however, there is a sufficiently strong argument favoring solid-state interpretations for detailed models to be proposed and analyzed. Three intuitively simple models have been considered. Each of them coheres satisfactorily with some data but not others, and it appears that a more sophisticated model is required, combining the most effective features of all three. What does seem clear is that all these models provide more satisfactory accounts of our current knowledge than any diffusionist alternative; but the diffusionist perspective remains valuable as a source of criticism for any proposed model.

The matter is worth resolving, not least because of the potential role of the transport machinery in regulating cellular activities. In principle, mRNA transport could be regulated in at least three (not necessarily independent) general ways. First, the substrate molecule might be altered, e.g., by differential splicing or by changing the structure of the 5′ cap or the 3′ tail, changing the ability of the RNA to bind to different parts of the transport system. Second, some parts of the transport machinery might be modified, e.g., by phosphorylation of one or more components of the pore complex (or, if the solid-state view is valid, the nucleoskeleton or cytoskeleton), with concomitant alteration of RNA affinity. Third, there might be a change in the quantity or activity of a factor that promotes or inhibits mRNA association with parts of the transport apparatus, by analogy with the factors that regulate protein transport. The potential of such mechanisms for the regulation of cellular activity cannot adequately be investigated until a satisfactory transport model is established, and a key part of this development will be fuller characterization of the relationship between pre-mRNA and the nucleoskeleton.

Regarding this central problem, the balance of evidence at present indicates that the nucleoskeleton has at least some role in mRNA transport, and that this role is more likely to be positive (in accord with the solid-state concept) than negative (in accord with the diffusionist concept). However, the question is far from closed. Some predictions and suitable tests have been suggested in this chapter, in the contexts of model evaluation and the explanation for the effects of viral infection on messenger RNA transport; but the rapidly expanding literature in these fields is likley to indicate profitable new directions in the near future.

## References

Adam, S. A., Lobl, T. J., Mitchell, M. A., and Gerace, L. (1989). Identification of specific binding proteins for a nuclear location sequence. *Nature* (*London*) **337,** 276–279.

Adams, R. I., and Pollard, T. D. (1986). Propulsion of organelles isolated from Acanthamoeba along actin filaments by myosin-I. *Nature* **322,** 754–756.

Agutter, P. S. (1980). Nucleocytoplasmic mRNA transport: A plea for methodological dualism. *Trends Cell Biol.* **4,** 278–279.

Agutter, P. S. (1984). Nucleocytoplasmic RNA transport. *Subcell. Biochem.* **10,** 281–357.

Agutter, P. S. (1988). Nucleocytoplasmic transport of mRNA: Its relationship to RNA metabolsim, subcellular structures and other nucleocytoplasmic exchanges. *Prog. Mol. Subcell. Biol.* **10,** 15–96.

Agutter, P. S. (1991). "Between Nucleus and Cytoplasm." Chapman Hall, London.

Agutter, P. S. (1994). Models for solid-state transport: messenger RNA movement from nucleus to cytoplasm. *Cell Biol. Int. Rep.* **18,** 849–58.

Agutter, P. S., and Birchall, K. (1979). Functional differences betwen mammalian nuclear protein matrices and pore-lamina complex laminae. *Exp. Cell Res.* **124,** 453–460.

Agutter, P. S., and Prochnow, D. (1994). Nucleocytoplasmic transport. *Biochem. J.* **300,** 609–618.

Agutter, P. S., Malone, P. C., and Wheatley, D. N. (1995). *J. Theoret. Biol.* (submitted for publication).

Alberts, B., Bray, D., Lewis, J., Raff, M., Roberts, K., and Watson, J. D. (1988). "Molecular Biology of the Cell," 2nd ed. Garland, New York.

Alonso-Caplan, F. V., Nemeroff, M. E., Qiu, Y., and Krug, R. M. (1992). Nucleocytoplasmic transport: The influenza virus NS1 protein regulates the transport of spliced NS2 mRNA and its precursor NS1 mRNA. *Genes Dev.* **6,** 255–267.

Ambron, R. T., Schmied, R., Huang, C. C., and Smedman, M. (1992). A signal sequence mediates the retrograde transport of proteins from the axon periphery to the cell body and then into the nucleus. *J. Neurosci.* **12,** 2813–2818.

Baglia, F. A., and Maul, G. G. (1982). Nuclear ribonucleoprotein release and NTPase activity are inhibited by antibodies directed against one nuclear matrix glycoprotein. *Proc. Natl. Acad. Sci. U.S.A.* **80,** 2285–2288.

Barrer, R. M. (1968). *In* "Diffusion in Polymers" (J. Crank and G. S. Park, eds.), pp. 165–217. Academic Press, New York.

Bataille, N., Helser, T., and Fried, H. M. (1990). Cytoplasmic transport of ribosomal subunits microinjected into the *Xenopus laevis* oocyte nucleus: A generalized, facilitated process. *J. Cell Biol.* **111,** 1571–1582.

Beltz, G. A., and Flint, S. J. (1979). Inhibition of HeLa cell protein synthesis during adenovirus infection. Restriction of cellular messenger RNA sequences to the nucleus. *J. Mol. Biol.* **131,** 353–373.

Benditt, J. O., Meyer, C., Fasold, H., Barnard, F. C., and Riedel, N. (1989). Interaction of a nuclear location signal with isolated nuclear envelopes and identification of signal-binding proteins by photoaffinity labeling. *Proc. Natl. Acad. Sci. U.S.A.* **86,** 9327–9331.

Berezney, R. (1980). Fractionation of the nuclear matrix. I. Partial separation into matrix protein fibrils and a residual ribonucleoprotein fraction. *J. Cell. Biol.* **85,** 641–650.

Bernd, A., Schroder, H. C., Zahn, R. K., and Muller, W. E. (1982). Modulation of the nuclear-envelope nucleoside triphosphatase by poly(A)-rich mRNA and by microtubule protein. *Eur. J. Biochem.* **129,** 43–49.

Berrios, M., and Fisher, P. A. (1986). A myosin heavy-chain-like polypeptide is associated with the nuclear envelope in higher eukaryotic cells. *J. Cell Biol.* **103,** 711–724.

Berrios, M., Osterhoff, N., and Fisher, P. A., (1986). *In situ* localization of DNa topoisomerase II, a major polypeptide component of the *Drosophila* nuclear matrix. *Proc. Natl. Acad. Sci. U.S.A.* **82,** 4142–4146.

Bers, D. M., and Peskoff, A. (1991). Diffusion around a cardiac calcium channel and the role of surface bound calcium. *Biophys. J.* **59,** 703–721.

Blobel, G. (1985). Gene gating: An hypothesis. *Proc. Natl. Acad. Sci. U.S.A.* **82,** 8527–8529.

Brasch, K. (1982). Fine structure and localization of the nuclear matrix *in situ. Exp. Cell Res.* **140,** 161–172.

Breeuwar, M., and Goldfarb, D. S., (1990). Facilitated nuclear transport of histone H1 and other small nucleophilic proteins. *Cell (Cambridge, Mass.)* **60,** 999–1008.

Carlslaw, H. S., and Jaeger, J. C. (1959). "Conduction of Heat in Solids." Oxford Univ. Press (Clarendon), Oxford.

Carter, K. C., Bowman, D., Carrington, W., Fogarty, K., McNeil, J. A., Fay, F. S., and Lawrence, J. B. (1993). A three-dimensional view of precursor messenger RNA metabolism within the mammalian nucleus [see comments]. *Science* **259,** 1330–1335.

Cervera, M., Dreyfuss, G., and Penman, S. (1981). Messenger RNA is translated when associated with the cytoskeletal framework in normal and vesicular stomatitis virus infected HeLa cells. *Cell (Cambridge, Mass.)* **23,** 113–120.

# Color Plates

PLATE 1 (He *et al.*, Fig. 3). Cartoon showing possible structural roles of NuMA in the interphase nuclear matrix (A) and the mitotic spindle (B).

PLATE 2 (Moreno Diáz De La Espina, Fig. 4). Stereopair obtained after serial sectioning of the nuclear matrix shown in Fig. 5. To produce the stereopairs, the outlines of the nuclear matrix components were drawn onto transparent sheets (one for each micrograph), their ($X$, $Y$) coordinates were recorded with a digitizer and memorized in the computer. The data were computerized with a program that includes: calculation and plotting of "smooth" profiles from coordinates, rotation of the profiles at 5°, and re-plotting of new profiles from the rotated coordinates. Both originals and rotated profiles constitute a stereopair. The lamina is represented in black, the profiles of the nucleolar matrix are red, and those of the patches of the internal matrix are green. Only the profiles of the internal matrix around the nucleolar matrix have been represented in the stereopair.

PLATE 3 (Moreno Diáz De La Espina, Fig. 14). Localization of fibrillarin in the nucleolar matrix by confocal microscopy after immunolabeling with the S4 serum. Sixteen consecutive optical sections (0.4 $\mu$m) through a nuclear matrix including the nucleolar area shown after deblurring to eliminate the out-of-focus fluorescence. The labeling is not randomly distributed but accumulates on discrete domains in the central part of the nucleolar matrix (large arrow). The boundary of the nuclear matrix is outlined by its faint fluorescence only visible in some middle sections (small arrows). Bars B–E = 10$\mu$m.

PLATE 4 (Panté and Aebi, Fig. 1). Schematic representation of the different components of the nuclear envelope (NE) together with the molecular trafficking occurring across the NE through the nuclear pore complexes (NPCs). The NE consists of an inner and an outer nuclear membrane enclosing the perinuclear space. The outer nuclear membrane is continuous with the endoplasmic reticulum (ER), so that the perinuclear space of the NE is contiguous with the lumen of the ER. The inner nuclear membrane is lined by the nuclear lamina, a near-tetragonal meshwork made of intermediate filament-like proteins called nuclear lamins. The NPCs are interposed at irregular intervals between the inner and outer nuclear membrane. Bidirectional molecular trafficking between the nucleus and the cytoplasm occurs through the NPCs which allow passive diffusion of ions and small molecules and energy-dependent transport of nuclear proteins, RNAs, and RNPs.

PLATE 5 (Panté and Aebi, Fig. 4). Surface renderings of the 52-MDa basic framework three dimensionally reconstructed from negatively stained NPCs released from the NE upon detergent treatment. (a) Slightly tilted view of the basic framework of the NPC which consists of eight multidomain spokes and exhibits strong 822 symmetry thus indicating that it is built of two identical halves relative to the central plane of the nuclear envelope. Note that due to its irreproducible appearance, the central plug or channel complex has been omitted in this reconstruction. When the pore membrane is positioned in this map of the basic framework of the NPC, eight ~10-nm-diameter peripheral channels are created between two adjacent spokes and the pore membrane border at a radius of ~40 nm. (b) Three different side views of one multidomain spoke cut out from the basic framework of the NPC as shown in (a). Each half-spoke is built from four distinct morphological domains termed annular (a), column (c), ring (r), and lumenal (l) domain. This figure has been adapted from Hinshaw *et al.* (1992).

PLATE 6 (Panté and Aebi, Fig. 7). Schematic representation of a consensus model of the membrane-bound NPC. Its major structural components include the basic framework (i.e., the spoke complex as shown in Fig. 4a), the central plug or channel complex, the cytoplasmic and nuclear rings, and the cytoplasmic filaments and nuclear basket. The 52-MDa basic framework of the NPC has been adapted from the 3-D reconstruction of negatively stained detergent-released NPCs (Hinshaw *et al.*, 1992). The cytoplasmic filaments and nuclear basket have been modeled based on EM data obtained by Ris (1991), Jarnik and Aebi (1991), and Goldberg and Allen (1992) (see also Fig. 5). In this consensus model of the NPC, we have also pictured a cytoplasmic and nuclear ring in addition to the two tenuous rings defined by the ring domain of the spoke complex (see Fig. 4a). The central plug or channel complex has been modeled as a transparent ellipsoidal particle to emphasize the fact that its definite structure remains to be determined.

PLATE 7 (Panté and Aebi, Fig. 9). Schematic diagram of the nuclear import pathway of proteins through the NPC. In the first step, the nuclear protein to be transported associates with an NLS receptor, this complex then docks to the cytoplasmic periphery (i.e., to the cytoplasmic filaments or the cytoplasmic ring) of an NPC from where it is delivered to the central plug or channel complex for translocation. Each one of these steps appears to be mediated by the action of one or several cytosolic factors. Once in the nucleus, the protein–receptor complex dissociates, and the receptor may be recycled for another round of transport.

PLATE 8 (Bastos *et al.*, Fig. 2). Diagram of the domain architecture of peripheral membrane proteins of vertebrate NPCs deduced from their predicted amino acid sequences. Depending on the presence of O-linked *N*-acetylglucosamine (*O*-GlcNac) residues, two groups of peripheral membrane proteins of the NPC have been distinguished. (a) Members of the O-linked glycoprotein family which in addition *O*-GlcNAc contain multiple copies of a more or less degenerate pentapeptide repeat XFXFG. POM121 (shown in Fig. 3) may also be classified as a member of this family. Other features of individual proteins are as follows: p62 contains a COOH-terminal α-helical coiled-coil domain, and NUP153 harbors four zinc finger motifs. In the case of CAN/NUP214, the repetitive XFXFG motif alternates with SVFG, FGQ, and FGG sequences. A leucine zipper is also evident in this protein. (b) Non-O-linked proteins that contain neither *O*-GlcNAc nor repetitive sequence motifs. The amino acid sequences of the three members of this group, NUP107, NUP155, and Tpr/p265, are quite unrelated. In the case of Tpr/p265, its major feature is an ~1600-residue-long α-helical coiled-coil domain beginning close to the $NH_2$-terminus. NUP107 is distinguished by a COOH-terminal leucine zipper. For additional information concerning these proteins, see Table I and references therein.

PLATE 9 (Bastos *et al.*, Fig. 3). Diagram of the domain architecture and membrane topology of both vertebrate and yeast integral membrane proteins of the NPC. The orientations shown are based upon deduced amino acid sequences as well as upon direct experimental evidence (see text). Three transmembrane NPC glycoproteins, gp210, POM121, and the yeast POM152p, have been cloned so far and sequenced. The translated sequences all reveal a distinct stretch of hydrophobic residues that is predicted to traverse the NPC membrane. Gp210 contains a large $NH_2$-terminal domain residing in the perinuclear space (PNS), the lumen of the nuclear envelope, and a small extralumenal COOH-terminal tail. In contrast, POM121 exhibits of a short $NH_2$-terminal tail residing in the PNS, and a large extralumenal COOH-terminal domain. The latter contains a number of XFXFG pentapeptide repeats clustered near the COOH-terminus. Like gp210, most of the mass of yeast POM152p is predicted to reside in the PNS. In this case, however, the orientation of the polypeptide is reversed such that it is the COOH-terminus that lies in the PNS. The lumenal domains of both gp210 and POM152p each possess consensus sites for *N*-glycosylation, close to the membrane-spanning segment, at least some of which are utilized (Greber *et al.*, 1990; Wozniak *et al.*, 1994). In addition, the amino acid sequence of POM152p contains eight repetitive segments, within the lumenal domain, each 24 residues long, with the consensus sequence C-G----V---L-G--PF---Y. The functions of these repeats are unknown. In the case of each of these proteins, the extralumenal domain is thought to be engaged in the inner region of the NPC.

PLATE 10 (Bastos *et al.*, Fig. 4). Diagram of the domain architecture of yeast NPC proteins deduced from their amino acid sequences. Depending on the occurrence of highly repeated motifs in their amino acid sequences, three groups of yeast NPC proteins have been distinguished. (a) The XFXFG family, NSP1p, NUP1p, and NUP2p, members of which exhibit numerous copies of a more or less degenerate pentapeptide motif XFXFG clustered in the central part of each molecule. (b) The GLFG family, NUP49p, NUP57p (not shown), NUP100p, NUP116p, and NUP145p, members of which contain several copies of a more or less degenerate tetrapeptide GLFG within their $NH_2$-terminal domains. Three members of this family—NUP100, NUP116, and NUP145—contain additional related domains including an RNA-binding motif. (c) The third group by default, includes the yeast nucleoporin interacting component NIC96p, SRP1p (not shown), and NUP133p (not shown). None of these proteins exhibits any obvious interrelationships and contains no repetitive sequence elements in the vein of XFXFG or GLFG referred to above. NIC96p may be recovered in a complex with NSP1p, NUP57p, and NUP49p (Grandi *et al.*, 1993). Interactions between these four proteins most likely involves the coiled-coil helical domains apparent in each. For more information about these proteins, see Table I and references therein.

PLATE 11 (Bastos *et al.*, Fig. 5). Diagram summarizing the immunolocalization of characterized NPC protein epitopes within the 3-D architecture of the consensus model of the NPC. The major structural components of the NPC include the basic framework (i.e., the spoke complex), the central plug or channel complex, the cytoplasmic and nuclear rings, and the cytoplasmic filaments and nuclear basket. The basic framework of the NPC has been adapted from a conical tilt reconstruction of negatively stained detergent-released NPCs (Hinshaw *et al.*, 1992). The cytoplasmic filaments and nuclear basket have been modeled based on EM data obtained by Jarnik and Aebi (1991), Ris (1991), and Goldberg and Allen (1992). The central plug or channel complex has been modeled as a transparent ellipsoidal particle to indicate the fact that its definite structure remains elusive (see chapter by Panté and Aebi in this volume). Tpr/p265, CAN/NUP214, and NUP180 exhibit epitopes residing in the cytoplasmic filaments (Wilken *et al.*, 1993; Byrd *et al.*, 1994; Panté *et al.*, 1994), whereas NUP153 COOH-terminal domain epitopes have been localized at the terminal ring of the nuclear basket (Panté *et al.*, 1994). p62 epitopes are exposed at both the cytoplasmic and nuclear periphery of the central plug or channel complex (Guan *et al.*, 1995). The transmembrane glycoprotein gp210 exhibits several epitopes in the lumen of the NE (Greber *et al.*, 1990) where based on its topology most of its mass resides (see Fig. 3; Greber *et al.*, 1990). Epitopes for the transmembrane glycoproteins POM121 and POM152 are also indicated. However, these latter two should not be taken literally since their exact localization awaits experimental determination.

PLATE 12 (Leonhardt and Cardoso, Fig. 1). Targeting of β-galactosidase to replication foci (A–C) or to the speckled compartment (D–F) by fusion to the DNA methyltransferase targeting sequence (Leonhardt *et al.*, 1992) or the *suppressor-of-white-apricot* RS domain (Li and Bingham, 1991), respectively. Mouse fibroblasts were transfected with the fusion constructs. Fusion proteins were detected with a monoclonal antibody to the β-galactosidase epitope (red; A, D) and compared to the DNA replication foci visualized with an antibody to DNA methyltransferase (green; B, E). Methyltransferase was previously shown to colocalize with sites of DNA replication in S phase cells (Leonhardt *et al.*, 1992). Double exposures are shown to better illustrate colocalization of the DNA MTase fusion at sites of DNA replication (yellow in C) and the different localization of the RS domain fusion (red and green in F).

PLATE 13 (Leonhardt and Cardoso, Fig. 4). Specific localization of cyclin A (red) at subnuclear DNA replication foci visualized by BrdU incorporation and costaining with BrdU-specific antibodies (green) in pulse labeled cultures of retrodifferentiating murine myotubes. Colocalization (yellow–orange color as a result of the double exposure) is seen in all the ring- and loop-like structures within the nuclei as well as in the dots near the nuclear periphery. For further details see Cardoso *et al.* (1993).

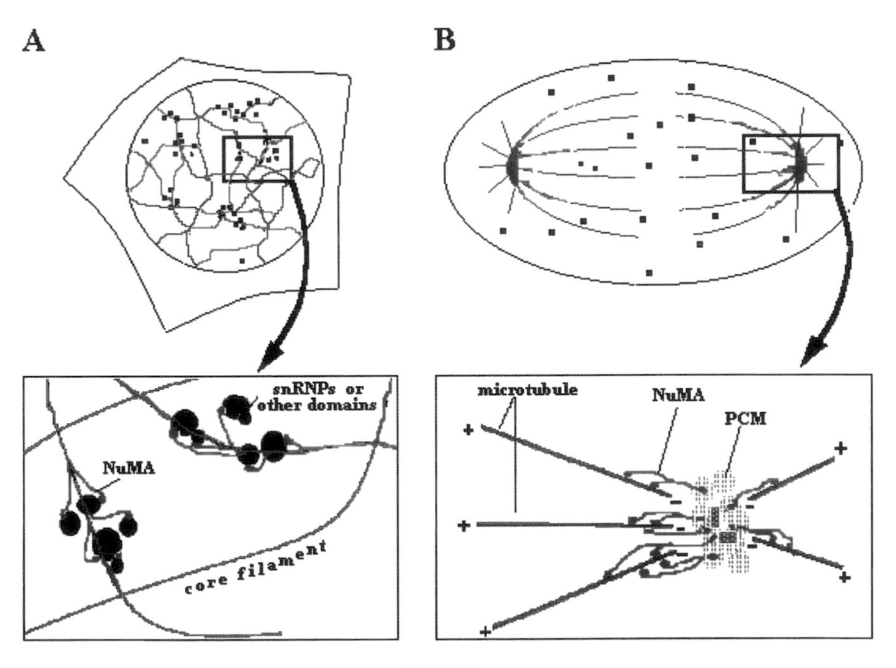


PLATE 1

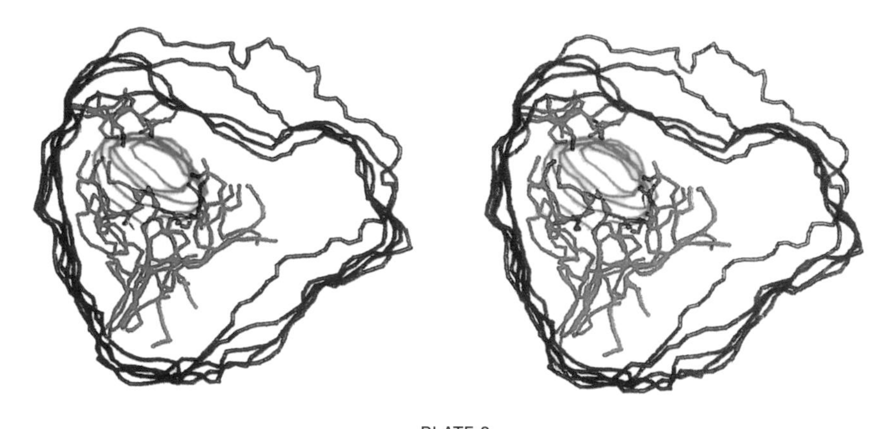

PLATE 2

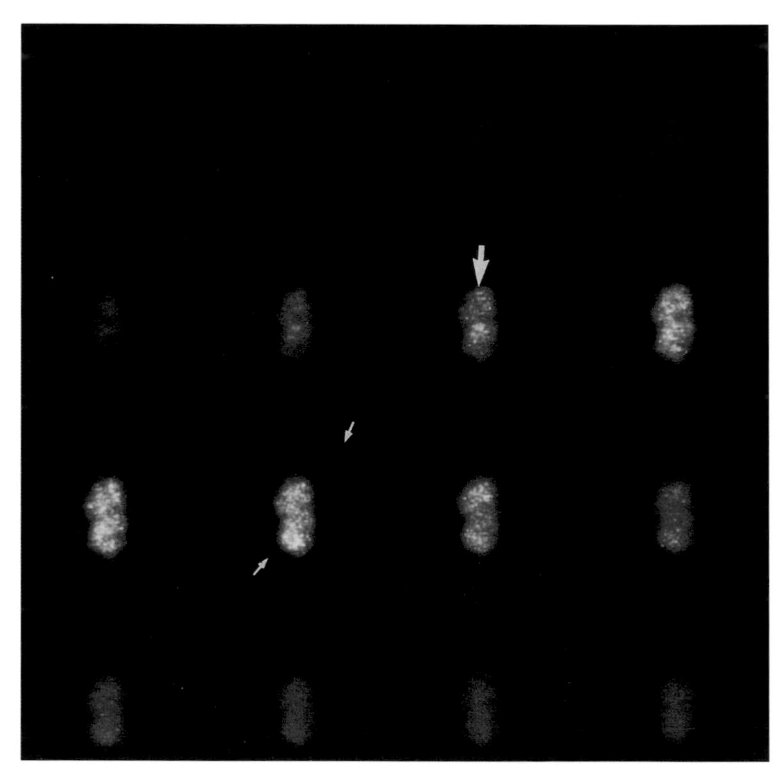

PLATE 3

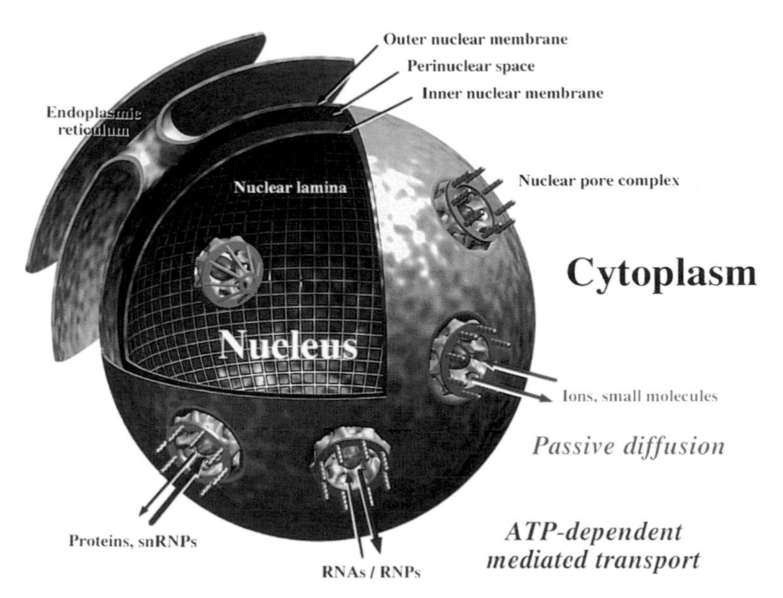


PLATE 4

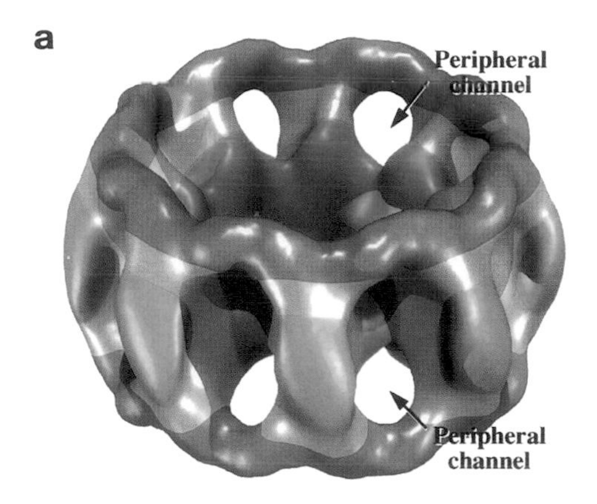


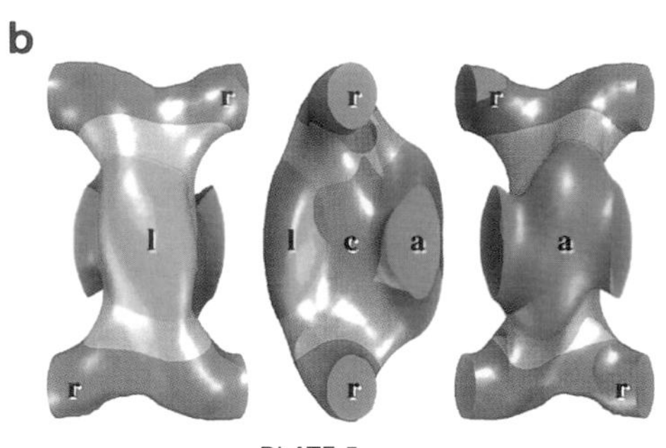


PLATE 5

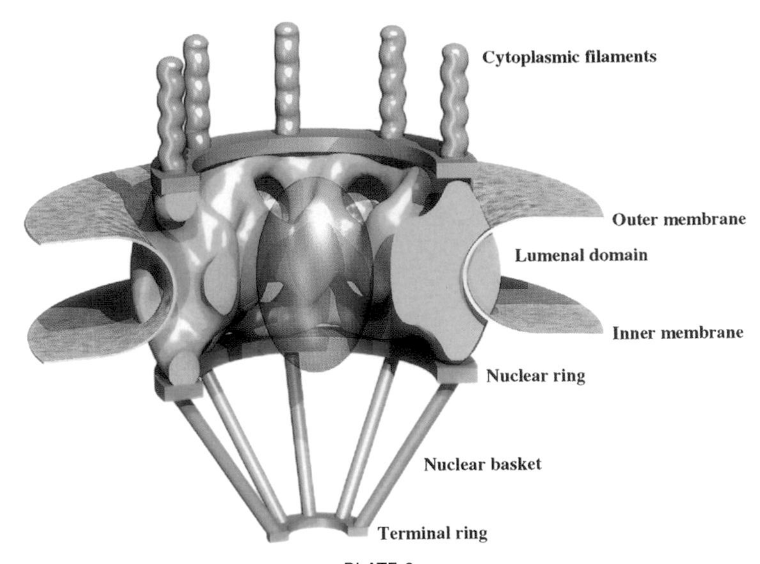


PLATE 6

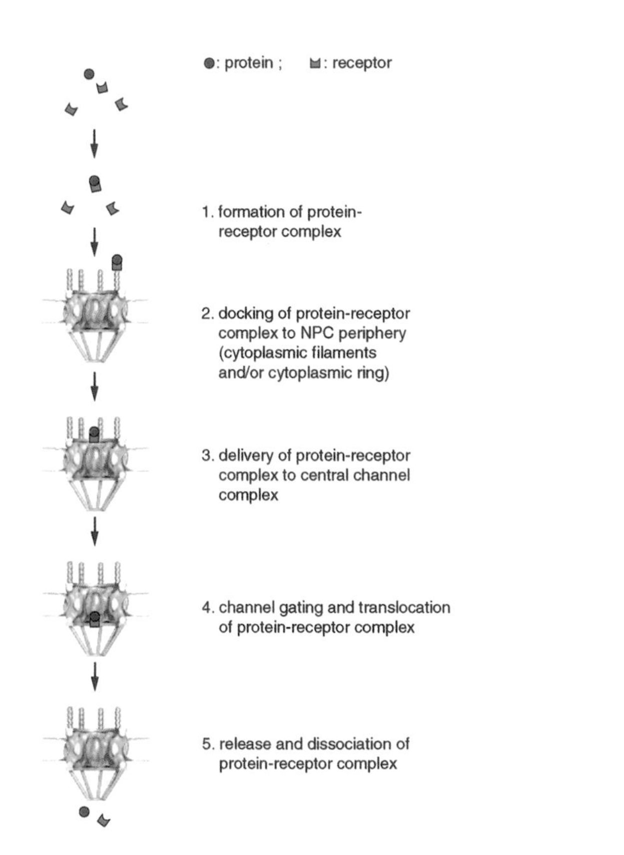


PLATE 7

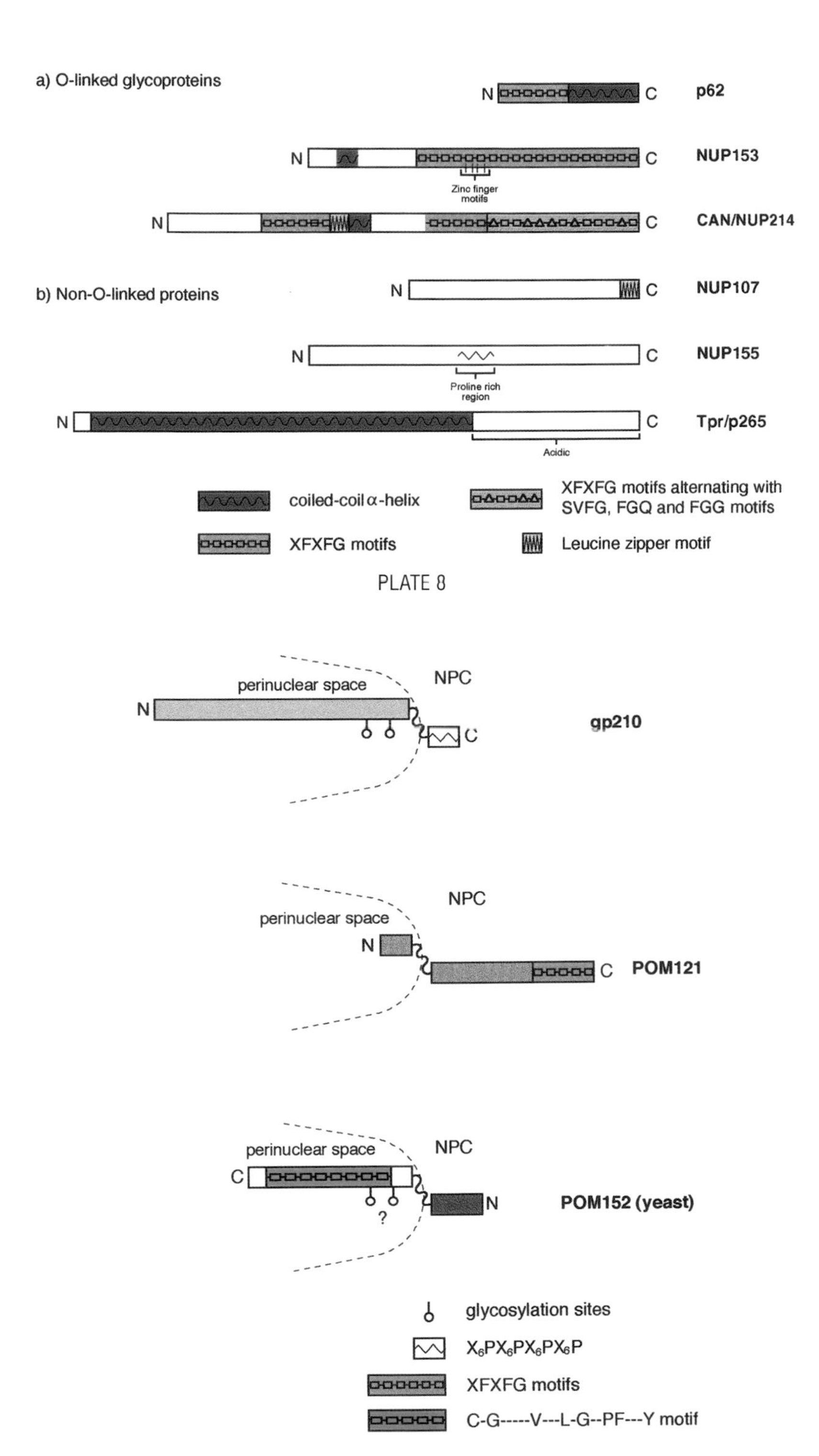


PLATE 8

PLATE 9

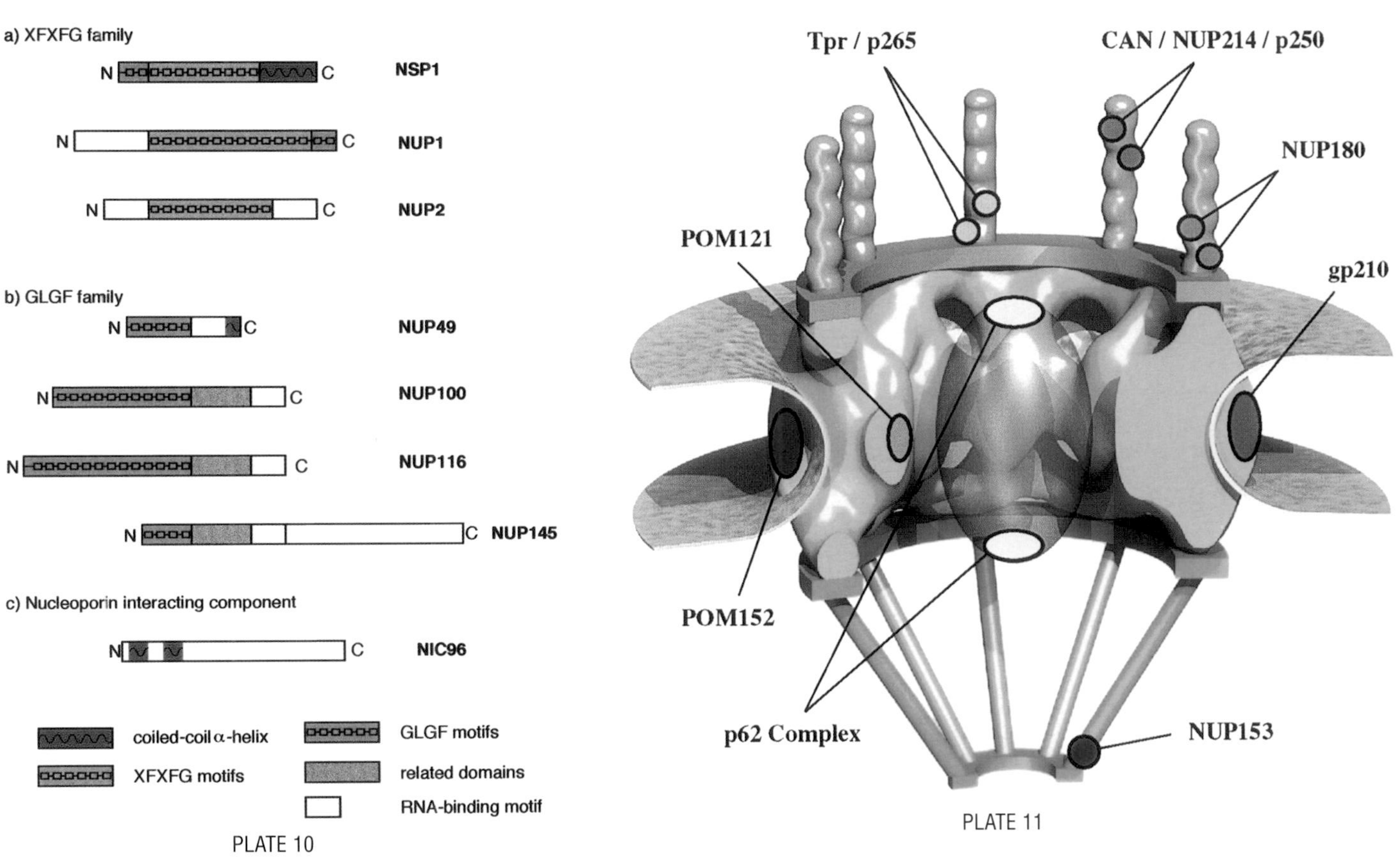


PLATE 10

PLATE 11

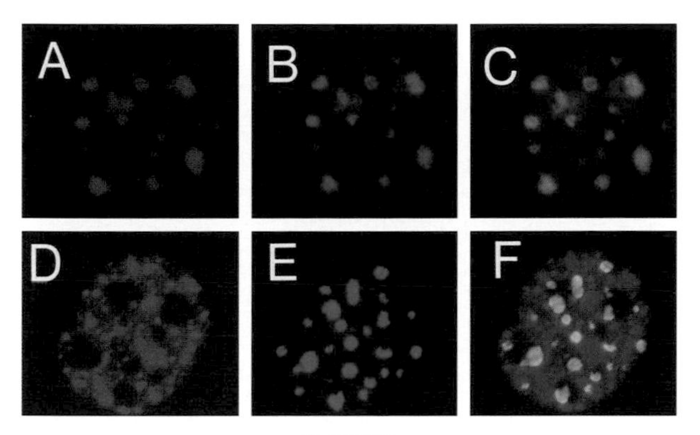


PLATE 12

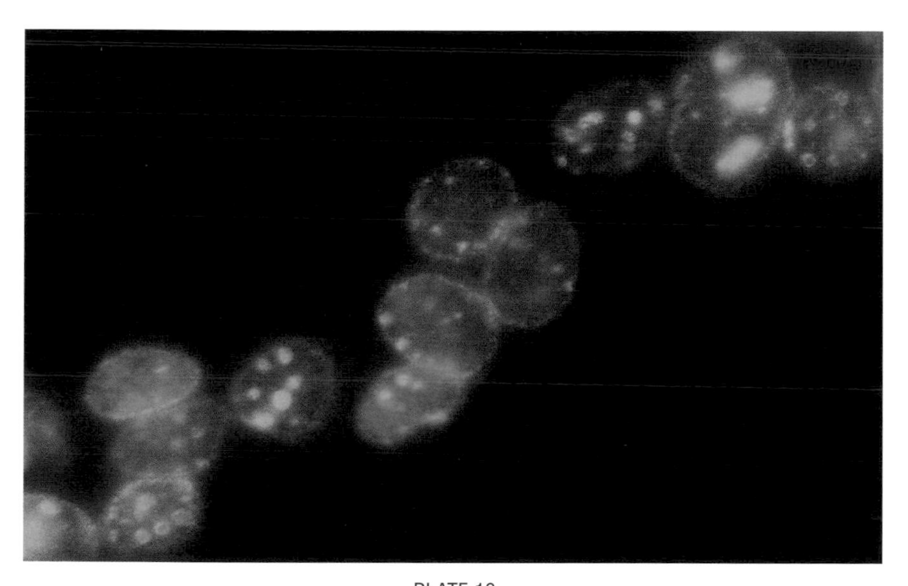

PLATE 13

Chang, D. D., and Sharp, P. A. (1989). Regulation by HIV Rev depends upon recognition of splice sites. *Cell (Cambridge, Mass.)* **59,** 789–795.

Cheng, H., and Bjerknes, M. (1989). Asymmetric distribution of actin mRNA and cytoskeletal pattern generation in polarized epithelial cells. *J. Mol. Biol.* **210,** 541–549.

Chew, E. C., Cheng, H., Chew, S. B., DeHarven, E., and Liew, C. C. (1992). Distribution of a novel nuclear protein in normal and regenerating liver cells. *In Vivo* **6,** 97–102.

Clawson, G. A., Woo, C. H., Button, J., and Smuckler, E. A. (1984). Photo affinity labelling of the major nucleoside triphosphatase of rat liver nuclear envelope. *Biochemistry* **23,** 3501–3507.

Clawson, G. A., Song, Y. L., Schwartz, A. M., Shukla, R. R., Patel, S. G., Connor, L., Blankenship, L., Hatem, C., and Kumar, A. (1991). Interaction of human immunodeficiency virus type I Rev protein with nuclear scaffold nucleoside triphosphatase activity. *Cell Growth Differ.* **2,** 575–582.

Cochrane, A. W., Chen, C. H., and Rosen, C. A. (1990a). Specific interaction of the human immunodeficiency virus Rev protein with a structured region in the env mRNA. *Proc. Natl. Acad. Sci. U.S.A.* **87,** 1198–1202.

Cochrane, A. W., Perkins, A., and Rosen, C. A. (1990b). Identification of sequences important in the nucleolar localization of human immunodeficiency virus Rev: Relevance of nucleolar localization to function. *J. Virol.* **64,** 881–885.

Comerford, S. A., Agutter, P. S., and McLennan, A. G. (1986). *In* "Nuclear Structures: Their Isolation and Characterization" (A. J. MacGillivray and G. D. Birnie, eds.), pp. 1–13. Butterworth, London.

Coutavas, E., Ren, M., Oppenheim, J. D., D'Eustachio, P., and Rush, M. G. (1993). Characterization of proteins that interact with the cell-cycle regulatory protein Ran/TC4. *Nature (London)* **366,** 585–587.

Crank, J. (1975). "The Mathematics of Diffusion." Oxford Univ. Press (Clarendon), Oxford.

Cullen, B. R., and Malim, M. H. (1991). The HIV-1 Rev protein: Prototype of a novel class of eukaryotic post-transcriptional regulators. *Trends Biochem. Sci.* **16,** 346–350.

Cullen, B. R., Hauber, J., Campbell, K., Sodroski, J. G., Haseltine, W. A., and Rosen, C. A. (1988). Subcellular localization of the human immunodeficiency virus trans-acting art gene product. *J. Virol.* **62,** 2498–2501.

Daefler, S., Klotman, M. E., and Wong, F. (1990). Trans-activating rev protein of the human immunodeficiency virus 1 interacts directly and specifically with its target RNA. *Proc. Natl. Acad. Sci. U.S.A.* **87,** 4571–4575.

Daly, T. J., Cook, K. S., Gray, G. S., Maione, T. E., and Rusche, J. R. (1989). Specific binding of HIV-1 recombinant Rev protein to the Rev-responsive element in vitro. *Nature (London)* **342,** 816–819.

d'Angelo, E. G. (1946). Micrurgical studies on Chironomus salivary gland chromosomes. *Biol. Bull. (Woods Hole, Mass.)* **90,** 71–87.

Dargemont, C., and Kuhn, L. C. (1992). Export of mRNA from microinjected nuclei of *Xenopus laevis* oocytes. *J. Cell. Biol.* **118,** 1–9.

Davis, L. I., and Blobel, G. (1986). Identification and characterization of a nuclear pore complex protein. *Cell (Cambridge, Mass.)* **50,** 699–709.

Davis, L. I., and Ish-Horowicz, D. (1991). Apical localization of pair-rule transcripts requires 3' sequences and limits protein diffusion in the *Drosophila* blastoderm embryo. *Cell (Cambridge, Mass.)* **67,** 927–940.

Dingwall, C., and Laskey, R. A. (1986). Protein import into the cell nucleus. *Annu. Rev. Cell Biol.* **2,** 367–390.

Divecha, N., Banfic, H., and Irvine, R. F. (1991). The polyphosphosinositide cycle exists in the nuclei of Swiss 3T3 cells under the control of a receptor (for IGF-I) in the plasma membrane, and stimulation of the cycle increases diacylglycerol and apparently induces translocation of protein kinase C to the nucleus. *EMBO J.* **10,** 3207–3214.

Donnan, F. G. (1927). Concerning the applicability of thermodynamics to the phenomena of life. *J. Gen. Physiol.* **8,** 685–688.

Dworetzky, S. I., and Feldherr, C. M. (1988). Translocation of RNA-coated gold particles through the nuclear pores of oocytes. *J. Cell Biol.* **106,** 575–584.

Eckner, R., Ellmeier, W., and Birnstiel, M. L. (1991). Mature mRNA 3′ end formation stimulates RNA export from the nucleus. *EMBO J.* **10,** 3513–3522.

Eckstein, E. C., and Belgacem, F. (1991). Model of platelet transport in flowing blood with drift and diffusion terms. *Biophys. J.* **60,** 53–69.

Einstein, A. (1905). Von der molecularkinetischen theorie der wärme gefordete bewegung von in ruhenden flüssigkeiten teilchen. *Ann. Phys. (Leipzig)* **17,** 549–554.

Emerman, M., Vazeux, R., and Peden, K. (1989). The rev gene product of the human immunodeficiency virus affects envelope-specific RNA localization. *Cell* (*Cambridge, Mass.*) **57,**1155–1165.

Felber, B. K., Hadzopoulou-Cladaras, M., Cladaras, C., Copeland, T., and Pavlakis, G. N. (1989). Rev protein human immunodeficiency virus type 1 affects the stability and transport of the viral mRNA. *Proc. Natl. Acad. Sci. U.S.A.* **86,** 1495–1499.

Feldherr, C. M. (1980). Ribosomal RNA synthesis and transport following disruption of the nuclear envelope. *Cell Tissue Res.* **205,** 157–162.

Feldherr, C. M., ed. (1992). "Nuclear Trafficking." Academic Press, San Diego, CA.

Fey, E. G., Wan, K. M., and Penman, S. (1984). Epithelial cytoskeletal framework and nuclear matrix/intermediate filament scaffold: Three-dimensional organization and protein composition. *J. Cell Biol.* **98,** 1973–1984.

Fick, A. (1855). Über Diffusion. *Ann. Phys.* (*Leipzig*) **94,** 59–86.

Fulton, A. B. (1982). How crowded is the cytoplasm? *Cell* (*Cambridge, Mass.*) **30,** 345–347.

Garcia-Bustos, J. F., Wagner, P., and Hall, M. N. (1991). Nuclear import substrates compete for a limited number of binding sites. Evidence for different classes of yeast nuclear import receptors. *J. Biol. Chem.* **266,** 22303–22306.

Garner, C. C., Tucker, R. P., and Matus, A. (1988). Selective localization of messenger RNA for cytoskeletal protein MAP2 in dendrites. *Nature* (*London*) **336,** 674–677.

Gershon, N., Porter, K., and Trus, B. (1983). *In* "Biological Structures and Coupled Flows" (A. Oplatka and M. Balaban, eds.), pp. 377–380. Academic Press, New York.

Getzenberg, R. H., Pienta, K. J., Ward, W. S., and Coffey, D. S. (1991). Nuclear structure and the three-dimensional organization of DNA. *J. Cell. Biochem.* **47,** 289–299.

Goldfarb, D. S. (1991). *Curr. Biol.* **1,** 212–214.

Goldfarb, D. S. (1992). Are the cytosolic components of the nuclear, ER, and mitochondrial import apparatus functionally related? *Cell* (*Cambridge, Mass.*) **70,** 185–188.

Gruss, P., Lai, C. J., Dhar, R., and Khoury, G. (1979). Splicing as a requirement for biogenesis of functional 16S mRNA of simian virus 40. *Proc. Natl. Acad. Sci. U.S.A.* **76,** 4317–4321.

Hadzopoulou-Cladaras, M., Felber, B. K., Cladaras, C., Athanassopoulos, A., Tse, A., and Pavlakis, G. N. (1989). The rev (trs/art) protein of human immunodeficiency virus type 1 affects viral mRNA and protein expression via a cis-acting sequence in the env region. *J. Virol.* **63,** 1265–1274.

Halling, P. J. (1989). Do the laws of chemistry apply to living cells? [see comments]. *Trends Biochem. Sci.* **14,** 317–318.

Hamm, J., and Mattaj, I. W. (1990). Monomethylated cap structures facilitate RNA export from the nucleus. *Cell* (*Cambridge, Mass.*) **63,** 109–118.

Hamm, J., Darzynkiewicz, E., Tahara, S. M., and Mattaj, I. W. (1990). The trimethylguanosine cap structure of U1 snRNA is a component of a bipartite nuclear targeting signal. *Cell* (*Cambridge, Mass.*) **62,** 569–577.

He, X. P., Bataille, N., and Fried, H. M. (1994). Nuclear export of signal recognition particle RNA is a facilitated process that involves the Alu sequence domain. *J. Cell Sci.* **107,** 903–912.

Heaphy, S., Finch, J. T., Gait, M. J., Karn, J., and Singh, M. (1991). Human immundeficiency virus type 1 regulator of virion expression, rev, forms nucleoprotein filaments after binding to a purine-rich "bubble" located within the rev-responsive region of viral mRNAs. *Proc. Natl. Acad. Sci. U.S.A.* **88,** 7366–7370.

Hesketh, J. E., and Pryme, I. E. (1988). Evidence that insulin increases the proportion of polysomes that are bound to the cytoskeleton in 3T3 fibroblasts. *FEBS Lett.* **231,** 62–66.

Hoock, T. C., Newcomb, P. M., and Herman, I. M. (1991). Beta-actin and its mRNA are localized at the plasma membrane and the regions of moving cytoplasm during the cellular response to injury. *J. Cell Biol.* **112,** 653–664.

Horowitz, S. B., and Miller, D. S. (1986). *NATO ASI Ser., A* **127,** 79–85.

Huang, S., and Spector, D. L. (1991). Nascent pre-mRNA transcripts are associated with nuclear regions enriched in splicing factors. *Genes Dev.* **5,** 2288–2302.

Imamoto, N., Matsuoka, Y., Kurihara, T., Kohno, K., Miyagi, M., Sakiyama, F., Okada, Y., Tsunasawa, S., and Yoneda, Y. (1992). Antibodies against 70-kD heat shock cognate protein inhibit mediated nuclear import of karyophilic proteins. *J. Cell Biol.* **119,** 1047–1061.

Imamoto-Sonobe, N., Matsuoka, Y., Semba, T., Okada, Y., Uchida, T., and Yoneda, Y. (1990). A protein recognized by antibodies to Asp-Asp-Asp-Glu-Asp shows specific binding activity to heterogenous nuclear transport signals. *J. Biol. Chem.* **265,** 16504–16508.

Jackson, D. A., and Cook, P. R. (1985). Transcription occurs at a nucleoskeleton. *EMBO J.* **4,** 919–925.

Jackson, D. A., Hassan, A. B., Errington, R. J., and Cook, P. R. (1993). Visualization of focal sites of transcription within human nuclei. *EMBO J.* **12,** 1059–1065.

Jeffrey, W. R. (1989). Localized mRNA and the egg cytoskeleton. *Int. Rev. Cytol.* **119,** 151–193.

Jones, N. L., and Kirkpatrick, B. A. (1987). The effects of human cytomegalovirus infection on cytoskeleton-associated polyribosomes. *Eur. J. Cell Biol.* **46,** 565–575.

Katze, M. G., Chen, Y. T., and Krug, R. M. (1984). Nuclear-cytoplasmic transport and VAI RNA-independent translation of influenza viral messenger RNAs in late adenovirus-infected cells. *Cell (Cambridge, Mass.)* **37,** 483–490.

Kim, S. Y., Byrn, R., Groopman, J., and Baltimore, D. (1989). Temporal aspects of DNA and RNA synthesis during human immunodeficiency virus infection: Evidence for differential gene expression. *J. Virol.* **63,** 3708–3713.

Kjems, J., Brown, M., Chang, D. D., and Sharp, P. A. (1991). Structural analysis of the interaction between the human immunodeficiency virus Rev protein and the Rev response element. *Proc. Natl. Acad. Sci. U.S.A.* **88,** 683–687.

Kleiman, R., Banker, G., and Steward, O. (1990). Differential subcellular localization of particular mRNAs in hippocampal neurons in culture. *Neuron* **5,** 821–830.

Korn, E. D., and Hammer, J. A. (1988). Myosins of nonmuscle cells. *Annu. Rev. Biophys. Biophys. Chem.* **17,** 23–45.

Krug, R. M. (1993). The regulation of export of mRNA from nucleus to cytoplasm. *Curr. Biol.* **5,** 944–949.

Lawrence, J. B., and Singer, R. H. (1986). Intracellular localization of messenger RNAs for cytoskeletal proteins. *Cell (Cambridge, Mass.)* **45,** 407–415.

Lawrence, J. B., Singer, R. H., and Marselle, L. M. (1989). Highly localised tracks of specific transcripts within interphase nuclei visualised by *in situ* hybridization. *Cell (Cambridge, Mass.)* **57,** 493–502.

Lee, W. C., and Malesi, T. (1989). Identification and characterization of a nuclear localization sequence-binding protein in yeast. *Proc. Natl. Acad. Sci. U.S.A.* **86,** 8808–8812.

Lenk, R., Ransom, L., Kaufmann, Y., and Penman, S. (1977). A cytoskeletal structure with associated polyribosomes obtained from HeLa cells. *Cell (Cambridge, Mass.)* **10,** 67–78.

Leppard, K. N., and Shenk, T. (1989). The adenovirus E1B 55 kd protein influences mRNA transport via an intranuclear effect on RNA metabolism. *EMBO J.* **8,** 2329–2336.

Long, B. H., Huang, C. Y., and Pogo, A. O. (1979). Isolation and characterization of the nuclear matrix in Friend erythroleukemia cells: Chromatin and hnRNA interactions with the nuclear matrix. *Cell (Cambridge, Mass.)* **18,** 1079–1090.

Lu, X. B., Heimer, J., Rekosh, D., and Hammarskjold, M. L. (1990). U1 small nuclear RNA plays a direct role in the formation of a rev-regulated human immunodeficiency virus env mRNA that remains unspliced. *Proc. Natl. Acad. Sci. U.S.A.* **87,** 7598–7602.

Luby-Phelps, K. (1994). Physical properties of cytoplasm. *Curr. Opin. Cell Biol.* **6,** 3–9.

Lydersen, B. K., and Pettijohn, D. E. (1980). Human-specific nuclear protein that associates with the polar region of the mitotic apparatus: Distribution in a human/hamster hybrid cell. *Cell (Cambridge, Mass.)* **22,** 489–499.

Malim, M. H., Hauber, J., Le, S. Y., Maizel, J. V., and Cullen, B. R. (1989). The HIV-1 rev *trans*-activator acts through a structured target sequence to activate nuclear export of unspliced viral mRNA. *Nature (London)* **338,** 254–257.

Maul, G. G. (1982). *In* "The Nuclear Envelope and the Nuclear Matrix" (G. G. Maul, ed.), pp. 1–11. Alan R. Liss, New York.

McDonald, J. R., and Agutter, P. S. (1980). The relationship between polyribonucleotide binding and the phosphorylation and dephosphorylation of nuclear protein. *FEBS Lett.* **116,** 145–148.

Meier, U. T., and Blobel, G. (1992). Noppl40 shuttles on tracks between nucleolus and cytoplasm. *Cell (Cambridge, Mass.)* **70,** 127–138.

Michaud, N., and Goldfarb, D. S. (1991). Multiple pathways in nuclear transport: The import of U2 snRNP occurs by a novel kinetic pathway. *J. Cell Biol.* **112,** 215–223.

Moon, R. T., Nicosia, R. F., Olsen, C., Hille, M. B., and Jeffrey, W. R. (1983). The cytoskeletal framework of sea urchin eggs and embryos: Developmental changes in the association of messenger RNA. *Dev. Biol.* **95,** 447–458.

Moore, M. S., and Blobel, G. (1992). The two steps of nuclear import, targeting to the nuclear envelope and translocation through the nuclear pore, require different cytosolic factors. *Cell (Cambridge, Mass.)* **69,** 939–950.

Muller, W. E., Agutter, P. S., Bernd, A., Bachmann, M., and Schroder, H.-C. (1985). *In* "Thresholds in Aging; 1984 Sandoz Lectures on Gerontology" (M. Bergener, M. Erminci, and H. B. Stahlin, eds.), pp. 21–58. Academic Press, London.

Muller,W. E., Okamoto, T., Reuter, P., Ugarković, D., and Schroder, H. C. (1990). Functional characterization of Tat protein from human immunodeficiency virus. Evidence that Tat links viral RNAs to nuclear matrix. *J. Biol. Chem.* **265,** 3803–3808.

Newmeyer, D. D., and Forbes, D. J. (1988). Nuclear import can be separated into distinct steps in vitro: Nuclear pore binding and translocation. *Cell (Cambridge, Mass.)* **52,** 641–653.

Newmeyer, D. D., and Forbes, D. J. (1990). An *N*-ethylmaleimide-sensitive cytosolic factor necessary for nuclear protein import: Requirement in signal-mediated binding to the nuclear pore. *J. Cell. Biol.* **110,** 547–557.

Nickerson, J. A., and Penman, S. (1992). Localization of nuclear matrix core filament proteins at interphase and mitosis. *Cell Biol. Int. Rep.* **16,** 811–826.

Ornelles, D. A., and Shenk, T. (1991). Localization of the adenovirus early region 1B 55-kilodalton protein during lytic infection: Association with nuclear viral inclusions requires the early region 4 34-kilodalton protein. *J. Virol.* **65,** 424–429.

Paine, P. L. (1984). Diffusive and non-diffusive proteins *in vivo. J. Cell Biol.* **99,** 188s–195s.

Paine, P. L. (1993). Nuclear protein accumulation by facilitated transport and intranuclear binding. *Trends Cell Biol.* **3,** 325–329.

Peters, R. (1986). Fluorescence microphotolysis to measure nucleocytoplasmic transport and intracellular mobility. *Biochim. Biophys. Acta* **864,** 305–359.

Peters, R. A. (1930). Surface structure in cell activity. *Trans. Faraday Soc.* **26,** 797–807.

Pfeifer, K., Weiler, B. E., Ugarković, D., Bachmann, M., Schroder, H. C., and Muller, W. E. (1991). Evidence for a direct interaction of Rev protein with nuclear envelop mRNA-translocation system. *Eur. J. Biochem.* **199,** 53–64.

Pienta, K. J., and Ward, W. S. (1994). An unstable nuclear matrix may contribute to genetic instability. *Med. Hypotheses* **42,** 45–52.

Pilder, S., Moore, M., Logan, J., and Shenk, T. (1986). The adenovirus E1B-55K transforming polypeptide modulates transport or cytoplasmic stabilization of viral and host cell mRNAs. *Mol. Cell. Biol.* **6,** 470–476.

Pinol-Roma, S., and Dreyfuss, G. (1992). Shuttling of pre-mRNA binding proteins between nucleus and cytoplasm. *Nature* (*London*) 355, 730–732.

Pomeroy, M. E., Lawrence, J. B., Singer, R. H., and Billings-Gaguardi, S. (1991). Distribution of myosin heavy chain mRNA in embryonic muscle tissue visualized by ultrastructural *in situ* hybridization. *Dev. Biol.* **143,** 58–67.

Prather, R. S., and Schatten, G. (1992). Construction of the nuclear matrix at the transition from maternal to zygotic control of development in the mouse: An immunocytochemical study. *Mol. Reprod. Dev.* **32,** 203–208.

Prochnow, D., Thomson, M., Schroder, H. C., Muller, W. E., and Agutter, P. S. (1994). Efflux of RNA from resealed nuclear envelope ghosts. *Arch. Biochem. Biophys.* **312,** 579–587.

Rao, M. V. N., and Prescott, D. M. (1967). Return of RNA into the nucleus after mitosis. *J. Cell Biol.* **35,** 109a.

Remenyik, C. J., and Kellermeyer, M. (1978). A fluid mechanical hypothesis for macromolecular transport in living cells: I. *Physiol. Chem. Phys.* **10,** 107–113.

Richardson, W. D., Mills, A. D., Dilworth, S. M., Laskey, R. A., and Dingwall, C. (1988). *Cell* (*Cambridge, Mass.*) **52,** 655–664.

Riedel, N., and Fasold, H. (1992). *In* "Nuclear Trafficking" (C. M. Feldherr, ed.), pp. 231–290. Academic Press, San Diego, CA.

Rosbash, M., and Singer, R. H. (1993). RNA travel: Tracks from DNA to cytoplasm. *Cell* (*Cambridge, Mass.*) **75,** 399–401.

Rosen, C. A. (1991). Regulation of HIV gene expression by RNA–protein interactions. *Trends Genet.* **7,** 9–14.

Rosen, C. A., Terwilliger, E., Dayton, A., Sodroski, J. G., and Haseltine, W. A. (1988). Intragenic *cis*-acting art gene-responsive sequences of the human immunodeficiency virus. *Proc. Natl. Acad. Sci. U.S.A.* **85,** 2071–2075.

Sahlas, D. J., Milankov, K., Park, P. C., and De Boni, U. (1993). Distribution of snRNPs, splicing factor SC-35 and actin in interphase nuclei: Immunocytochemical evidence for differential distribution during changes in functional states. *J. Cell Sci.* **105,** 347–357.

Sarnow, P., Hearing, P., Anderson, C. W., Halbert, D. N., Shenk, T., and Levine, A. J. (1984). Adenovirus early region 1B 58,000-dalton tumor antigen is physically associated with an early region 4 25,000-dalton protein in productively infected cells. *J. Virol.* **49,** 692–700.

Saxton, M. J. (1982). Lateral diffusion in an archipelago. Effects of impermeable patches on diffusion in a cell membrane. *Biophys. J.* **39,** 165–173.

Scherrer, K. (1990). Prosomes, subcomplexes of untranslated mRNP. *Mol. Biol. Rep.* **14,** 1–9.

Schindler, M., and Jiang, L.-W. (1986). Nuclear actin and myosin as control elements in nucleocytoplasmic transport. *J. Cell Biol.* **102,** 859–862.

Schindler, M., and Jiang, L. W. (1987). Epidermal growth factor and insulin stimulate nuclear pore-mediated macromolecular transport in isolated rat liver nuclei. *J. Cell Biol.* **104,** 849–853.

Schroder, H.-C., Rottmann, M., Bachmann, M., Muller, W. E., McDonald, A. R., and Agutter, P. S. (1986). Proteins from rat liver cytosol which stimulate mRNA transport. *Eur. J. Biochem.* **159,** 51–59.

Schroder, H.-C., Bachmann, M., Diehl-Seifert, B., and Muller, W. E. (1987a). Transport of mRNA from nucleus to cytoplasm. *Prog. Nucleic Acids Res. Mol. Biol.* **34,** 89–142.

Schroder, H.-C., Trolltsch, D., Friese, U., Bachmann, M., and Muller, W. E. G. (1987b). Mature mRNA is selectively released from the nuclear matrix by an ATP/dATP-dependent mechanism sensitive to topoisomerase inhibitors. *J. Biol. Chem.* **262,** 8917–8925.

Schroder, H.-C., Trolltsch, D., Wenger, R., Bachmann, M., and Muller, W. E. G. (1987c). Cytochalasin B selectively releases ovalbumin mRNA precursors but not the mature ovalbumin mRNA from hen oviduct nuclear matrix. *Eur. J. Biochem.* **167,** 239–245.

Schroder, H.-C., Diehl-Seifert, B., Rottmann, M., Aitken, S. J. M., Bryson, B. A., Agutter, P. S., and Muller, W. E. G. (1988). Functional dissection of the nuclear envelope mRNa translocation system: Effects of phorbol ester and a monoclonal antibody recognizing cytoskeletal structures. *Arch. Biochem. Biophys.* **261,** 394–404.

Schroder, H.-C., Friese, U., Bachmann, M., Zaubitzer, T., and Muller, W. E. G. (1989). Energy requirement requirement and kinetics of transport of poly(A)-free histone mRNA compared to poly(A)-rich mRNA from isolated L-cell nuclei. *Eur. J. Biochem.* **181,** 149–158.

Schroder, H. C., Wenger, R., Ugarković, D., Friese, K., Bachmann, M., and Muller, W. E. (1990). Differential effect of insulin and epidermal growth factor on the mRNA translocation system and transport of specific poly(A+) mRNA and poly(A−) mRNA in isolated nuclei. *Biochemistry* **29,** 2368–2378.

Schumm, D. E., Hanausek-Walasek, M., Yannarel, A., and Webb, T. E. (1977). Changes in nuclear RNA transport incident to carcinogenesis. *Eur. J. Cancer* **13,** 139–142.

Schweiger, A., and Kostka, G. (1984). Concentration of particular high molecular mass phosphoproteins in rat liver nuclei and nuclear matrix decreases following inhibition of RNA synthesis by alpha-amanitin. *Biochim. Biophys. Acta* **782,** 262–268.

Shen, W. C., Selvakumar, D., Standford, D. R., and Hopper, A. K. (1993). The *Saccharomyces cerevisiae* LOS1 gene involved in pre-tRNA splicing encodes splicing encodes a nuclear protein that behaves as a component of the nuclear matrix. *J. Biol. Chem.* **268,** 19436–19444.

Shi, Y., and Thomas, J. O. (1992). The transport of proteins into the nucleus requires the 70-kilodalton heat shock protein or its cytosolic cognate. *Mol. Cell. Biol.* **12,** 2186–2192.

Sibon, O. C., Cremers, F. F., Boonstra, J., Humbel, B. M., and Verkleij, A. J. (1993). Localisation of EGF-receptor mRNA in the nucleus of A431 cells by light microscopy. *Cell Biol. Int. Rep.* **17,** 1–11.

Singer, R. H. (1992). The cytoskeleton and mRNA localization. *Curr. Opin. Cell Biol.* **4,** 15–19.

Sippel, A. E., Hynes, N., Groner, B., and Schutz, G. (1977). Frequency distribution of messenger sequences within polysomal mRNA and nuclear RNA from rat liver. *Eur. J. Biochem.* **77,** 141–151.

Srivastava, D. K., and Bernhard, S. A. (1986). Metabolite transfer via enzyme-enzyme complexes. *Science* **234,** 1081–1086.

Sterne-Marr, R., Blevitt, J. M., and Gerace, L. (1992). O-linked glycoproteins of the nuclear pore complex interact with a cystolic factor required for nuclear protein import. *J. Cell. Biol.* **116,** 271–280.

Stochaj, U., and Silver, P. A. (1992). A conserved phosphoprotein that specifically binds nuclear localization sequences is involved in nuclear import. *J. Cell Biol.* **117,** 473–482; erratum: **118**(1), 215 (1992).

Sundell, C. L., and Singer, R. H. (1990). Actin mRNA localizes in the absence of protein synthesis. *J. Cell Biol.* **111,** 2397–2403.

Sundell, C. L., and Singer, R. H. (1991). Requirement of microfilaments in sorting of actin messenger RNA. *Science* **253,** 1275–1277.

Tawfic, S., Goueli, S. A., Olson, M. O., and Ahmed, K. (1993). Androgenic regulation of the expression and phosphorylation of prostatic nucleolar protein B23. *Cell. Mol. Biol. Res.* **39,** 43–51.

Ueyama, H., Nakayasu, H., and Ueda, K. (1987). Nuclear actin and transport of RNA. *Cell Biol. Int. Rep.* **11,** 671–677.

van Driel, R., Humbel, B., and de Jong, L. (1991). The nucleus: A black box being opened. *J. Cell. Biochem.* **47,** 311–316.

van Eekelen, C. A. G., and van Venrooij, W. J. (1981). HnRNA and its attachment to a nuclear matrix. *J. Cell Biol.* **88,** 554–563.

van Venrooij, W. J., Sillekens, P. T. G., van Eekelen, C. A. G., and Reinders, R. T. (1981). On the association of mRNA with the cytoskeleton in uninfected and adenovirus infected human KB cells. *Exp. Cell Res.* **135,** 79–91.

Vedeler, A., Pryme, I. F., and Hesketh, J. E. (1990). Insulin and step-up conditions cause a redistribution of polysomes among free, cytoskeletal-bound and membrane-bound fractions in Krebs II ascites cells. *Cell Biol. Int. Rep.* **14,** 211–218.

Vedeler, A., Pryme, I. F., and Hesketh, J. E. (1991). The characterization of free, cytoskeletal and membrane-bound polysomes in Krebs II ascites and 3T3 cells. *Mol. Cell. Biochem.* **100,** 183–193.

Venkatesan, S., Gerstberger, S. M., Park, H., Holland, S. M., and Nam, Y. (1992). Human immunodeficiency virus type 1 Rev activation can be achieved without Rev-responsive element RNA if Rev is directed to the target as a Rev/MS2 fusion protein which tethers the MS2 operator RNA. *J. Virol.* **66,** 7469–7480.

von Smoluchowski, M. (1906). Zur kinetischen Theore der Brownschen Molekulabewegung und der Suspensionen. *Ann. Phys.* (*Leipzig*) **B21,** 756–780.

Wachsberger, P. R., and Cos, R. A. (1993). Vedelations in nuclear matrix ultrastructure of G1 mammalian cells following heat shock: Resinless section electron microscopy, biochemical, and immunoflourescence studies. *J. Cell. Physiol.* **155,** 615–634.

Walter, P. Gilmore, R., and Blobel, G. (1984). Protein transloation across the endoplasmic reticulum. *Cell (Cambridge, Mass.)* **38,** 5–8.

Wang, J., Cao, L. G., Wang, Y. L., and Pederson, T. (1991). Localization of pre-messenger RNA at discrete nuclear sites. *Proc. Natl. Acad. Sci. U.S.A.* **88,** 7391–7395.

Warrick, H. M., and Spudlich, J. A. (1987). Protein translocation across the endoplasmic reticulum. *Annu. Rev. Cell Biol.* **3,** 379–421.

Welch, G. R., and Clegg, J. S. (1987). "The Organization of Cell Metabolism," NATO ARI Ser. A, B127. Plenum, New York.

Wojcieszyn, J. W., Schlegel, R. A., Wu, E. S., and Jacobson, K. A. (1981). Diffusion of injected macromolecules within the cytoplasm of living cells. *Proc. Natl. Acad. Sci. U.S.A.* **78,** 4407–4410.

Wolosewick, J. J., and Porter, K. R. (1976). Stereo high-voltage electron microscopy of whole cells of the human diploid line, WI-38. *Am. J. Anat.* **147,** 303–323.

Xing, Y., and Lawrence, J. B. (1991). Preservation of specific RNA distribution within the chromatin-depleted nuclear substructure demonstrated by *in situ* hybridization coupled with biochemical fractionation. *J. Cell Biol.* **112,** 1055–1064.

Yamasaki, L., and Lanford, R. E. (1992). Nuclear transport: A guide to import receptors. *Trends Cell Biol.* **2,** 123–127.

Yamasaki, L., Kanda, P., and Lanford, R. E. (1989). Identification of four nuclear transport signal-binding proteins that interact with diverse transport signals. *Mol. Cell. Biol.* **9,** 3028–3036.

Yang, F., Demma, M., Warren, V., Dharmawardhane, S., and Condeelis, J. (1990). Identification of an actin-binding protein from *Dyctostelium* as elongation factor 1. *Nature* (*London*) **347,** 494–496.

Yisraeli, J. K., Sokol, S., and Melton, D. A. (1990). A two-step model for the localization of maternal mRNA in *Xenopus* oocytes. *Development* (*Cambridge, UK*) **108,** 289–298.

Zachar, Z., Kramer, J., Mims, I. P., and Bingham, P. M. (1993). Evidence for channeled diffusion of pre-mRNAs during nuclear RNA transport in metazoans. *J. Cell Biol.* **121,** 729–742.

Zambetti, G., Fey, E. G., Penman, S., Stein, J., and Stein, G. (1990). Multiple types of mRNA–cytoskeleton interactions. *J. Cell. Biochem.* **44,** 177–187.

Zeitlin, S., Parent, A., Silverstein, S., and Efstratiadis, A. (1987). Pre-mRNA splicing and the nuclear matrix. *Mol. Cell. Biol.* **7,** 111–120.

Zeng, C., He, D., Berget, S. M., and Brinkley, B. R. (1994). Nuclear-mitotic apparatus protein: A structural protein interface between the nucleoskeleton and RNA splicing. *Proc. Natl. Acad. Sci. U.S.A.* **91,** 1505–1509.

Zini, N., Martelli, A. M., Sabatelli, P., Santi, S., Negri, C., Astaldi-Ricotti, G. C., and Maraldi, N. M. (1992). The 180-kDa isoform of topoisomerase II is localized in the nucleolus and belongs to the structural elements of the nucleolar remnant. *Exp. Cell Res.* **200,** 460–466.

Zwanzig, R., and Szabo, A. (1991). Time dependent rate of diffusion-influenced ligand binding to receptors on cell surfaces. *Biophys. J.* **60,** 671–678.

# Toward a Molecular Understanding of the Structure and Function of the Nuclear Pore Complex

Nelly Panté* and Ueli Aebi*,†
*M. E. Müller Institute for Microscopy, Biozentrum, University of Basel, CH-4056 Basel, Switzerland, and †Department of Cell Biology and Anatomy, The Johns Hopkins University School of Medicine, Baltimore, Maryland 21205

Molecular trafficking between the nucleus and cytoplasm of interphase cells occurs via the nuclear pore complexes (NPCs), ~120-MDa supramolecular assemblies embedded in the double-membraned nuclear envelope. Significant progress has been made in elucidating the 3-D architecture of the NPC, and in identifying and characterizing NPC proteins, and cloning and sequencing their genes. Several of these proteins have now been localized within the 3-D structure of the NPC. Over the past few years there have also been some advances in establishing the signals, receptors, and factors mediating import of nuclear proteins and snRNP particles, and nuclear export of RNAs and RNP particles. Nevertheless, relatively little is known about the molecular mechanisms underlying nucleocytoplasmic transport through the NPC. Here we review recent advances toward the 3-D structure and molecular architecture of the NPC, and the molecular basis of nucleocytoplasmic transport of proteins, RNAs, and RNP particles through the NPC.

**KEY WORDS:** Nuclear envelope, Nuclear pore complex, Nuclear pore complex proteins, Nucleocytoplasmic transport.

## I. Introduction

As illustrated schematically in Fig. 1 (color plate 4), in interphase eukaryotic cells the nucleus is separated from the cytoplasm by the nuclear envelope (NE), which consists of a double membrane enclosing a lumen, the "perinuclear space." The outer nuclear membrane faces the cytoplasm and is continuous with the endoplasmic reticulum (ER), so that the perinuclear space is contiguous with the lumen of the ER. On its cytoplasmic surface the outer nuclear membrane is often studded with ribosomes, as is the rough ER. The inner nuclear membrane faces the nucleoplasm and at its nucleoplasmic surface is lined by the nuclear lamina, a polymer of intermediate filament-like proteins, the nuclear lamins (Aebi *et al.,* 1986; Gerace and Burke, 1988). The nuclear lamina provides a general framework for NE structure and is an anchoring site for interphase chromatin. Interposed at irregular intervals between the inner and outer nuclear membranes, where the two membranes are fused to form the "pore membrane," are large supramolecular assemblies called nuclear pore complexes (NPCs). As indicated in Fig. 1 (color plate 4), the NPCs are the sites for bidirectional exchange of material between the nucleus and the cytoplasm. The NPC allows passive diffusion of ions and small molecules, but molecules larger than 9 nm in diameter do not traverse the NPC freely. These are selectively imported into the nucleus by a signal-requiring and ATP-dependent mechanism. Macromolecular assemblies such as ribonucleoprotein (RNP) particles containing mRNA are also selectively exported from the nucleus by active transport. Despite significant progress in understanding the structure and function of the NPC, the molecular mechanism of nucleocytoplasmic transport through the NPC has remained elusive. Here we review recent advances made toward the elucidation of the molecular architecture of the NPC and toward the understanding of the mechanism of mediated transport through the NPC.

## II. Architecture of the Nuclear Pore Complex

The structure of the NPC has been extensively investigated by different electron microscopy (EM) specimen preparation methods and imaging techniques (Panté and Aebi, 1993, 1994a), including atomic force microscopy (AFM) in physiological buffer environment (Panté and Aebi, 1993; Braunstein and Spudich, 1994; Goldie *et al.,* 1994; Oberleithner *et al.,* 1994). Most of these structural studies have been performed using nuclei from *Xenopus laevis* oocytes, which have the advantage that their NEs can readily be manually isolated and yield high-density NPCs. As illustrated in Fig. 2a, when intact *Xenopus* oocyte NEs are spread on an EM grid and viewed in a transmission electron microscope after negative staining, extensive arrays of NPCs are revealed. In these images the NPCs appear as round particles with a diameter of ~125 nm and a distinct eightfold rotational symmetry (Unwin and Milligan, 1982; Akey, 1989; Reichelt *et al.,* 1990; Jarnik and Aebi, 1991). However, as documented in Figs. 2b and 2c, depend-

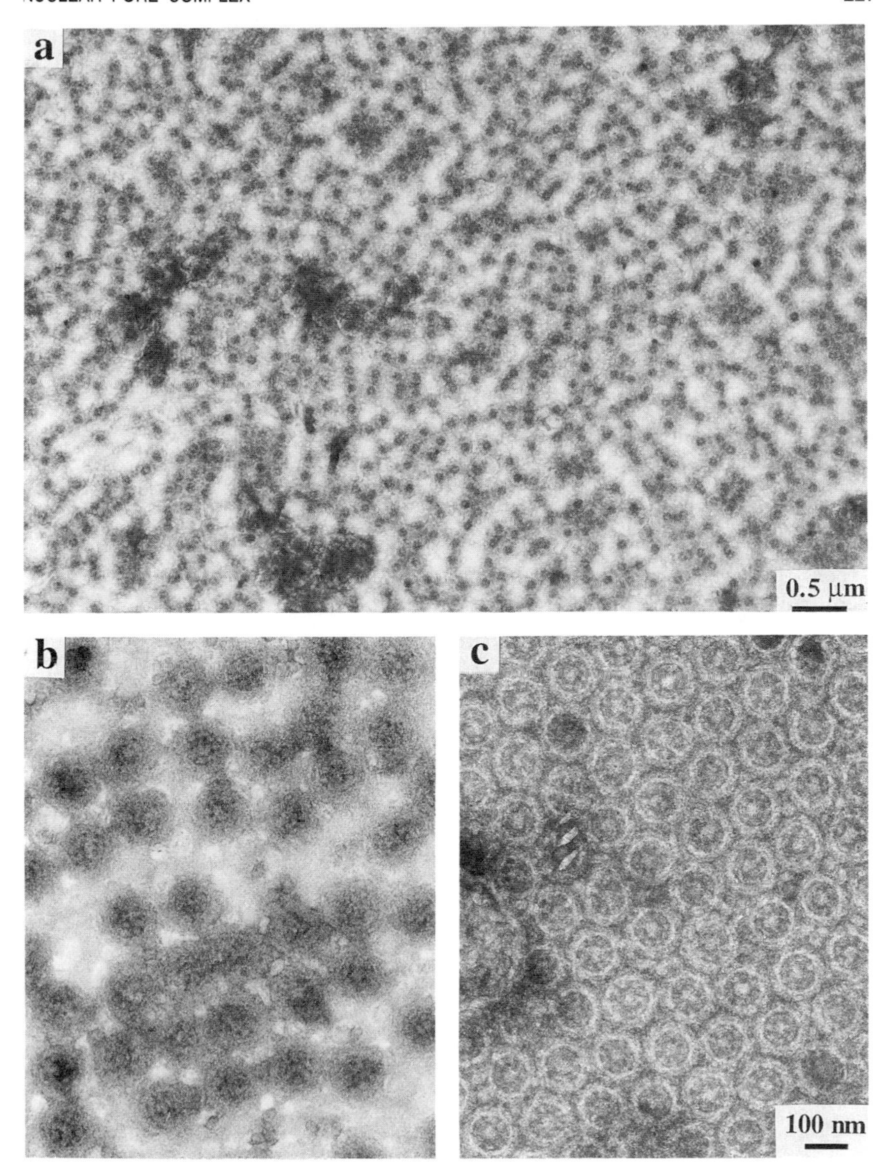


FIG. 2 Electron micrographs of negatively stained *Xenopus* oocyte nuclear envelope (NE). (a) Low magnification view, (b) nuclear, and (c) cytoplasmic face of negatively stained intact NEs. Depending on which face of the NE adsorbed to the EM grid, the NPCs revealed a different morphology, thus revealing the highly asymmetric architecture of the NPC with regard to its nuclear (b) and cytoplasmic (c) periphery. Bars: 0.5 $\mu$m (a) and 100 nm (b, c).

ing on the face of the NE that is adsorbed to the EM support film, the NPCs reveal different morphologies: if the NE adsorbs to the EM film with its cytoplasmic face so that its nuclear face is exposed, the NPCs appear irregularly stained and poorly preserved (Fig. 2b). Whereas when the NE adsorbs to the EM film with its nuclear face, the NPCs appear well preserved and exhibit a distinct eightfold rotational symmetry (Fig. 2c). As will be discussed in Section II,C, this distinct appearance of the nuclear and cytoplasmic faces of the NPC is due to the presence of different peripheral NPC components.

As illustrated in Fig. 3, when the nuclear membranes have been solubilized by treatment with nondenaturing detergents, three distinct NPC components are revealed: (i) the "spoke complex" or basic framework of the NPC which consists of eight "spokes" embracing a "central pore" (Fig. 3b), (ii) the cytoplasmic ring (Fig. 3c), and (iii) the nuclear ring (Fig. 3d). Thus traditionally the NPC has been described as a tripartite assembly consisting of the spoke complex sandwiched between a cytoplasmic and a nuclear ring (Unwin and Milligan, 1982; Akey, 1989; Reichelt *et al.,* 1990; Jarnik and Aebi, 1991). Often the pore of the NPC is "plugged" with a "central channel complex" (also called "central plug" or "transporter") which, depending on the isolation and/or specimen preparation, has a highly variable appearance (see Section II,D). More recently, the use of more elaborate EM specimen preparations and imaging techniques has further documented the existence of distinct filamentous components attached to both the cytoplasmic and nuclear ring (see Section II,C; Jarnik and Aebi, 1991; Ris, 1991; Goldberg and Allen, 1992; Goldie *et al.,* 1994). In the following sections, we first present mass measurements of the entire NPC and its different components before we discuss various structural aspects of the different NPC components in greater detail.

## A. STEM Mass Analysis

The mass of the intact NPC and its different structural components has been determined directly by quantitative scanning transmission EM (STEM) (Reichelt *et al.,* 1990). Accordingly, as summarized in Table I, the membrane-bound intact NPC has a mass of 124 MDa (i.e., $124 \times 10^6$ daltons), and the spoke complex or basic framework of the NPC without the central channel complex 52 MDa. The membrane-bound NPC without the central channel complex has a mass of 112 MDa, thus the central plug or channel complex has a mass on the order of 12 MDa. Two types of rings were distinguished in detergent-treated NEs: heavy, 32-MDa, and light, 21-MDa rings. Since the 32-MDa rings were also observed as "footprints" when isolated nuclei were rolled back and forth on EM grids, these must represent

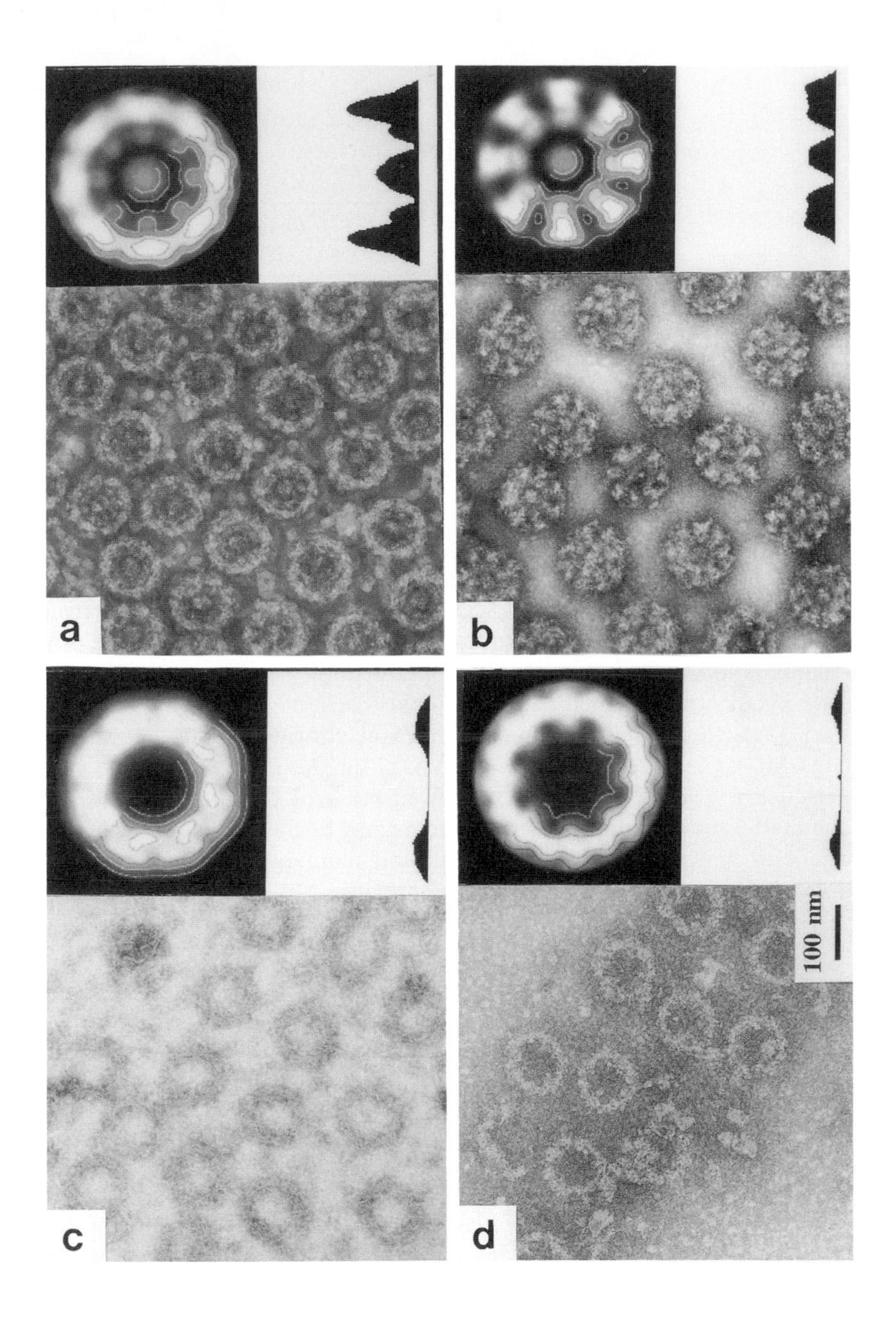


FIG. 3 Negatively stained intact NPCs and distinct NPC components. Electron micrographs, correlation averages (always including 20 eightfold rotationally symmetrized NPCs; upper insets), and radial mass density profiles (lower insets) of intact NPCs (a), and distinct NPC components yielded after treatment of spread NEs with 0.1% Triton 100 (b–d ). (a) Membrane-bound, intact NPCs; (b) plug–spoke complexes; (c) cytoplasmic rings that are also left behind after rolling intact nuclei on a carbon-coated EM grid; and (d) nuclear rings. Bar: 100 nm (a–d ). Adapted from Jarnik and Aebi (1991).

TABLE I
Masses of the Intact NPC and Its Major Structural Components

| | Mass in MDa[a] |
|---|---|
| Nuclear pore complex | 124.0 ± 11.0 |
| Basic framework | 51.7 ± 5.3 |
| Cytoplasmic ring | 32.0 ± 5.5 |
| Nuclear ring | 21.1 ± 3.7 |
| Central channel complex | 12.0 ± 1.1 |

[a] Determined by quantitative scanning transmission EM and adapted from Reichelt *et al.* (1990).

the cytoplasmic rings with the remnants of one to several collapsed cytoplasmic filaments attached (see Section II,C). The 21-MDa nuclear rings frequently revealed some mass in the center (see Fig. 3d), which due to its variation among rings was excluded from the mass measurements. The sum of the masses of its principal components—i.e., the basic framework, the central plug, the cytoplasmic ring, and the nuclear ring—yields a total mass of 117 MDa (i.e., 52 + 12 + 32 + 21 MDa) for the intact NPC. The small difference (~6%) between the measured mass of the intact membrane-bound NPC and the sum of its components may be due to loss of some of its peripheral components (i.e., the cytoplasmic filaments and nuclear basket) during detergent treatment and/or specimen preparation.

## B. 3-D Architecture of the Basic Framework of the NPC

The architecture of the basic framework of the NPC (i.e., the spoke complex) has recently been revealed in 3-D reconstructions of both negatively stained (Hinshaw *et al.,* 1992) and frozen hydrated (Akey and Radermacher, 1993) NPCs obtained after detergent treatment of *Xenopus* NEs. Accordingly, each spoke is built of two approximately identical halves; hence, the entire spoke complex yields 822 symmetry with one half-spoke representing the asymmetric unit. Since the basic framework of the NPC has a mass of 52 MDa (Reichelt *et al.,* 1990; see also Table I), the mass of one half-spoke is ~3.3 MDa, i.e., on the order of a ribosome. As documented in Fig. 4 (color plate 5), in the 3-D mass map of negatively stained detergent-treated NPCs each half-spoke is built from four distinct morphological domains termed "annular," "column," "ring," and "lumenal" domain (Hinshaw *et al.,* 1992). When the nuclear membranes are positioned in this 3-D map, the lumenal domain of each spoke extends into the lumen of the NE and eight ~10-nm-diameter "peripheral channels" are created between two adjacent spokes and the pore membrane border (see Fig. 4a, color plate 5). As speculated by Hinshaw *et al.* (1992), these peripheral channels may represent sites for passive diffusion of ions and small molecules, and they may also facilitate import of inner nuclear membrane proteins (Soullan and Worman, 1993; reviewed by Wiese and Wilson, 1993).

The 3-D reconstruction of frozen hydrated detergent-treated NPCs (Akey and Radermacher, 1993) has revealed a similar 3-D map of the basic framework of the NPC. However, these authors included the central plug or channel complex in their reconstruction, thus the spoke complex embraces an elaborate, barrel-like central channel complex. Also by including the central channel complex, this reconstruction yields eight internal channels between two adjacent spokes and the central channel complex. However, since this tomographic reconstruction is based on a relatively large missing tilt cone of information that affects the final 3-D map predominantly at low radii, the central mass of their 3-D model may have been overestimated despite application of a solvent flattening procedure to the reconstruction (Akey and Radermacher, 1993). In addition, as we will discuss in Section II,C, other NPC components (i.e., the nuclear basket) may contribute significantly to what in ice-embedded NPC images appears as the central plug. Moreover, since the abundance and morphology of the central plug or channel complex depends on the isolation and preparation conditions employed (see Section II,D), it is necessary to develop a procedure to control the reproducible appearance of the central plug among NPCs before computing 3-D maps of an irreproducible structure.

### C. Peripheral NPC Components

As documented in Fig. 3, in addition to revealing the basic framework (i.e., the spoke complex), detergent treatment of NEs releases two types of rings: (i) cytoplasmic rings, which are predominantly positively stained by uranyl salts, and are also yielded by rolling intact nuclei on an EM grid (Jarnik and Aebi, 1991); and (ii) nuclear rings, which are negatively stained and appear less massive than the cytoplasmic rings both by comparison of their radial mass density profiles (Figs. 3c and 3d), and from direct STEM mass measurements (see Section II,A and Table I; Reichelt *et al.*, 1990). However, it has been argued that these rings represent an integral part of the basic framework of the NPC (Hinshaw *et al.,* 1992; Akey and Radermacher, 1993). Indeed, the 3-D map of the basic framework yields two tenuous rings at the cytoplasmic and nuclear face of the NPC (see Fig. 4a, color plate 5) that are defined by the ring domain of the multidomain spokes (see Fig. 4b, color plate 5). Since according to these authors the basic framework of the NPC exhibits good 822 symmetry, these two tenuous rings should be the same or at least very similar. In contrast, the cytoplasmic and nuclear rings released upon detergent treatment of *Xenopus* oocyte NE preparations yield distinct masses (see Section II,A and Table I; Reichelt *et al.,* 1990). Moreover, the basic framework of the NPC—without the central channel complex—reconstructed from negatively stained NPCs after detergent treatment (Hinshaw *et al.,* 1992) has a mass of 52 MDa. Hence, there must be additional components associated with the basic framework of the NPC to account for the ~110-MDa mass of the intact, unplugged NPC (Reichelt *et al.,* 1990). Interestingly, the masses of the 32-MDa cytoplasmic and 21-MDa nuclear ring amount to just about what has to be added to the 52-MDa mass of the basic framework (i.e., 105 MDa) to arrive at a mass of approximately 110 MDa for the intact, unplugged NPC. Therefore, it is conceivable that the cytoplasmic and nuclear rings are not—or only partially—represented by the two tenuous rings of the 3-D mass density maps of the basic framework of detergent-released NPCs (see Fig. 4a, color plate 5).

As illustrated in Fig. 5a, distinct filamentous structures are associated with both the cytoplasmic and nuclear rings. These were first observed in embedded/thin-sectioned NEs but were believed to represent a specimen preparation artifact since they were not depicted by other EM preparation techniques (Franke and Scheer, 1970a,b; Franke, 1974). Recently, the use of more elaborate specimen preparations and imaging techniques has further documented the existence of these peripheral filamentous components of the NPC (Jarnik and Aebi, 1991; Ris, 1991; Goldberg and Allen, 1992; Panté and Aebi, 1993; Goldie *et al.,* 1994). As documented in Figs. 5b and 5c, when *Xenopus* oocyte NEs are visualized in the EM after quick freezing/freeze drying/rotary metal shadowing, the cytoplasmic face of the NPC looks distinct from the nuclear face (see also Jarnik and Aebi, 1991). Accordingly, the cytoplasmic face of the NPC is topped with a 100- to 110-nm outer diameter ring from which eight more or less kinky filaments protrude (see Fig. 5b, arrowheads) which have a tendency to collapse onto themselves and thus often appear as "short cylinders" or "cigars." The presence of these cytoplasmic filaments is best documented in situations where they have bent to the side and adhered to cytoplasmic filaments of adjacent NPCs, thus appearing as "NPC-connecting fibrils" (see Fig. 5b, small arrows). As shown in Fig. 5c, the nuclear face of the NPC accommodates a more tenuous, 90- to 100-nm outer diameter ring from which eight thin, 50- to 100-nm-long filaments emanate and are joined distally by a 30- to 50-nm-diameter terminal ring, thus forming a "basket" or "fishtrap." These cytoplasmic filaments and nuclear baskets make the NPC periphery distinctly asymmetric.

High-resolution scanning EM of critical point-dried/metal-sputtered isolated NEs has revealed similar structures (Ris, 1991; Goldberg and Allen, 1992). In addition, in *Triturus cristatus*—but not *Xenopus*—this technique has depicted the existence of an ordered fibrous nuclear lattice, termed the "NE lattice" or "NEL" that is connected to the nuclear baskets via their terminal rings (Goldberg and Allen, 1992). The chemical composition and function of this NEL remains to be established. Remnants of such a lattice or filament system connecting adjacent baskets has also been observed in *Xenopus* oocyte NEs in the form of "basket-connecting filaments" (Jarnik and Aebi, 1991; Ris, 1991).

More recently, the native cytoplasmic and nuclear periphery of the NPC has been visualized by atomic force microscopy of spread *Xenopus* oocyte NEs kept in physiological buffer (Panté and Aebi, 1993; Goldie *et al.,* 1994). In agreement with the results revealed by dehydrated specimens (Jarnik and Aebi, 1991; Ris, 1991; Goldberg and Allen, 1992), the corresponding AFM topographs revealed a high degree of asymmetry between the nuclear and cytoplasmic periphery of the NPC. As documented in Figs. 5d and 5e, by AFM the cytoplasmic face of the NPC appears "donut-like," whereas the nuclear face exhibits a "dome like" appearance. However, since the resolution in these AFM images is insufficient to resolve individual NPC-associated filaments, the *in vivo* conformation of these cytoplasmic and nuclear filaments has remained uncertain. For example, the cytoplasmic filaments have been described as "granules" (Stewart and Whytock, 1988), short cylinders (Jarnik and Aebi, 1991), and "T-shaped" particles (Goldberg and Allen, 1992). At this stage it is difficult to determine which of these conformations—if any—represents the native conformation or to which extent these different conformations may merely represent preparation artifacts. It is conceivable that at least some of these conformations may represent distinct functional states of the cytoplasmic filaments.

As to their possible functional role, it has been proposed that the cytoplasmic filaments might be the "docking sites" for proteins to be imported into the nucleus, and might deliver the docked material to the central channel complex for active translocation (Gerace, 1992). Since the integrity of the nuclear baskets depends on divalent cations (i.e., they disassemble upon removal of divalent cations and reassemble upon their addition; Jarnik and Aebi, 1991), these might be directly involved in the active transport of proteins, RNAs, or RNP particles through the NPC.

## D. The Central Plug or Channel Complex

Often a massive, ~12-MDa particle that has been termed central plug, central channel complex, or transporter, resides in the central pore of the

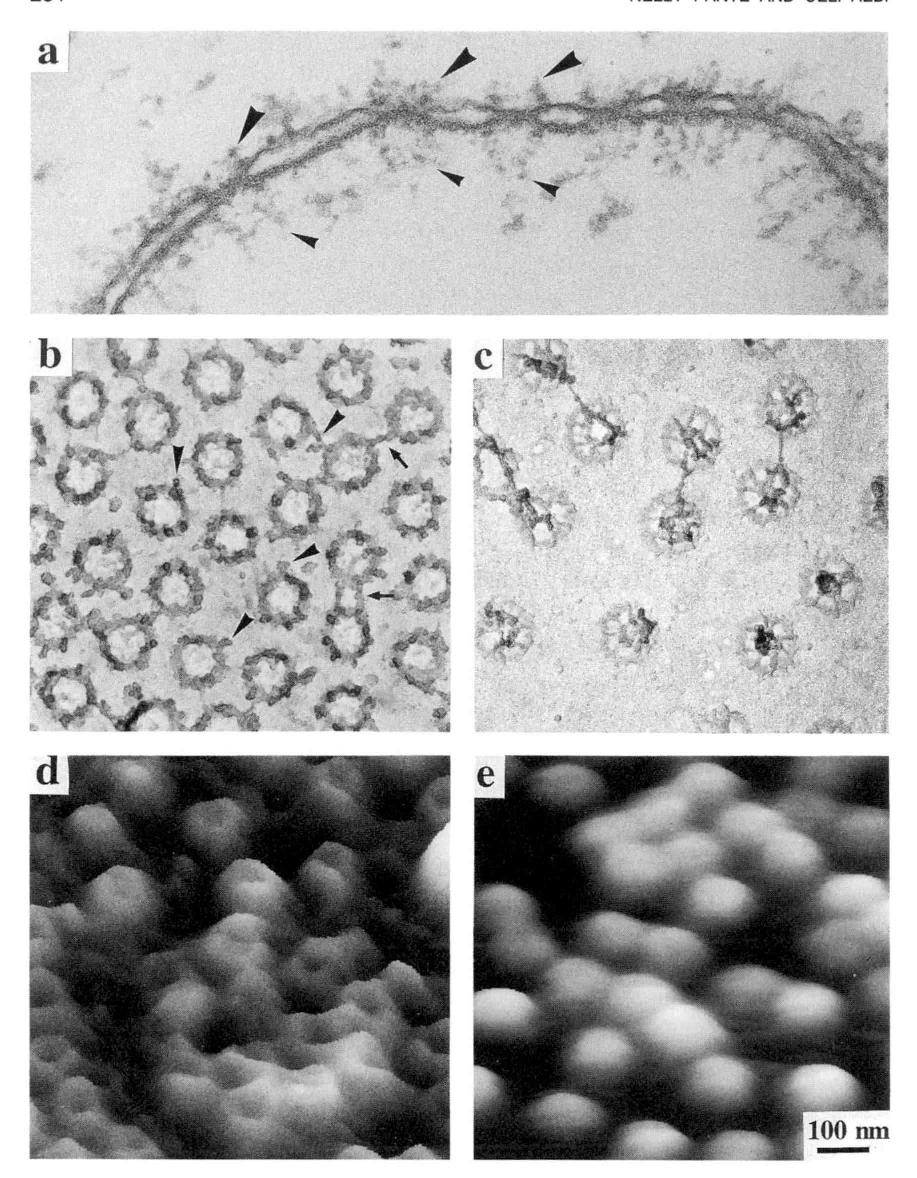


FIG. 5 The NPC is highly asymmetric with regard to its cytoplasmic and nuclear periphery. (a) Cross section of Epon-embedded NE revealing both cytoplasmic (large arrowheads) and nuclear (small arrowheads) filamentous structures associated with the NPC periphery. (b) Cytoplasmic and (c) nuclear face of NPCs revealed by transmission EM of quick-frozen/freeze-dried/rotary metal-shadowed intact *Xenopus* oocyte NEs. (d) Cytoplasmic and (e) nuclear topography of NPCs revealed by atomic force microscopy (AFM) of intact *Xenopus* oocyte NEs kept in physiological buffer. Relatively short cytoplasmic filaments (b; arrowheads), and nuclear baskets (c) are revealed by quick freezing/freeze-drying/rotary

NPC (see Figs. 2c and 3a). The frequent lack of this central structure and its highly variable appearance within a given NPC population has led to suggestions that at least in part it may represent material in transit (i.e., ribonucleoprotein particles) rather than an integral component of the NPC (e.g., Jarnik and Aebi, 1991; Gerace, 1992). Nevertheless, Akey (1990) has sorted several thousand NPC images from frozen–hydrated NEs and classified them in four groups which after applying image analysis techniques revealed four different conformations for the central particle that were related to different transport states of the NPC and termed "closed," "docked," "open/in transit" and "open." Based on this computer classification, Akey (1990) has proposed this central structure to represent the actual transporter and modeled it as a double-iris arrangement that can assume several distinct configurations as it actively transports molecules and particles through the NPC. Recently, the central plug as it appears in frozen–hydrated NPCs after detergent treatment has been reconstructed in 3-D (Akey and Radermacher, 1993). Accordingly, it has a rather complex, hourglass-like structure ~62-nm long with distal and central diameters of ~42 and ~32 nm, respectively. However, as we have discussed in Section II,B, reconstruction errors—due to sampling and limited tilt range—accumulate predominantly at low radii (i.e., in the center of the 3-D mass map), thus the central mass in this 3-D reconstruction may be overestimated. Furthermore, this representation of the central plug is not entirely consistent with the double iris-like model previously proposed by Akey (1990).

In view of the recently described peripheral components of the NPC (see Section II,C and Fig. 5), it is also conceivable that a substantial fraction of what in projection appears as the central plug in fact represents remnants of the nuclear basket (including the terminal ring) which may have been squashed into the pore upon embedding the NPC in a thin layer of negative stain or a thin ice film. In support of this notion, the mass density profile of negatively stained NPCs in the presence of EDTA—a condition that causes disassembly of the nuclear baskets (see above; Jarnik and Aebi, 1991)—is significantly attenuated when compared with that of NPCs isolated in the presence of m*M* amounts of divalent cations. Thus to more systematically investigate the nature and 3-D structure of the central plug

metal-shadowing. The arrowheads in (b) point to cytoplasmic filaments protruding from the cytoplasmic ring of the NPC, whereas the small arrows in (b) mark filaments that have bent to the side and thereby adhered to filaments of adjacent NPCs thus appearing as NPC connecting fibrils. The resolution of the AFM images (d and e) is insufficient to resolve individual NPC-associated filaments, thus the cytoplasmic face of the NPC appears donut-like (d), whereas the nuclear face exhibits a dome-like appearance (e). Bar: 100 nm (a–e).

or channel complex, it is first necessary to develop a method that yields a reproducible appearance of this NPC component, and at the same time controls the structural integrity of the nuclear baskets. Toward achieving this goal, we have been exploring a number of different buffers, incubation conditions, and chemical fixation protocols. As illustrated in Fig. 6, we have found that ~95% of the NPCs harbor a massive central plug when during isolation the NEs are stabilized with Cu-*ortho*-phenanthroline, an oxidizing agent causing intra- and intermolecular S–S bridge formation.

Based on the similarity of its overall size and shape, octagonal symmetry, and mass, it has been proposed that the central plug may represent a "vault" ribonucleoprotein particle (Kedersha *et al.,* 1991), with the vaults acting as "transport vehicles," i.e., carrying nuclear proteins as their cargo. Recent evidence for vault immunoreactivity at the NPC (Chugani *et al.,* 1993) supports this idea. However, a more specific and comprehensive characterization of the central plug in terms of its molecular structure and association with the basic framework of the NPC must still be achieved before it can be identified with any known particle such as vaults. Similarly, any such candidate particles have to be subjected to a more stringent analysis regarding their possible interaction or association with the NPC.

### E. A Consensus Model for the NPC

The large amount of structural studies (reviewed above) have elucidated the architecture of some of the components of the NPC. In the consensus model presented in Fig. 7 (color plate 6) we have identified the distinct structural components of the NPC discussed above. Accordingly, the intact membrane-embedded NPC consists of a basic framework (i.e., the spoke complex shown in Fig. 4a, color plate 5) sandwiched between a cytoplasmic and a nuclear ring. These two rings are in addition to the two tenuous rings defined by the ring subunits of the spokes (see Fig. 4a, color plate 5). The cytoplasmic ring is decorated with eight cytoplasmic filaments, and the nuclear ring is topped with a basket-like filamentous assembly. The central pore often harbors a central plug or channel complex whose definite structure and functional role (i.e., to act as a transporter) remain to be established.

## III. Toward the Molecular Details of the NPC

In contrast to the relatively large amount of structural studies (see Section II), less is known about the chemical composition and molecular architec-

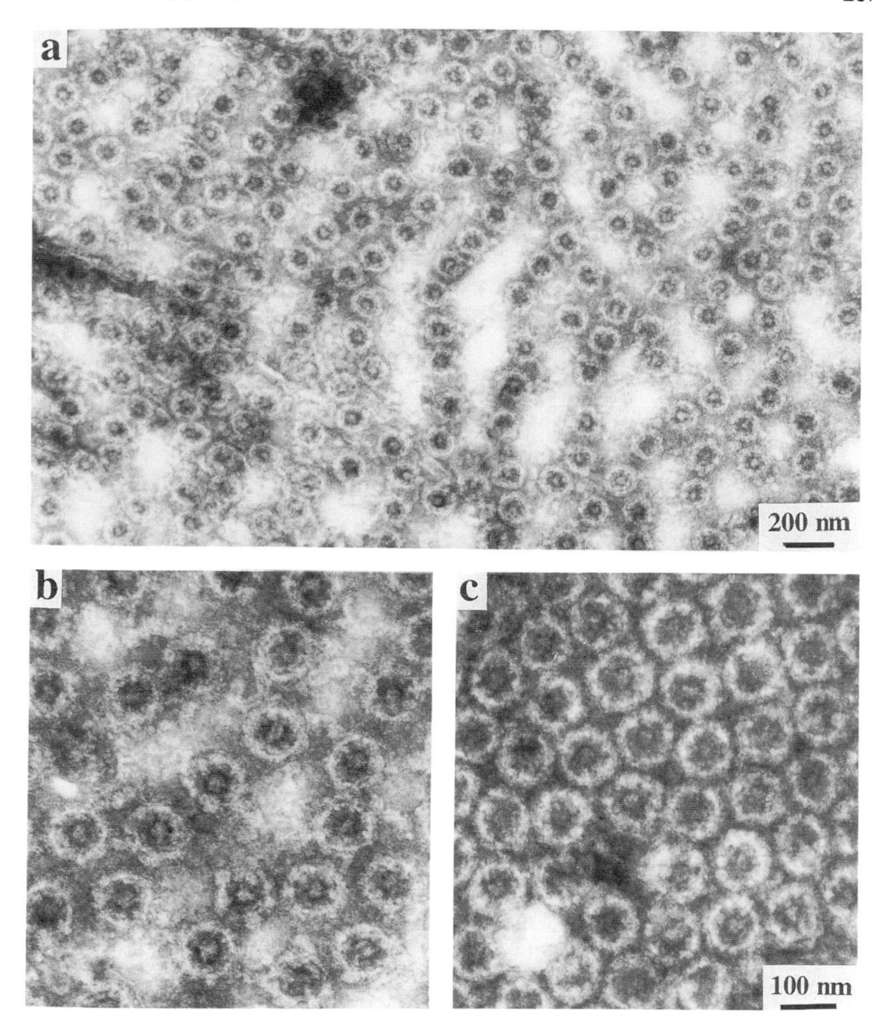


FIG. 6 Chemical manipulation of isolated NEs changes both the abundance and appearance of the central plug or channel complex. (a, b) Negatively stained *Xenopus* oocyte NEs that have been isolated in the presence of 0.1 m*M* Cu–*ortho*-phenanthroline, an oxidizing agent causing intra- and intermolecular S–S bridge formation. (c) Control, prepared as in (a) but in the absence of Cu–*ortho*-phenanthroline. Accordingly, ~95% of the NPCs yield a massive central plug when stabilized with Cu–*ortho*-phenanthroline during isolation. Bar: 200 nm (a) and 100 nm (b, c).

ture of the NPC. Based on its molecular mass of about 120 MDa (Reichelt *et al.,* 1990), it is believed that the NPC is composed of multiple copies (i.e., 8 or 16) on the order of ~100 different polypeptides. For several years, gp210 and p62 have been the only two well-characterized NPC polypeptides. However, the production of specific antibodies, the use of molecular genetics, and the development of an isolation procedure for NPCs from yeast (Rout and Blobel, 1993) have recently enabled identification, characterization, and cloning and sequencing of an increasing number of NPC proteins (reviewed by Fabre and Hurt, 1994; Panté and Aebi, 1994b; Rout and Wente, 1994). Moreover, the combination of well-characterized antibodies with different EM specimen preparation methods has allowed localization of several of these proteins to distinct structural components of the NPC. Thus the *molecular* architecture of the NPC is now on the way to be elucidated. In this section, we will focus on the recent progress made toward the localization of NPC proteins within the 3-D structure of the NPC, and the isolation and characterization of NPC subcomplexes. For more information about the presently identified NPC proteins the reader is referred to the chapter on NPC proteins by Bastos *et al.* in this volume.

### A. 3-D Localization of Specific Proteins to Distinct NPC Components

As documented in Fig. 8 (see also Fig. 4 of Bastos *et al.,* in this volume), epitopes of five different NPC proteins have recently been localized within the 3-D NPC architecture. Three of these proteins—p62, NUP153 and CAN/NUP214/p250—are members of the O-linked glycoprotein family that are modified at up to 10–20 sites with O-linked *N*-acetylglucosamine (GlcNac) (Davis and Blobel, 1987; Holt *et al.,* 1987; Snow *et al.,* 1987). Due to the relatively strong cross-reactivity of some of the antibodies raised against NPC proteins, localization of p62—the first O-linked NPC protein studied in some detail (see chapter by Bastos *et al.,* in this volume)—has remained ambiguous (Cordes *et al.,* 1991; Panté and Aebi, 1993). To resolve this ambiguity, we have recently produced a monoclonal antibody, RL31, that reacts specifically with rat p62 (Guan *et al.,* 1995). As illustrated in Fig. 8a, RL31 labels both the nuclear and cytoplasmic periphery of the central plug or channel complex of rat liver NPCs, albeit the labeling at the nuclear periphery is more frequent. This localization is consistent with the presumed involvement of the p62 complex in the nuclear import of proteins (Finlay *et al.,*1991) and may indicate that the p62 complex associates with material that is actively transported through the NPC. The fact that the p62 complex is located at both the cytoplasmic and nuclear periphery of the central channel complex (Fig. 8a) suggests that it might be shuttled back and forth from the cytoplasmic to the nuclear periphery of the central channel complex.

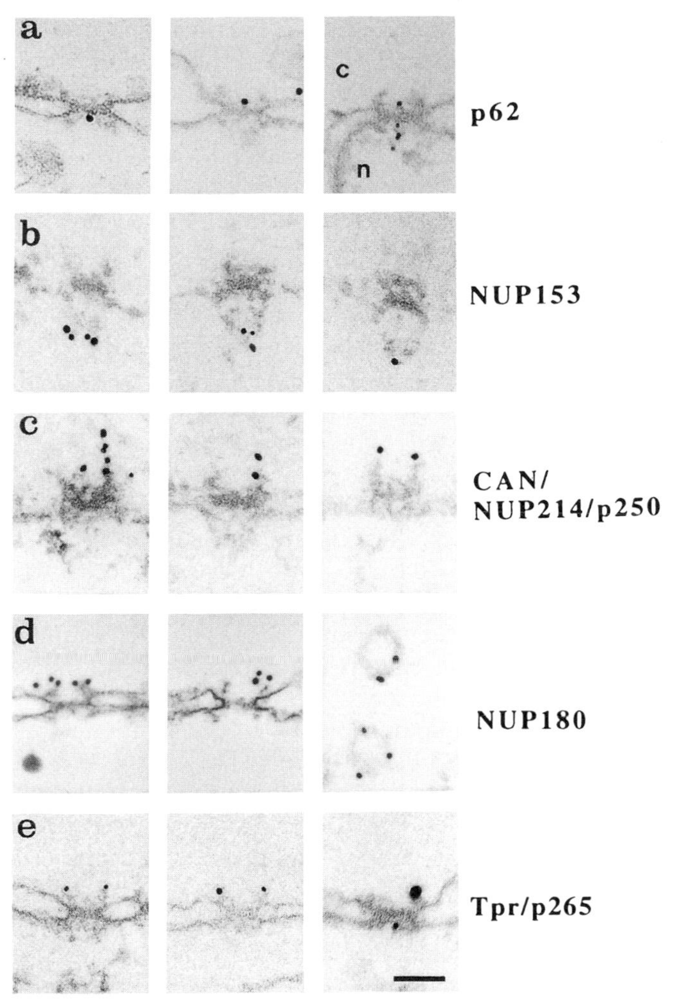


FIG. 8 Immunolocalization of NPC proteins. Selected examples of labeled NPCs seen in cross sections revealing the localization of epitopes of (a) p62 at both the cytoplasmic and nuclear periphery of the central plug or channel complex (using the monoclonal RL31 antibody; Guan *et al.,* 1995); (b) NUP153 close to or at the terminal ring of the nuclear basket (using an antipeptide antibody; Panté *et al.,* 1994); (c) CAN/NUP214/p250 at the cytoplasmic filaments (using a polyclonal anti-p250 antibody; Panté *et al.,* 1994); (d) NUP180 at the cytoplasmic ring or cytoplasmic filaments (using an affinity-purified antibody from a serum of a patient with overlap connective tissue disease; Wilken *et al.,* 1993); and (e) Tpr/p265 at the cytoplasmic filaments (using the monoclonal RL30 antibody; Byrd *et al.,* 1994). In all cases, the antibodies were directly conjugated to 8-nm colloidal gold, and the NEs were labeled prior to embedding and thin sectioning. The example in the right panel of (e) is an NPC that has been double-labeled with RL31 (anti-p62) conjugated to 8-nm colloidal gold and RL30 (anti-Tpr/p265) conjugated to 14-nm colloidal gold. (a) and (e) rat liver NEs; (b), (c) and (d) *Xenopus* oocyte NEs; c, cytoplasmic, and n, nuclear side of the NE. Bar: 100 nm (a–e).

Using a polyclonal antibody raised against a fusion protein expressed from a NUP153 cDNA construct, NUP153 has been unequivocally localized to the nuclear periphery of the NPC (Sukegawa and Blobel, 1993). However, in this labeling study NUP153 could not be identified with a particular NPC component(s). Recently, more specific localization of NUP153 has been achieved. Using an antibody raised against an extract of nuclear matrix proteins that by immunoblotting recognizes NUP153, Cordes *et al.,* (1993) have localized this protein to intranuclear NPC-attached filaments which, among other structures, may represent nuclear baskets that have been disrupted during sample preparation. More specifically, Panté *et al.* (1994) have identified NUP153 as a constituent of the nuclear basket with at least one of its epitopes residing in the terminal ring (see Fig. 8b). An interesting feature of this protein is its four zinc finger motifs, each containing two pairs of cysteine residues ($Cys_2$–$Cys_2$) (Sukegawa and Blobel, 1993; McMorrow *et al.,* 1994). The localization of NUP153 at the nuclear baskets (see Fig. 8b) together with its four zinc finger motifs is consistent with the stabilizing effect of $Zn^{2+}$ on the nuclear baskets (Jarnik and Aebi, 1991): in the presence of 0.5 m*M* $ZnCl_2$ well-formed baskets are observed, whereas when divalent cations are chelated by 2 m*M* EDTA or EGTA, the nuclear baskets become destabilized and are disrupted. Surprisingly, if after destabilization by EDTA or EGTA divalent cations are added back, the nuclear baskets reform—most efficiently with $Zn^{2+}$ (Jarnik and Aebi, 1991). These findings indicate that the zinc finger motifs of NUP153, in addition to binding DNA or RNA (Sukegawa and Blobel, 1993), may have a role in maintaining the structural integrity of the nuclear baskets, and therefore could be directly involved in the active transport of proteins, RNAs, or RNP particles through the NPC. We are currently testing this hypothesis using an antipeptide antibody against the zinc finger motifs.

CAN/NUP214 has also been expressed as a fusion protein, and a polyclonal antibody raised against this fusion protein labeled the cytoplasmic periphery of the NPC (Kraemer *et al.,* 1994). However, this labeling was not specific enough to locate CAN/NUP214 to a distinct NPC components(s). Using a monoclonal antibody called QE5, Panté *et al.* (1994) have identified an ~250-kDa O-linked NPC glycoprotein, termed p250, in extracts of BHK cells. As illustrated in Fig. 8c, a polyclonal antibody raised against p250 specifically labeled the cytoplastic filaments of *Xenopus* oocyte NPCs at multiple sites. On immunoblots, p250 is also recognized by the monoclonal RL1 antibody used by Snow *et al.* (1987) (B. Burke and R. Bastos, personal communication). In addition, p250 is recognized by polyclonal antibodies raised against $NH_2$- and COOH-terminal peptides synthesized based on the deduced amino acid sequence of cloned human CAN (B. Burke and R. Bastos, personal communication). Taken together, these results indicate that p250 corresponds to the originally identified ~210-

kDa O-linked NPC glycoprotein (Snow *et al.,* 1987) that has recently been demonstrated to represent a homologue of human CAN and termed CAN/NUP214 (Kraemer *et al.,* 1994).

Epitopes of at least two other—both non-O-linked—NPC proteins have been localized to the cytoplasmic filaments of the NPC: (i) NUP180 (Fig. 8d; Wilken *et al.,* 1993), and (ii) Tpr/p265 (Fig. 8e; Byrd *et al.,* 1994). Based on the ~1600-residue-long $\alpha$-helical coiled-coil domain of Tpr/p265 (Mitchell and Cooper, 1992), it is conceivable that this protein, together with CAN/NUP214/p250 (see Fig. 8c; Panté *et al.,* 1994) and NUP180 (see Fig. 8d; Wilken *et al.,* 1993), forms the backbone of the cytoplasmic filaments.

As can be appreciated in Fig. 8 (see also Fig. 4 of Bastos *et al.,* in this volume), the epitopes of the five NPC proteins that have thus far been identified with distinct structural components of the NPC are localized either at the cytoplasmic or nuclear periphery of the NPC. Therefore the molecular constituents of the 52-MDa basic framework of the NPC (i.e., the spoke complex) remain to be identified. Thus far, gp210—a transmembrane glycoprotein bearing N-linked high mannose oligosaccharides with most of its mass residing in the perinuclear space of the NE (Gerace *et al.,*1982; Wozniak *et al.,* 1989; Greber *et al.,* 1990)—is the only NPC protein that has been identified as a constituent of the basic framework of the NPC. gp210 has been proposed to be a component of the radially outermost domain of the spoke complex, i.e., the lumenal domain, which extends into the perinuclear space (Hinshaw *et al.,* 1992; Akey and Radermacher, 1993) and appears as distinct "knobs" in thin sections (Jarnik and Aebi, 1991). Hence, it has been proposed that gp210 may act as a membrane anchor for the NPC, and it may also play a topogenic role during NPC assembly (Greber *et al.,* 1990; Gerace, 1992).

## B. Isolation and Characterization of Distinct Subcomplexes of the NPC

When assembled in the NPC, several NPC proteins may mutually interact to form distinct subcomplexes. In vertebrate species, it was first reported that some of the soluble NPC proteins contained in *in vitro* nuclear reconstitution extracts from *Xenopus* oocytes form a supramolecular complex with an estimated molecular mass of 254 kDa, which contains p68, the *Xenopus* homolog of rat p62, together with several other NPC proteins (Dabauvalle *et al.,* 1990). The rat homolog of this supramolecular complex (i.e., the p62 complex) has also been isolated and characterized at the molecular level (Finlay *et al.,* 1991; Kita *et al.,* 1993; Buss and Stewart, 1995). It consists of p62 interacting with two other proteins of molecular mass 58 (p58) and

54 (p54) kDa. The estimated molecular mass of the p62 complex ranges between 200 and 600 kDa with no consensus on its subunit stoichiometry (Finlay *et al.,* 1991; Kita *et al.,* 1993; Buss and Stewart, 1995). To resolve these ambiguities, we have developed a modified procedure to isolate the p62 complex from rat liver NEs (Guan *et al.,* 1995). In addition to p58 and p54, p62—in this supramolecular complex—is associated with a 45-kDa NPC protein having a peptide map similar to that of p58. However, p45 is only revealed when extreme caution is taken to avoid proteolysis during isolation, so it is unlikely that p45 is a proteolytic fragment of p58 produced during isolation of the p62 complex. When examined in the electron microscope after negative staining or glycerol spraying/rotary metal shadowing, this complex appears as a ~16-nm-diameter donut-like particle (Guan *et al.,* 1995). We are currently investigating its molecular mass and subunit stoichiometry by quantitative scanning transmission EM, and whether this particle represents an oligomer of the p62 complex.

Using the monoclonal QE5 antibody, Panté *et al.* (1994), in addition to the p62 complex, have identified two distinct NPC subcomplexes in extracts of BHK cells. While this antibody recognizes p62, NUP153, and CAN/NUP214/p250 on Western blots, it immunoprecipitates three additional polypeptides, p54, p58, and p75, which were found to be associated with p62 (p54 and p58) and with CAN/NUP214/p250 (p75). Furthermore, in these extracts NUP153 existed as a homo-oligomer of ≥1 MDa, thus most likely representing an octamer. As several of its epitopes have been located to the terminal ring of the nuclear baskets (see Fig. 8b; Panté *et al.,* 1994), it is conceivable that the octameric NUP153 complex represents the basic framework of the eightfold symmetric terminal ring.

Hurt (1990) has expressed the $\alpha$-helical COOH-terminal domain of the yeast NPC protein NSP1 (Hurt, 1988) as a fusion protein with a cytosolic protein, and found that this protein was targeted to yeast NEs. Based on this observation, he proposed that NSP1 might specifically interact with several other yeast NPC proteins (Hurt, 1990). Indeed, NSP1 has now been shown to form a complex with three other yeast NPC proteins, NIC96, NSP49, and a novel 54-kDa protein that has not yet been cloned and sequenced (Grandi *et al.,* 1993). Recently, Bélanger *et al.* (1994) trying to identify and characterize subcomplexes of yeast NPCs have found that the nucleoporins NUP1 and NUP2 interact with SRP1, the product of a gene originally identified as a suppressor of temperature-sensitive mutations of RNA polymerase I in yeast, which has been localized to the NE by immunofluorescence microscopy (Yano *et al.,* 1992). Since SRP1 binds either NUP1 or NUP2 but not both simultaneously, Bélanger *et al.* (1994) have proposed the existence of two distinct NPC subcomplexes in yeast (i.e., SRP1–NUP1 and SRP1–NUP2). However, based on significant similarity between the amino acid sequence of SRP1 and that of armadillo/plakoglobin/β-catenin—proteins linking cadherins to the cytoskeleton at intercellular junctions—it has been speculated that SRP1 might link NPCs to the nucleoskeleton (Yano *et al.*, 1992, 1994; Bélanger *et al.*, 1994). Most recently, *Xenopus* importin—a cytosolic protein mediating binding of nuclear proteins to the NPC—has been cloned and sequenced, and the resulting amino acid sequence is 44% identical with SRP1 (Görlich *et al.*, 1994; see Section IV,B). Thus SRP1 may be the yeast homolog of importin. The latter possibility is supported by (i) the fact that SRP1—in addition to being located at the NE—is present in a soluble pool (Yano *et al.*, 1992), and (ii) depletion as well as mutations of SRP1 yield distinct phenotypes related to nuclear functions (e.g., fragmentation of the nucleolus, inhibition of transcription, defects in nuclear division/segregation; Yano *et al.*, 1994) that might be due to nucleocytoplasmic transport defects (Görlich *et al.*, 1994).

## IV. Molecular Trafficking through the NPC

As illustrated schematically in Fig. 1 (color plate 4), two different types of nucleocytoplasmic transport through the NPC occur: (1) passive diffusion of ions and small molecules through an aqueous channel with a physical diameter of ~9 nm (Paine *et al.*, 1975), and (2) mediated transport of proteins, RNAs, and ribonucleoprotein particles through a gated channel with a functional diameter of up to 26 nm (Felherr *et al.*, 1984; Dworetzky and Feldherr, 1988). Although mediated nucleocytoplasmic transport of different substrates may use the same transport machinery, depending on the substrate it appears to occur via different signal pathways. Some of the signals, receptors, and factors mediating nuclear import of proteins or RNPs and nuclear export of RNA have begun to be identified and isolated. In this section we focus on some recent advances made toward identification and characterization of these signals and factors. For background information and further details, the reader is referred to some recent reviews covering this topic (Forbes, 1992; Gerace, 1992; Izaurralde and Mattaj, 1992; Mattaj *et al.*, 1993; Newmeyer, 1993).

### A. Passive Diffusion

Early microinjection experiments of dextrans have indicated that the NPC has the properties of a "molecular sieve" (Paine *et al.*, 1975). Accordingly, molecules larger than ~9 nm in diameter are excluded from the nucleus, while smaller molecules can passively diffuse through the NPC with a rate inversely proportional to their size. The recent 3-D reconstruction of the

NPC (see Fig. 4, color plate 5; Hinshaw *et al.*, 1992) from negatively stained preparations of detergent-treated *Xenopus* oocyte NEs (see Fig. 3b) has indicated that there may exist eight ~10-nm-diameter slightly kinked channels at a radius of ~40 nm that are located between two adjacent spokes and the pore membrane (see Section II,B and Fig. 4a, color plate 5). These peripheral channels have been proposed to be sites for passive diffusion of ions and small molecules through the NPC (Hinshaw *et al.*, 1992). In contrast, the 3-D reconstruction of ice-embedded NPCs (Akey and Radermacher, 1993) has revealed eight ~10-nm-diameter channels at a radius of ~32 nm located between two adjacent spokes and the central channel complex that have also been speculated to represent sites for passive diffusion (Akey and Radermacher, 1993). Thus location of the diffusional channels within the NPC remains controversial.

## B. Import of Nuclear Proteins

Import of nuclear proteins through the NPC is highly selective, energy dependent, and requires cytosolic factors (reviewed by Forbes, 1992; Gerace, 1992; Newmeyer, 1993). *In vitro* studies have demonstrated that this process can be experimentally dissected into two distinct steps: (1) docking of the nuclear protein to the NPC, and (2) translocation of the NPC-bound protein through a gated channel (Newmeyer and Forbes, 1988; Richardson *et al.*, 1988). The first step does not require ATP and is temperature independent, whereas the second step depends on ATP hydrolysis. Accordingly, two cytoplasmic fractions have been identified, one required for ATP independent docking of the nuclear protein to the NPC, and a second for active translocation of the nuclear protein through the NPC (Newmeyer and Forbes, 1990). To more systematically identify and characterize specific cytosolic factors both in mammalian cells and *Xenopus* oocyte extracts, the development of a digitonin-permeabilized cell system has become a powerful tool to study mediated nuclear protein import *in vitro* (Adam *et al.*, 1990).

Targeting of nuclear proteins to the NPC is specified by short amino acid segments, called nuclear localization sequences (NLSs), on the protein to be transported (reviewed by Dingwall and Laskey, 1991; Garcia-Bustos *et al.*, 1991). Two types of NLSs have been identified: (1) the Simian virus 40 (SV40) large T antigen type of NLS, which consists of a single segment of seven basic amino acid residues (Chelsky *et al.*, 1989); and (2) the *Xenopus* nucleoplasmin type of bipartite NLS, which contains two interdependent segments of basic amino acids (three to four residues each) separated by 10 intervening "spacer" residues (Robbins *et al.*, 1991).

A number of NLS binding proteins have been identified (reviewed by Yamasaki and Lanford, 1992), but thus far only two NLS binding proteins of 54 and 56 kDa isolated from mammalian cells have been demonstrated to have the properties of a functional receptor (Adam *et al.,* 1989; Adam and Gerace, 1991). However, in addition to this 54/56-kDa "NLS receptor," cytosolic factors are required in order to stimulate mediated nuclear protein import *in vitro* (Adam and Gerace, 1991). As it binds to NLS peptides and is required for import of nuclear proteins, the ubiquitous cellular protein hsc70 also has been implicated in the mediated transport of nuclear proteins through the NPC (Imamoto *et al.,* 1992; Shi and Thomas, 1992). Since in other systems this protein acts as a chaperone mediating protein folding, transport, and assembly (reviewed by Ellis and van der Vies, 1991; Gething and Sambrook, 1992), it has been suggested that hsc70 might facilitate the interaction between the NLS-bearing proteins and the NLS receptor (Imamoto *et al.,* 1992). Another alternative is that hsc70 might facilitate the release of the NLS receptor at a later stage of nuclear import (Newmeyer, 1993).

Since the O-linked glycoproteins of the NPC have been implicated in mediated import of nuclear proteins through the NPC (Finlay *et al.,* 1987; Dabauvalle *et al.,* 1988a,b; Featherstone *et al.,* 1988; Finlay and Forbes, 1990), Sterne-Marr *et al.,* (1992) have investigated whether these interact with cytosolic factor(s). Accordingly, incubation of transport-competent cytosol with O-linked glycoproteins reduced its ability to support nuclear import of proteins. To this end, Adam and Adam (1994) have identified and purified a 97-kDa protein from bovine erythrocytes that together with the NLS receptor is able to reconstitute the binding of the nuclear protein–-receptor complex to the NPC in the digitonin-permeabilized cell assay (see above; Adam *et al.,* 1990) in an ATP- and temperature-independent way, but does not mediate the translocation of the complex into the nucleus. Since the p97-facilitated binding of the NLS receptor to the NPC was inhibited by the lectin wheat germ agglutinin (WGA; Adam and Adam, 1994), these authors proposed that an initial docking site for the NLS receptor might be an O-linked glycoprotein. Based on the recent localization of the O-linked glycoprotein CAN/NUP214/p250 as a constituent of the cytoplasmic filaments (see Section II,A, Fig. 8; Panté *et al.,*1994), it is tempting to speculate that NUP214 might represent an initial docking site at the NPC for nuclear proteins destined for import. As other O-linked glycoproteins (e.g., p62; Finlay and Forbes, 1990) also have been demonstrated to be involved in nuclear protein import, it is conceivable that some of these proteins might represent site(s) for the protein–receptor complex while it is transported through the NPC.

Fractionation of cytosol from both mammalian cells and *Xenopus* oocyte extracts with respect to import activity, and the finding that nuclear import of proteins is inhibited by GTP-γ-S and other nonhydrolyzable GTP analogs have led to the identification of the small GTPase Ran/TC4 as a cytosolic factor required for the translocation step during nuclear import of proteins (Melchior *et al.,* 1993; Moore and Blobel, 1993). Accordingly, addition of purified Ran/TC4 or recombinant Ran/TC4 to Ran/TC4-depleted cytosol in the presence of GTP increases significantly the level of nuclear import of NLS-bearing proteins in a digitonin-permeabilized cell assay. However, when added with the *Xenopus* cytosol fraction, termed fraction A by Moore and Blobel (1992), which is required for binding of NLS-bearing proteins to the NPC but not for the translocation step, neither purified Ran/TC4 nor human recombinant Ran/TC4 were sufficient to mediate import of nuclear proteins (Moore and Blobel, 1993). Thus additional cytosolic factors provided by the Ran/TC4-depleted cytosol are required in this *in vitro* import assay. Recently, Moore and Blobel (1994) purified a *Xenopus* 10-kDa protein that together with Ran/TC4 and fraction A is able to reconstitute nuclear protein import using digitonin-permeabilized BRL cells. Therefore, these authors conclude that p10 and Ran/TC4 are the two cytosolic factors required for the actual translocation step during nuclear protein import through the NPC. However, it remains to be determined which components of the *Xenopus* cytosolic fraction A mediate the docking of NLS-bearing proteins to the NPC, and whether these are identical to the mammalian 54/56-kDa NLS receptor and p97 (see above).

In the search for additional cytosolic factors required for nuclear protein import, Görlich *et al.* (1994) have recently identified and purified a 60-kDa protein from *Xenopus* oocytes, termed importin, that mediates docking of NLS-bearing proteins to the NPC but is not involved in the translocation step. This protein was cloned and sequenced, and its amino acid sequence yielded 44% identity to SRP1 (Görlich *et al.,* 1994), the product of a gene previously identified as a suppressor of mutants defective in RNA polymerase I (Yano *et al.,* 1992; see Section III,B) and recently demonstrated to be associated with yeast NPCs (Bélanger *et al.,* 1994). Since it is purified from the same source, importin is probably a component of the *Xenopus* oocyte fraction A defined by Moore and Blobel (1992; see above). However, in contrast to Moore and Blobel (1993) who were unable to reconstitute nuclear protein import with only fraction A and Ran/TC4, Görlich *et al.* (1994) succeeded in producing nuclear import of BSA–NLS conjugate comparable to the level achieved with unfractionated cytosol employing importin purified from *Xenopus* oocyte together with recombinant Ran/TC4 in the absence of any other cytosolic fraction. Nevertheless, it is conceivable that at least trace amounts of additional cytosolic factors were present in their import assay. In support of this possibility, whereas recombinant importin was able to mediate both NPC docking and translocation of BSA–NLS conjugate, albeit less efficiently than purified importin, when recombinant importin was added with importin-depleted cytosol, nuclear import of BSA–NLS conjugate was comparable to that achieved with authentic importin (Görlich *et al.,* 1994). Based on the comparison of the isolation procedure and the sensitivity to the sulfhydryl alkylating reagent *N*-ethylmaleimide (NEM), it was concluded that importin is different from the mammalian 54/56-kDa NLS receptor purified by Adam and Gerace (1991). It remains to be determined whether importin binds directly to NLS-bearing proteins, thus being a *Xenopus* NLS receptor homolog, or whether importin represents the *Xenopus* homolog of the binding factor p97 identified by Adam and Adam (1994; see above).

In summary, the above reviewed studies have documented the requirement for multiple cytosolic factors in NPC-mediated nuclear protein import, which is clearly a multistep pathway. As illustrated in Fig. 9 (color plate 7), based on all the presently available data, at least five steps might be distinguished: (1) While residing in the cytoplasm, the NLS-bearing protein destined for nuclear import binds to a soluble receptor (e.g., the mammalian p54/p56-kDa NLS receptor) via its specific NLS (Newmeyer and Forbes, 1988; Adam and Gerace, 1991); this interaction may be stabilized by some cytosolic factor(s) such as hsc70 (Imamoto *et al.,* 1992; Shi and Thomas, 1992). (2) Depending on the action of additional cytosolic factors (e.g., p97; Adam and Adam, 1994), this protein–receptor complex then docks to an NPC by specific binding to some "peripheral" NPC component such as the cytoplasmic ring or the cytoplasmic filaments (Richardson *et al.,* 1988; Sterne-Marr *et al.,* 1992; Adam and Adam, 1994). (3) From this peripheral docking site the protein–receptor complex is next delivered to specific site(s) at the central channel complex [e.g., some of the O-linked glycoproteins (Sterne-Marr *et al.,* 1992)—most probably involving the p62 complex] which harbors the actual transport machine. (4) Active translocation of the protein–receptor complex through the central channel complex occurs after channel gating to accomodate the particular size and shape of the protein–receptor complex; this step is energy dependent and requires the small GTPase Ran/TC4 and hydrolyzable GTP (Melchior *et al.,* 1993; Moore and Blobel, 1993). (5) During or after release into the nucleus, the protein–receptor complex dissociates, and the receptor may be recycled for further rounds of transport (Adam *et al.,* 1989; Adam and Gerace, 1991).

Several aspects of these five distinct steps specifying nuclear protein import are as yet hypothetical or remain elusive; for example: (1) the site(s) and mechanism of ATP utilization; (2) the site(s) and mechanism of action of the different cytosolic factors; (3) the NPC ligands, and the role of specific NPC components involved in the different transport steps; and (4) the nature and molecular mechanism of the gated channel. Furthermore, nuclear import of proteins has been shown and is now generally accepted to be an ATP-dependent process (Newmeyer and Forbes, 1988; Richardson *et al.,* 1988). However, the recent involvement of the Ran/TC4 GTPase and the inhibition of nuclear import by GTP-$\gamma$-S and other nonhydrolyzable GTP analogues (Melchior *et al.,* 1993; Moore and Blobel, 1993) open the possibility that GTP—rather than ATP—hydrolysis may be the primary energy source sustaining active nuclear protein import *in vivo.* Consistent with this idea, Pruschy *et al.* (1994) have recently reported that the mediated—since it is inhibited by WGA and chilling—nuclear import of calmodulin does not require ATP hydrolysis *in vitro.*

## C. Export/Import of RNAs and RNP Particles

In addition to nuclear proteins, the NPC also imports ribonucleoprotein complexes into the nucleus. The import of U-rich small nuclear ribonucleoprotein (U snRNP) particles that contain a 5′-trimethyl G cap and are complexed with proteins termed Sm has been studied in some detail (reviewed by Izaurralde and Mattaj, 1992; Mattaj *et al.,* 1993). The import of one of these, the U6 snRNP particles, is inhibited by proteins bearing the SV40 T antigen NLS (Michaud and Goldfard, 1991), thus import of U6 snRNP particles seems to occur by a mechanism identical—or at least very similar—to that of import of nuclear proteins (see above). However, in the case of U1, U2, U3, U4, and U5 snRNP particles, their nuclear import does not compete with NLS-bearing proteins, but it requires both the 5′-trimethyl G cap and binding of Sm proteins (Fischer and Lührmann, 1990; Hamm *et al.,* 1990; Michaud and Goldfard, 1992; Fischer *et al.,* 1993). Moreover, neither proteins bearing an NLS nor an excess of free 5′ trimethyl G cap competes with nuclear import of the U3 snRNP particle (Michaud and Goldfard, 1992). Thus there appear to exist at least three distinct signaling pathways for import of snRNP particles.

Recently, Marshallsay and Lührmann (1994), using the digitonin-permeabilized cell system developed by Adam *et al.* (1990), have demonstrated that nuclear import of RNP particles requires some cytosolic factors. The identity of these cytosolic factors and whether they are the same as those that mediate nuclear import of proteins remain to be determined (see Section VI,B). Nevertheless, this approach now opens the possibility to identify and characterize both the specific signals and factors mediating the nuclear import of RNP particles.

Compared with the nuclear import of proteins and RNP particles, the export of RNAs and RNP particles is still rather poorly understood. Several classes of RNAs, including snRNAs, mRNAs, and tRNAs, are exported through the NPC. Since these RNAs are packaged with proteins into RNP complexes, they are probably exported in the form of RNP particles. Indeed, export of Balbiani ring granules—premessenger RNP particles in the salivary glands of *Chironomus*—through the NPC has been visualized by EM, and it appears to be a polar process in that the 5′ end of the RNA exits the nucleus first (Mehlin *et al.,* 1992).

Similar to import of nuclear proteins, export of RNAs from the nucleus also is a signal-dependent, receptor-mediated process that requires energy in the form of ATP hydrolysis (Zasloff, 1983; Bataillé *et al.,* 1990; Dargemont and Kühn, 1992). However, since the export of RNAs may be controlled at multiple levels within the nucleus, it has been difficult to determine the signals involved in this process and the nuclear site(s) where these signals exert their effect. Nevertheless, some of the signals and factors mediating nuclear export of RNAs and RNP particles are starting to emerge. In the case of mRNAs and snRNAs, it has been shown that the monomethylated RNA cap structures facilitate their nuclear export; thus the signal(s) may reside within the primary structure of the RNA (Hamm and Mattaj, 1990). Moreover, a nuclear cap-binding protein that might play a role similar to the NLS receptor in nuclear protein import (see Section IV,B) has been identified (Izaurralde *et al.,* 1992). Most recently, Jarmolowski *et al.* (1994) have demonstrated that nuclear export of different classes of RNAs is mediated by specific rather than common factors. Thus export of RNAs and RNP particles may occur by different signaling pathways than does import of nuclear proteins and RNP particles (see above).

## V. Concluding Remarks

Significant progress has been made over the past few years toward understanding the 3-D architecture and molecular composition of the NPC. As a consequence, the 3-D structure of its basic framework has now been determined to a resolution of just under 10 nm (see Fig. 4, color plate 5). Also, a model of its different components has emerged (see Fig. 7, color plate 6), and epitopes of a number of NPC proteins have been localized to these components by immuno-EM (see Fig. 8). Thus the *molecular* architecture of the NPC is definitely taking shape. Nevertheless, to date only about two dozen NPC proteins have been identified, cloned, and characterized, a number accounting for ~15% of the NPC mass. Interestingly, almost all of these appear to be constituents of the peripheral components (i.e., the cytoplasmic or nuclear rings, the cytoplasmic filaments, nuclear baskets) or the central channel complex, with only few constituents of the basic framework of the NPC (see Fig. 8). Therefore, we have to go a long way before the complete NPC architecture will be unveiled at molecular detail. Toward this goal, the recent success to bulk-isolate NPCs from yeast (Rout and Blobel, 1993) has opened the possibility to more systematically identify the protein constituents of yeast NPCs. However, to date little is known about the mammalian or *Xenopus* homologues corresponding to the increasing number of cloned and characterized yeast nucleoporins, which will be important if yeast NPC should become a model for NPC structure and function. Moreover, to eventually reconstitute functional NPCs *in vitro,* we also have to mass-isolate and molecularly characterize the different subcomplexes or components of the NPCs (e.g., the spoke complex, the cytoplasmic and nuclear rings, the cytoplasmic filaments, the nuclear basket, and the central plug or channel complex; see Section III,B), determine their molecular architecture, and decipher their functional roles in mediated nucleocytoplasmic transport.

As a first step to more systematically investigate the molecular mechanisms underlying mediated import of nuclear proteins *in vitro,* the development of a digitonin-permeabilized cell system (Adam *et al.,* 1990) has led to the identification of a number of signals, receptors, and factors involved in nuclear import of proteins. However, the site(s) on the NPC where these exert their function and their mechanism of action remain to be established. Similarly, the signals, receptors, and factors that mediate nuclear import of RNP particles and nuclear export of RNAs have just begun to be elucidated, but again the nuclear site(s) where they exert their effect remain to be determined. Nevertheless, these advances have revealed the existence of multiple nuclear import–export pathways that might be specific for certain cell types, or stages of development of differentiation. Finally, the recent success to image and at the same time manipulate NPCs in their native buffer environment by atomic force microscopy (see Figs. 5d and 5e) provides us with the exciting possibility to directly correlate NPC structure with function.

## Acknowledgments

The authors are indebted to C. Henn for designing and preparing Figs. 1, 4, and 7. We thank Dr. R. Milligan (Scripps Research Institute, La Jolla, CA) who provided the data of the 3-D reconstruction of negatively stained detergent-released NPCs that enabled us to produce Figs. 4 and 7. We are grateful to Dr. B. Burke (Harvard Medical School, Boston, MA) and Dr. L. Gerace (Scripps Research Institute, La Jolla, CA) for providing us with several anti-NPC antibodies. We thank K. N. Goldie for providing the micrographs for Figs. 5d and 5e. Ms. H. Frefel and Ms. M. Zoller are thanked for their expert photographic work. This work was supported by the M. E. Müller Foundation of Switzerland, by grants from the Swiss National Foundation, and the Human Frontier Science Program (HFSP).

## References

Adam, J. H., and Adam, S. A. (1994). Identification of cytosolic factors required for nuclear localization sequence-mediated binding to the nuclear envelope. *J. Cell Biol.* **125,** 547–555.

Adam, S. A., and Gerace, L. (1991). Cytosolic proteins that specifically bind nuclear localization signals are receptors for nuclear import. *Cell (Cambridge, Mass.)* **66,** 837–847.

Adam, S. A., Lobl, T. J., Mitchell, M. A., and Gerace, L. (1989). Identification of specifically binding proteins for a nuclear location sequence. *Nature (London)* **337,** 276–279.

Adam, S. A., Sterne-Marr, R., and Gerace, L. (1990). Nuclear protein import in permeabilized mammalian cells requires soluble cytoplasmic factors. *J. Cell Biol.* **111,** 807–816.

Aebi, U., Cohn, J., Buhle, L., and Gerace, L. (1986). The nuclear lamina is a meshwork of intermediate-type filaments. *Nature (London)* **323,** 560–564.

Akey, C. W. (1989). Interactions and structure of the nuclear pore complex revealed by cryo-electron microscopy. *J. Cell Biol.* **109,** 955–970.

Akey, C. W. (1990). Visualization of transport-related configurations of the nuclear pore transporter. *Biophys. J.* **58,** 341–355.

Akey, C. W., and Radermacher, M. (1993). Architecture of the *Xenopus* nuclear pore complex revealed by three-dimensional cryo-electron microscopy. *J. Cell Biol.* **122,** 1–19.

Bataillé, N., Helser, T., and Fried, H. M. (1990). Cytoplasmic transport of ribosomal subunits microinjected into the *Xenopus laevis* oocyte nucleus: A generalized, facilitated process. *J. Cell Biol.* **111,** 1571–1582.

Bélanger, K. D., Kenna, M. A., Wei, S., and Davis, L. (1994). Genetic and physical interactions between Srp 1p and nuclear pore complex proteins NUP1p and NUP2p. *J. Cell Biol.* **126,** 619–630.

Braunstein, D., and Spudich, A. (1994). Structure and activation dynamics of RBL-2H3 cells observed with scanning force microscopy. *Biophys. J.* **66,** 1717–1725.

Buss, F., and Stewart, M. (1995). Macromolecular interactions in the nucleoporin p62 complex of rat nuclear pores: Binding of nucleoporin p54 to the rod domain of p62. *J. Cell Biol.* **128,** 251–261.

Byrd, D., Sweet, D. J., Panté, N., Konstantinov, K. N., Guan, T., Saphire, A. C. S., Mitchell, P. J., Cooper, C. S., Aebi, U., and Gerace, L. (1994). Tpr, a large coiled coil protein whose amino terminus is involved in activation of oncogenic kinases, is localized to the cytoplasmic surface of the nuclear pore complex. *J. Cell Biol.* **127,** 1515–1526.

Chelsky, D., Ralph, R., and Jonak, G. (1989). Sequence requirements for synthetic peptide-mediated translocation to the nucleus. *Mol. Cell. Biol.* **9,** 2487–2492.

Chugani, D. C., Rome, L. H., and Kedersha, N. L. (1993). Localization of vault particles to the nuclear pore complex. *J. Cell Sci.* **106,** 23–29.

Cordes, V., Waizenegger, I., and Krohne, G. (1991). Nuclear pore complex glycoprotein p62 of *Xenopus laevis* and mouse: cDNA cloning and identification of its glycosylation region. *Eur. J. Cell Biol.* **55,** 31–47.

Cordes, V., Reidenbach, S., Köhler, A., Stuurman, N., van Driel, R., and Franke, W. W. (1993). Intranuclear filaments containing a nuclear pore complex protein. *J. Cell Biol.* **123,** 1333–1344.

Dabauvalle, M.-C., Benevente, R., and Chaly, N. (1988a). Monoclonal antibodies to a $M_r$ 68,000 pore complex protein interfere with nuclear protein uptake in *Xenopus* oocytes. *Chromosoma* **97,** 193–197.

Dabauvalle, M.-C., Schultz, B., Scheer, U., and Peters, R. (1988b). Inhibition of nuclear accumulation of karyophilic proteins by microinjection of the lectin WGA. *Exp. Cell Res.* **174,** 291–296.

Dabauvalle, M.-C., Loos, K., and Scheer, U. (1990). Identification of a soluble precursor complex essential for nuclear pore assemble *in vitro. Chromosoma* **100,** 56–66.

Dargemont, C., and Kühn, L. C. (1992). Export of mRNA from microinjected nuclei of *Xenopus laevis* oocytes. *J. Cell Biol.* **118,** 1–9.

Davis, L. I., and Blobel, G. (1987). The nuclear pore complex contains a family of glycoproteins that includes p62: Glycosylation through a previously unidentified cellular pathway. *Proc. Natl. Acad. Sci. U.S.A.* **84,** 7552–7556.

Dingwall, C., and Laskey, R. A. (1991). Nuclear targeting sequences—a consensus? *Trends Biochem. Sci.* **16,** 478–481.
Dworetzky, S. I., and Feldherr, C. M. (1988). Translocation of RNA-coated gold particles through the nuclear pores of oocytes. *J. Cell Biol.* **106,** 575–584.
Ellis, R. J., and van der Vies, S. M. (1991). Molecular chaperones. *Annu. Rev. Biochem.* **60,** 321–347.
Fabre, E., and Hurt, E. C. (1994). Nuclear transport. *Curr. Opin. Cell Biol.* **6,** 335–342.
Featherstone, C., Darby, M. K., and Gerace, L. (1988). A monoclonal antibody against the nuclear pore complex inhibits nucleocytoplasmic transport of protein and RNA *in vivo*. *J. Cell Biol.* **107,** 1289–1297.
Feldherr, C. M., Kallenbach, E., and Schultz, N. (1984). Movement of a karyophilic protein through the nuclear pores of oocytes. *J. Cell Biol.* **99,** 2216–2222.
Finlay, D. R., and Forbes, D. J. (1990). Reconstitution of biochemically altered nuclear pores: Transport can be eliminated and restored. *Cell (Cambridge, Mass.)* **60,** 17–29.
Finlay, D. R., Newmeyer, D. D., Price, T. M., and Forbes, D. J. (1987). Inhibition of *in vitro* nuclear transport by a lectin that binds to nuclear pores. *J. Cell Biol.* **104,** 189–200.
Finlay, D. R., Meier, E., Bradley, P., Horecka, J., and Forbes, D. J. (1991). A complex of nuclear pore proteins required for pore function. *J. Cell Biol.* **114,** 169–183.
Fischer, U., and Lührmann, R. (1990). An essential signaling role for the $m^3G$ cap in the transport of U1 snRNPs to the nucleus. *Science* **249,** 786–790.
Fischer, U., Sumpter, V., Sekine, M., Satoh, T., and Lührmann, R. (1993). Nucleo-cytoplasmic transport of U snRNPs: definition of a nuclear localization signal in the Sm core domain that binds a transport receptor independently of the $m^3G$ cap. *EMBO J.* **12,** 573–583.
Forbes, D. J. (1992). Structure and function of the nuclear pore complex. *Annu. Rev. Cell Biol.* **8,** 495–527.
Franke, W. W. (1974). Structure, biochemistry and functions of the nuclear envelope. *Int. Rev. Cytol., Suppl.* **4,** 71–236.
Franke, W. W., and Scheer, U. (1970a). The ultrastructure of the nuclear envelope of amphibian oocytes: A reinvestigation. I. The mature oocyte. *J. Ultrastruct. Res.* **30,** 288–316.
Franke, W. W., and Scheer, U. (1970b). The ultrastructure of the nuclear envelope of amphibian oocytes: A reinvestigation. II. The immature oocyte and dynamic aspects. *J. Ultrastruct. Res.* **30,** 317–327.
Garcia-Bustos, J., Heitman, J., and Hall, M. N. (1991). Nuclear protein localization. *Biochim. Biophys. Acta* **1071,** 83–101.
Gerace, L. (1992). Molecular trafficking across the nuclear pore complex. *Curr. Opin. Cell Biol.* **4,** 637–645.
Gerace, L., and Burke, B. (1988). Functional organization of the nuclear envelope. *Annu. Rev. Cell Biol.* **4,** 335–374.
Gerace, L., Ottaviano, Y., and Kondor-Koch, C. (1982). Identification of a major polypeptide of the nuclear pore complex. *J. Cell Biol.* **95,** 826–837.
Gething, M.-J., and Sambrook, J. (1992). Protein folding in the cell. *Nature (London)* **355,** 33–44.
Goldberg, M. W., and Allen, T. D. (1992). High resolution scanning electron microscopy of the nuclear envelope: Demonstration of a new, regular, fibrous lattice attached to the baskets of the nucleoplasmic face of the nuclear pores. *J. Cell Biol.* **119,** 1429–1440.
Goldie, K. N., Panté, N., Engel, A., and Aebi, U. (1994). Exploring native nuclear pore complex structure and conformation by scanning force microscopy in physiological buffers. *J. Vac. Sci. Technol., B* **12,** 1482–1485.
Görlich, D., Prehn, S., Laskey, R., and Hartmann, E. (1994). Isolation of a protein that is essential for the first step of nuclear protein import. *Cell (Cambridge, Mass.)* **79,** 767–778.
Grandi, P., Doye, V., and Hurt, E. C. (1993). Purification of NSP1 reveals complex formation with 'GLFG' nucleoporins and a novel nuclear pore protein NIC96. *EMBO J.* **12,** 3061–3071.

Greber, U. F., Senior, A., and Gerace, L. (1990). A major glycoprotein of the nuclear pore complex is a membrane-spanning polypeptide with a large lumenal domain and a small cytoplasmic tail. *EMBO J.* **9,** 1495–1502.

Guan, T., Müller S., Kleir, G., Panté, N., Blevitt, J. M., Häner, M., Paschal, B., Aebi, U., and Gerace, L. (1995). Structural analysis of the p62 complex, an assembly of O-linked glycoproteins that localizes near the central gated channel of the nuclear pore complex. Submitted for publication.

Hamm, J., and Mattaj, I. W. (1990). Monomethylated cap structures facilitate RNA export from the nucleus. *Cell (Cambridge, Mass.)* **63,** 109–118.

Hamm, J., Darzynkiewicz, E., Tahara, S. M., and Mattaj, I. W. (1990). The trimethyguanosine cap structure of U1 snRNA is a component of a bipartite nuclear targeting signal. *Cell (Cambridge, Mass.)* **62,** 569–577.

Hinshaw, J. E., Carragher, B. O., and Milligan, R. A. (1992). Architecture and design of the nuclear pore complex. *Cell (Cambridge, Mass.)* **69,** 1133–1141.

Holt, G. D., Snow, C. M., Senior, A., Haltiwanger, R. S., Gerace, L., and Hart, G. W. (1987). Nuclear pore complex glycoproteins contain cytoplasmically disposed O-linked N-acetylglucosamine. *J. Cell Biol.* **104,** 1157–1164.

Hurt, E. C. (1988). A novel nucleoskeletal-like protein located at the nuclear periphery is required for the life cycle of *Saccharomyces cerevisiae. EMBO J.* **7,** 4323–4334.

Hurt, E. C. (1990). Targeting of a cytosolic protein to the nuclear periphery. *J. Cell Biol.* **111,** 2829–2837.

Imamoto, N., Matsuoka, Y., Kurihara, T., Kohno, K., Miyagi, M., Sakiyama, F., Okada, Y., Tsunasawa, S., and Yoneda, Y. (1992). Antibodies against 70-kD heat shock cognate protein inhibit mediated nuclear import of karyophilic proteins. *J. Cell Biol.* **119,** 1047–1061.

Izaurralde, E., and Mattaj, I. W. (1992). Transport of RNA between nucleus and cytoplasm. *Semin. Cell Biol.* **3,** 279–288.

Izaurralde, E., Stepinski, J., Darzynkiewicz, E., and Mattaj, I. W. (1992). A cap binding protein that may mediate nuclear export of RNA polymerase II-transcribed RNAs. *J. Cell Biol.* **118,** 1287–1295.

Jarmolowski, A., Boelens, W. C., Izaurralde, E., and Mattaj, I. W. (1994). Nuclear export of different classes of RNA is medicated by specific factors. *J. Cell Biol.* **124,** 627–635.

Jarnik, M., and Aebi, U. (1991). Toward a more complete 3-D structure of the nuclear pore complex. *J. Struct. Biol.* **107,** 291–308.

Kedersha, N. L., Heuser, J. E., Chugani, D. C., and Rome, L. H. (1991). Vaults. III. Vault ribonucleoprotein particles open into flower-like structures with octagonal symmetry. *J. Cell Biol.* **112,** 225–235.

Kita, K., Omata, S., and Horigome, T. (1993). Purification and characterization of a nuclear pore glycoprotein complex containing p62. *J. Biochem. (Tokyo)* **113,** 377–382.

Kraemer, D., Wozniak, R. W., Blobel, G., and Radu, A. (1994). The human CAN protein, a putative oncogene product associated with myeloid leukemogenesis, is a nuclear pore complex protein that faces the cytoplasm. *Proc. Natl. Acad. Sci. U.S.A.* **91,** 1519–1523.

Marshallsay, C., and Lührmann, R. (1994). In vitro nuclear import of snRNPs: Cytosolic factors mediate $m^3$G-cap dependence of U1 and U2 snRNP transpport. *EMBO J.* **13,** 222-231.

Mattaj, I. W., Boelens, W., Izaurralde, E., Jarmolowski, A., and Kambach, C. (1993). Nucleocytoplasmic transport and snRNP assembly. *Mol. Biol. Rep.* **18,** 79–83.

McMorrow, I. M., Bastos, R., Horton, R., and Burke, B. (1994). Sequence analysis of a cDNA encoding a human nuclear pore complex protein, hnup153. *Biochim. Biophys. Acta* **1217,** 219–223.

Mehlin, H., Daneholt, B., and Skoglund, U. (1992). Translocation of a specific premessenger ribonucleoprotein particle through the nuclear pore studied with electron microscope tomography. *Cell (Cambridge, Mass.)* **69,** 605–613.

Melchior, F., Paschal, B., Evans, J., and Gerace, L. (1993). Inhibition of nuclear protein import by nonhydrolyzable analogues of GTP and identification of the small GTPase Ran/TC4 as an essential transport factor. *J. Cell Biol.* **123,** 1649–1659.

Michaud, N., and Goldfard, D. (1991). Multiple pathways in nuclear transport; the import of U2 snRNP occur by a novel kinetic pathway. *J. Cell Biol.* **112,** 215–223.

Michaud, N., and Goldfard, D. (1992). Microinjected U snRNAs are imported to oocyte nuclei via the nuclear pore complex by three distinguishable targeting pathways. *J. Cell Biol.* **116,** 851–861.

Mitchell, P. J., and Cooper, C. S. (1992). Nucleotide sequence analysis of human *tpr* cDNA clones. *Oncogene* **7,** 383–388.

Moore, M. S., and Blobel, G. (1992). The two steps of nuclear import, targeting to the nuclear envelope and translocation through the nuclear pore, require different cytosolic factors. *Cell (Cambridge, Mass.)* **68,** 939–950.

Moore, M. S., and Blobel, G. (1993). The GTP-binding protein Ran/TC4 is required for protein import into the nucleus. *Nature (London)* **365,** 661–663.

Moore, M. S., and Blobel, G. (1994). Purification of a Ran-interacting protein that is required for protein import into the nucleus. *Proc. Natl. Acad. Sci. U.S.A.* **91,** 10212–10216.

Newmeyer, D. D. (1993). The nuclear pore complex and nucleocytoplasmic transport. *Curr. Opin. Cell Biol.* **5,** 395–407.

Newmeyer, D. D., and Forbes, D. J. (1988). Nuclear import can be separated into distinct steps *in vitro:* Nuclear pore binding and translocation. *Cell (Cambridge, Mass.)* **52,** 641–653.

Newmeyer, D. D., and Forbes, D. J. (1990). An N-ethylmaleimide-sensitive cytosolic factor necessary for nuclear protein import: Requirement in signal-mediated binding to the nuclear pore. *J. Cell Biol.* **110,** 547–557.

Oberleithner, H., Brinckmann, E., Schwab, A., and Krohne, G. (1994). Imaging nuclear pores of aldosterone-sensitive kidney cells by atomic force microscopy. *Proc. Natl. Acad. Sci. U.S.A.* **91,** 9784–9788.

Paine, P. L., Moore, L. C., and Horowitz, S. B. (1975). Nuclear envelope permeability. *Nature (London)* **254,** 109–114.

Panté, N., and Aebi, U. (1993). The nuclear pore complex. *J. Cell Biol.* **122,** 977–984.

Panté, N., and Aebi, U. (1994a). Towards understanding the 3-D structure of the nuclear pore complex at the molecular level. *Curr. Opin. Struct. Biol.* **4,** 187–196.

Panté, N., and Aebi, U. (1994b). Toward the molecular details of the nuclear pore complex. *J. Struct. Biol.* **113,** 179–189.

Panté, N., Bastos, R., McMorrow, I., Burke, B., and Aebi, U. (1994). Interactions and three-dimentional localization of a group of nuclear complex proteins. *J. Cell Biol.* **126,** 603–617.

Pruschy, M., Ju, Y., Spitz, L., Carafoli, E., and Goldfarb, D. S. (1994). Facilitated nuclear transport of calmodulin in tissue culture cells. *J. Cell Biol.* **127,** 1527–1536.

Reichelt, R., Holzenburg, A., Buhle, E. L., Jarnik, M., Engel, A., and Aebi, U. (1990). Correlation between structure and mass distribution of the nuclear pore complex, and of distinct pore complex components. *J. Cell Biol.* **110,** 883–894.

Richardson, W. D., Mills, A. D., Dilworth, S. M., Laskey, R. A., and Dingwall, C. (1988). Nuclear protein migration involves two steps: Rapid binding at the nuclear envelope followed by slower translocation through the nuclear pores. *Cell (Cambridge, Mass.)* **52,** 655–664.

Ris, H. (1991). The 3-D structure of the nuclear pore complex as seen by high voltage electron microscopy and high resolution low voltage scanning electron microscopy. *EMSA Bull.* **21,** 54–56.

Robbins, J., Dilworth, S. M., Laskey, R. A., and Dingwall, C. (1991). Two interdependent basic domains in nucleoplasmin nuclear targeting sequence: Identification of a class of bipartite nuclear targeting sequence. *Cell (Cambridge, Mass.)* **64,** 615–623.

Rout, M. P., and Blobel, G. (1993). Isolation of the yeast nuclear pore complex. *J. Cell Biol.* **123,** 771–783.

Rout, M. P., and Wente, S. R. (1994). Pores for thought: Nuclear pore complex proteins. *Trends Cell Biol.* **4,** 357–365.

Shi, Y., and Thomas, J. O. (1992). The transport of proteins into the nucleus requires the 70-kilodalton head shock protein or its cytoplasmic cognate. *Mol. Cell Biol.* **12,** 2186–2192.

Snow, C. M., Senior, A., and Gerace, L. (1987). Monoclonal antibodies identify a group of nuclear pore complex glycoproteins. *J. Cell Biol.* **104,** 1143–1156.

Soullan, B., and Worman, H. J. (1993). The amino-terminal domain of the lamin B receptor is a nuclear envelope targeting signal. *J. Cell Biol.* **120,** 1093–1100.

Sterne-Marr, R., Blevitt, J. M., and Gerace, L. (1992). O-linked glycoproteins of the nuclear pore complex interact with a cytosolic factor required for nuclear protein import. *J. Cell Biol.* **116,** 271–280.

Stewart, M., and Whytock, S. (1988). The structure and interactions of components of nuclear envelopes from *Xenopus* oocyte germinal vesicles observed by heavy metal shadowing. *J. Cell Sci.* **90,** 409–423.

Sukegawa, J., and Blobel, G. (1993). A nuclear pore complex protein that contains zinc finger motifs, binds DNA, and faces the nucleoplasm. *Cell (Cambridge, Mass.)* **72,** 29–38.

Unwin, P. N. T., and Milligan, R. A. (1982). A large particle associated with the perimeter of the nuclear pore complex. *J. Cell Biol.* **93,** 63–75.

Wiese, C., and Wilson, K. L. (1993). Nuclear membrane dynamics. *Curr. Opin. Cell Biol.* **5,** 387–394.

Wilken, N., Kossner, U., Senécal, J.-L., Scheer, U., and Dabauvalle, M.-C. (1993). Nup180, a novel nuclear pore complex protein localizing to the cytoplasmic ring and associated fibrils. *J. Cell Biol.* **123,** 1345–1354.

Wozniak, R. W., Bartnik, E., and Blobel, G. (1989). Primary structure analysis of an integral membrane glycoprotein of the nuclear pore. *J. Cell Biol.* **108,** 2083–2092.

Yamasaki, L., and Lanford, R. E. (1992). Nuclear transport: A guide to import receptors. *Trends Cell Biol.* **2,** 123–127.

Yano, R., Oakes, M., Yamaghishi, M., Dodd, J. A., and Nomura, M. (1992). Cloning and characterization of SRP1, a suppressor of temperature-sensitive RNA polymerase I mutations, in *Saccharomyces cerevisiae. Mol. Cell Biol.* **12,** 5640–5641.

Yano, R., Oakes, M. L., Tabb, M. M., and Nomura, M. (1994). Yeast Srp1p has homology to armadillo/plakoglobin/β-catenin and participates in apparently multiple nuclear functions including the maintenance of the nucleolar structure. *Proc. Natl. Acad. Sci. U.S.A.* **91,** 6880–6884.

Zasloff, M. (1983). tRNA transport form the nucleus in a eukaryotic cell: Carrier-mediated translocation process. *Proc. Natl. Acad. Sci. U.S.A.* **80,** 6436–6440.

# Nuclear Pore Complex Proteins

Ricardo Bastos,* Nelly Panté,** and Brain Burke*
*Department of Cell Biology, Harvard Medical School, Boston, Massachusetts 02115, and **M. E. Müller Institute for Microscopy, Biozentrum, University of Basel, CH-4056 Basel, Switzerland

The nuclear envelope forms the boundary between the nucleus and the cytoplasm and as such regulates the exchange of macromolecules between the two compartments. The channels through the nuclear envelop that actually mediate this macromolecular traffic are the nuclear pore complexes. These are extremely elaborate structures which in vertebrate cells exhibit a mass of approximately 120 MDa. They are thought to be composed of as many as 100 distinct polypeptide subunits. A major challenge in the field of nucleocytoplasmic transport is to identify these subunits and to determine their functions and interactions in the context of the three-dimensional structure of the nuclear pore complex. It is the aim of this review to summarize what is currently known of the 20 or so nuclear pore complex proteins that have been described in either vertebrate or yeast cells.

**KEY WORDS:** Nuclear envelope, Nuclear pore complex, Nucleoporin, Nucleocytoplasmic transport, Protein traffic, Mitosis.

## I. Introduction

The nuclear envelope is a membranous organelle that separates the nuclear and cytoplasmic compartments and as such is a defining feature of eukaryotic cells. A consequence of this separation is that the nucleus contains a unique spectrum of proteins required for the organization and utilization of the genome. All of these proteins must be synthesized in the cytoplasm, the exclusive location of the protein synthetic machinery, a situation that obviously demands continuous macromolecular trafficking between the two compartments. This development of complexity by higher cells has a number of implications, first and foremost of which is that the nuclear envelope provides opportunities for new levels of cellular regulation that are not available in prokaryotes. For instance, transcription, and therefore many aspects of cell physiology, can be modulated at least in part by controlling acces of certain key transcription factors to the nucleus (Steward, 1989; Ghosh and Baltimore, 1990; Nasmyth *et al.*, 1990). On perhaps a more dramatic note, the orchestration of DNA replication and by implication, the basic eukaryotic cell cycle, has been suggested to depend upon the exclusion of cardinal regulatory molecules from the interphase nucleus (Blow and Laskey, 1988). It is evident from this that the delineation of nuclear and cytoplasmic compartments and the parallel development of mechanisms to move macromolecules between them is of profound evolutionary significance. Because of the growing appreciation of its central role in eukaryote metabolism, in many respects no less significant than the processes of transcription and translation, the mechanics of nucleocytoplasmic transport have become an important focus of cell biology research. As a consequence, the control of macromolecular traffic across the nuclear envelope has been the subject of a number of recent reviews and commentaries (Forbes, 1992; Gerace, 1992; Newmeyer, 1993; Panté and Aebi, 1993, 1994; Fabre and Hurt, 1994). Therefore, rather than provide unnecessary duplication by covering the entire field, it is the goal of this article to provide the reader with a comprehensive and up-to-date description of the properties and interactions of protein subunits of the channels through the nuclear envelope that mediate molecular communication between the nucleus and cytoplasm.

## II. Nucleocytoplasmic Transport

Because the nuclear envelope separates two biochemically distinct compartments, it follows that it should be both structurally and functionally asymmetric. The functional asymmetry is partly reflected in the nuclear envelope's role as a selective barrier in that it mediates both the export of mature RNA molecules from the nucleus and the import of nuclear proteins and ribonucleoprotein particles. At the same time, unprocessed RNAs as well as large cytoplasmic macromolecules across the nuclear envelope has been a topic of intense investigation. Early cell physiological studies showed that the nuclear envelope has the properties of a molecular sieve (Paine, 1975; Paine *et al.*, 1975). These studies were originally performed using sized tritiated dextrans (Paine *et al.*, 1975) as well as various proteins and small molecules (Paine, 1975) microinjected into living cells (insect salivary gland and amphibian oocytes), and indicated that relatively small molecules may exchange by diffusion between the nucleus and cytoplasm through what appear to be 9 to 10-nm-diameter aqueous channels. Larger molecules were found to remain in the compartment into which they were injected with no exchange across the nuclear envelope, even after extended periods. For a globular protein this limiting size for passive diffusion would correspond to about 50–60 kDa. It has subsequently become clear that macromolecules above this 9–10-nm size limit must present specific signals in order to traverse the nuclear envelope. Evidence supporting such a selective transport model has been reviewed by (Dingwall and Laskey, 1986).

In this section we will summarize briefly the current state of knowledge of nuclear import/export signals and their receptors. However, for a more in-depth discussion of this topic we refer the reader to the following excellent reviews: Goldfarb and Michaud, 1991; Silver, 1991; Davis, 1992; Forbes, 1992. It is clear that many macromolecules too large to be accommodated by a 10-nm channel may traverse the nuclear envelope in processes that require specific signals and are energy dependent. The best characterized of these processes is that involved in nuclear protein import where specific nuclear localization sequences (NLSs) contained within nuclear transport substrates have been defined (Dingwall and Laskey, 1986, 1991). The prototype NLS was described for SV40 large T antigen, which consists of a short, basic peptide PKKKRKV (Kalderon *et al.,* 1984a,b). Subsequently a somewhat more complex NLS was described for the nuclear protein nucleoplasmin (Dingwall *et al.,* 1988) which contains two basic segments separated by a small spacer, $KRX_{10}KKKK$ (X being an unspecified amino acid residue.) The majority of NLSs that have now been described appear to conform more closely to the dibasic nucleoplasmin type (Robbins *et al.,* 1991) and the general consensus is that even the SV40 NLS may simply represent one end of a continuum of spacer lengths (Dingwall and Laskey, 1991). These basic domain NLSs are extremely variable in terms of location within polypeptide chains and are not preferentially located at either terminus (Richardson *et al.,* 1986). The SV40 T antigen NLS, for instance, extends from positions 126 to 132 in the wild-type molecule but can actually function at several different locations. Indeed the SV40 NLS does not actually have to form part of the polypeptide chain (Goldfarb *et al.,* 1986), since the NLS peptide can be chemically crosslinked to a wide variety of proteins and will cause them to localize to the nucleus when microinjected into living cells (Goldfarb *et al.,* 1986) or when employed in *in vitro* nuclear import assays (Newmeyer and Forbes, 1988). While the NLS will not function in certain contexts (Roberts *et al.,* 1987) at least one important requirement appears, not surprisingly, to be that it is exposed at the surface of the molecule (Nelson and Silver, 1989). An obvious and experimentally confirmed implication of the plasticity of NLS locations is that they are not cleaved during the import process and are therefore available for reuse. This is not a trivial feature since many nuclear proteins are released into the cytoplasm during mitosis and must then reaccumulate in the daughter nuclei following nuclear envelope reformation. Notwithstanding this, the function of an NLS may still be modulated by, for instance, phosphorylation of adjacent residues (Rihs and Peters, 1989; Rihs *et al.*, 1991), tethering to cytoplasmic proteins (Roth *et al.*, 1989; Steward, 1989; Ghosh and Baltimore, 1990), or conformational masking (Picard and Yamamoto, 1987; Picard *et al.*, 1988).

Receptors for basic domain NLSs have yet to be characterized in detail. However, one widely held view is that initial recognition of the NLS by its receptor occurs in the cytoplasm and it is a soluble NLS–receptor complex which in fact docks at the nuclear envelope prior to translocation into the nucleus (Silver, 1991; Forbes, 1992; Gerace, 1992). A possible candidate for an NLS receptor has been identified in *Xenopus* egg extracts (Görlich *et al.*, 1994). Named importin, this 60-kDa protein is the *Xenopus* homolog of the yeast protein SRP1p with which it shares 44% sequence identity. In *in vitro* transport assays importin has been shown to be essential for the docking of NLS-containing transport substrates at the nuclear periphery. Together with a second soluble import factor, the ras-related GTP-binding protein Ran/TC4 (Moore and Blobel, 1994), and in the presence of ATP and GTP, importin can replace cytosol in promoting signal-mediated nuclear protein import *in vitro*. These two proteins will be discussed in more detail in subsequent sections.

There are at least two other kinetically distinct nuclear import pathways in addition to the basic domain NLS-dependent pathway. These have been described for subsets of U RNPs (Michaud and Goldfarb, 1991, 1992). One of these is dependent upon the integrity of 3Me-G cap of the U RNA and since it is not competed by basic domain NLSs, presumably involves a distinct receptor system. Similarly RNA export appears to involve multiple pathways (Jarmolowski *et al.*, 1994), one of which involves recognition of the 7Me-G cap structure. Recently a cap-binding protein has been identified that appears to be involved in the export process (Izaurralde *et al.*, 1992).

## III. Nuclear Pore Complexes

The channels that mediate the exchange of macromolecules between the nucleus and cytoplasm are the nuclear pore complexes (NPCs). These are extremely elaborate multiprotein assemblies that span both the inner and outer nuclear membranes in places where the two phospholipid bilayers are in continuity. They are anchored on the nucleoplasmic side to the nuclear lamina, a filamentous protein meshwork closely associated with the nucleoplasmic face of the inner nuclear membrane (Gerace and Burke,

1988). The interrelationships of these structures are shown in Fig. 1. A typical mammalian somatic cell may possess 2000–4000 NPCs per nucleus representing a density of about 2–5 NPCs per $\mu m^2$. NPC densities as high as 20 per $\mu m^2$ may be observed in certain cells such as *Xenopus laevis* oocytes.

Direct evidence for the role of the NPC in nucleocytoplasmic transport was provided in a series of pioneering experiments by Feldharr and colleagues (1984) where colloidal gold particles coated with the nucleophilic protein nucleoplasmin were microinjected into the cytoplasm of *Xenopus* oocytes. Subsequent electron microscopic analysis of the oocytes revealed gold particles passing through the central region of all identifiable NPCs en route to the nucleus. Gold particles lacking nucleoplasmin or coated with an irrelevant protein remained in the cytoplasm and showed no association with NPCs. Similar experiments were carried out employing gold particles coated with tRNA microinjected into the oocyte nucleus (Dworetzky and Feldherr, 1988). As before, gold particles were found exiting the nucleus via all identifiable NPCs. As well as demonstrating the role of NPCs in nucleocytoplasmic transport, these experiments also indicated that each NPC can support bidirectional movement, both import and export. Finally, by the use of gold particles of increasing size it was possible to demonstrate that the NPCs could accommodate particles with diameters up to 26 nm, two to three times the limit for free diffusion (Dworetzky *et al.,* 1988).

This section will give a brief overview of NPC structure and organization. For an in-depth treatment of this topic the reader is referred to the chapter by Panté and Aebi in this volume. High-resolution electron microscopy (EM) of manually isolated amphibian oocyte nuclear envelopes has shown NPCs to possess eightfold radial symmetry about an axis perpendicular to the plane of the nuclear envelope (Gall, 1967; Hinshaw *et al.,* 1992; Unwin and Milligan, 1982; Akey and Radermacher, 1993). This eightfold symmetry is particularly evident at the level of the nuclear membranes in the equatorial region of the NPC that appears as a ring–spoke complex about 145–150 nm in diameter and 75 nm deep. A significant portion of the mass of this structure resides within the perinuclear space and in this way is firmly anchored within the nuclear membranes (Akey, 1989; Hinshaw *et al.,* 1992; Akey and Radermacher, 1993). The ring–spoke complex embraces a large central channel, frequently appearing occluded by a central plug that may form part of the transport apparatus (Akey and Radermacher, 1993), as well as eight peripheral channels (Hinshaw *et al.,* 1992). The latter have been suggested to allow the free exchange of small molecules (less than 10 nm in diameter) across the nuclear envelope (Hinshaw *et al.,* 1992). In contrast, it is the central channel that appears to be gated and which may mediate translocation of larger particles (with diameters up to 26 nm) provided that they bear specific nuclear import/export signals. In this way the sieve-like properties of the nuclear envelope as a whole may be ac-

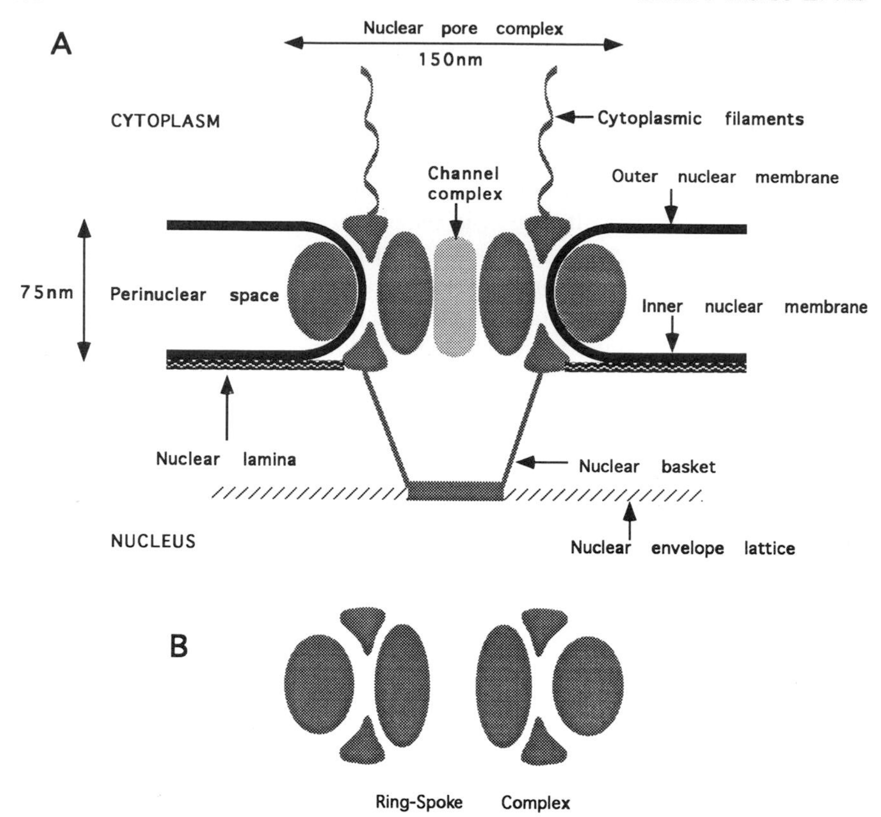


FIG. 1 Highly stylized diagram showing the interrelationships of the major components of the nuclear envelope. These include the inner and outer nuclear membranes that are joined in regions where they are penetrated by nuclear pore complexes (A). Four major elements of the NPC are distinguished in the diagram: the cytoplasmic filaments, the central plug or channel complex, the nuclear basket, and the ring–spoke complex. The latter, as referred to for the purpose of this review, comprises the major structural framework of the NPC and includes cytoplasmic and nuclear ring subunits and the spoke assembly. The ring–spoke complex is shown isolated in (B) simply to indicate that it includes features that extend into the perinuclear space. Other features in (A) include the nuclear lamina, which is closely associated with both the inner nuclear membrane and underlying chromatin and which attaches to the NPCs at an ill-defined site near the base of the nuclear baskets. The nuclear envelope lattice connects the distal rings of adjacent nuclear baskets and in certain cell types may be extremely elaborate. While major dimensions are indicated in the diagram, it has not been drawn strictly to scale.

counted for in the structure and organization of the NPCs themselves (Figs. 1 and 5) (Color plate 11).

In addition to the ring–spoke complex, NPCs also possess extensive peripheral structures extending into both the cytoplasm and the nuclear interior (references in Franke, 1974; Panté and Aebi, 1994). On the cytoplasmic face of the NPC there are eight short (~100 nm), kinky filaments, each attached at one end to the ring–spoke complex (Jarnik and Aebi, 1991). It has been suggested that a protein whose destination is the nucleus and which bears a nuclear localization signal may initially dock at these filaments as a complex with a cytosolic NLS receptor, prior to translocation across the NPC (Richardson *et al.,* 1988; Gerace, 1992). On the nucleoplasmic face of the NPC there are additional filamentous structures consisting of eight apparently rigid filaments anchored proximally to the ring–spoke complex and joined at their distal ends by a 50-nm-diameter ring. The overall appearance is of a basket or fish trap-like structure extending about 100 nm into the nucleoplasm (Jarnik and Aebi, 1991; Ris, 1991; Goldberg and Allen, 1992). In certain cell types the distal rings of many baskets are observed to be linked by an oftentimes extensive planar meshwork of filaments (Goldberg and Allen, 1992). This meshwork is quite distinct from the other, better characterized filament network associated with the nuclear envelope, the nuclear lamina (Gerace and Burke, 1988), being separated from it by at least 50–100 nm. The nuclear lamina itself is tightly associated with the inner nuclear membrane (Gerace and Burke, 1988) and provides an additional anchorage for each NPC, which it contacts at the base of the basket structure in the vicinity of the ring–spoke complex (Goldberg and Allen, 1992). The major structural features of the NPC are summarized in Figs. 1 and 5 (color plate 11).

A major recent advance has been the bulk isolation of yeast NPCs (Rout and Blobel, 1993), a feat that has so far not been achieved in vertebrate systems (see below). The yeast NPCs appear structurally very similar, although somewhat smaller than their vertebrate counterparts having a diameter of about 100 nm versus the 145 nm for *Xenopus* oocyte NPCs (Akey, 1989; Hinshaw *et al.,* 1992; Akey and Radermacher, 1993; see also Panté and Aebi, 1994). However, the diameter of the central channel appears to be conserved between yeast and vertebrates. Whether the peripheral NPC structures consisting of the nuclear basket and cytoplasmic filaments are present in yeast NPCs has yet to be ascertained. Biophysical (hydrodynamic) measurements indicate a mass of about 66 MDa for the yeast NPC (Rout and Blobel, 1993) compared with a mass of about 125 MDa for the oocyte NPC (Reichelt *et al.,* 1990). The latter value was obtained by quantitative scanning transmission electron microscopy. To put these mass measurements in perspective, a eukaryotic ribosome, with a major axis of about 32 nm, has a mass of about 4 MDa and contains about 80 polypeptides and 4 RNAs! Analysis of the polypeptide composition of yeast NPC preparations reveals a total of 60–80 potential NPC proteins in a range of relative abundances (Rout and Blobel, 1993). This figure is not inconsistent with the suggestion, frequently heard, that the vertebrate NPC may contain as many as 100 different polypeptides (Reichelt *et al.,* 1990). One should, however, be aware that some NPC proteins may be present in very high copy number. It is quite possible therefore that a significant portion of the NPC mass could be accounted for by a relatively small number of proteins.

## IV. Nuclear Pore Complex Proteins

In recent years considerable effort has been applied to the identification of NPC components in both vertebrates and in yeast and increasing numbers are now coming to light. Despite this progress, and even assuming very generous stoichiometries, it is still only possible to account for about 15–30% of the vertebrate NPC mass with known NPC proteins or nucleoporins (abbreviated as nups). It seems likely therefore that these inventory-type investigations will continue at least into the near future. Notwithstanding this, it is now possible in the light of a number of cDNA sequences to place many of these NPC proteins into family groupings on the basis of common structural features present in both higher and lower organisms. It is the aim of this section to summarize what is known of vertebrate and yeast nucleoporins.

### A. Vertebrate Nucleoporins

Identification and characterization of vertebrate NPC proteins has been bedeviled by the fact that it has not so far proven possible to purify morphologically identifiable NPCs. This failure is primarily a consequence of the tight association of NPCs with the relatively insoluble nuclear lamina (Gerace and Burke, 1988). Conditions that disassemble the lamina or remove NPCs from the lamina also result in disruption of the NPCs (Gerace *et al.,* 1984). Because a direct analysis of NPC polypeptide composition is ruled out, rather more circumspect methods have been used to identify NPC components. Two approaches have yielded the majority of vertebrate nucleoporins. The first has involved generation of monoclonal antibodies against nuclear envelopes or nuclear envelope extracts followed by identification of those antibodies that recognize NPC components. The second approach is based upon the isolation of cDNAs encoding candidate nucleoporins followed by the demonstration that they are indeed NPC subunits.

This has generally been achieved by localizing these proteins to the NPC by EM–immunocytochemistry employing antibodies raised against fusion proteins produced in bacteria, or by expressing epitope-tagged proteins in mammalian cells and again using immunocytochemistry to localize these tagged proteins to the NPC. Nucleoporins have also been identified by raising antibodies specifically against candidate nuclear envelope proteins as well as by screening autoimmune sera. Those vertebrate nucleoporins and their families that have been characterized to date are described in the following sections. Their structure and properties are summarized in Table I and in Figs. 2 and 3 (color plates 8 and 9).

### 1. The O-Linked Glycoprotein Family

A family of 10–20 nucleoporins has been identified employing a variety of monoclonal antibodies raised against detergent extracts of purified rat liver nuclear envelopes (Davis and Blobel, 1986; Snow *et al.,* 1987). This family of proteins, which range in size from about 45 to 230 kDa exhibit an impressive degree of immunological cross-reactivity and are united in the property of binding to the lectin wheat germ agglutinin (WGA) (Davis and Blobel, 1987; Holt *et al.,* 1987; Park *et al.,* 1987). WGA binding is a result of extensive modification with single O-linked *N*-acetylglucosamine moieties (O-GlcNAC, Torres and Hart, 1984; Holt *et al.,* 1987). This modification, which has now been found on a large number of intracellular proteins (Hart, 1992), is somewhat unusual. In contrast to most other forms of glycosylation, *O*-GlcNAc is found on proteins or protein domains that are exposed to the cytosol and its addition is catalyzed by a cytosolic *N*-acetylglucosaminyl transferase that utilizes UDP-*N*-acetylglucosamine as the sugar donor (Haltiwanger *et al.,* 1992). Other more familiar forms of glycosylation, involving chains of sugars, occur on membrane, secretory, and lysosomal proteins that are exposed to the extracytoplasmic space. The sugar transferases in these cases are invariably membrane proteins. An additional feature of *O*-GlcNAc is that it turns over (Kearse and Hart, 1991), its removal being mediated by a recently described cytosolic *N*-acetylglucosaminidase (Yong and Hart, 1994). The dynamic nature of this modification has led to the view that it has more in common with phosphorylation than glycosylation. Indeed it has been suggested that the two modifications might actually act antagonistically (Hart, 1992). Proteins that bear *O*-GlcNAc are invariably phosphoproteins and one function of this modification might be to block potential phosphorylation sites (Hart, 1992).

The use of WGA coupled to colloidal gold particles in electron microscopic localization studies has revealed that the WGA-binding proteins are located on both faces of the NPC in the central region of the ring–spoke

TABLE I
Vertebrate NPC Proteins

| Name | Isolation | Properties and possible functions | Location within the NPC | References |
|---|---|---|---|---|
| O-linked glycoprotein family | | | | |
| p62 (62 kDa) | Biochemical | XFXFG repeats (N-terminal) C-terminal domain with heptad repeats/coiled-coil structure. Multiple O-linked GlcNAc and putative phosphorylation sites. Forms a complex with p58 and p54. Essential for NPC function | Central region of both cytoplasmic and nuclear NPC faces | Davis and Blobel (1986); Snow *et al.* (1987); Starr *et al.* (1990) |
| POM121 (121 kDa) | Biochemical | C-terminal XFXFG repeats. Integral membrane protein. Multiple O-linked GlcNAc and putative phosphorylation sites. May anchor NPC components to the pore membrane | Pore membrane domain | Hallberg *et al.* (1993) |
| NUP153 (153 kDa) | Immunological (mAbs) | C-terminal XFXFG repeats. Four $C_2$–$C_2$ zinc fingers in the central region of the molecule. Phosphoprotein containing multiple O-linked GlcNAc. N-terminal domain required for targeting to the NPC. Forms homooligomers. Organizes chromatin around NPC? | Terminal ring of the nuclear basket. Filaments? | Cordes *et al.* (1993); Sukegawa and Blobel (1993); McMorrow *et al.* (1994); Panté *et al.* (1994) |

| | | | | |
|---|---|---|---|---|
| CAN/NUP214 (214 kDa) | Genetic and biochemical | C-terminal XFXFG repeats, central leucine zipper. Forms a complex with a 75-kDa NPC protein. Putative oncogene product associated with myeloid leukemogenesis. Docking site for incoming nuclear proteins? | Cytoplasmic filaments | von Lindern *et al.* (1992b); Kraemer *et al.* (1994); Panté *et al.* (1994) |
| Others | | | | |
| gp210 (210 kDa) | Biochemical | Type I integral membrane protein. Contains high mannose N-linked oligosaccharides on large lumenal domain. Thought to anchor the NPC in the nuclear membranes | Pore membrane domain | Gerace *et al.* (1982); Wozniak *et al.* (1989); Greber *et al.* (1990); Wozniak and Blobel (1992) |
| NUP107 (107 kDa) | Biochemical | Leucine zipper at C-terminus | Not established | Radu *et al.* (1994) |
| NUP155 (155 kDa) | Biochemical | Multiple potential phosphorylation sites | Exposed on both the nuclear and cytoplasmic faces of the NPC | Radu *et al.* (1993) |
| Tpr (265 kDa) | Genetic and biochemical | Extended central coiled-coil domain. Very sensitive to proteolysis. Involved in activation of oncogenic kinases | Cytoplasmic face/filaments of the NPC | Byrd *et al.* (1994) |
| NUP180 (180 kDa) | Immunological (autoimmune serum) | Sequence not yet established. Conserved in vertebrates. May correspond to a proteolytic product of Tpr | Cytoplasmic ring and cytoplasmic filaments | Wilken *et al.* (1993) |

complex as well as in the nuclear basket structure and cytoplasmic filaments (Akey and Goldfarb, 1989; Panté *et al.,* 1994). In a variety of functional studies both *in vitro* and *in vivo,* the latter by microinjection, WGA has been shown to inhibit signal-dependent nucleocytoplasmic transport (Finlay *et al.,* 1987; Yoneda *et al.,* 1987; Dabauvalle *et al.,* 1988b; Wolff *et al.,* 1988) as does a monoclonal antibody that recognizes the majority of the NPC O-linked glycoproteins (Featherstone *et al.,* 1988). Additional studies employing *in vitro* nuclear assembly systems have demonstrated the requirement for the WGA-binding proteins in the formation of functional NPCs (Dabauvalle *et al.,* 1990; Finlay and Forbes, 1990; Finlay *et al.,* 1991). Furthermore, one or more of these proteins binds and essential cytosolic factor required for nuclear protein import in an *in vitro* transport assay (Sterne-Marr *et al.,* 1992). While the O-linked glycoproteins are clearly intimately involved in nucleocytoplasmic transport, the role of the *O*-GlcNAc modification itself is less clear. In fact *in vitro* galactosylation (Miller and Hanover, 1994) studies and the finding that it is not present in yeast (Radu *et al.,* 1994) would argue that it plays no essential role in the transport process per se. cDNAs encoding four members (CAN/NUP214, NUP153, POM121, and p62) of the O-linked glycoprotein family of nucleoporins from various organisms have now been isolated. The derived amino acid sequences reveal that each family member contains multiple (up to 34) copies of the pentapeptide repeat XFXFG, where X corresponds to any amino acid with a small or polar side chain. Searches of sequence data banks reveal no other vertebrate proteins in which this pentapeptide is iterated more than twice. There are, however, several yeast proteins that contain numerous copies of this motif, and all of them are NPC components. The occurrence of this motif in these various yeast and vertebrate proteins almost certainly accounts for their extensive immunological cross-reactivities.

***a. p62*** p62 (Davis and Blobel, 1986; Snow *et al.,* 1987) was the first member of the family of vertebrate O-linked NPC glycoproteins whose primary structure was elucidated (Starr *et al.,* 1990). Sequences are now available for the rat (Starr *et al.,* 1990), mouse (Cordes *et al.,* 1991), human (Carmo-Fonseca *et al.,* 1991), and *Xenopus* (Cordes *et al.,* 1991) proteins. The basic features of the rat p62 amino acid sequence are shown in Fig. 2. Cursory inspection indicates that the protein falls into two obvious but distinct halves separated by a centrally located tract of 24 serine and threonine residues. The amino-terminal half contains all of the 15 XFXFG pentapeptide repeats. In addition, it is also rich in serine and threonine residues and has been shown to contain all of the *O*-GlcNAc present on the molecule, possibly as many as 22 residues in the case of mouse p62 (Cordes *et al.,* 1991). Based on circular dichroism measurements as well as on sequence-based secondary structure predictions, this region of p62 is thought to be composed of a series of alternating $\beta$-sheets and loops (Buss *et al.,* 1994). The carboxy-terminal half of the molecule, in contrast, exhibits heptad repeats and is believed to form an $\alpha$-helical coiled-coil domain involved in self-association, a view that is reinforced by the finding that recombinant p62 produced in *Escherichia coli* forms rod-shaped dimers about 35 nm in length (Buss *et al.,* 1994).

Mammalian p62 can be eluted from intact NPCs in a large subcomplex or particle containing at least two other less well-characterized NPC proteins, p58 and p54 (Finlay *et al.,* 1991; Kita *et al.,* 1993; Panté *et al.,* 1994). Mass measurements for this particle range from 231 kDa (Kita *et al.,* 1993) to about 600 kDa (Finlay *et al.,* 1991; Panté *et al.,* 1994). The lower figure takes into account the fact that the particle appears somewhat asymmetric (with a frictional ratio of 2.0) and is therefore likely to lie closer to reality. It has been estimated that rat liver NPCs may each contain about 19 (most likely 16 or 24 given the NPC symmetry) copies of this particle (Kita *et al.,* 1993). Similarly, in *Xenopus* eggs, in which NPCs are disassembled, the p62 homolog (p68) has been detected in a discreet soluble particle with a mass in the region of about 250 kDa (Dabauvalle *et al.,* 1990). In a series of *in vitro* reconstitution experiments the p62 particle has been shown to be essential for the formation of functional NPCs (Dabauvalle *et al.,* 1990; Finlay *et al.,* 1991). These findings extend the observations that antibodies against p62 block nuclear protein import when microinjected into living cells (Dabauvalle *et al.,* 1988a; Benavente *et al.,* 1989a,b). The exact location of p62 within the NPC has until recently remained somewhat ambiguous. This is at least partly a consequence of the high degree of immunological cross-reactivity associated with the O-linked NPC glycoproteins. The emerging view, however, is that p62, and by implication p58 and p54, is located in the central region of the NPC, most likely associated with the central channel complex or plug, and accessible on both faces (Cordes *et al.,* 1991; Guan *et al.,* 1995), a localization consistent with the functional studies referred to above (Fig. 5, color plate 11).

***b. CAN/NUP214*** CAN (214 kDa) was originally identified as the product of a putative oncogene (*can*) associated with leukemogenesis (von Lindern *et al.,* 1992b). In certain types of acute myeloid leukemia a chromosomal translocation results in the formation of a hybrid gene consisting of the 5′ end of the *dek* gene fused to the 3′ end of the *can* gene (von Lindern *et al.,* 1992a). Since the fusion occurs within an intron in both genes the result is a new inframe DEK–CAN fusion protein of 165 kDa (von Lindern *et al.,* 1992a). A second translocation involving the *set* and *can* genes and which is associated with acute undifferentiated leukemia gives rise to a SET–CAN fusion protein of 155 kDa (von Lindern *et al.,* 1992b). While the functions of DEK and SET remain unknown, protein and cDNA sequence analysis has shown that CAN corresponds to a high-molecular-weight member of the WGA-binding nucleoporin family and which has been named NUP214 (Kraemer *et al.,* 1994). An overview of the CAN/NUP214 amino acid sequence is shown in Fig. 2. Its most notable features are a centrally located region bearing the hallmarks of a leucine zipper and in which the leukemia-associated breakpoint is located, and a number of XFXFG type repeats located predominantly in the C-terminal half of the molecule and interspersed with 12 copies of the tripeptide FGQ. Multiple copies of this tripeptide are also seen in the yeast NPC proteins NUP100 and NUP116 where it overlaps with a tetrapeptide repeat, GLFG (see below). This C-terminal segment of CAN/NUP214 is also rich in serine and threonine residues and may represent a region of the molecule that is modified with *O*-glcNAc. It has recently been shown that CAN/NUP214 is tightly associated with a novel 75-kDa NPC protein (Panté *et al.,* 1994). The latter is not a member of the O-linked glycoprotein family nor does it contain XFXFG repeats (R. Bastos and B. Burke, unpublished observations). Its notable features are an acidic amino terminal domain and a potentially helical carboxy-terminal domain, with a cysteine/histidine-rich central region that could function in metal binding. These two proteins, CAN/NUP214 and p75, may be eluted directly from NPCs as a discreet high Stoke's radius complex (Panté *et al.,* 1994). Depending upon conformation, the mass of this complex could be as high as 1–2 MDa. High-resolution immunoelectron microscopic studies have recently shown that CAN/NUP214, and by implication p75, is a component of the short filaments on the cytoplasmic face of the NPC (Panté *et al.,* 1994). These filaments have been strongly implicated as the initial sites of docking of NLS receptor/nuclear protein complexes prior to translocation across the NPC (Gerace, 1992). This raises the possibility that CAN/NUP214, or a protein with which it is associated, could be involved in NLS receptor recognition by the NPC. In any event, all of these observations open up new avenues of investigation of CAN/NUP214 function. As for its involvement in leukemogenesis, the role of CAN/NUP214 is still not clear. It is possible that DEK and SET are the real culprits and that the CAN/NUP214 C-terminal sequences causes these proteins to mislocalize. Alternatively, it may cause malfunction of a regulatory segment or perhaps even provide these proteins with a dimerization domain leading to the "activation."

***c. NUP153*** The complete cDNA sequence has been obtained for both rat (Sukegawa and Blobel, 1993) and human (McMorrow *et al.,* 1994) NUP153. This protein is unique among nucleoporins characterized to date in that it contains a central zinc finger domain of about 250 amino acids. Within this domain are four apparent $Cys_2$—$Cys_2$ zinc fingers with the consensus sequence $WXCX_2CX_3NX_6CX_2C$, and which at least *in vitro* are capable of binding DNA in a zinc-dependent manner (Sukegawa and Blobel, 1993). The spacing within these zinc fingers is somewhat unusual and there are only four other proteins in the data banks that exhibit such a motif, namely mdm-2 (Fakharzadeh *et al.,* 1991) and its human homolog and the products of the *Drosophila sol* gene (Delaney *et al.,* 1991) and the yeast *ARP* gene (Wehner *et al.,* 1993). The latter two are of unknown function. The carboxy-terminal domain of NUP153 contains numerous (about 30) XFXFG repeats and is extremely rich in hydroxyl amino acids (32%). It would seem likely, by analogy with p62, that this region of the molecule contains *O*-GlcNAc addition sites. The 600 residue amino terminal domain contains a small number (7) of degenerate XFXFG repeats but is otherwise not related in any obvious way to other proteins in the data banks. This domain is, however, essential for assembly into the NPC and appears to function at least in part in the formation of NUP153 homo-oligomers possibly involving the short potential helical segment indicated in Fig. 2 (Lin *et al.,* 1995). Such oligomers, which can be readily eluted directly from intact NPCs, behave as discreet particles with a Stoke's radius 9.7 nm (Panté *et al.,* 1994). This would be consistent with a mass of up to 1 MDa depending upon conformation. Immuno-EM studies have shown that NUP153 is localized exclusively to the nucleoplasmic face of the NPC (Cordes *et al.,* 1993; Sukegawa and Blobel, 1993; Panté *et al.,* 1994) where it is a component of the distal ring of the nuclear baskets (Panté *et al.,* 1994). The function of NUP153 is still a matter of speculation. However, it is conceivable given its DNA binding properties that it is involved in the organization of chromatin in the vicinity of the NPC. The location of NUP153 in the nuclear baskets is also very suggestive in light of the observation that these structures are destabilized by EGTA treatment, but reform upon addition of divalent cations, with $Zn^{2+}$ being the most efficacious (Jarnik and Aebi, 1991).

***d. POM121*** POM121 was identified as an integral membrane protein of the nuclear envelope that bound to WGA (Halberg *et al.,* 1993). Like NUP153, POM121 contains multiple (23) XFXFG pentapeptide repeats clustered toward the carboxy-terminus of the molecule, a region that is also rich in serine and threonine residues and as with p62 is thought to represent the location of most of the O-glcNAc. POM121 differs from other members of the O-linked glycoprotein family in that it possesses a 44-residue hydrophobic segment, beginning at position 28, which is flanked by basic residues. This segment of the molecule may form either one or two membrane-spanning $\alpha$-helices. In the latter case the helices would lie antiparallel leaving both the amino- and carboxy-terminal on one side of the membrane and a small polypeptide loop on the other side. In either case, the bulk of the molecule, greater than 90%, is predicted to be exposed to the cytoplasm with only a small segment extending into the perinuclear space (Fig. 3). The reverse situation is observed with another NPC transmembrane protein gp210, described below. Immunocytochemical studies clearly show that POM121 is a component of the NPC and must logically be localized in the NPC membrane domain (Fig. 5, color plate 11). While the precise function of POM121 is unknown, it most likely plays at least some role in anchoring the NPC in the nuclear membranes.

### 2. Other Vertebrate Nucleoporins

Sequence information is available for four other nucleoporins at the time of writing, gp210, NUP107, NUP155, and Tpr/p265. None of these proteins contains O-linked GlcNAc nor do they exhibit the XFXFG pentapeptide repeat or the GLFG repeats observed in yeast nucleoporins (below).

***d. Gp210*** Gp210 (Fig. 3) was the first NPC-specific protein to be identified (initially referred to as gp190) and was originally detected as an integral membrane protein of the nuclear envelope that bound to the lectin concanavalin A by virtue of endoplasmic reticulum (ER)-type high mannose N-linked oligosaccharides (Gerace *et al.,* 1982). Its location within the NPC membrane domain has been demonstrated by EM–immunocytochemistry (Gerace *et al.,* 1982). cDNA sequencing and subsequent topological analyses indicated that it is a conventional type I transmembrane glycoprotein with a single membrane-spanning domain close to the carboxy-terminus (Wozniak *et al.,* 1989). Its orientation is such that the bulk of the molecule comprising about 1800 amino acid residues, is located in the perinuclear space (PNS, Fig. 5, color plate II) with only a small, proline-rich cytoplasmic tail consisting of 58 amino acids at the carboxy-terminus (Greber *et al.,* 1990). The lumenal domain may well correspond to the "knobs," sometimes observed in ultrathin sections, extending into the PNS at the equator of the NPC (Jarnik and Aebi, 1991). The large size and abundance of gp210 (Gerace *et al.,* 1982), about 25 copies per NPC representing about 5–6 MDa, suggests that this protein alone contributes a significant fraction of that portion of NPC mass that is located within the PNS (at a radius of about 60–70 nm). The suggested role of gp210 is as an anchor for the NPC in the nuclear membranes, with its small carboxy-terminal domain presumably being engaged in the cytoplasmic region of the ring–spoke complex. In this respect it resembles a viral envelope spike glycoprotein such as the Semliki Forest Virus $E_2$ where a small cytoplasmic tail interacts with an underlying nucleocapsid (Zhao *et al.,* 1994). A second 23-amino acid residue hydrophobic segment is present in the large lumenal domain. However, this does not appear to be stably integrated into the membrane and in some respects is reminiscent of certain viral fusion peptides (White,

1990). A surprising feature of gp210 is that it is the transmembrane domain that determines sorting to the NPC membrane from its site of synthesis on the endoplasmic reticulum and outer nuclear membranes (Wozniak and Blobel, 1992). Presumably it is involved in lateral interactions (both homotypic and heterotypic) with other transmembrane helices. The cytoplasmic domain, it turns out, is only a weak sorting determinant. This property of the gp210 transmembrane domain brings to mind the sorting by kin-recognition proposed for certain integral membrane proteins of the Golgi apparatus (Nilsson and Warren, 1994). While the lumenal domain of gp210 seems to be relatively unimportant in terms of assembly at the NPC, this domain does seem to have some influence on the function of the NPC in terms of nuclear protein import. Microinjection into fibroblasts of mRNA encoding the heavy and light chains of a monoclonal antibody directed against a lumenal epitope of gp210 was found to inhibit both NLS-dependent transport and diffusion into the nucleus (Greber and Gerace, 1992). Presumably functional antibody assembled in the lumen of the ER and perinuclear space and was able to bind to the gp210 lumenal domain. Binding of antibody must then cause some form of conformational alteration, rotation about the axis of the transmembrane domain for instance, which might in turn modify the interactions of the cytoplasmic domain.

***b. NUP107 and NUP155*** NUP 107 (Radu *et al.,* 1994) and NUP155 (Radu *et al.,* 1993) are members of a group of about 30 proteins that can be extracted from nuclear envelopes in 2 *M* urea and do not bind to WGA (Radu *et al.,* 1993). Both have been shown to be NPC proteins by immunogold labeling of ultrathin cryosections and appear to be located somewhere in the central region of the NPC with NUP155 seemingly exposed on both nuclear and cytoplasmic faces. Neither protein exhibits any of the features, XFXFG repeats for instance, that are found in other nucleoporins. NUP107 is noteworthy in that it contains a leucine zipper at its carboxy-terminus. Whether this is involved in self-association or interactions with other proteins has yet to be resolved. At present there are no clues as to the function of either of these two proteins.

***c. Tpr/p265 and NUP180*** Tpr, short for translocated promoter region, was identified several years ago as a human gene that was rearranged in a number of tumors and transformed cells (Park *et al.,* 1986). It encodes a very large protein of about 265 kDa (Mitchell and Cooper, 1992). The most striking feature of this protein is a large extended $\alpha$-helical coiled-coil domain of about 1600 amino acids (similar in size to the myosin II rod domain) flanked by a small amino-terminal domain and a larger acidic carboxy-terminal domain. Tpr has been implicated in the activation of a number of proto-oncogenes, most notably *met* (Park *et al.,* 1986), *trk* (Greco *et al.,* 1992), and *raf* (Ishikawa *et al.,* 1987). This activation in each case involves fusion of a 5′ sequence of the Tpr gene to a segment of the proto-oncogene resulting in the formation of a hybrid gene encoding a chimeric protein. In the case of *raf* activation, for instance, the chimera produced consists of the N-terminal 232 amino acids of Tpr fused to the *raf* serine kinase domain. Employing a monoclonal antibody, RL30, raised against rat liver nuclear envelope extracts, it has recently been shown that there is a large nonglycosylated 265-kDa protein located on the cytoplasmic face of the NPC that is associated with the cytoplasmic filaments (Byrd *et al.,* 1994). In a rather remarkable parallel with CAN/NUP214, this protein has been shown by limited protein sequence analysis to be none other than the rat homologue of Tpr (Byrd *et al.,* 1994). So rearrangements in two quite distinct genes encoding very different NPC proteins are both associated with oncogenesis. The significance of this finding is at present completely obscure. As suggested for CAN/NUP214, the Tpr N-terminal segment may cause mislocalization of the various oncoproteins with which it has been shown to be associated. Alternatively it may interfere with regulatory sequences or it may provide them with a possible dimerization domain leading to inappropriate activation. The extensive coiled-coil structure of Tpr would be consistent with the latter possibility. In this regard it may be more than just a coincidence that the original identification of activated human *trk* revealed a hybrid protein consisting of the *trk* tyrosine kinase domain fused to tropomyosin, another coiled-coil polypeptide (Marin-Zanka *et al.,* 1986).

The Tpr/p265 protein has been found to be extremely sensitive to proteolysis, one of its breakdown products being a 175 kDa polypeptide (Byrd *et al.,* 1994). It has been suggested that this may well correspond to a 180-kDa protein (NUP180) identified by Wilken *et al.* (1993) using a human autoantibody from a patient suffering from an overlap connective tissue disease. NUP180, which is itself very sensitive to proteolysis, may be extracted from nuclear envelopes in 2 *M* urea and is therefore like Tpr/p265, classified as a peripheral membrane protein. It does not bind WGA so it is not a member of the O-linked glycoprotein family. Immunogold labeling of *Xenopus* oocyte nuclear envelopes has localized this protein to the short filaments on the cytoplasmic face of the NPC (Fig. 5, color plate 11). In terms of function, *in vivo* and *in vitro* experiments have shown no effect on anti-NUP180 antibodies on nucleocytoplasmic transport. Wilken *et al.* (1993) speculate that NUP180 may form an often suggested link between the NPC cytoplasmic filaments and the intermediate filament network (Carmo-Fonseca *et al.,* 1987). Alternatively, if p180 is indeed a breakdown product of Tpr/p265, it could as Byrd *et al.* (1994) suggest form the core of the NPC cytoplasmic filaments.

## B. Yeast Nucleoporins

The search for yeast nucleoporins initially followed strategies similar to those employed to identify the vertebrate proteins described in the previous section. In fact one of the first yeast nucleoporins to be described, NUP1p, was identified and its gene cloned using monoclonal antibodies raised against rat nucleoporins (Davis and Fink, 1990). In recent years, however, important advances have been made on two additional fronts. The first of these, largely practical in nature, has been the isolation of the yeast NPC in milligram quantities (Rout and Blobel, 1993). The enrichment achieved is 607-fold with a recovery of 45%. Both microscopically and biochemically these NPCs are roughly 90% pure. The significance of this achievement cannot be overstated since for the first time it has provided us with direct insight into the composition of the NPC in all its complexity. In addition to all but one (NUP2p) of the known yeast nucleoporins, the NPC preparation contains on the order of 80 uncharacterized proteins, each of which may be a nucleoporin. These can now be subjected to limited amino acid sequence analysis as a prelude to the isolation of their respective genes. The isolation of the yeast NPC may also provide material for high-resolution structural studies that will hopefully lead to the precise localization of yeast nucleoporins within the 3-D structure of the NPC.

The second advance is more conceptual, taking advantage of yeast genetics to identify new NPC proteins on the basis of their functional or physical interaction with known yeast nucleoporins. Such screens are founded on the premise that mutations in two genes that are not individually lethal, may in combination generate a lethal phenotype if their encoded proteins are required to associate or if they share some form of functional overlap or interdependent activity (Koshland *et al.*, 1985; Bender and Pringle, 1991). This phenomenon is referred to as synthetic lethality. This approach has been employed successfully to identify new nucleoporins as well as to establish interactions between previously identified proteins (Wimmer *et al.*, 1992; Loeb *et al.*, 1993; Bélanger *et al.*, 1994).

At the time of writing there are 12 yeast nucleoporins that have been characterized at the molecular level with several more soon to be described (summarized in Table II and Figs. 3 and 4, color plates 9 and 10). As with the vertebrate nucleoporins, family relationships are immediately obvious. Three of the yeast nucleoporins contain the type of XFXFG repeats that are characteristic of the family of NPC O-linked glycoproteins of vertebrates. The yeast proteins, however, appear not to be modified with *O*-GlcNAc (Kalinich and Douglas, 1989) and consequently do not bind to WGA. This finding is consistent with the fact that, in contrast to the situation in vertebrates, WGA does not block nuclear protein import in yeast cell-

TABLE II
Yeast NPC Proteins

| Name | Isolation | Sequence motifs | Properties and possible functions | References |
|---|---|---|---|---|
| XFXFG family | | | | |
| NSP1p (86 kDa) | Immunological (cloned by expression) | Central domain containing KPAFSFGAK repeats. Degenerate GLFG repeats in N-terminal domain. Essential C-terminal domain with heptad repeats indicating coiled-coil structure. Possible calcium-binding function | Essential NPC protein. Forms quaternary complex with NUP49p, NUP57p, and NIC96p | Hurt (1988); Nehrbass *et al.* (1990); Wimmer *et al.* (1992); Grandi *et al.* (1993) |
| NUP1p (113 kDa) | Immunological (cloned by expression) | Central domain containing XFXFG repeats | Essential NPC protein. Interacts with SRP1p | Davis and Fink (1990); Bélanger *et al.* (1994); Bogerd *et al.* (1994) |
| NUP2p (78 kDa) | Immunological (cloned by expression) | Central domain containing XFXFG repeats | Nonessential. Null mutant exhibits synthetic lethality with *NSP1* and *NUP1* alleles. Interacts with SRP1p | Belanger *et al.* (1994); Loeb *et al.* (1993) |
| GLFG family | | | | |
| NUP49p (49 kDa) | Immunological (cloned by expression), genetic (synthetic lethal screen) | N-terminal GLFG repeats. C-terminal helical domain | Essential. Forms subcomplex with NSP1p, NUP57p, and NIC96p. Gene exhibits synthetic lethal interaction with *nsp1ts* | Wente *et al.* (1992); Wimmer *et al.* (1992) |

| | | | | |
|---|---|---|---|---|
| NUP57p (57 kDa) | Genetic (synthetic lethal screen) | N-terminal GLFG repeats. C-terminal helical domain | Essential. Forms subcomplex with NSP1p, NUP49p, and NIC96p. Gene exhibits synthetic lethal interaction with *nsp1ts* | Grandi *et al.* (1993, 1995) |
| NUP100p (100 kDa) | Immunological (cloned by expression) | GLFG repeats in N-terminal domain. C-terminal region displays homologies with NUP116p and NUP145p | Nonessential. Synthetic lethal interaction with *nup116ts* | Wente *et al.* (1992) |
| NUP116p (116 kDa) | Immunological (cloned by expression), genetic (synthetic lethal screen) | N-terminal GLFG repeats. C-terminal region displays homologies with NUP100p and NUP145p in sequences comprising the nucleoporin RNA-binding motif (NRM) | Null mutant is thermosensitive and results in structural changes in the nuclear membranes. It is synthetic lethal with *nsp1*, *nup100*, and *nup145* alleles | Wente *et al.* (1992); Wimmer *et al.* (1992) |
| NUP145p (145 kDa) | Immunological (cloned by expression), genetic (synthetic lethal screen) | N-terminal GLFG repeats. Centrally located NRM (also found in NUP100p and NUP116p) | Essential. Binds RNA *in vitro*. Synthetic lethal interaction with *nsp1* alleles. N-terminal truncation gives rise to a nuclear membrane structural defect | Fabre *et al.* (1994); Wente and Blobel (1994) |
| Membrane proteins | | | | |
| POM152p (152 kDa) | Biochemical | Type II integral membrane protein with a large lumenal domain and small cytoplasmic tail | Nonessential. Role in anchoring NPC in the nuclear membranes? | Wozniak *et al.* (1994) |

(*continued*)

TABLE II (*continued*)

| Name | Isolation | Sequence motifs | Properties and possible functions | References |
|---|---|---|---|---|
| Others | | | | |
| NIC96p (96 kDa) | Biochemical | Helical N-terminal domain | Essential. Forms complex with NSP1p, NUP49p, and NUP57p. Synthetic lethal relationship with *nsp1ts*. Depletion of NIC96p results in inhibition of nuclear protein import | Grandi *et al.* (1993, 1995) |
| SRP1p (67 kDa) | Genetic | Central region contains eight *arm* repeats of 42 residues. Highly charged terminal domains | Essential. *srp1* alleles are suppressors of RNA pol1. Exhibits synthetic lethal interaction with *nup1* alleles. Associates with both NUP1p and NUP2p. Homolog of *Xenopus* "importin," an essential nuclear protein import factor | |
| NUP133p (133 kDa) | Genetic (synthetic lethal screen) | No striking features or homologies | Nonessential. Implicated in RNA export. Synthetic lethal interaction with *nup49ts*. NUP133p null cells exhibit NPC clustering | Doye *et al.* (1994) |

free systems (Kalinich and Douglas, 1989). A second family of yeast nucleoporins is characterized by the repeat GLFG, a motif that has not yet been observed in any of the vertebrate NPC proteins. However, vertebrate GLFG nucleoporins almost assuredly exist since certain of the yeast GLFG proteins were identified using a monoclonal antibody (MAb192) raised against rat liver nuclear envelope preparations and which recognizes vertebrate nucleoporins (Wente *et al.,* 1992). There at present only two yeast NPC-associated proteins that appear to have known vertebrate counterparts. The first is NSP1p (below) which bears some resemblance to p62 (described above). The second is the recently described *Xenopus* nuclear transport factor, importin, which is clearly a homolog of SRP1p (below). The fact that so few strong relationships have so far been established should not be viewed as evidence that the yeast and vertebrate NPCs are fundamentally different. More likely it is a reflection of the complexity of the NPC. To make a simple analogy, we currently are fishing in two pools containing many varieties of fish and using different bait. Perhaps it is not too surprising that at this relatively early stage we have yet to catch exactly the same species in both pools!

### 1. The XFXFG Family of Yeast Nucleoporins

***a. NUP1p*** NUP1p was identified using monoclonal antibodies (MAb306 and MAb350) raised against rat liver nucleoporins (Davis and Fink, 1990). On immunoblots of rat liver nuclear envelopes, MAb350 recognizes many members of the O-linked glycoprotein family of nucleoporins, its specificity resembling that of WGA. On blots of yeast nuclei a 130-kDa protein is recognized, albeit weakly, as well as a doublet at about 100-kDa plus several other smaller polypeptides. The 130-kDa protein is NUP1p while the pair at 100-kDa correspond to NSP1p and NUP2p (below). Cloning of the *NUP1* gene by expression using MAb350 revealed a protein with a calculated molecular weight of 113-kDa (Fig. 4). The most striking feature of the predicted amino acid sequence is a central region of about 600 amino acid residues containing 24 XFXFG repeats, which clearly brand this protein as a relative of the vertebrate O-linked glycoproteins. The average spacer length between each repeat is 13 residues, each of which contains multiple serines and threonines with the majority also exhibiting three or more charged residues. This central repeat region is flanked by a highly charged ~330-residue amino terminal domain, and a smaller, virtually uncharged, ~120-residue carboxy-terminal domain.

Disruption experiments indicate that *NUP1* is an essential gene (Davis and Fink, 1990). Similarly, overexpression of *NUP1* results in an arrest in cell growth. *nup1* mutant cells exhibit both import and export defects as well as gross changes in nuclear envelope morphology (Bogerd *et al.,* 1994).

Strains carrying *nup1* alleles consisting of partial truncation of either the N- or C-terminus are viable but grow more slowly than cells carrying the wild-type gene (Loeb *et al.,* 1993). These alleles exhibit synthetic lethality with a gene encoding another XFXFG protein, NUP2p, an observation thought to reflect partial functional overlap (Loeb *et al.,* 1993; see below). The *nup1* allele truncated at the carboxy-terminus (lacking the last 35 residues) also displays synthetic lethality with another gene SRP1 (Bélanger *et al.,* 1994; see below). Deletion of most of the C-terminal domain (88 residues) renders the protein nonfunctional (Bogerd *et al.,* 1994). At present the precise function of *NUP1* is unknown and its location within the NPC has yet to be determined.

***b. NSP1p*** NSP1p was originally identified as a nucleoskeletal-like protein (Hurt, 1988) and subsequently found, by immunocytochemistry, to be a component of the NPC (Nehrbass *et al.,* 1990). Cloning of the *NSP1* gene by expression using antibodies prepared against the yeast nuclear envelope revealed an encoded protein with a calculated mass of 86 kDa (Hurt, 1988). Like NUP1p, NSP1p exhibits a central core of multiple XFXFG repeats (22 in all). However, in contrast to all of the other XFXFG nucleoporins, both vertebrate and yeast, the NSP1p repeats extend up to 9 amino acids, and 14 out of the 22 show perfect sequence conservation of the nonapeptide KPAFSFGAK. Of the 21 spacer peptides, 19 are precisely 10 residues in length with charged and hydroxyl amino acid residues predominating. This central repeat region is flanked by an amino terminal domain resembling the end domains of cytokeratins and containing a series of degenerate GLFG repeats (below), and a carboxy-terminal domain which is partly homologous to the calcium-binding proteins parvalbumin (Coffee and Bradshaw, 1973; Jauregui-Adell and Pechere, 1978) and oncomodulin (MacManus *et al.,* 1983). These latter two proteins are members of the EF-hand family of calcium-binding proteins (Kretsinger, 1980). Whether NSP1p is a calcium-binding protein has yet to be resolved. However, residues in parvalbumin that coordinate $Ca^{2+}$ are conserved in NSP1p.

Like *NUP1, NSP1* is an essential gene, strong overexpression of which also results in growth arrest (Hurt, 1988). Depletion of NSP1p causes an inhibition of nuclear protein import and a concomitant decline in the density of NPCs in the nuclear envelope (Mutvei *et al.,* 1992). Similarly, cells harboring a temperature-sensitive allele of *NSP1* fail to import NLS-bearing proteins into the nucleus at the nonpermissive temperature (Nehrbass *et al.,* 1993). The essential function of NSP1p resides exclusively in its carboxy-terminal domain stretching from residues 603 to 823 (Nehrbass *et al.,* 1990). The remainder of the molecule consisting of the amino terminal domain and the entire central repeat domain appears to be completely dispensable. Cells that synthesize only the carboxy-terminal domain of NSP1p grow essentially normally. However, *NSP1* truncated in this manner exhibits synthetic lethality with deletion of the nonessential gene, *NUP2* (Loeb *et al.,* 1993), implying that certain functions of one may be duplicated in the other, i.e., that there is functional redundancy. Perhaps not surprisingly, this functional C-terminal domain of *NSP1* can cause relocation of a heterologus cytosolic protein to the nuclear envelope (Hurt, 1990).

A number of single amino acid changes in a predicted α-helical segment of the carboxy-terminal domain of NSP1p results in the acquisition of a ts phenotype (Nehrbass *et al.,* 1990). Certain of these ts alleles are synthetic lethal in combination with mutations in several additional nucleoporin genes, *NUP49* (Wimmer *et al.,* 1992), *NUP57* (Grandi *et al.,* 1995), *NUP116* (Wimmer *et al.,* 1992), *NUP145* (Fabre *et al.,* 1994), and *NIC96* (Grandi *et al.,* 1995), all of which are described in following sections. A number of other synthetic lethal mutants await characterization. Affinity chromatography employing the NSP1p carboxy-terminal domain fused to protein A has revealed the existence of an essential NPC subcomplex containing in addition to NSP1p, three other NPC proteins, NUP49p, NUP57p, and NIC96p (Grandi *et al.,* 1993). The discovery of this subcomplex provides an obvious basis for the synthetic lethality of *nsp1ts* observed in combination with *nup49, nup57,* and *nic96* alleles. At present little is known of the precise location of the subcomplex within the NPC. However, there is some evidence to suggest that NSP1p is exposed on the cytoplasmic face (Schlenstedt *et al.,* 1993).

***c. NUP2p*** The most recent member of the yeast XFXFG family of nucleoporins to be described is NUP2p (Loeb *et al.,* 1993). Like *NUP1,* the *NUP2* gene was obtained in an expression screen employing MAb350. The *NUP2* gene encodes a 720 residue protein with a predicted mass of 78 kDa. It is clearly related to NUP1p and NSP1p in that it exhibits a central domain containing 16 XFXFG repeats (Fig. 4). NUP2p is highly charged (~29% charged residues) but with a roughly neutral p*I.* Neither the amino- or carboxy-terminal domains, each of about 175 residues, shows any obvious resemblance to those in NUP1p or NSP1p. The NUP2p carboxy-terminal domain does, however, exhibit some sequence similarity to the human Ran-binding protein-1 (Ran BP1, Coutavas *et al.,* 1993), amounting to 25% identity with an additional 49% of conservative changes out of a total of 162 residues (B. Burke, unpublished observations). Ran BP1 is a small (24 kDa, 203 residue) protein that associates with the GTP-bound form of the small *ras*-related GTP binding protein Ran/TC4 (Drivas *et al.,* 1990; Bischoff and Ponstingl, 1991). RanBP1 does not appear to possess GAP activity and is thought to be a Ran/TC4 effector molecule. While the precise function of Ran/BP1 is not known, there are indications from its sequence that it may be capable of binding RNA. Interestingly, the highest degree of similarity with NUP2p is in the region of a putative RNA-binding motiff. The significance of all of this is that Ran/TC4 has been clearly shown to play a role in the process of nucleocytoplasmic transport (Melchior *et al.,* 1993; Moore and Blobel, 1993). This obviously raises the possibility that *NUP2* might actually interact with the yeast Ran/TC4 homolog.

Whatever the function of NUP2p, Ran/TC4 effector or not, it is clearly redundant since yeast strains in which *NUP2* is deleted exhibit normal growth rates with no apparent morphological abnormalities or any other discernible phenotype (Loeb *et al.,* 1993). Furthermore, overexpression of *NUP2* causes only a partial growth inhibition, compared with the complete arrest in growth induced by excess NUP1p. *NUP2* deletion is, however, lethal when combined with nonlethal mutations in certain other nucleoporin genes, most notably *NUP1* and *NSP1* (Loeb *et al.,* 1993). This can be at least partly accounted for by a certain degree of functional overlap. For instance, *nup1* mutants yielding either amino-terminal or carboxy-terminal truncations exhibit synthetic lethality with a *nup2* null mutant. However, at least with the amino-terminally truncated NUP1p (lacking residues 4–141), the lethal phenotype can be rescued by expression of the first 174 amino acid residues of NUP2p, indicating that the amino-terminal regions of both of these molecules must be functionally redundant. Similarly, cells that express NSP1p-C (consisting only of amino acid residues 606–823, and which lack the amino terminal and central repeat domains, residues 1–605) are inviable in combination with a *nup2* null mutation, but are rescued by expression of amino acid residues 1–174 of NUP2p, again suggesting functional overlap of the amino-terminal regions. Finally, in common with NUP1p, NUP2p has been found to interact with another nucleoporin, SRP1p (Bélanger *et al.,* 1994; see below).

### 2. The GLFG Family of Yeast Nucleoporins

At present there are five known yeast nucleoporins characterized by numerous copies of the tetrapeptide, GLFG. Two complementary approaches have been used in the identification of these proteins. The first has involved employing a monoclonal antibody (MAb192) raised against rat nucleoporins to isolate genes encoding cross-reacting yeast proteins (Wente *et al.,* 1992). On immunoblots of yeast nuclei it recognizes a number of polypeptides with apparent molecular weights ranging from 49 to 118 kDa. The second approach has involved screening for mutant genes that are lethal only in combination with a nonlethal *nsp1* mutant (Wimmer *et al.,* 1992). As is the case with the XFXFG family, GLFG nucleoporins exhibit a certain degree of functional redundancy.

***a. NUP49p*** The *NUP49* gene was identified by expression using the MAb192 monoclonal antibody (Wente *et al.,* 1992), as well as appearing in a synthetic lethal screen in combination with a temperature-sensitive a mutant of *NSP1* (Wimmer *et al.,* 1992). The *NSP1* mutation employed in this screen results a single amino acid substitution (leucine to serine at position 640) in the essential carboxy-terminal domain, a region of the NSP1p molecule now known to interact with NUP49p, as well as NIC96p and NUP57p (Grandi *et al.,* 1993).

As predicted from the DNA sequence, *NUP49* encodes a protein of 472 amino acid residues with a calculated molecular weight of 49 kDa (Fig. 4). Inspection of the amino acid sequence reveals that the amino-terminal half contains 14 repeats with the sequence GLFG or some conservative variation thereof. This region of the molecule contains only a few (7) basic amino acids and is completely devoid of acidic residues. It also contains multiple epitopes recognized by MAb192 (Wente *et al.,* 1992). The carboxy-terminal half of the molecule contains a far higher charge density and exhibits a series of heptad repeats consistent with a coiled-coil conformation. Genetic evidence would suggest that it is this region of the molecule that interacts with the NSP1p carboxy-terminal domain.

Gene disruption experiments have demonstrated that NUP49p is an essential component of the NPC (Wente *et al.,* 1992; Wimmer *et al.,* 1992), and while it has not yet been assigned a specific function, *nup49 ts* mutants have been identified that display differential inhibition of both nuclear protein import (*nup49–313*) and RNA export (*nup49–316*). The latter mutant has recently been employed in a synthetic lethal screen leading to the identification of an additional NPC protein NUP133p, which also appears to have a role in the RNA export process (Doye *et al.,* 1994; see below).

***b. NUP57p*** NUP57p was originally described as a 54 kDa nup recognized by MAb192 (Wente *et al.,* 1992) and which could be recovered in a quaternary complex (described above) with NSP1p, NUP49p, and NIC96p (Grandi *et al.,* 1993). The gene was finally isolated in a synthetic lethal screen with *nsp1ts* (Grandi *et al.,* 1995). Sequence analysis predicts a protein of 541 residues (57 kDa). The amino-terminal half of the protein contains three XFXFG repeats followed by nine GLFG repeats with spacers of up to 33 residues. The presence of the GLFG repeats is consistent with the reactivity with MAb192. The carboxy-terminal half of the molecule contains a series of heptad repeats indicating that this region of NUP57p may form a coiled-coil secondary structure, and suggests a role in oligomerization. Gene disruption experiments indicate that NUP57p is essential. However, its vital functions seem to reside in the C-terminal domain, since cells expressing only a 20-kDa C-terminal segment are viable albeit temperature sensitive.

***c. NUP100p*** *NUP100* was identified by screening a yeast expression library with MAb192 (Wente *et al.,* 1992). It encodes a protein of 959 amino acid residues with a calculated mass of 100 kDa and a p*I* predicted to be about 9.4 (Fig. 4). Like NUP49p and NUP57p, the sequence can be divided into two obvious regions. The amino-terminal segment consisting of about 600 residues contains a total of 29 GLFG-type repeats, as well as a number of copies (14) of the dipeptide FG, most of the latter being clustered within the first 250 residues. This GLFG region of the molecule is largely charge free with only a few basic residues and one acidic residue (an aspartic acid). The carboxy-terminal domain is highly charged in comparison and contains segments (extending from residue 799 to the carboxy-terminus) that exhibit significant homology to sequences in NUP1 16p (Wente *et al.,* 1992; Wimmer *et al.,* 1992) and NUP145p (Fabre *et al.,* 1994; see below). At least in the case of the latter two nucleoporins, these sequences have been shown to possess the capacity to bind RNA, and are referred to as nucleoporin RNA-binding motifs (RNMs, Fabre *et al.,* 1994).

*NUP100* is a nonessential gene, since null cells exhibit no apparent growth defects nor is there evidence of gross morphological changes (Wente *et al.,* 1992). The *nup100* null mutant is, however, lethal in combination with a temperature-sensitive *nup116* null mutant (Wente *et al.,* 1992). The inference is that NUP100p and NUP116p share some degree of functional overlap. The *nup100* deletion also intensifies a phenotype involving structural changes in the nuclear envelope, which is observed with an N-terminal deletion of NUP145p (below) and also renders these cells temperature sensitive (Wente and Blobel, 1994).

***d. NUP116p*** Like NUP49p, NUP116p was identified using MAb192 (Wente *et al.,* 1992) as well as appearing independently in the *nsp1* synthetic lethal screen (Wimmer *et al.,* 1992). In terms of its organization, the 1113 residue NUP116p polypeptide (predicted p*I* ~9.3) closely resembles NUP100p (Fig. 4). The region of the molecule consisting of residues 200–715 contains all of the 33 GLFG repeats and is largely devoid of charge. There are also 17 FG dipeptides, more that half of which are contained within the first 200 residues. The amino acid sequence consisting of the ~150 residues extending to the carboxy-terminus exhibits a high degree of homology to the carboxy-terminal region of NUP100p as well as to a centrally located segment of NUP145p (below) comprising the nucleoporin RNA-binding motif. This segment of NUP116p has been shown to bind RNA *in vitro* with a preference for poly(G) (Fabre *et al.,* 1994). It does not bind single- or double-stranded DNA.

Deletion of *NUP116* renders cells temperature sensitive. At 30°C *nup116* null cells grow about two to three times more slowly than their wild-type counterparts while at 37°C the deletion is lethal (Wente *et al.,* 1992; Wente and Blobel, 1993; Fabre *et al.,* 1994). The double deletion of *NUP116* and the nonessential *NUP100* is apparently lethal at all temperatures, indicative of some level of functional overlap of their encoded proteins. The temperature-sensitive *nup116* deletion mutant (*nup116Δ*) exhibits a remarkable phenotype. Upon shift of the *nup116Δ* cells to the restrictive temperature there is an immediate decline in RNA export. Electron microscopic examination of the nuclear envelopes both before and after the temperature shift provides a clear structural basis for this change in RNA transport. At the permissive temperature, essentially normal NPCs are observed as well as NPCs contained in intranuclear annulate lamellae extending from the inner nuclear membrane. Following incubation at 37°C, however, the NPCs become physically occluded. Each NPC appears sealed on the cytoplasmic face with a double membrane segment that is continuous with the nuclear membranes. Wente and Blobel (1993) refer to these very aptly as sealed herniated NPCs. These herniated NPCs are apparently still transport competent since export material accumulates over time between the cytoplasmic face of the NPC and the occluding membranes. These observations suggest that NUP116p must be somehow involved in maintaining the stability of the NPC within the nuclear membranes, although it is not itself a membrane protein.

***e. NUP145p*** As was the case with NUP49p and NUP116p, NUP145p, the fifth member of the GLFG family, was also identified both immunologically and in a synthetic lethal screen (Fabre *et al.,* 1994; Wente and Blobel, 1994). *NUP145* encodes a protein of 1314 residues, the first 220 of which contain 12 GLFG repeats with only a few basic residues and none that are acidic. There is a preponderance of acidic residues in the carboxy-terminal region of the protein which endows it with a calculated p*I* of 5.45, similar to that of NUP49p. NUP145p contains an NRM homologous to that found at the carboxy-terminus of both NUP100p and NUP116p. However, the NUP145p NRM is positioned centrally between residues 453 and 604 leaving a large ~700-residue carboxy-terminal domain of unique sequence. Like the NUP116p NRM, that found in NUP14p binds RNA *in vitro* with a preference for poly(G) (Fabre *et al.,* 1994).

Gene disruption experiments have demonstrated that *NUP145* encodes an essential protein (Fabre *et al.,* 1994). Repression of synthesis *in vivo* is accompanied by a decline in RNA export followed by a similar decline in nuclear protein import (Fabre *et al.,* 1994). Whether the decline in import is secondary to the export defect has yet to be resolved. What is clear, however, is that the essential function of NUP145p does not reside in the NRM, since this region of the molecule can be deleted without compromising cell viability (Fabre *et al.,* 1994). This is also true in double mutants involving similar NMR deletions of NUP100p and/or NUP116p. Cells harboring the triple NRM deletion (in NUP100p, NUP116p, and NUP145p) are still viable but only at reduced temperature (Fabre *et al.,* 1994).

A *nup145* mutant (*nup145Δ N*) encoding an N-terminally truncated protein lacking the entire GLFG domain as well as 50% of the NRM gives rise to cells with a nuclear envelope structural defect similar to that observed with *nup116Δ* involving herniated NPCs (Wente and Blobel, 1994). In this case, however, the herniated NPCs appear in lobulated clusters and furthermore the defect is not lethal, doubling time of cells harboring the mutant gene being only slightly longer than that of wild-type cells. Since exported material does not seem to accumulate in the herniations, it is inferred that the NPCs in the *nup145ΔN* cells are not completely occluded (Wente and Blobel, 1994). The *nup145ΔN* mutant is a synthetic lethal with *nup116Δ* (which alone renders cells temperature sensitive). Deletion of the nonessential *NUP100* renders *nup145ΔN* temperature-sensitive lethal. At the permissive temperature in these cells harboring the double mutation, the *nup145ΔN* structural phenotype appears more severe (Wente and Blobel, 1994). It is clear from the foregoing discussion that *NUP100, NUP116,* and *NUP145* all interact genetically. Furthermore *NUP116* and *NUP145* must interact in some manner with *NSP1,* since a synthetic lethal screen employing *nsp1ts* turned up both of these genes (Wimmer *et al.,* 1992; Fabre *et al.,* 1994). At present, however, the molecular basis of any of these interactions has yet to be fully understood.

### 3. Nucleoporins That Interact with XFXFG and GLFG Family Members

Three additional yeast NPC proteins have been identified on the basis of both physical and genetic interaction with different members of the XFXFG and GLFG nucleoporin families. None of these proteins contains sequence motifs characteristic of other nucleoporins.

***a. NIC96p*** NIC96 (short for nucleoporin interacting component of 96 kDa) is a polypeptide of 839 amino acids that contains neither the XFXFG nor the GLFG repeats characteristic of many other nups (Fig. 4). It was originally identified by affinity chromatography using the essential carboxy-terminal domain of NSP1p and forms a quarternary complex with, in addition to NSP1p, two GLFG proteins, NUP49p and NUP57p (Grandi *et al.,* 1993). *NIC96* is an essential gene, alleles of which are, not surprisingly, synthetic lethal of *nsp1ts* (Grandi *et al.,* 1995). Depletion of *NIC96,* which encodes a very abundant NPC component (Rout and Wente, 1994), was found to result in failure to import nuclear proteins, while at the same time RNA export was unaffected (Grandi *et al.,* 1995).

The amino terminal region of NIC96p extending to residue 150 exhibits a series of heptad repeats with hydrophobic residues at positions 1 and 4, a characteristic of polypeptides that form $\alpha$-helical coiled coils (Grandi *et al.,* 1993). The finding that each of the other components of the NSP1p subcomplex (NSP1p, NUP49p, and NUP57p) contains a similar domain suggests that this may form the basis for their mutual interactions. In the case of NIC96p there is now some experimental evidence based on analysis of N- and C-terminal deletions to support such a notion (Grandi *et al.,* 1995). The central region of the NIC96p molecule extending from residues 151 to 530 contains three hydrophobic segments, with the most striking lying between residues 322 and 342. Deletion of this 20-residue segment renders the molecule nonfunctional, while certain point mutations in this region result in the acquisition of a temperature-sensitive phenotype (Grandi *et al.,* 1995). These hydrophobic stretches are each somewhat reminiscent of the membrane-spanning domain of transmembrane proteins, although there is no evidence to suggest such a role in NIC96p. This central region of the molecule also contains a putative NLS (526–530). The C-terminal domain of NIC96p consisting of about 300 residues is not essential for cell viability. However, the defects that give rise to synthetic lethality with *nsp1* and *nup57* alleles are located in this region of the molecule. The inference is that while the C-terminal domain of NIC96p is not required for the association with NSP1p/NUP49p/NUP57p, it becomes essential when other members of the subcomplex contain nonlethal defects.

At the time of writing, neither the function nor the precise location of these four proteins (NIC96p, NSP1p, NUP49p, NUP57p) within the NPC have been determined. However, the fact that deletion of any one of them is lethal suggests that the subcomplex in which they are associated has both a unique and essential role in the activity of the NPC. Based upon the similarities of NSP1p with vertebrate p62 it has been suggested that the NSP1p/NUP49p/NUP57p/NIC96p subcomplex may have a function related to that of the vertebrate p62/58/54 subcomplex (Grandi *et al.,* 1995). If this is the case, by analogy with the latter, the yeast subcomplex (or at least NSP1p) may be located in the vicinity of the central channel and exposed on both faces of the NPC (see Fig. 5, color plate II).

***b. SRP1p*** An allele of *SRP1* was originally identified as a suppressor of temperature-sensitive RNA polymerase I mutations (Yano *et al.,* 1992). This study also demonstrated that *SRP1* is an essential gene and that the protein it encodes is associated with the nuclear envelope in large macromolecular complexes. The gene subsequently reappeared in a synthetic lethal screen utilizing a *nup1* mutant encoding a C-terminally truncated protein (Bélanger *et al.,* 1994). The deduced amino acid sequence reveals of 542-residue polypeptide with three obvious regions. The amino-

and carboxy-terminal regions of the molecule (120 and 80 residues, respectively) are highly charged (~25% charged residues). In contrast, the central region is more hydrophobic containing eight repeats, each of 42 amino acids, that display significant similarity to the *"arm"* motif found in the *D. melanogaster armadillo* segment polarity gene product as well as β-catenin and plakoglobin. These are proteins that link cadherins to the cytoskeleton at intracellular junctions. The *arm* motif is also found in the exchange factor for Ras-related small G proteins that might have some significance in terms of SRP1p function (below).

SRP1p has recently been shown to be a yeast homologue of the *Xenopus* nuclear protein import factor, importin, with which it shares 44% sequence identity, as well as the human Rch1 protein (Görlich *et al.,* 1994). Importin was identified as a component of *Xenopus* egg cytosol essential for the import of basic domain NLS bearing proteins into nuclei *in vitro.* Importin functions at an early step in the transport process involving energy-independent docking of the transport substrate at the nuclear periphery. Subsequent translocation requires Ran/TC4 plus ATP and GTP. Combined, these two proteins, in the presence of nucleotides, can replace cytosol in *in vitro* nuclear import assays. From these results it is evident that importin has characteristics expected of an NLS receptor. It has not, however, been demonstrated that importin actually binds NLSs.

The relationship between SRP1p and importin raises an apparent paradox. SRP1p has been shown by immunofluorescence microscopy to be associated with the nuclear periphery, almost certainly in NPCs (Yano *et al.,* 1992). Indeed, direct physical interaction with two nucleoporins has been convincingly documented (below). Importin on the other hand has been suggested to be largely cytosolic (Görlich *et al.,* 1994). Certainly the nuclear envelopes of permeabilized cells have insufficient associated importin to provide activity in *in vitro* nuclear protein import assays (Görlich *et al.,* 1994). The resolution is probably that importin/SRP1p is only relatively loosely associated with the NPC, cycling on and off during the transport process. In fact some soluble SRP1p was observed upon fractionation of yeast cells (Yano *et al.,* 1992), and it may well be that importin is easily lost under the conditions employed to obtain transport-competent nuclei in digitonin-permeabilized cells.

Biochemical analyses have demonstrated a direct association between NUP1p and SRP1p that occurs in part via the NUP1p carboxy-terminal domain (Bélanger *et al.,* 1994). A direct interaction has also been demonstrated between SRP1p and NUP2p. However, the interactions with NUP1p and with NUP2p are mutually exclusive, a single SRP1 polypeptide associates with one or the other, not both simultaneously. There are at least two interpretations for these observations. The first is that these two different complexes may perform redundant functions. An alternative, however, is that they may portray two dynamic transport states of the NPC. For instance it is not beyond the bounds of possibility that during the transport cycle of the NPC, SRP1p may hand off from NUP1p to NUP2 or vice versa. The appearance of these two complexes might then represent our first glimpse of the NPC at a mechanistic level. It must be admitted, however, that the first interpretation involving functional redundancy is more easily reconciled with the finding that NUP2p is nonessential. As a piece of pure speculation, the interaction between NUP2p and SRP1p could be modulated by the yeast Ran/TC4 homologue, in light of the sequence similarities of NUP2p and SRP1p with RanBP-1 and the GTP–GDP exchange factor for Ras-related small G proteins, respectively. It is also possible that it is the interaction of SRP1p with transport substrates that is under G protein control. If, as suggested for importin, SRP1p does indeed interact with proteins containing a basic domain NLS (perhaps as a receptor), at some point during the transport process the NLS-bearing protein must be released, while at the same time SRP1/importin must be cycled back to the cytosol. This event could be mediated by Ran/TC4.

*c.* ***NUP133p*** *NUP133* was identified in a synthetic lethal screen employing a temperature-sensitive allele of *NUP49* (*nup49–316,* Doyle *et al.*, 1994). Cells harboring *nup49–316* are defective in RNA export at the restrictive temperature. Sequence analysis of *NUP133* indicates that it encodes an acidic protein of 1157 amino acid residues with a predicted molecular weight of 133 kDa. It contains two short (16–18 residue) hydrophobic stretches of undefined function, but otherwise exhibits no striking features or homologies. Immunofluorescence experiments suggest that NUP133p is indeed a component of the NPC although its precise localization within the structure remains unknown. Cells lacking NUP133p are viable but temperature sensitive. At 30°C they grow two or three times more slowly than wild-type cells, while at 37°C growth is completely arrested. At 18°C no growth defect is observed. The inhibition of growth at elevated temperatures is accompanied by the accumulation of nuclear poly(A)$^+$ RNA. In contrast, import of proteins into the nucleus appears unaffected. Ultrastructural examination of cells lacking NUP133p has shown that the NPCs are extensively clustered. Generally this clustering does not feature altered nuclear membrane structure of the type observed in *nup145ΔN* cells (Wente and Blobel, 1994; see above). In addition it appears to be independent of the export defect since it occurs even at 18°C at which temperature nuclear accumulation of RNA is not observed (Doye *et al.*, 1994).

The basis for the synthetic lethality when *nup133* and *nup49* mutants are combined is not clear. At present there is no evidence for a direct association between the two proteins. Doye *et al.* (1994) suggest that NUP133p and the

NSP1p/NUP49p/NUP57p/NIC96p subcomplex may possess overlapping or redundant functions with respect to mRNA export.

#### 4. Yeast NPC Membrane Proteins

To date only one integral membrane protein of the yeast NPC has been characterized. This protein, POM152p, was identified as a concanavalin A-binding glycoprotein that cofractionates with the yeast NPC (Wozniak *et al.,* 1994). Its location within the NPC has been confirmed by immunogold labeling of both ultrathin sections and isolated NPCs. Sequence analysis of the *POM152* gene predicts a protein of 1337 residues (Fig. 3) with a single potential membrane-spanning domain of 20 amino acid residues beginning at position 176 (Wozniak *et al.,* 1994). There are five consensus sequences for N-linked oligosaccharide addition, two on the amino-terminal side and three on the carboxy-terminal side of the transmembrane segment. Protein sequence data suggest that one of the latter sites, at position 280, is indeed glycosylated. This would indicate that POM152p has the orientation of a type II membrane protein with a carboxy-terminal domain of 1142 residues extending into the PNS and an amino-terminal domain of 175 residues on the pore side of the membrane. This orientation would be consistent with the fact that POM152p does not have a cleaved signal sequence. POM152p bears no similarity to any other proteins with the exception of a stretch of 19 residues in the amino-terminal domain adjacent to the membrane-spanning segment, which exhibits ~50% identify to a short sequence in the cytoplasmic domain of vertebrate POM121p (Hallberg *et al.,* 1993). The significance of this similarity is unclear. The only other notable feature of the POM152p sequence is the presence in the carboxy-terminal domain of eight repeats of ~24 amino acid residues. Each repeat is separated by a spacer region of ~100 residues. The function of these repeats is unknown.

A surprising finding is that *POM152* is nonessential. Indeed cells harboring a *pom152* disruption display no discernible phenotype or growth defect at a variety of temperatures (Wozniak *et al.,* 1994) despite the fact that the gene encodes a major (by mass) NPC component (Rout and Wente, 1994). However, overexpression of the gene results in an inhibition of cell growth. In terms of function, POM152p is likely to play at least some role in anchoring the NPC in the nuclear membranes, as is suggested for the vertebrate nucleoporins gp210 and POM121. Presumably the small cytoplasmic domain of POM152p engages in and forms part of the inner (i.e., nonlumenal) region of the ring–spoke complex. Finally, POM152p expressed in mammalian cells becomes correctly localized indicating that the sorting and assembly pathway has been conserved between yeast and mammals (Wozniak *et al.,* 1994).

## V. NPC Dynamics

During interphase in higher eukaryotes, and in particular during S-phase, the nuclear envelope surface area roughly doubles in preparation for the subsequent mitosis. Since the density of NPCs within the nuclear envelope remains roughly constant, new NPCs must be inserted *de novo* (Maul *et al.,* 1971, 1972; Maul, 1977). It is not hard to imagine that the biosynthetic assembly of an NPC, composed as it is of dozens of subunits, is a complex multistep process probably involving a number of assembly intermediates. During mitosis, the nuclear envelope is disassembled to allow chromosomes access to the mitotic apparatus. On completion of chromatid segregation, all of the disassembled nuclear envelope components are reutilized of two new nuclear envelopes, one in each daughter cell. Since morphologically identifiable NPCs cannot be seen in mitotic cells (Zeligs and Wollman, 1979), the inference is that like all other elements of the nuclear envelope, they too undergo a cycle of mitotic disassembly followed by reassembly. The implication of these observations, expanded upon below, is that there are two distinct assembly pathways for NPCs, one operative during interphase and the other operative at the end of mitosis.

### A. Biosynthesis and Assembly of Nucleoporins

The assembly of newly synthesized nucleoporins during interphase occurs in a preexisting closed structure, and as such raises some intriguing topological problems. NPCs surround aqueous channels through the nuclear envelope where the inner and outer membranes are joined (Figs. 1 and 5, color plate II). Therefore, if formation of a new NPC, including the aqueous channel, were to occur from scratch it must at some point involve a fusion event between the inner and outer nuclear membranes. Such a progress might involve a specific fusion protein (or complex of proteins) that could well be a component of the NPC. If an NPC component were to be involved it is likely to be an integral membrane protein, since the fusion event must be initiated at the membrane surfaces facing the perinuclear space. Possible candidates would be the nucleoporins gp210 or POM121. Gp210 in fact has certain appealing features for the performance of such a function, most notably a stretch of ~20 hydrophobic amino acids located within the large lumenal domain. As pointed out by Greber *et al.* (1990), it is somewhat reminiscent of the fusion peptide of certain viral membrane proteins (White, 1990). Such a function for gp210 is, however, highly speculative, and has yet to be addressed experimentally. An alternative scheme for interphase

NPC assembly could involve lateral division and growth of existing NPCs. This model, which would sidestep the issue of inner and outer nuclear membrane fusion, might involve incorporation of additional subunits into an NPC to give a rotational symmetry greater than eight followed by division along a plane perpendicular to the nuclear membranes. It is also possible that the division could occur first followed by incorporation of new subunits into the daughter NPCs. Either division scenario predicts that occasionally NPC-like structures should be observed that do not exhibit the typical eightfold symmetry. What events might drive incorporation of additional subunits or force division is not at all clear. It is worth pointing out that even this division model of NPC assembly must involve membrane fusion. However, in this case it would involve the nuclear membrane segment facing the NPC channel and which is exposed to the cytosol. This clearly raises the possibility of the involvement of well-described cytosolic membrane fusion factors in interphase NPC assembly (Rothman, 1994; Rothman and Warren, 1994).

Only a few studies have been carried out on the biosynthesis of NPC proteins, and the majority of these have concerned p62. Davis and Blobel (1987) originally showed that it is synthesized as a soluble precursor and is assembled into the nuclear envelope only very slowly with a halftime of several hours. It is certainly glycosylated prior to assembly since nonglycosylated p62 cannot be detected even after only a brief labeling period (Davis and Blobel, 1987). This observation is consistent with the findings that the GlcNAc transferase is a cytosolic enzyme (Haltiwanger *et al.,* 1992), and that *in vitro* synthesized p62 acquires GlcNAc very rapidly, either contranslationally or within 5 min of synthesis (Starr and Hanover, 1990; Cordes *et al.,* 1991). Whether *O*-GlcNAc plays any role in modulating *de novo* assembly has yet to be addressed. It is also not yet known whether p62 associates with p58 and p54 before or after incorporation into the NPC. However, another subcomplex of the NPC, CAN/NUP214-p75, forms very rapidly after synthesis (within 15 min, R. Bastos and B. Burke, unpublished observations), seemingly prior to assembly. The kinetics of assembly of the integral membrane proteins of the NPC have not been examined; however, at least gp210 (and probably the others) is synthesized on membrane-bound polysomes and inserted cotranslationally into the ER or outer nuclear membrane from where it would be free to diffuse laterally until captured at an NPC assembly site. It would seem likely that for a structure as complex as the NPC, both soluble and membrane-associated subcomplexes or modules would initially form and that it is these that ultimately associate to form a mature NPC.

Most *de novo* NPC assembly must occur during S-phase. However, it is not clear whether nucleoporin synthesis is restricted to this phase of the cell cycle. If in fact synthesis is not regulated in this manner, it would have to be argued that assembly was somehow tied to the progression of the cell cycle. Continuous synthesis with a restricted period of assembly might in part account for the protracted assembly halftime measured for p62 (Davis and Blobel, 1987).

### B. NPCs during Mitosis

The fate of NPCs during mitosis in higher eukaryotes has yet to be satisfactorily elucidated. However, it is clear that their disappearance from the nuclear membranes during prometaphase is a consequence of their partial disassembly. All NPC components that have been looked at to date become dispersed throughout the mitotic cytoplasm. There have been none convincingly documented that remain associated with chromosomes. The disassembly process most likely occurs to the level of subcomplexes and not to single polypeptides since it is clear that at least p62 remains associated with p58 and p54 in *Xenopus* eggs (Finlay *et al.,* 1991). NPC membrane components appear to be restricted to a subset of nuclear envelope-derived vesicles in mitotic cells. Vigers and Lohka showed that formation of nuclear envelopes *in vitro* requires two dissimilar nuclear envelop-derived vesicle populations, one of which is enriched in NPC components (Vigers and Lohka, 1991). More recently it has been demonstrated that an inner nuclear membrane protein, the lamin B receptor, resides in vesicles that are distinct from those that contain gp210 (Chaudhary and Courvalin, 1993). This finding is reinforced by the observation that reassembly of the lamin B receptor at the end of mitosis actually precedes that of the NPC membrane protein (Chaudhary and Courvalin, 1993).

Mitotic reassembly of NPCs does not pose the same topological problems encountered in *de novo* assembly since the NPC membrane components are located in discrete vesicles rather than in a single closed membrane surface. Reformation of an NPC at the end of mitosis simply involve docking of a group of such vesicles about the circumference of a central core formed from the soluble components of the ring–spoke complex. Fusion of adjacent vesicles would then regenerate an intact NPC. Other schemes involving preassembly of half pore complexes (cytoplasmic and nucleoplasmic) followed by their association to form a mature NPC have also been suggested (Sheehan *et al.,* 1988).

The question of the regulation of NPC dynamics during mitosis has yet to be addressed in detail. In all likelihood disassembly of the NPC during prometaphase will involve phosphorylation of key NPC components by mitotically activated kinases, followed by dephosphorylation during telophase. It is possible that, in the case of the vertebrate O-linked glycoproteins, phorphorylation of certain sites might require prior removal of *O*-

GlcNAc. In this way the presence of *O*-GlcNAc might impose an extra layer of control on the assembly status of the NPC. At present, however, such a notion remains purely speculative.

## VI. Future Directions

Considerable progress has been made in recent years in the study of nucleocytoplasmic transport. In particular new NPC components are being described at a remarkable rate. While the field is clearly still in what can only be described as its inventory phase, the tools and methodologies are now becoming available with which to address mechanistic questions. Genetic and biochemical analyses are beginning to define interactions among nucleoporins almost as fast as they are identified while high-resolution immunocytochemical studies are being perfomed to map individual nucleoporins as well as complexes of nucleoporins in the 3-D structure of the NPC. The development of a variety of *in vitro* transport systems is shedding new light on the role of soluble components of the transport process. The key to further understanding of nucleocytoplasmic transport is the convergence and refinement of these varied approaches in which structure and biochemistry are united. This is best appreciated if one recalls that to traverse the entire NPC a macromolecule must travel on the order of 250 nm in a multistep process involving among other things recognition by an NLS receptor, docking of the occupied NLS receptor at the NPC, ATP, and GTP hydrolysis, release from the NLS receptor, and gating of the NPC (not necessarily in this order). Because the NPC operates vectorially each of these (and probably numerous other) biochemically distinct events will likely be encountered or triggered sequentially as a transport substrate progresses across the NPC. Any models of NPC mechanics must therefore be able to accomodate this vectorial function in terms of appropriate localizations of effector molecules. With this in mind, the objectives for the next few years in the elucidation of NPC function are therefore likely to be threefold. (1) The ordering of discrete biochemical events in the translocation process utilizing both *in vitro* and *in vivo* experimental systems much as has been performed in the elucidation of the mechanisms of insertion of polypeptides into the ER or of intracellular vesicle trafficking. (2) Identification of individual nucleoporins or subcomplexes of nucleoporins with distinct biochemical events. This goal will most likely be accomplished utilizing both *in vitro* and genetic approaches and represents probably the major challenge in NPC research. (3) Mapping of individual nucleoporins or subcomplexes of nucleoporins within the refined three-dimensional structure of the NPC. The convergence of these three areas of research should begin to shed some light on the molecular mechanisms of signal-mediated translocation across the nuclear envelope.

## Acknowledgments

The authors are indebted to Mr. C. Henn (M. E. Müller Institute, University of Basel) for designing and preparing Fig. 5. We thank Dr. R. Milligan (Scripps Research Institute, La Jolla, CA) who provided the data of the 3-D reconstruction of negatively stained detergent-released NPCs that enabled us to produce Fig. 5. We also wish to thank Drs. P. Grandi, V. Doye, and E. Hurt (EMBL, Heidelberg) for providing preprints of their work describing NUP57p and NUP133p. B. B. was supported by grants from the National Institutes of Health and the Alberta Heritage Foundation for Medical Research. N.P. was supported by the M. E. Müller Foundation of Switzerland, and by a grant from the Human Frontier Science Program (HFSP).

## References

Akey, C. W. (1989). Interactions and structure of the nuclear pore complex revealed by cryo-electron microscopy. *J. Cell Biol.* **109,** 955–970.

Akey, C. W., and Goldfarb, D. S. (1989). Protein import through the nuclear pore complex is a multi-step process. *J. Cell Biol.* **109,** 971–982

Akey, C. W., and Radermacher, M. (1993). Architecture of the *Xenopus* nuclear pore complex revealed by three dimensional cryo-electron microscopy. *J. Cell Biol.* **122,** 1–20.

Bélanger, K. D., Kenna, M. A., Wei, S., and Davis, L. I. (1994). Genetic and physical interactions between srp1p and nuclear pore complex proteins nup1p and nup2p. *J. Cell Biol.* **126,** 619–630.

Benavente, R., Dabauvalle, M.-C., Scheer, U., and Chaly, N. (1989a). Functional role of newly formed pore complexes in postmitotic nuclear reorganization. *Chromosoma* **98,** 233–241.

Benavente, R., Scheer, U., and Chaly, N. (1989b). Nucleocytoplasmic sorting of macromolecules following mitosis: Fate of nuclear constituents after inhibition of pore complex function. *Eur. J. Cell Biol.* **50,** 209–219.

Bender, A., and Pringle, J. R. (1991). Use of a screen for synthetic lethal and multicopy suppressor mutants to identify new genes involved in morphogenesis in *Saccharomyces cerevisiae. Mol. Cell. Biol.* **11,** 1295–1305.

Bischoff, F. R., and Ponstingl, H. (1991). Mitotic regulator protein RCC1 is complexed with a nuclear ras-related polypeptide. *Proc. Natl. Acad. Sci. U.S.A.* **88,** 10830–10834.

Blow, J. J., and Laskey, R. A. (1988). A role of the nuclear envelope in controlling DNA replication within the cell cycle. *Nature* (*London*) **332,** 546–548.

Bogerd, A. M., Hoffman, J. A., Amberg, D. C., Fink, G. R., and Davis, L. I. (1994). nup1 mutants exhibit pleiotropic defects in nuclear pore complex function. *J. Cell Biol.* **127,** 319–332.

Buss, F., Kent, H., Stewart, M., Bailler, S. B., and Hanover, J. A. (1994). Role of different domain in the self-association of rat nucleoporin p62. *J. Cell Sci.* **107,** 631–638.

Byrd, D., Sweet, D. J., Panté, N., Konstantinov, K. N., Guan, T., Saphire, A. C. S., Mitchell, P. J., Cooper, C. S., Aebi, U., and Gerace, L. (1994). Tpr, a large coiled-coil protein whose amino-terminus is involved in activation of oncogenic kinases, is localized to the cytoplasmic surface of the nuclear pore complex. *J. Cell Biol.* **127,** 1515–1526.

Carmo-Fonseca, M., Cidadao, A. J., and David-Ferreira, J. F. (1987). Filamentous cross-bridges link intermediate filaments to the nuclear pore complexes. *Eur. J. Cell Biol.* **45,** 282–290.

Carmo-Fonseca, M., Kern, H., and Hurt, E. C. (1991). Human nucleoporin p62 and the essential yeast nuclear pore protein NSP1 show sequence homology and a similar domain organization. *Eur. J. Cell Biol.* **55,** 17–30.

Chaudhary, N., and Courvalin, J.-C. (1993). Stepwise reassembly of the nuclear envelope at the end of mitosis. *J. Cell Biol.* **122,** 295–306.

Coffee, C. C., and Bradshaw, R. A. (1973). Carp muscle calcium-binding protein. I. Characterization of the tryptic peptides and the complete amino acid sequence of component B. *J. Biol. Chem.* **248,** 3305–3312.

Cordes, V., Waizenegger, I., and Krohne, G. (1991). Nuclear pore complex glycoprotein p62 of *Xenopus laevis* and mouse: cDNA cloning and identification of its glycosylated region. *Eur. J. Cell Biol.* **55,** 31–47.

Cordes, V., Reidenbach, S., Köhler, A., Stuurman, N., van Driel, R., and Franke, W. W. (1993). Intranuclear filaments containing a nuclear pore complex protein. *J. Cell Biol.* **123,** 1333–1344.

Coutavas, E., Ren, M., Oppenheim, J. D., D'Eustachio, P., and Rush, M. G. (1993). Characterization of proteins which interact with the cell-cycle regulatory protein Ran/TC4. *Nature (London)* **366,** 585–587.

Dabauvalle, M.-C., Benevente, R., and Chaly, N. (1988a). Monoclonal antibodies to a $M_r$ 68,000 pore complex glycoproteininterfere with nuclear protein uptake in *Xenopus* oocytes. *Chromosoma* **97,** 193–197.

Dabauvalle, M.-C., Schulz, B., Scheer, U., and Peters, R. (1988b). Inhibition of nuclear accumulation of karyophilic proteins in living cells by microinjection of the lectin wheat germ agglutinin. *Exp. Cell Res.* **174,** 291–296.

Dabauvalle, M.-C., Loos, K., and Scheer, U. (1990). Identification of a soluble precursor complex essential for nuclear pore assembly in vitro. *Chromosoma* **100,** 56–66.

Davis, L. I. (1992). Control of nucleocytoplasmic transport. *Curr. Opin. Cell Biol.* **4,** 424–429.

Davis, L. I., and Blobel, G. (1986). Identification and characterization of a nuclear pore complex protein. *Cell (Cambridge, Mass.)* **45,** 699–709.

Davis, L. I., and Blobel, G. (1987). Nuclear pore complex contains a family of glycoproteins that includes p62: Glycosylation through a previously unidentified cellular pathway. *Proc. Natl. Acad. Sci. U.S.A.* **84,** 7552–7556.

Davis, L. I. and Fink, G. R. (1990). The NUP1 gene encodes an essential component of the yeast nuclear pore complex. *Cell (Cambridge, Mass.)* **61,** 965–976.

Delaney, S. J., Hayward, D. C., Barleben, F., Fishbach, K.-F., and Gabor Miklos, G. L. (1991). Molecular cloning and analysis of small optic lobes, a structural brain gene of *Drosophilla melanogaster. Proc. Natl. Acad. Sci. U.S.A.* **88,** 7214–7218.

Dingwall, C., and Laskey, R. A. (1986). Protein import into the cell nucleus. *Annu. Rev. Cell Biol.* **2,** 367–390.

Dingwall, C., and Laskey, R. A. (1991). Nuclear targeting sequences—a concensus? *Trends Biochem. Sci.* **16,** 478–481.

Dingwall, C., Robbins, J., Dilworth, S. M., Roberts, S. M., and Richardson, W. D. (1988). The nucleoplasmin nuclear localization sequence is larger and more complex than that of SV40 large T antigen. *J. Cell Biol.* **107,** 841–849.

Doye, V., Wepf, R., and Hurt, E. C. (1994). A novel nuclear pore protein Nup133p with distinct roles in poly(A)$^+$ RNA transport and nuclear pore distribution. *EMBO J.* **13,** 6062–6075.

Drivas, G. T., Shih, A., Coutavas, E., Rush, M. G., and D'Eustachio, P. (1990). Characterization of four novel ras-like genes expressed in a human teratocarcinoma cell line. *Mol. Cell. Biol.* **10,** 1793–1798.

Dworetzky, S. I., and Feldherr, C. M. (1988). Translocation of RNA coated gold particles through the nuclear pores of oocytes. *J. Cell Biol.* **106,** 575–584.

Dworetzky, S. I., Lanford, R. E., and Feldherr, C. M. (1988). The effects of variations in the number and sequence of targeting signals on nuclear uptake. *J. Cell Biol.* **107,** 1279–1287.

Fabre, E., and Hurt, E. C. (1994). Nuclear transport. *Curr. Opin. Cell Biol.* **6,** 335–342.

Fabre, E., Boelens, W. C., Wimmer, C., Mattaj, I. W., and Hurt, E. C. (1994). Nup145p is required for nuclear export of mRNA and binds homopolymeric RNA in vitro via a novel conserved motif. *Cell (Cambridge, Mass.)* **78,** 275–289.

Fakharzadeh, S. S., Trusko, S. P., and George, D. L. (1991). Tumorigenic potential associated with enhanced expression of a gene that is amplified in a mouse tumor cell line. *EMBO J.* **10,** 1565–1569.

Featherstone, C., Darby, M. K., and Gerace, L. (1988). A monoclonal antibody against the nuclear pore complex inhibits nucleocytoplasmic transport of protein and RNA in vivo. *J. Cell Biol.* **107,** 1289–1287.

Feldherr, C., Kallenbach, E., and Schultz, N. (1984). Movement of a karyophilic protein through the nuclear pores of oocytes. *J. Cell Biol.* **99,** 2216–2222.

Finlay, D. R., and Forbes, D. J. (1990). Reconstitution of biochemically altered nuclear pores: Transport can be eliminated and restored. *Cell (Cambridge, Mass.)* **60,** 17–29.

Finlay, D. R., Newmeyer, D. D., Price, T. M., and Forbes, D. J. (1987). Inhibition of in vitro nuclear transport by a lectin that binds to nuclear pores. *J. Cell Biol.* **104,** 189–200.

Finlay, D. R., Meier, E., Bradley, P., Horecka, J., and Forbes, D. J. (1991). A complex of nuclear pore proteins required for pore function. *J. Cell Biol.* **114,** 169–183.

Forbes, D. J. (1992). Structure and function of the nuclear pore complex. *Annu. Rev. Cell Biol.* **8,** 495–527.

Franke, W. W. (1974). Structure, function and biochemistry of the nuclear envelope. *Int. Rev. Cytol. Suppl.* **4,** 71–236.

Gall, J. G, (1967). Octagonal nuclear pores. *J. Cell Biol.* **32,** 391–399.

Gerace, L. (1992). Molecular trafficking across the nuclear pore complex. *Curr. Opin. Cell Biol.* **4,** 637–645.

Gerace, L., and Burke, B. (1988). Functional organization of the nuclear envelope. *Annu. Rev. Cell Biol.* **4,** 335–374.

Gerace, L., Ottaviano, Y., and Kondor-Koch, C. (1982). Identification of a major polypeptide of the nuclear pore complex. *J. Cell Biol.* **95,** 826–837.

Gerace, L., Comeau, C., and Benson, M. (1984). Organization and modulation of nuclear lamina structure. *J. Cell Sci. Suppl.* **1,** 137–160.

Ghosh, S., and Baltimore, D. (1990). Activation *in vitro* of NF-κB by phosphorylation of its inhibitor IκB *Nature (London)* **344,** 678–682.

Goldberg, M. W., and Allen, T. D. (1992). High resolution scanning electron microscopy of the nuclear envelope: Demonstration of a new, regular, fibrous lattice attached to the baskets of the nucleoplasmic face of the nuclear pores. *J. Cell. Biol.* **119,** 1429–1440.

Goldfarb, D., and Michaud, N. (1991). Pathways for the nuclear transport of proteins and RNAs. *Trends Cell Biol.* **1,** 20–24.

Goldfarb, R., Gariepy, J., Schoolnick, G., and Kornberg, R. (1986). Synthetic peptides as nuclear localization signals. *Nature (London)* **322,** 641–644.

Görlich, D., Prehn, S., Laskey, R. A., and Hartmann, E. (1994). Isolation of a protein that is essential for the first step of nuclear protein import. *Cell (Cambridge, Mass.)* **79,** 767–778.

Grandi, P., Doye, V., and Hurt, E. C. (1993). Purification of NSP1 reveals complex formation with GLFG nucleoporins and a novel nuclear pore protein NIC96. *EMBO J* **12,** 3061–3071.

Grandi, P., Schlaich, N., Tekotte, H., and Hurt, E. C. (1995). Functional interaction of Nic96p with a core nucleoporin complex consisting of Nsp1p, Nup49p and a novel protein Nup57p. *EMBO J.* **14** (in press).

Greber, U. F., and Gerace, L. (1992). Nuclear protein import is inhibited by an antibody to a lumenal epitope of a nuclear pore complex glycoprotein. *J. Cell Biol.* **116,** 15–30.

Greber, U. F., Senior, A., and Gerace, L. (1990). A major glycoprotein of the nuclear pore complex is a membrane-spanning polypeptide with a large lumenal domain and a small cytoplasmic tail. *EMBO J.* **9,** 1495–1502.

Greco, A., Pierotti, M. A., Bongarzone, I., Pagliardini, S., Lanzi, C., and Della Porta, G. (1992). TRK-T1 is a novel oncogene formed by the fusion of TPR and TRK genes in human papillary thyroid carcinoma. *Oncogene* **7,** 237–242.

Guan, T. *et al.,* (1995). In preparation.

Hallberg, E., Wozniak, R. W., and Blobel, G. (1993). An integral membrane protein of the pore membrane domain of the nuclear envelope contains a nucleoporin-like region. *J. Cell Biol.* **122,** 513–522.

Haltiwanger, R. S., Blomberg, M. A., and Hart, G. W. (1992). Glycosylation of nuclear and cytoplasmic proteins. Purification and characterization of a uridine diphospho-*N*-acetylglucosamine:polypeptide *β*-*N*-acetylglucosaminyltransferase. *J. Biol. Chem.* **267,** 9005–9013.

Hart, G. W. (1992). Glycosylation. *Curr. Opin. Cell Biol.* **4,** 1017–1023.

Hinshaw, J. E., Carragher, B. O., and Milligan, R. A. (1992). Architecture and design of the nuclear pore complex. *Cell (Cambridge, Mass.)* **69,** 1133–1141.

Holt, G. D., Snow, C. M., Senior, A., Haltiwanger, R. S., Gerace, L., and Hart, G. W. (1987). Nuclear pore complex glycoproteins contain cytoplasmically disposed O-linked *N*-acetylglucosamine. *J. Cell Biol.* **104,** 1157–1164.

Hurt, E. C. (1988). A novel nucleoskeletal-like protein located at the nuclear periphery is required for the life cycle of *Saccaromyces cerevisiae. EMBO J.* **7,** 4323–4334.

Hurt, E. C. (1990). Targeting of a cytosolic protein to the nuclear periphery. *J. Cell Biol.* **111,** 2829–2837.

Ishikawa, F., Takaku, F., Nagao, M., and Sugimura, T. (1987). Rat c-raf oncogene activation by a rearrangement that produces a fused protein. *Mol. Cell. Biol.* **7,** 1226–1232.

Izaurralde, E., Stepinski, J., Darzynkiewicz, E., and Mattaj, I. W. (1992). A cap-binding protein that may mediate export of RNA polymerase II-transcribed RNAs. *J. Cell Biol.* **118,** 1287–1295.

Jarmolowski, A., Boelens, W. C., Izaurralde, E., and Mattaj, I. W. (1994). Nuclear export of different classes of RNA is mediated by specific factors. *J. Cell Biol.* **124,** 627–635.

Jarnik, M., and Aebi, U. (1991). Toward a more complete 3-D structure of the nuclear pore complex. *J. Struct. Biol.* **107,** 291–308.

Jauregui-Adell, J., and Pechere, J.-F. (1978). Parvalbumins from coelocanth muscle. III. Amino acid sequence of the major component. *Biochim. Biophys. Acta* **536,** 275–282.

Kalderon, D., Richardson, W. D., Markham, A. F., and Smith, A. E. (1984a). Sequence requirements for nuclear location of simian virus 40 large-T-antigen. *Nature (London)* **311,** 33–38.

Kalderon, D., Roberts, B. L., Richardson, W. D., and Smith, A. E. (1984b). A short amino acid sequence able to specify nuclear location. *Cell (Cambridge, Mass.)* **39,** 499–509.

Kalinich, J. F., and Douglas, M. G. (1989). In vitro translocation through the yeast nuclear envelope: Signal-dependent transport requires ATP and calcium. *J. Biol. Chem* **264,** 17979–17989.

Kearse, K. P., and Hart, G. W. (1991). Lymphocyte activation induces rapid changes in nuclear and cytoplasmic glycoproteins. *Proc. Natl. Acad. Sci. U.S.A.* **88,** 1701–1705.

Kita, K., Omata, S., and Horigome, T. (1993). Purification and characterization of a nuclear pore glycoprotein complex containing p62. *J. Biochem. (Tokyo)* **113,** 377–382.

Koshland, D., Kent, J. C., and Hartwell, L. H. (1985). Genetic analysis of the mitotic transmission of minichromosomes. *Cell (Cambridge, Mass.)* **40,** 393–403.

Kraemer, D., Wozniak, R. W., Blobel, G., and Radu, A. (1994). The human CAN protein, a putative oncogene product associated with myeloid leukemogenesis, is a nuclear pore complex protein that faces the cytoplasm. *Proc. Natl. Acad. Sci. U.S.A.* **91,** 1519–1523.

Kretsinger, R. H. (1980). Structure and evolution of calcium modulated proteins. *CRC Crit. Rev. Biochem.* **8,** 119–174.

Lin, A., Bastos, R., and Burke, B. (1995). In preparation.

Loeb, J. D., Davis, L. I., and Fink, G. R. (1993). NUP2, a novel yeast nucleoporin, has functional overlap with other proteins of the nuclear pore complex. *Mol. Biol. Cell.* **4,** 209–222.

MacManus, R., Watson, J. P., and Yaguchi, M. (1983). The complete amino acid sequence of oncomodulin—a parvalbumin-like calcium-binding protein from Morris hepatoma 5123tc. *Eur. J. Biochem.*

Marin-Zanka, D., Hughes, S. H., and Barbacid, M. (1986). A human oncogene formed by the fusion of a truncated tropomyosin and protein tyrosine kinase sequences. *Nature (London)* **319,** 743–748.

Maul, G. G. (1977). Nuclear pore complexes: Elimination and reconstruction during mitosis. *J. Cell Biol.* **74,** 492–500.

Maul, G. G., Price, J. W., and Lieberman, M. W. (1971). Formation and distribution of nuclear pore complexes in interphase. *J. Cell Biol.* **51,** 405–418.

Maul, G. G., Maul, H., Scogna, J., Lieberman, Stein, G. *et al.* (1972). Time sequence of nuclear pore formation in phytohaemagglutinin-stimulated lymphocytes and in HeLa cells during the cell cycle. *J. Cell Biol.* **55,** 433–447.

McMorrow, I. M., Bastos, R., Horton, H., and Burke, B. (1994). Sequence analysis of a cDNA encoding a human nuclear pore complex protein, hnup153. *Biochim. Biophys. Acta* **1217,** 219–223.

Melchior, F., Paschal, B., Evans, J., and Gerace, L. (1993). Inhibition of nuclear protein import by non-hydrolyzable analogues of GTP and identification of the small GTPase Ran/TC4 as an essential transport factor. *J. Cell Biol.* **123,** 1649–1659.

Michaud, N., and Goldfarb, D. S. (1991). Multiple pathways in nuclear transport: The import of U2 snRNP occurs by a novel kinetic pathway. *J. Cell Biol.* **112,** 215–223.

Michaud, N., and Goldfarb, D. S. (1992). Microinjected U snRNAs are imported to oocyte nuclei via the nuclear pore complex by three distinguishable targeting pathways. *J. Cell Biol.* **116,** 851–861.

Miller, M. W., and Hanover, J. A. (1994). Functional nuclear pores reconstituted with $\beta$1-4 galactose-modified O-linked *N*-acetyglucosamine glycoproteins. *J. Biol. Chem.* **269,** 9289–9297.

Mitchell, P. J., and Cooper, C. S. (1992). The human tpr gene encodes a protein of 2094 amino acids that has extensive coiled-coil regions and an acidic C-terminal domain. *Oncogene* **7,** 2329–2333.

Moore, M. S., and Blobel, G. (1993). The GTP-binding protein Ran/TC4 is required for protein import into the nucleus. *Nature (London)* **365,** 661–663.

Moore, M. S., and Blobel, G. (1994). A G protein involved in nucleocytoplasmic transport: The role of Ran. *Trends Biochem. Sci.* **19,** 211–216.

Mutvei, A., Dihlmann, S., Herth, W., and Hurt, E. C. (1992). NSP1 depletion in yeast affects nuclear pore formation and nuclear accumulation. *Eur. J. Cell Biol.* **59,** 280–295.

Nasmyth, K., Adolf, G., Lydall, D., and Sheddon, A. (1990). The identification of a second cell cycle control on the HO promoter in yeast: Cell cycle regulation of SWI5 nuclear entry. Cell *(Cambridge, Mass.)* **62,** 631–647.

Nehrbass, U., Kern, H., Mutvei, A., Horstmann, H., Marshallsay, B., and Hurt, E. C. (1990). NSP1: A yeast nuclear envelope protein localized at the nuclear pores exerts it essential function by its carboxy-terminal domain. *Cell (Cambridge, Mass.)* **61,** 979–989.

Nehrbass, U., Fabre, E., Dihlmann, S., Herth, W., and Hurt, E. C. (1993). Analysis of nucleo-cytoplasmic transport in a thermosensitive mutant of nuclear pore protein NSP1. *Eur. J. Cell Biol.* **62,** 1–12.

Nelson, M., and Silver, P. (1989). Context affects nuclear protein localization in *Sacharomyces cerevisiae. Mol. Cell. Biol.* **9,** 384–389.

Newmeyer, D. D. (1993). The nuclear pore complex and nucleocytoplasmic transport. *Curr. Opin. Cell Biol.* **3,** 395–407.

Newmeyer, D. D., and Forbes, D. J. (1988). Nuclear import can be separated into distinct steps in vitro: Nuclear pore binding and translocation. *Cell* (*Cambridge, Mass.*) **52,** 641–653.

Nilsson, T., and Warren, G. (1994). Retention and retrieval in the endoplasmic reticulum and Golgi apparatus. *Curr. Opin. Cell Biol.* **6,** 517–521.

Paine, P. L. (1975). Nucleocytoplasmic movement of fluorescent tracers microinjected into living salivary gland cells. *J. Cell Biol.* **66,** 652–657.

Paine, P. L., Moore, L. C., and Horowitz, S. B. (1975). Nuclear envelope permeability. *Nature* (*London*) **254,** 109–114.

Panté, N., and Aebi, U. (1993). The nuclear pore complex. *J. Cell Biol.* **122,** 977–984.

Panté, N., and Aebi, U. (1994). Towards understanding the three-dimensional structure of the nuclear pore complex at the molecular level. *Curr. Opin. Struct. Biol.* **4,** 187–196.

Panté, N., Bastos, R., McMorrow, I., Burke, B., and Aebi, U. (1994). Interactions and three-dimensional localization of a group of nuclear pore complex proteins. *J. Cell Biol.* **126,** 603–617.

Park, M. K., Dean, M., Cooper, C. S., Schmidt, M., O'Brien, S. J., Blair, D. G., and Vande Woude, G. F. (1986). Mechanism of met oncogene activation. *Cell* (*Cambridge, Mass.*) **45,** 895–904.

Park, M. K., D'Onofrio, M., Willingham, M. C., and Hanover, J. A. (1987). A monoclonal antibody against a family of nuclear pore proteins (nucleoporins): O-linked *N*- acetylglucosamine is part of the immunodeterminant. *Proc. Natl. Acad. Sci. U.S.A.* **84,** 6462–6466.

Picard, D., and Yamamoto, K. R. (1987). Two signals mediate hormone-dependent nuclear localization of the glucocorticoid receptor. EMBO J. **6,** 3333–3340.

Picard, D., Salser, S. J., and Yamamoto, K. R. (1988). A movable and regulable inactivation function within the steroid binding domain of the glucocorticoid receptor. *Cell* (*Cambridge, Mass.*) **54,** 1073–1080.

Radu, A., Blobel, G., and Wozniak, R. W. (1993). Nup155 is a novel nuclear pore complex protein that contains neither repetitive sequence motifs nor reacts with WGA. *J. Cell Biol.* **121,** 1–9.

Radu, A., Blobel, G., and Wozniak, R. W. (1994). Nup107 is a novel nuclear pore complex protein that contains a leucine zipper. *J. Biol. Chem.* **269,** 17600–17605.

Reichelt, R., Holzenburg, A., Buhle, E. L., Jarnik, M., Engel, A., and Aebi, U. (1990). Correlation between structure and mass distribution of the nuclear pore complex and of distinct pore components. *J. Cell Biol.* **110,** 883–894.

Richardson, W. D., Roberts, B. L., and Smith, A. E. (1986). Nuclear location signals in polyoma virus large-T. *Cell* (*Cambridge, Mass.*) **44,** 77–85.

Richardson, W. D., Mills, A. D., Dilworth, S. M., Laskey, R. A., and Dingwall, C. (1988). Nuclear protein migration involves two steps: Rapid binding at the nuclear envelope followed by slower translocation through the nuclear pores. *Cell* (*Cambridge, Mass.*) **52,** 655–664.

Rihs, H.-P., and Peters, R. (1989). Nuclear transport kinetics depend on phosphorylation-site-containing sequences flanking the karyophilic signal of the Simian virus 40 T-antigen. *EMBO J.* **8,** 1479–1484.

Rihs, H.-P., Jans, D. A., Fan, H., and Peters, R. (1991). The rate of nuclear cytoplasmic protein transport is determined by the casein kinase II site flanking the nuclear localization sequence of the SV40 T-antigen. *EMBO J.* **10,** 633–639.

Ris, H. (1991). The 3-D structure of the nuclear pore complex as seen by high voltage electron microscopy and high resolution low voltage scanning electron microscopy. EMSA Bull. **21,** 54–56.

Robbins, J., Dilworth, S. M., Laskey, R. A., and Dingwall, C. (1991). Two interdependent basic domains in nucleoplasmin nuclear targeting sequence: Identification of a class of bipartite nuclear targeting sequence. *Cell* (*Cambridge, Mass.*) **64,** 615–623.

Roberts, B. L., Richardson, W. D., and Smith, A. E. (1987). The effect of protein context on nuclear location signal function. *Cell (Cambridge, Mass.)* **50,** 465–475.

Roth, S., Stein, D. and Nüsslein-Volhard, C. (1989). A gradient of nuclear localization of the dorsal protein determines dorsoventral pattern in the Drosophila embryo. *Cell (Cambridge, Mass.)* **59,** 1189–1202.

Rothman, J. E. (1994). Mechanisms of intracellular protein transport. *Nature (London)* **372,** 55–63.

Rothman, J. E., and Warren, G. (1994). Implications of the SNARE hypothesis for intracellular membrane topology and dynamics. *Curr. Biol.* **4,** 220–233.

Rout, M. P., and Blobel, G. (1993). Isolation of the yeast nuclear pore complex. *J. Cell Biol.* **123,** 771–783.

Rout, M. P., and Wente, S. R. (1994). Pores for thought: nuclear pore complex proteins. *Trends Cell Biol.* **4,** 357–365.

Schlenstedt, G., Hurt, E., Doye, V., and Silver, P. A. (1993). Reconstitution of nuclear protein transport with semi-intact yeast cells. *J. Cell Biol.* **123,** 785–798.

Sheehan, M. A., Mills, A. D., Sleeman, A. M., Laskey, R. A., and Blow, J. J. (1988). Steps in the assembly of replication-competent nuclei in a cell-free system. *J. Cell Biol.* **106,** 1–12.

Silver, P. A. (1991). How proteins enter the nucleus. *Cell (Cambridge, Mass.)* **64,** 489–497.

Snow, C. M., Senior, A., and Gerace, L. (1987). Monoclonal antibodies identify a group of nuclear pore complex glycoproteins. J. Cell Biol. **104,** 1143–1156.

Starr, C. M., and Hanover, J. A. (1990). Glcosylation of nuclear pore protein p62: Reticulocyte lysate catalyzes O-linked *N*-acetylglucosamine addition in vitro. *J. Biol. Chem.* **265,** 6868–6873.

Starr, C. M., D'Onofrio, M., Park, M. K., and Hanover, J. A. (1990). Primary sequence and heterologous expression of nuclear pore glycoprotein p62. *J. Cell Biol.* **110,** 1861–1871.

Sterne-Marr, R., Blevitt, J. M., and Gerace, L. (1992). O-linked glycoproteins of the nuclear pore complex interact with a cytosolic factor required for nuclear protein import. *J. Cell Biol.* **116,** 271–280.

Steward, R. (1989). Relocalization of the dorsal protein from the cytoplasm to the nucleus correlates with its function. *Cell (Cambridge, Mass.)* **59,** 1179–1188.

Sukegawa, J., and Blobel, G. (1993). A nuclear pore complex protein that contains zinc fingers and faces the nucleoplasm. *Cell (Cambridge, Mass.)* **72,** 29–38.

Unwin, P. N. T., and Milligan, R. A. (1982). A large particle associated with the perimeter of the nuclear pore complex. *J. Cell Biol.* **93,** 63–75.

Vigers, G. P. A., and Lohka, M. J. (1991). A distinct vesicle population targets membranes and pore complexes to the nuclear envelope in *Xenopus* eggs. *J. Cell Biol.* **112,** 545–556.

von Lindern, M., Fornerod, M., van Baal, S., Jaegle, M., de Wit, T., Buijs, A., and Grosveld, G. (1992a). The translocation (6;9), associated with a specific subtype of acute myeloid leukemia, results in the fusion of two genes, *dek* and *can,* and the expression of a chimeric, leukemia-specific *dek-can* mRNA. *Mol. Cell. Biol.* **12,** 1687–1697.

von Lindern, M., van Baal, S., Wiegant, J., Raap, A., Hagemijer, A., and Grosveld G. (1992b). *can,* a putative oncogene associated with myeloid leukemogenesis, may be activated by fusion of its 3′ half to different genes: Characterization of the *set* gene. *Mol. Cell. Biol.* **12,** 3346–3355

Wehner, E. P., Rao, E., and Brendel, M. (1993). Molecular structure and genetic regulation of SFA, a gene responsible for resistance to formaldehyde in *Saccharomyces cerevisiae,* and characterization of its protein product. *Mol. Gen. Genet.* **237,** 351–358.

Wente, S. R., and Blobel, G. (1993). A temperature-sensitive NUP116 null mutant forms a nuclear envelope seal over the nuclear pore complex thereby blocking nucleocytoplasmic traffic. *J. Cell Biol.* **123,** 275–284.

Wente, S. R., and Blobel, G. (1994). NUP145 encodes a novel yeast glycine-leucine-phenylalanine-glycine (GLFG) nucleoporin required for nuclear envelope structure. *J. Cell Biol.* **125,** 955–969.

Wente, S. R., Rout, M. P., and Blobel, G. (1992). A new family of yeast nuclear pore complex proteins. *J. Cell Biol.* **119,** 705–723.

White, J. M. (1990). Viral and cellular membrane fusion proteins. *Annu. Rev. Physiol.* **52,** 675–697.

Wilken, N., Kossner, U., Senécal, J.-L., Scheer, U., and Dabauvalle, M.-C. (1993). Nup180, a novel nuclear pore complex protein localizing to the cytoplasmic ring and associated fibrils. *J. Cell Biol.* **123,** 1345–1354.

Wimmer, C., Doye, V., Grandi, P., Nehrbass, and Hurt, E. C. (1992). A new subclass of nucleoporins that functionally interact with nuclear pore protein NSP1. *EMBO J.* **11,** 5051–5061.

Wolff, B., Willingham, M. C., and Hanover, J. A. (1988). Nuclear protein import: Specificity for trasnport across the nuclear pore. *Exp. Cell Res.* **178,** 318–334.

Wozniak, R. W., and Blobel, G. (1992). The single transmembrane segment of gp210 is sufficient for soring to the pore membrane domain of the nuclear envelope. *J. Cell Biol.* **119,** 1441–1449.

Wozniak, R. W., Bartnik, E., and Blobel, G. (1989). Primary structure analysis of an integral membrane glycoprotein of the nuclear pore. *J. Cell Biol.* **108,** 2083–2092.

Wozniak, R. W., Blobel, G., and Rout, M. P. (1994). POM152 is an integral protein of the pore membrane domain of the yeast nuclear envelope, *J. Cell Biol.* **125,** 31–42.

Yano, R., Oakes, M., Yamaghishi, M., Dodd, J. A., and Nomura, M. (1992). Cloning and characterization of SRP1, a suppressor of temperature-sensitive RNA polymerase 1 mutations, in *Saccaromyces cerevisiae. Mol. Cell Biol.* **12,** 5640–5651.

Yoneda, Y., Imamoto-Sonobe, N., Yamaizumi, M., and Uchida, T. (1987). Revesible inhibition of protein import into the nucleus by wheat germ agglutinin injected into cultured cells. *Exp. Cell Res.* **173,** 586–595.

Yong, D. L.-Y., and Hart, G. W. (1994). Purification and characterization of an O-glcNAc selective *N*-acetyl-β-D-glucosaminidase from rat spleen cytosol. *J. Biol. Chem.* **269,** 19321–19330.

Zelings, J. D., and Wollman, S. H. (1979). Mitosis in rat thyroid epithelial cells in vivo: Ultrastructural changes in cytoplasmic organelles during the mitotic cycle. *J. Ultrastruc. Res.* **66,** 53–77.

Zhao, H., Lindqvist, B., Garoff, H., von Bonsdorff, C.-H., and Liljeström, P. (1994). A tyrosine-based motif in the cytoplasmic domain of the alphavirus envelope protein is essential for budding. *EMBO J.* **13,** 4204–4211.

# Targeting and Association of Proteins with Functional Domains in the Nucleus: The Insoluble Solution

Heinrich Leonhardt and M. Cristina Cardoso
Humboldt Universität Berlin, Franz-Volhard-Klinik am Max-Delbrück-Centrum für Molekulare Medizin, Department of Nephrology, Hypertension, and Genetics, 13122 Berlin, Germany

The mammalian nucleus is highly organized into distinct functional domains separating different biochemical processes such as transcription, RNA processing, DNA synthesis, and ribosome assembly. A number of proteins known to participate in these processes were found to be specifically localized at their corresponding functional domains. A distinct targeting sequence, necessary and sufficient for the localization to DNA replication foci, was identified in the N-terminal, regulatory domain of DNA methyltransferase and DNA ligase I and might play a role in the coordination of DNA replication and DNA methylation. The fact that the targeting sequence is absent in lower eukaryotic and prokaryotic DNA ligase I homologs suggests that "targeting" is a rather recent development in evolution. Finally, targeting sequences have also been identified in some splicing factors and in viral proteins, which are responsible for their localization to the speckled compartment and to the nucleolus, respectively. These higher levels of organization are likely to contribute to the regulation and coordination of the complex and interdependent biochemical processes in the mammalian nucleus.

**KEY WORDS:** DNA methyltransferase, DNA ligase I Cyclin A, cdk2, Replication protein A, Targeting sequence, RS domain, Nucleolus, Coiled body, DNA replication, DNA methylation, RNA splicing, Cell cycle.

## I. Introduction

The goal of this chapter is to outline the functional organization of mammalian nuclei with emphasis on protein targeting. The term "protein targeting" in a broad sense refers to mechanisms that cause proteins to end up in specific cellular compartments. The term was initially coined in the field of intracellular protein traffic and was used to refer to the controlled and energy-dependent transport of proteins across membranes and the transport between intracellular compartments in vesicles. Starting with the synthesis at the ribosomes in the cytoplasm, a new protein has several options regarding the cellular localization. Short signal or presequences, usually at the N-terminus, determine whether a protein is directly translocated into the endoplasmic reticulum, imported into mitochondria or into the nucleus, or whether it only remains in the cytoplasm. All these transport or targeting processes are energy dependent and, except for nuclear transport, the signal or presequences are cleaved off during the transport. The nuclear import is special because the nucleus breaks down during each cell cycle and most proteins must be reimported after the nucleus is rebuilt (for a detailed discussion on the nuclear import process and the function of the nuclear pore see chapters by Agutter; Panté and Aebi; and Bastos and Burke).

For a long time the nucleus was considered to be the final destination of protein targeting. This is true as there are no organizing or dividing membranes within the nucleus, but nonetheless the nucleus is anything but a homogeneous cellular compartment with freely diffusible components. Even by regular light microscopy the nucleolus can be identified as a distinct subnuclear structure. Electron micrography clearly showed that there are no membranes around the nucleolus, but still a number of proteins were found to be specifically localized in the nucleolus while other nuclear proteins were excluded. A detailed analysis led to the identification of nucleolar targeting sequences, which are necessary and sufficient for the nucleolar localization just like the above nuclear localization sequences are required for nuclear uptake (see Section II). In addition to the nucleolus there are several other subnuclear structures, which are not visible by regular light microscopy. Three distinct RNA-containing structures including interchromatin granules (IGs), perichromatin fibrils (PFs), and coiled bodies (CBs) were identified by immunofluorescence and electron microscopy. The targeting of proteins to these "compartments" and their role in transcription and RNA processing is discussed in Section III. Finally, DNA replication also occurs in distinct nuclear foci, and proteins directly or indirectly involved in DNA replication are localized and targeted to these foci (see Section IV).

The specific enrichment of proteins in distinct nuclear structures or compartments where the respective nuclear processes occur is often referred to as "functional organization of the nucleus" and obviously raises a number of interesting questions. First of all, what forces are responsible for this organization into functional domains? In case of the nuclear localization the basic principle seems to be simple; there is a separating membrane and only proteins with a nuclear localization sequence (NLS) are recognized and imported through the nuclear pore complex. Other proteins lacking the NLS are either small enough to passively diffuse through the nuclear pore or remain outside unless they are associated with other NLS-containing factors. As mentioned above, there are no organizing membranes in the nucleus and—apart from nuclear components like chromosomes, proteins, and RNA limiting the free diffusion path—imported proteins seem to be per se freely diffusible. Hence the specific enrichment of proteins at, e.g., replication foci must be based on association and retention at these structures. In several cases targeting sequences were identified, which are necessary and sufficient for localization at these nuclear structures. Examples for targeting to nucleoli, speckled compartment, coiled bodies, and replication foci are reviewed throughout this chapter.

The functional organization of the nucleus also has consequences for the regulation of biological processes. The concentration of factors, which are involved in different steps of a complex process, in functional nuclear foci and the organization of the process in an assembly line fashion is expected to greatly increase the overall efficiency, fidelity, and specificity. Over the past few years progress has been made toward the "appreciation" of these higher order structures. After a decade of dispute over potential artifactual experimental conditions, several different experimental approaches are now in place to examine the architecture of these structures and to analyze their contribution to the regulation and coordination of biological processes throughout the cell cycle.

## II. Targeting of Proteins to the Nucleolus

In the interphase eukaryotic nucleus, the most conspicuous morphological structure seen in the light microscope is the nucleolus. Even though it has been a preferred target of cytologists for centuries, it was only in the 1960s that findings derived by electron microscopic autoradiography of [$^3$H]uridine incorporation established its major function in rRNA synthesis and processing and ribosome biogenesis (Granboulan and Granboulan, 1965).

The electron microscope unveils three major ultrastructural components within the nucleolus: pale-staining fibrillar centers (FC) surrounded by dense fibrillar components (DFC), which are themselves embedded in a granular component (GC). The size of the nucleolus changes according to the metabolic state of the cell and it is the contraction or expansion of the granular component that is mostly responsible for these differences. The nucleolus is usually surrounded by a layer of condensed chromatin that is continuous with the intranucleolar chromatin in the FCs. The major components and biochemical functions of the different intranucleolar compartments were essentially laid out in the mid-1960s but the details of where in the nucleolus ribosomal genes are transcribed remained highly controversial. Recently, the discrepancies in the mapping of the transcriptional sites within the nucleolus have been reevaluated by the development of a nonisotopic method based on the incorporation of 5-bromouridine-5′-triphosphate (BrUTP) into RNA in streptolysin O-permeabilized mammalian cells (Dundr and Raska, 1993). This new technique provided decisive evidence that the site of nucleolar transcription is the DFC including its boundary to the FC (Hozák *et al.,* 1994; Hozák, 1995). The FCs contain DNA from the nucleolar organizer region (NOR) of a chromosome (i.e., the region of the chromosome containing tandem copies of rRNA genes separated by nontranscribed spacer DNA) and proteins such as RNA polymerase I and others needed for rRNA synthesis (e.g., DNA topoisomerase I and the transcription factor UBF). The DFC surrounding the FC contains the pre-rRNA transcripts with their 5′ tails encased by a protein-rich granule and U3 snoRNA (Mougey *et al.,* 1993). The latter is the most abundant nucleolar-specific snRNA and it has been implicated in the early pre-rRNA processing steps (Kass *et al.,* 1990; Savino and Gerbi, 1990; Hughes and Ares, 1991). Upon hypotonic treatment, each of these actively transcribing units shows a "Christmas tree"-like appearance, with the tip of the tree representing the point on the DNA where transcription begins. These transcripts are then processed and converted into preribosomal particles forming the GC. It is not yet clear whether these various processing and assembly steps occur at particular sites within the nucleolus or whether each pre-rRNA carries its private processing machinery, and maturation, therefore, takes place independently of the location within the nucleolus. In any event, after these steps, ribosomal precursors are released from the nucleolus into the nucleoplasm and are transported to the cytoplasm where the final steps of ribosome maturation take place. Detailed information on nucleolar transcription and structure can be found in the chapters by Jackson and Cook, van Driel *et al.,* and Moreno Diaz de la Espina.

This vectorial process involves an unusual degree of nucleocytoplasmic interaction, with precursor rRNA being synthesized in the nucleolus and being soon associated with proteins synthesized in the cytoplasm. These include most ribosomal proteins and, in addition, nonribosomal nucleolar proteins as well as small RNA molecules (nucleolar snRNAs) that are involved in pre-rRNA processing and preribosome assembly. The two last are recycled in the nucleolus when the preribosomal particles are exported to the cytoplasm. Among the nonribosomal nucleolar proteins identified so far, some may represent structural proteins forming a nucleolar skeleton (Franke *et al.,* 1981), others are transiently associated with preribosomes and are involved in processing of rRNA (e.g., fibrillarin; Tollervey *et al.,*1991), and others shuttle between the nucleolus and the cytoplasm (e.g.,

C23/nucleolin and B23/NO38, Borer *et al.,* 1989; Nopp140, Meier and Blobel, 1992; NAP57, Meier and Blobel, 1994) (see Table I).

To establish and maintain this functional nonmembranous intranuclear compartment where rDNA transcription, pre-rRNA processing, and preribosome assembly take place, a very intricate set of protein–protein and protein–nucleic acid interactions are required at each mitotic cycle. During prophase of mitosis, the nucleolus disintegrates. In the telophase nucleus prenucleolar bodies (PNB) appear and, when rRNA transcription resumes, they fuse at the NORs forming the interphase nucleolus (Scheer and Benavente, 1990; Scheer *et al.,* 1993). This complex process of nucleolar reformation is called nucleogenesis. The nucleolus can be used as a paradigm of how function relates to spatial and temporal organization, and indeed it uncovers, at its own scale, the underlying principles of the functional organization of the nucleus within which it is formed. Although it has no separating membranes, specific proteins have been localized to this highly specialized region of the nucleus, and some have even been assigned to certain ultra-

TABLE I
Cellular Proteins[a] Localized to the Nucleolus

| Protein | Reference |
|---|---|
| RNA polymerase I | Scheer and Rose (1984); Raska *et al.* (1989) |
| UBF transcription factor | Rendon *et al.* (1992); Roussel *et al.* (1993) |
| Topoisomerase I | Zini *et al.* (1994) |
| Topoisomerase II $\alpha$ and $\beta$ | Zini *et al.* (1994) |
| NOP1/fibrillarin | Ochs *et al.* (1985) |
| GAR1 | Girard *et al.* (1992) |
| B23/NO38/nucleophosmin | Ochs *et al.* (1985) |
| C23/nucleolin | Ochs *et al.* (1985); Pfeifle *et al.* (1986) |
| Nopp140 | Meier and Blobel (1992) |
| NSR1 | Yan and Mélèse (1993) |
| NAP57 | Meier and Blobel (1994) |
| HSP70 | Pelham (1984) |
| PCNA | Waseem and Lane (1990) |
| Annexin V | Sun *et al.* (1992) |
| Cyclin B–cdc2 complex | Ookata *et al.* (1992); Gallagher *et al.* (1993) |
| Angiogenin | Moroianu and Riordan (1994) |

[a] Ribosomal proteins not included.

structural compartments within the nucleolus. A summary of cellular proteins (ribosomal proteins not included) that have been localized to the nucleolus is shown in Table I, although these proteins are not necessarily exclusively present in the nucleolus (see Warner, 1990).

One clue on how nucleolar function might be coordinated during cell cycle progression comes from the observation that a subset of active cyclin B–cdc2 kinase complex is relocated to the nucleolus during meiosis in starfish oocytes (Ookata *et al.*, 1992) and during mitosis in fission yeast (Gallagher *et al.*, 1993). These results provide the cytological framework and are in good agreement with the identification of nucleolar proteins NO38 and nucleolin as possible mitotic substrates for phosphorylation by cyclin B–cdc2 (Belenguer *et al.*, 1990; Peter *et al.*, 1990). The existence of such a complex structure as the nucleolus raises a number of interesting questions. How is this structure assembled and disassembled at each mitotic cycle? What are the molecular interactions involved? How does the structure relate to the function? How is access to this compartment denied or granted?

Some of these questions were answered by studies on the molecular determinants responsible for the localization/recruitment of key regulatory proteins of human retroviruses HTLV-1 (Rex) and HIV-1 (Rev and Tat) to the nucleolus of infected cells (Hatanaka, 1990). Nucleolar targeting sequences, designated NOS, were mapped to the first 19 amino acids (aa) of HTLV-1 Rex protein (Siomi *et al.*, 1988) and to amino acids 35–50 and 48–57 in the middle of HIV-1 Rev (Kubota *et al.*, 1989) and Tat (Dang and Lee, 1989; Endo *et al.*, 1989) proteins, respectively. The mapping of the HIV-1 Rev nucleolar targeting sequence generated some controversy in that another group reported that sequences extending towards the N-terminus (amino acids 1–58) are necessary to direct a Rev–$\beta$-galactosidase fusion protein to the nucleolus (Cochrane *et al.*, 1990). Furthermore, nucleolar localization is required for the ability of Rev to regulate HIV gene expression, since nonnucleolar forms of the protein showed reduced function in transactivation assays, even though they were still imported into the nucleus (Kubota *et al.*, 1989; Cochrane *et al.*, 1990). Since the identification of HTLV-1 Rex NOS, other nucleolar targeting sequences have been mapped in viral proteins and are listed in Table II.

The identification of short basic sequences able to target reporter proteins to the nucleolus prompted a series of searches for such sequences in cellular nucleolar proteins. Early work identified in the human heat shock protein (HSP) 70 a sequence (amino acids 250–267) that resembles the Tat NOS. Indeed, when fused to a reporter protein, this sequence could mediate nucleolar localization, even though not as efficiently as the Tat NOS. This was interpreted as reflecting the fact that HSP70 nucleolar targeting sequence is cryptic at physiological temperature and is exposed under stressed conditions (Dang and Lee, 1989). Deletion analysis of the mouse nucleolar

TABLE II
Nucleolar Targeting Sequences in Viral Proteins

| Virus | Protein | Targeting sequence | Reference |
|---|---|---|---|
| HTLV-1 | Rex | aa 1–19 | Siomi *et al.* (1988) |
| HIV-1 | Tat | aa 48–57 | Dang and Lee (1989)<br>Endo *et al.* (1989) |
| HIV-1 | Rev | aa 35–50<br>aa 1–58 | Kubota *et al.* (1989)<br>Cochrane *et al.* (1990) |
| SFV | nsP2 | aa 470–539 | Rikkonen *et al.* (1992) |
| HSV-1 | ICP27 | aa 110–152 | Mears *et al.* (1995) |

transcription factor UBF indicated that a wide region, including the HMG-box 1 (crucial for rDNA binding) and the COOH-terminal acidic tail, is required for nucleolar accumulation (Maeda *et al.,* 1992). Similarly, deletion of 24-amino acids near the carboxy-terminus of nucleolar shuttling protein NO38 abrogates nucleolar localization, but fusion of this domain to a reporter protein did not target the hybrid protein to the nucleolus (Peculis and Gall, 1992). Extensive mutational analysis of nucleolin, the most abundant nucleolar protein in vertebrate cells, established the requirement of both the RNA-binding domains and the GAR domain for nucleolar localization (Créancier *et al.,* 1993; Messmer and Dreyer, 1993; Schmidt-Zachmann and Nigg, 1993). Similar experiments using NSR1, a yeast nucleolar protein, demonstrated that either the amino terminus or the RNA-binding domains are sufficient to target a heterologous protein to the nucleolus, and the GAR domain, even though not sufficient, facilitates nucleolar accumulation (Yan and Mélèse, 1993). More recently, a central domain (aa 21–124) was identified in the yeast nucleolar protein GAR1, which is sufficient for both nuclear uptake and nucleolar localization of a reporter protein. This protein, unlike the previous ones, does not contain a functional classical NLS nor a bipartite NLS. The authors propose that the protein could be transported to the nucleolus passively by binding to another nucleolar component (for instance, as a part of a ribonucleoprotein complex) or, alternatively, GAR1 nucleolar transport could occur via a pathway different from the other nucleolar proteins (Girard *et al.,* 1994).

Indeed, the NOS signals of Rev, Rex and Tat also share little homology, apart from a stretch of basic amino acids. Clusters of basic amino acids may suggest binding to nucleic acids (e.g., rRNA) or to acidic proteins but, per se, are not sufficient for nucleolar accumulation since other proteins share this characteristic (e.g., histones) but do not localize to the nucleolus. The fact that there is no common nucleolar targeting signal suggests that nucleolar recruitment is the result of functional interactions with other molecules already present at the nucleolus. In fact, the NOS domain of HTLV-1 Rex specifically interacts with the nucleolar shuttling protein B23, which is thought to play a role in ribosome assembly (Adachi *et al.,* 1993). Moreover, B23 and other nucleolar shuttling proteins such as nucleolin (Borer *et al.,* 1989), Noppl40 (Meier and Blobel, 1992), and NAP57 (Meier and Blobel, 1994) possibly play a dual role, in ribosome assembly (consistent with their ultrastructural localization) and in the nucleolocytoplasmic transport of ribosomal proteins into the nucleolus and preribosomal particles to the cytoplasm.

Altogether, these experiments illustrate multiple ways by which proteins (and, possibly also nucleolar snRNAs and 5S rRNA) are recruited and retained in the nucleolus dependent upon specific functional interactions with different nucleolar components. These multiple interactions establish the foundations of this biochemical and structural entity where ribosomes are formed. How is this structure reformed at each mitotic cycle? The nucleating sites at telophase are the rDNA loci (NORs) and these, upon transcriptional activation, recruit fibrillar structures (PNBs) that provide essential proteins and RNAs to assemble a functional nucleolus. PNBs are formed from material surrounding chromosomes that is either segregated to the daughter cells or synthethized very early after anaphase (Scheer and Benavente, 1990; Scheer *et al.,* 1993; see also Jiménez-Garcia *et al.,* 1994). Certain aspects of this molecular assembly process can benefit from the availability of cell-free systems for its study, one being the assembly of PNBs in *Xenopus* egg extracts (Bell *et al.,* 1992; Wu *et al.,* 1993), and another, the observation that solubilized amoeba nucleolar components can reassociate *in vitro* into nucleoli-like bodies with similar protein composition and ultrastructure (Trimbur and Walsh, 1993). A detailed analysis of the macromolecular interactions leading to this coordinated assembly should elucidate how the functional compartmentalization of the nucleolus is achieved and how this architecture enhances the efficiency of the process of ribosome biosynthesis.

## III. Sites and Organization of RNA Processing

Over the past two decades a vast amount of information was gathered about the transcriptional activation of individual genes and the mechanism of the subsequent RNA processing, but our understanding of the overall coordination of genome activity and the spatial organization of transcription, splicing, and mRNA export is still at an early stage. In addition to the nucleolus (see Section II) there are at least three other distinct RNA-containing nuclear structures identified by electron microscopy, which are generally referred to as perichromatin fibrils, interchromatin granules, and coiled bodies (Fakan and Puvion, 1980).

The characterization of these structures was greatly facilitated by the identification of antisera from autoimmune disease patients that react with splicing factors or small nuclear ribonucleoprotein particles (snRNP). These antisera as well as a variety of other antibodies and *in situ* hybridization probes against snRNP components give a speckled staining pattern, also referred to as "speckled compartment," which typically includes 20 to 50 bright spots (interchromatin granules) that seem to be connected by a less bright reticulum (perichromatin fibrils) (Spector, 1993). Furthermore, one to five nuclear foci are stained, which correspond to coiled bodies (Raska *et al.,* 1991; Carmo-Fonseca *et al.,* 1992; Huang and Spector, 1992). In addition to snRNP components, non-snRNP splicing factors including SC35 and SF2/ASF also were identified at these nuclear speckles (Fu and Maniatis, 1990; Spector *et al.,* 1991).

These three structures—interchromatin granules, perchromatin fibrils, and coiled bodies—are morphologically distinct and also differ in their composition. While snRNP components are mostly concentrated at interchromatin granules and coiled bodies, hnRNP proteins are mostly associated with nascent RNA transcripts at perichromatin fibrils (Pinol-Roma and Dreyfuss, 1993). Interestingly, the essential splicing factor U2AF was detected at coiled bodies but not at the other nuclear speckles (Carmo-Fonseca *et al.,* 1991; Zamore and Green, 1991). Coiled bodies can be specifically visualized by immunofluorescence or immunoelectron microscopy using antibodies against p80-coilin (Andrade *et al.,* 1991; Raska *et al.,* 1991). Furthermore, they contain fibrillarin, which is otherwise found only in the nucleolus (Raska *et al.,* 1990, 1991; Andrade *et al.,* 1991).

Since coiled bodies do not contain DNA, hnRNP, nor the SC35 and SF2/ASF splicing factors, it was argued that they are most likely not directly involved in transcription and pre-mRNA splicing (Lamond and Carmo-Fonseca, 1993).

Ever since these snRNP-containing nuclear structures were identified, their role in transcription and splicing has been discussed. The analysis of the spatial organization of RNA processing is hampered by the lack of methods to directly visualize the process of RNA splicing itself. *In situ* autoradiography of [$^3$H]uridine incorporation studies showed rapid labeling at perchromatin fibrils but little or no labeling at interchromatin granules or coiled bodies (Fakan *et al.,* 1976, 1984; Fakan and Nobis, 1978; Fakan and Puvion, 1980; Spector, 1990; Fakan, 1994). Moreover, recent studies using DNA, intron, and splice-junction probes, which specifically recognize pre-mRNA but not spliced mRNA, indicated that splicing occurs cotranscriptionally at the location of the gene and neither spliced mRNA nor introns were detected at interchromatin granules or coiled bodies (Zhang *et al.,* 1994). In other words, the structures with the highest concentration of splicing factors did not appear to be the sites where splicing occurs. In view of these results, it was speculated that interchromatin granules and coiled bodies might be nuclear domains for the preassembly, recycling, and perhaps storage of splicing factors (Mattaj, 1994). This hypothesis is somewhat reminiscent of nucleoli, which are also enriched in ribosomal proteins and RNA and yet are not sites of translation and mainly represent assembly sites. The problem with these studies, as mentioned before, is that they do not directly visualize the process of splicing itself; instead they identify the sites where selected mRNA splicing products and/or substrates are enriched. Previous studies examining other mRNAs have reached different conclusions, namely that fibronectin RNA tracks (representing the site of both transcription and splicing) are associated with SC35-rich subnuclear domains (Xing *et al.,* 1993). It remains to be decided what is the exception and what is the rule.

Whatever the function of these structures may be, it is intriguing that most snRNPs and splicing factors are concentrated at these sites, which raises again the question of whether they are localized at these structures or concentrated in these domains. In other words, whether they bind to some organizing scaffold or whether these domains are merely open spaces surrounded by dense chromatin. The most direct link between the nuclear matrix and RNA processing came from the identification of several nuclear matrix antigens that were localized to the speckled compartment and appeared to be involved in splicing. Two of these nuclear matrix antigens (B4A11 and B1C8) were related to the family of SR proteins (Blencowe *et al.,* 1994). Since these nuclear matrix antigens are part of the insoluble matrix and bind to splicing complexes at the same time, they could be good candidates for mediating the association of the splicing machinery with the nuclear matrix and may thus contribute to the coordination and regulation of RNA processing.

The assembly of the splicing machinery also seems to involve some sort of protein targeting. The localization of the *Drosophila* splicing regulators, *transformer* (*tra*) and *suppressor-of-white-apricot* [$su(w^a)$], to nuclear speckles was in both cases shown to be caused by an about 120 amino acid arginine/serine-rich (RS) domain. Interestingly, these RS domains could be successfully swapped between the two unrelated factors *tra* and $su(w^a)$ and seemed to be dispensible for enzyme activity but required for efficiency (Li and Bingham, 1991). It should be helpful for our understanding of the formation and function of these structures to determine what these RS domains bind to, whether it is some organizing scaffold, like perhaps the fibrils seen by electron microscopy, or just another splicing factor, which would then just lead back to the initial question of what this second factor binds to. It would be interesting to compare the RS domains from these splicing regulators with the ones of the above described nuclear matrix antigens. Considering the two lines of evidence it seems reasonable to speculate that SR proteins like *tra* and *su*($w^a$) might exert their regulatory role in part by coupling splicing complexes to the nuclear matrix.

Cell division poses several interesting problems—not only the chromosomes but also the nuclear matrix and its associated structures have to be duplicated and distributed to the two daughter cells—and involves a transient and at least partial disassembly of these otherwise insoluble structures during mitosis. It has been shown before that the speckled compartment breaks up during mitosis and splicing factors are dispersed throughout the cytoplasm (Spector and Smith, 1986). Recently, a serine kinase (SPRK1) was identified, which phosphorylates serine residues of serine/arginine-rich domain of SR proteins and regulates their subnuclear localization. Phosphorylation causes disassembly of the speckled compartment upon entry into mitosis and inhibits splicing *in vitro* indicating an important role of SRPK1 in the regulation of RNA processing and in the control of the dynamic organization of nuclear structures (Gui *et al.,* 1994a,b; Spector, 1994). Another example of subnuclear protein targeting is p80-coilin. The first 102 amino acids of p80-coilin are necessary and sufficient for targeting to sphere organelles, which appear to be the amphibian homolog of mammalian coiled bodies (Wu *et al.,* 1994).

Finally, there also seems to be a potential role for pRb at snRNP-containing nuclear structures. Several independent studies established that the early $G_1$, hypophosphorylated form of the retinoblastoma protein (pRb) is tightly bound to the insoluble nuclear framework (Mittnacht and Weinberg, 1991; Templeton *et al.,* 1991; Templeton, 1992; Mancini *et al.,* 1994; Mittnacht *et al.,* 1994). Hypophosphorylated pRb binds viral oncogenes and, for this reason, is thought to be the form active in growth suppression (DeCaprio *et al.,* 1988; Ludlow *et al.,* 1989; Imai *et al.,* 1991; Templeton *et al.,* 1991). In concert with these results, microinjection of the serine/threonine protein phosphatases types 1 (PP1) and 2A (PP2A) leads to increased nuclear affinity of pRb and to cell cycle arrest (Alberts *et al.,* 1993). Many cellular proteins have been identified that interact with pRb (Riley *et al.,* 1994; Wang *et al.,* 1994), but they are easily extractable soluble proteins and therefore not likely to play a role in "docking" to the nuclear matrix. Recently, in a screen for pRb-interacting cellular proteins using the yeast two-hybrid system and the N-terminal 300 amino acids of pRb as a bait, one gene was isolated encoding a novel nuclear matrix protein, p84, which localizes to subnuclear regions associated with RNA processing (Durfee *et al.,* 1994). Also, p84 interacts via its carboxy-terminal half with the hypophosphorylated form of pRb that is present during $G_1$ and, by immunofluorescence and confocal microscopy of preextracted cells, p84 intranuclear speckles colocalize extensively (even though not totally) with pRb, Sm antigens, and another recently identified nuclear matrix protein B1C8 (Durfee *et al.,* 1994). These results suggest a connection between the nuclear matrix (including B1C8), pRb, and RNA metabolism. Genetic evidence pointing to a functional role of the N-terminus of pRb in tumor suppression has recently been obtained from a mutation isolated from a retinoblastoma tumor, which is predicted to remove only exon 4, coding for amino acids 126–166 (Dryja *et al.,* 1993; Hogg *et al.,* 1993). The association of pRb with both matrix-bound and soluble components of the nucleus might represent a novel mechanism by which pRb exerts its growth-suppressive activity under physiological conditions and it would be interesting to examine whether pRb or its associated protein p84 have a direct role in the organization and regulation of RNA metabolism. In summary, all of these recently discovered interactions surrounding the nuclear matrix and the splicing machinery are still at a very speculative stage but they open exciting new avenues of research.

## IV. DNA Replication Foci in Mammalian Nuclei

### A. Visualization of DNA Replication Foci and Changes during S-Phase

The successful reproduction of viral replication in the test tube with isolated soluble components (Li and Kelly, 1984; Stillman and Gluzman, 1985; Wobbe *et al.,* 1985) suggested that DNA replication would occur evenly distributed throughout the nucleus. However, over the last two decades several methods for the visualization of cellular sites of DNA synthesis have been developed that clearly show that viral and cellular DNA replication occurs in discrete foci. The first methods were based on the incorporation of radioactively labeled [$^{3}$H]thymidine, which can be detected by autoradiography (Huberman *et al.,* 1973; Fakan and Hancock, 1974). More recently, a number of nonradioactive labeling techniques combined with immunofluorescence detection were developed that utilize thymidine analogs like 5-bromo-2′-deoxyuridine (BrdU) (Gratzner, 1982; Nakamura *et al.,* 1986; van Dierendonck *et al.,* 1989; Mazzotti *et al.,* 1990; Fox *et al.,* 1991) or iododeoxyuridine (IdU) and chlorodeoxyuridine (CldU) (Aten *et al.,* 1992; Manders *et al.,* 1992). Alternatively, biotin–11-dUTP labeling was used with permeablilized cells (Banfalvi *et al.,* 1989; Mills *et al.,* 1989; Nakayasu and Berezney, 1989). Most recently, fluorescently labeled nucleotides also were used for direct visualization of DNA synthesis (Hassan and Cook, 1993). It is reassuring that all these different approaches lead to the same basic conclusion that DNA replication occurs in discrete nuclear foci and the pattern of these replication foci changes in a characteristic manner throughout S-phase (for an extensive review on this subject see chapter by Berezney *et al.*).

The fact that the pattern of replication foci changes throughout S-phase can be exploited to address basic questions concerning the regulation of DNA replication. Yanishevsky and Prescott (1978) could show in cell fusion experiments that late S-phase cells induce early S-phase DNA labeling patterns in $G_1$ nuclei. These results indicate a preexisting regulatory mechanism in $G_1$ nuclei that determines the spatial and temporal order of DNA replication and is prematurely triggered by a diffusible factor from S-phase cells. Using a retrodifferentiation cellular system, Cardoso *et al.* (1993) could show that transient expression of SV40 large T antigen induces terminally differentiated myotubes to reenter S-phase with the exact same patterns of replication foci as, e.g., cycling myoblasts, indicating that this regulatory mechanism also persists in terminally differentiated cells that are permanently withdrawn from the cell cycle. Moreover, in these induced myotubes, neighboring nuclei, which share the same cytoplasm and should, therefore, have equal access to cytoplasmic factors, show very different replication patterns ranging from typical early to late S-phase patterns and include even $G_1$ nuclei with no detectable DNA synthesis. These results suggest that the nucleus represents an autonomous unit and may at least in part control the timing of S-phase (Cardoso *et al.,* 1993).

## B. Proteins Present at Replication Foci

The first step toward an understanding of molecular events involved in the exact duplication of DNA molecules was done in the late 1950s, when the group of Arthur Kornberg isolated from extracts of *Escherichia coli* an enzyme that could synthesize DNA and was accordingly named DNA polymerase. This first enzyme had, however, only a miniscule activity, which was only little over background levels and was clearly orders of magnitude below rates observed *in vivo.* During the following decades other forms of DNA polymerase and, most importantly, other subunits and accessory proteins were isolated that improved the rate and fidelity of DNA replication *in vitro* and gradually helped to narrow the gap between replication *in vitro* and *in vivo.* This development is also reflected in the terminology, starting with DNA replication enzymes over replication complexes to replication machinery. This inflation in the terminology recently culminated in the term "replication factory" coined by the group of Peter Cook (Hozák *et al.,* 1993).

While research in prokaryotes has progressed to a point where an understanding of the basic process and its control mechanisms appears to be within reach, there is not even an *in vitro* system for cellular DNA replication in mammals available and it is still debated what an origin of chromosomal DNA replication is and whether they exist in mammals to begin with. Even with all the proteins identified and characterized by genetic and biochemical approaches it remains impossible to imagine how they can work together to achieve as precise and well-coordinated a round of DNA replication as observed in mammalian cells. It remains to be explained how $3 \times 10^9$ base pairs per human cell get replicated with extremely high fidelity in a matter of hours involving up to 60,000 origins of DNA replication, which are activated at specific times during S-phase and yet each segment of the 46 chromosomes is replicated once and only once per cell cycle. A first clue about how to close the gap between *in vitro* replication results and observations in mammalian cells came from early experiments showing an association of newly synthesized DNA with the nuclear matrix (Berezney and Coffey, 1975). Later biochemical fractionation experiments showed that DNA polymerase activity was specifically associated with the insoluble fraction of the nucleus (Smith and Berezney, 1980; Noguchi *et al.,* 1983; Jackson and Cook, 1986; Tubo and Berezney, 1987). Over the past few years it has become clear that the association of DNA replication with insoluble nuclear structures and its organization into megacomplexes of up to a micrometer in size may provide the higher order structure, which make the precision, efficiency, and complex coordination of mammalian DNA replication possible. The role of the nuclear matrix in the organization and regulation of DNA replication is discussed in the chapter by Berezney *et al.* This section focuses on the view of cellular DNA replication through the microscope. As discussed in Section IV,A, DNA replication occurs at discrete foci in mammalian nuclei and can easily be visualized by pulse labeling with either biotinylated dUTP or BrdU. It is therefore possible with standard immunofluorescence microscopy to determine whether a given protein is present at sites of DNA replication *in vivo* (Cardoso and Leonhardt, 1995). The first protein to be identified at replication foci was PCNA, which shows dispersed nucleoplasmic distribution throughout the cell cycle and only in S-phase relocates to replication foci and then shows the same characteristic and changing replication patterns (Bravo and Macdonald-Bravo, 1985, 1987; Celis and Celis, 1985, Celis *et al.,* 1986). Since then a number of other proteins directly or indirectly involved in DNA replication were found to be specifically localized at these replication foci (see Table III).

In addition to PCNA DNA polymerase $\alpha$ (Hozák *et al.,* 1993) and RPA70 (Cardoso *et al.,* 1993) also were shown to be localized at sites of cellular DNA replication. RPA70 is the 70-kDa subunit of the single-stranded DNA

TABLE III
Cellular Proteins Identified at Replication Foci

| Protein | Reference |
|---|---|
| PCNA (proliferating cell nuclear antigen) | Bravo and Macdonald-Bravo (1987) |
| DNA methyltransferase | Leonhardt *et al.* (1992) |
| DNA polymerase $\alpha$ | Hozák *et al.* (1993) |
| Replication factor A, 70-kDa subunit (RPA70) | Cardoso *et al.* (1993) |
| cdk2 (cyclin-dependent kinase) | Cardoso *et al.* (1993) |
| Cyclin A | Cardoso *et al.* (1993) |
| DNA ligase I | Cardoso *et al.* (1995) |

binding replication factor A. Interestingly, the 34-kDa subunit, RPA34, was not detected at replication foci but instead appeared to be associated with condensed chromosomes during mitosis. These results may indicate a transient interaction of RPA34 or might be due to a specific distortion or masking of the epitope at replication foci, since a monoclonal antibody was used (Cardoso *et al.,* 1993). Most recently, DNA ligase I, which is involved in joining Okazaki fragments during lagging strand synthesis, was also shown to be localized at nuclear replication foci (Cardoso *et al.,* 1995). In addition to these replication factors several other proteins not directly involved in DNA replication like DNA methyltransferase also were found at replication foci (Leonhardt *et al.,* 1992). DNA methyltransferase adds methylgroups to the newly synthesized DNA strand at hemimethylated CpG sites after DNA replication and thus maintains the cellular DNA methylation pattern. In the course of experiments on the control of DNA replication the cell cycle regulators cyclin A and cdk2 also were identified at sites of cellular DNA replication (Cardoso *et al.,* 1993; for a discussion see Section IV,D). This localization of cyclin A during S-phase was also shown in HeLa cells by means of immunoelectron microscopy (Sobczak-Thepot *et al.,* 1993). Finally, even lamin B was shown to be associated with sites of DNA replication during mid to late S-phase, suggesting a potential role in the organization of replicating chromatin (Moir *et al.,* 1994). These analyses screening for proteins specifically localized at replication foci have already provided several interesting clues about the organization and regulation of nuclear replication foci and further insights are to be expected for the coming years as more proteins are screened.

Equally interesting are questions concerning the formation of these replication foci, meaning which proteins are present, in what phase of the cell cycle, and in what order do they bind. First evidence for the possible existence of such prereplication foci or centers was provided by microinjection experiments showing PCNA containing foci about 15 min before DNA synthesis was detected (Kill *et al.,* 1991). Similar conclusions were also derived from *in vitro* experiments with *Xenopus* egg extracts showing RPA70 containing foci during $G_1$-phase (Adachi and Laemmli, 1992). The dynamics of the nuclear organization also becomes evident upon viral infection, which not only forces reprogramming of the transcriptional and translational machinery, but also causes an extensive reorganization of nuclear structures (Quinlan *et al.,* 1984; de Bruyn Kops and Knipe, 1988; Walton *et al.,* 1989; for a detailed discussion of "nuclear matrix and virus function" see chapter by Deppert). A number of cellular proteins were reported to be recruited to sites of viral DNA replication including replication protein A (RPA), PCNA, retinoblastoma protein, p53, DNA ligase I, and DNA polymerase $\alpha$ (Wilcock and Lane, 1991). Interestingly, the retinoblastoma protein was not detected at sites of cellular DNA replication (H. Leonhardt and M. C. Cardoso, unpublished results). It will be interesting to compare the structure of viral and cellular replication foci. The changes induced by viral infection may reveal some of the organizing principles of mammalian nuclei, much like it has been used to elucidate cell cycle regulation.

## C. Targeting of Proteins to Replication Foci

### 1. DNA Methyltransferase

In Section IV,B a number of proteins were described that are specifically localized at nuclear replication foci (see Table III). In the absence of dividing membranes, which could explain this compartmentalization, it seems reasonable to assume that this localization is caused by binding to some sort of insoluble structures. This hypothesis was first put to a test in the case of the mammalian DNA methyltransferase (DNA MTase), which is localized at sites of DNA replication (Leonhardt *et al.,* 1992). Mammalian DNA is methylated in both strands at about 50–80% of all CpG dinucleotides. After DNA replication, the newly synthesized DNA strand is at first unmethylated resulting in hemimethylated CpG sites, which are then specifically recognized and methylated by DNA MTase, precisely propagating the previous DNA methylation pattern. Already partial inactivation of DNA MTase causes embryonic lethality demonstrating the importance of the precise maintenance of methylation patterns for differentiation and mammalian development in general (Li *et al.,* 1992, 1993).

The mammalian DNA MTase has a C-terminal, catalytic domain of about 500 amino acids and an N-terminal regulatory domain of about 1000 amino acids (Bestor *et al.,*1988; Leonhardt and Bestor, 1993). The enzyme has a strong preference for hemimethylated CpG sites, which is consistent with a role in the maintenance of methylation patterns (Gruenbaum *et al.,* 1982; Bestor and Ingram, 1983). The specific localization of DNA MTase at replication foci could, therefore, be caused by high substrate concentration, since hemimethylated sites are generated by DNA replication and should be most abundant at these foci. Alternatively, the specific localization could be achieved by a targeting sequence that concentrates the enzyme at these replication foci by some sort of docking process. To test this alternative, various parts of the DNA MTase gene were fused to the totally unrelated $\beta$-galactosidase gene from *E. coli* and expressed in mouse fibroblast cells. The distribution of the fusion protein can be determined with antibodies recognizing an epitope in the $\beta$-galactosidase part and can be compared to the distribution of the endogenous enzyme or—for that matter—any other cellular components that can be visualized by fluorescent microscopy techniques. With this experimental approach a targeting sequence was identified, which is necessary and sufficient for targeting to replication foci and by itself causes the localization of a protein as unrelated as the bacterial $\beta$-galactosidase to replication foci (Fig. 1A–1C, color plate 12). For comparison, targeting to the speckled compartment using an RS domain fusion is shown in Fig. 1D–1F, color plate 12). The targeting sequence of the DNA MTase is located in the N-terminal, regulatory domain and is dispensible for enzyme activity *in vitro* (Leonhardt *et al.,* 1992; Fig. 2).

The DNA MTase targeting sequence is surprisingly hydrophobic and large (about 250 amino acids) suggesting that most of it has a function in the globular folding of the enzyme placing short and discontinuous stretches of amino acids into the right three-dimensional configuration to form the binding interface. Searches of protein and DNA databases failed to identify other proteins with sequence similarity suggesting that the DNA MTase targeting sequence is rather unique and not a member of a general new type of protein motif. The definition of this targeting sequence should now make it possible to identify the component(s) of the replication foci that DNA MTase binds to and should contribute to our understanding of the architecture of replication factories. It is tempting to speculate that the targeting sequence plays an important role in the coordination of DNA replication and methylation *in vivo* by placing the DNA MTase at the right position in the replication factory. The efficient recognition and methylation of hemimethylated sites would then not simply rely on diffusion and random encounters but would be achieved by one processive replication-methylation machinery or assembly line (see model, Fig. 3).

## 2. DNA Ligase I

The identification of a targeting sequence in case of the mammalian DNA MTase raised the question whether other proteins also use a similar princi-

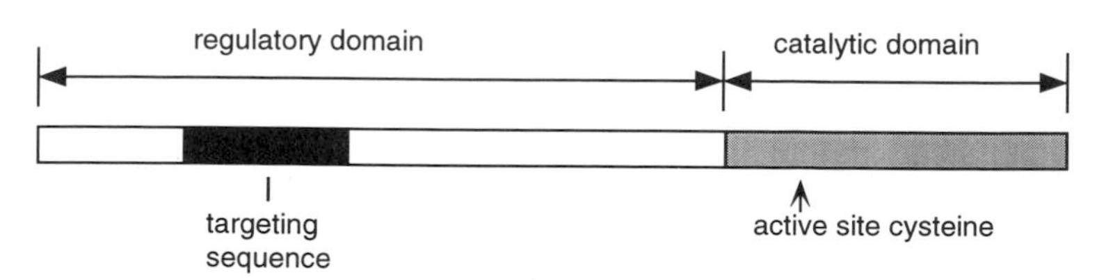


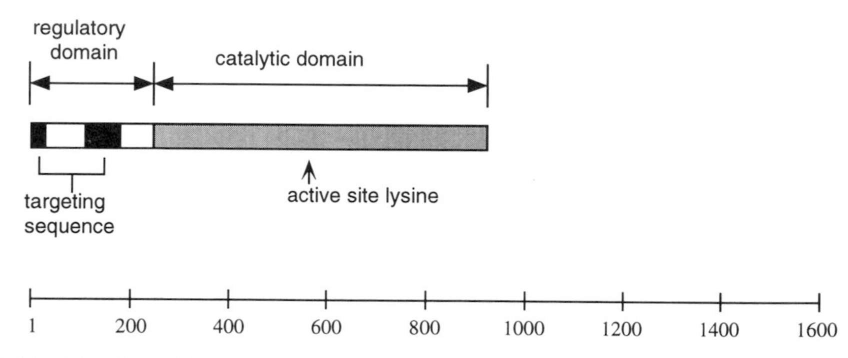


FIG. 2 Mapping of the targeting sequence in the mouse DNA methyltransferase and human DNA ligase I proteins. Drawings are to scale and numbers at the bottom represent amino acids.

ple for localization at replication foci. A good candidate was the DNA ligase I, which is involved in joining Okazaki fragments during lagging-strand synthesis and in excision-repair (Barnes *et al.,* 1992). The human DNA ligase I has 919 amino acids and the catalytic center is around the active site lysine residue at position 568 (Barnes *et al.,* 1990). The enzyme has, like DNA MTase, a protease-sensitive N-terminal domain of 249 amino acids, which has a negative regulatory function and is dispensable for enzyme activity *in vitro* and for complementation of yeast and bacterial ligase mutants (Kodama *et al.,* 1991; see Fig. 2). DNA ligase is activated at the onset of S-phase by phosphorylation of the N-terminal regulatory domain (Prigent *et al.,* 1992). Biochemical fractionation experiments showed that DNA ligase I is part of a large protein complex of replication factors (Malkas *et al.,* 1990) and is localized at replication foci *in vivo* (Cardoso *et al.,* 1995). To study the basis of this subnuclear localization pattern the same approach was used that had led to the identification of the DNA MTase targeting sequence (Leonhardt *et al.,* 1992; see Section IV,C,1). The analysis of a set of epitope-tagged deletion mutants identified a bipartite targeting sequence within the N-terminal, regulatory domain of the human

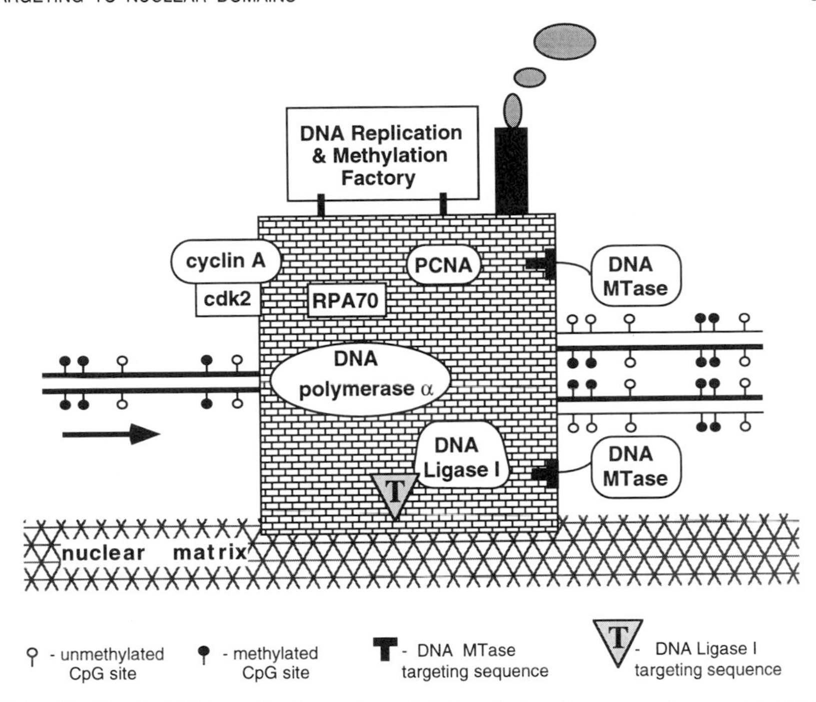


FIG. 3 Model of a DNA replication and methylation factory in mammalian nuclei. DNA is reeled through the factory, which is attached to the nuclear matrix. In this process DNA is duplicated and the hemimethylated newly synthesized DNA is methylated in an assembly-line fashion. The proteins identified to date at subnuclear replication foci are named and the remaining building blocks of this factory (in the form of bricks) represent all the other proteins necessary for DNA replication and methylation. Targeting sequences responsible for association with these foci are depicted for DNA methyltransferase and DNA ligase I.

DNA ligase I (Cardoso *et al.,* 1995; see also Fig. 2). This targeting sequence is in functional terms very similar to that of the DNA MTase, as it is dispensable for enzyme activity *in vitro,* can by itself direct unrelated proteins like $\beta$-galactosidase to replication foci, and is necessary and sufficient for targeting. A direct comparison of both sequences, however, failed to detect any similarities and in fact they are quite different. The DNA ligase I targeting sequence is very hydrophilic, while the DNA MTase sequence is on the contrary very hydrophobic. These results suggest that the two enzymes bind to different components of the replication factory, which makes sense in light of their different enzymatic activities. A comparison with other DNA ligases showed that the targeting sequence is highly conserved in mammals (mouse and human) but is absent in lower eukaryotes like yeast, suggesting that targeting is a rather recent but important development in evolution. It is noteworthy that both enzymes are involved in cellular processes where efficiency is critical, DNA MTase has to methylate in the order of $10^8$ hemimethylated sites and DNA ligase has to join about $10^6$–$10^7$ Okazaki fragments per cell cycle (assuming that Okazaki fragments are 200 to 1000 bp in size). In both cases it is crucial for the viability of the cell that these processes are totally completed; any omissions might have serious consequences. The targeting of DNA ligase I to replication foci might be a mechanism used by mammalian cells to ensure this efficiency (see model, Fig. 3).

### 3. Other Candidates

In addition to proteins already identified at replication foci (see Table III), there are many more proteins involved in SV40 replication that are, therefore, also expected to be involved in cellular DNA replication. Complete SV40 replication can be obtained with SV40 large T antigen, replication protein A, DNA topoisomerase I and II, DNA polymerase $\alpha$-primase, replication factor C, PCNA, DNA polymerase $\delta$, maturation factor I, and DNA ligase I (Waga *et al.,* 1994). Most—if not all—of these proteins together with still unidentified factors are expected to be localized at replication foci. It should, therefore, be interesting to determine what other factors are localized at replication foci and whether distinct targeting sequences can be identified using the same approach as described above. Obviously, not all of these proteins are expected to have distinct targeting sequence. Especially smaller proteins, e.g., cdk2, which are altogether as big (or small) as the targeting sequence of the DNA MTase, are not likely to have a separate targeting sequence. Nonetheless, the analysis of epitope-tagged deletion mutants should identify regions that are necessary for localization at replication foci, even if they are part of the catalytic domain. It would then be a semantic question whether it can still be called targeting sequence. In any event, the identification of such sequences can be used to screen for interacting proteins, which should lead to the identification of other known and unknown proteins. These results would then outline a network of interactions and should definitely help to elucidate the architecture of replication foci *in vivo.*

Finally, an interesting candidate for having a targeting sequence is DNA topoisomerase I. This enzyme has, like DNA ligase I and DNA MTase, a protease-sensitive N-terminal domain, which is dispensable for enzyme activity *in vitro* (Bjornsti and Wang, 1987; D'Arpa *et al.,* 1988; Alsner *et al.,* 1992) and is required for complete SV40 DNA replication (Yang *et al.,* 1987).

## D. A Link between Cell Cycle Regulation and DNA Replication

Our understanding of cell cycle regulation is continuously challenged by the ever-increasing number of new cyclins, cyclin-dependent kinases, and other regulatory proteins, such as the recently identified family of cyclin-cdk inhibitors. The most critical step in cell cycle progression is the control of S-phase entry. Biochemical analyses showed that the phosphorylation of RPA34 is a key event at the initiation of DNA replication (Din *et al.*, 1990). The problem is that, as is the case with other putative substrates for phosphorylation by cyclin–cdk complexes, RPA34 can be phosphorylated *in vitro* by several distinct complexes including cyclins A, B, or E and cdc2 or cdk2 (Fotedar and Roberts, 1991; Dutta and Stillman, 1992; Elledge *et al.*, 1992). Do these different cyclin–cdk complexes have overlapping substrate specificity *in vivo*? Or, alternatively, do they carry out specific tasks *in vivo* that are nonredundant? If so, how can the physiologically meaningful complex be sorted out?

Overexpression and inhibition experiments suggest that different cyclins and cdks target different substrates and act at distinct times during the cell cycle. Immunofluorescence analysis in mammalian cells induced to reenter S-phase showed specific colocalization of cyclin A and cdk2, but not of cyclin B and cdc2, with subnuclear sites of ongoing DNA synthesis (Fig. 4; color plate 13) (Cardoso *et al.*, 1993). Similar results were obtained for cyclin A by immunoelectronmicroscopy (Sobczak-Thepot *et al.*, 1993). Colocalization was observed with DNA replication patterns typical of early, mid, and late S-phase suggesting that cyclin A–cdk2 acts throughout S-phase and not only at the beginning. This cell cycle-dependent redistribution of cyclin A and cdk2 perfectly matches the one of replication protein PCNA (Cardoso *et al.*, 1993). Localization of cyclin–cdk complexes to specific domains might explain how they can have overlapping substrate specificity *in vitro* but carry out specific and nonredundant tasks *in vivo*. Obviously, colocalization by itself does not prove any involvement but it forms a consistent picture, together with biochemical data, showing that cyclin A and cdk2 are at the right place at the right time to "trigger" DNA replication. These results demonstrate how the functional organization of the mammalian nucleus can provide important clues about protein function *in vivo* and strongly suggest that cyclin A–cdk2 is the "S phase-promoting factor" or "replication factor S" postulated by Roberts and D'Urso (1988; Reed, 1991), providing a direct link between cell cycle regulation and DNA replication.

## V. Insoluble Solutions Derived from Functional Domains in the Nucleus

The most direct way to regulate an enzyme is through its abundance by either synthesis or degradation. Alternatively, the subcellular localization of enzymes may be used to control their biological activity. A very obvious case is the cytoplasmic versus nuclear localization and well-known examples for this type of regulation are NF-κB, Rel, and dorsal, which were shown to be regulated at the level of nuclear uptake (Schmitz *et al.,* 1991). The functional organization of the nucleus represents another potential level for regulating the activity of nuclear proteins. The activity of proteins may be positively regulated by targeting a protein to the appropriate subnuclear compartment containing the respective substrate. The local enzyme and substrate concentration is thereby dramatically increased and consequently also the rate of catalysis. The increase in velocity due to protein targeting and integration into complex protein machineries can only be estimated at this moment. On the other hand, preventing targeting of proteins to a particular subnuclear compartment may be as effective as preventing its nuclear uptake. Interestingly, several smaller replication foci appeared to have lower or no detectable levels of DNA MTase, which could have simple technical reasons or might point to a replicon-specific regulation of DNA methylation at the level of targeting (Leonhardt *et al.,* 1992). The identification of these intranuclear targeting sequences makes it possible to screen for interacting factors and post-translational modification which may control protein targeting within the nucleus. The recently discovered SRPK1 kinase phosphorylating the RS domain of SR proteins and thus causing the break up of the speckled compartment (Gui *et al.,* 1994a; see Section III) may fall into this category. A particular subnuclear site may also constitute a "protective environment" shielding proteins from degradation or modifying factors, which would lead to their inactivation. Finally, there are additional subnuclear compartments of still unknown function like the nuclear dots and PIKA (Ascoli and Maul, 1991; Saunders *et al.,* 1991, respectively), which contribute to the complex organization of mammalian nuclei.

Almost by definition, the functional organization provides valuable clues about protein function and, hence, may help with model building. Since, e.g., several replication proteins were found to be specifically localized to nuclear replication foci (Table III) it seems reasonable to expect that other proteins identified at these foci might be involved in some aspects of DNA replication, just as the nuclear localization of proteins is usually taken as an indication for a role in the nucleus. One example for this approach is the identification of cyclin A and cdk2 at replication foci providing a link between cell cycle regulation and DNA replication (Cardoso *et al.,* 1993; Section IV,D). Obviously, localization by itself is not a proof and proteins caught "in flagranti" might have actually been "innocent bystanders," but these localization studies complement biochemical and genetic data and can provide compelling evidence about protein function *in vivo.* The integration of various enzyme activities into complex and insoluble macromolecular machineries is likely to provide the efficiency, specificity, and coordination of biological processes *in vivo* and is discussed throughout this chapter (see also Fig. 3). These higher order structures, however, defy the principle of traditional biochemical analysis in two ways. First, biochemistry mostly deals with the soluble fraction of the cell and hence misses all insoluble structures. Second, the traditional biochemical approach is to purify cellular components and characterize them *in vitro,* but this takes them out of their "natural environment" or context and may identify affinities to other cellular components that they never encounters *in vivo.*

For many years, research on the nuclear matrix and its associated components has been hampered by technical difficulties (Cook, 1988). It is now that a number of experimental approaches are in place to explore the functional organization of the nucleus and to determine its role in the regulation of biological processes. On the biochemical side, methods have been developed to prepare nuclear matrix samples for functional assays and matrix components are being identified by monoclonal antibodies, cloned and sequenced. Moreover, the association of nuclear components with functional domains can be studied by immunofluorescence microscopy and deletion analyses. Most recently, a method for the direct visualization of the functional organization of the nucleus also was developed. The green fluorescent protein (GFP; Chalfie *et al.,* 1994) was fused with the targeting sequence of the DNA MTase and expressed in mammalian cells. The fusion protein, which is by itself fluorescent, is targeted to nuclear replication foci and can be followed in living cells allowing the direct visualization of nuclear structures *in vivo* and to follow its dynamic reorganization during the cell cycle and in response to external stimuli (Leonhardt *et al.,* 1995). It is now feasible to study the functional organization of the nucleus in real time, altogether avoiding potential fixation and staining artifacts. Furthermore, this approach can easily be adapted for the analysis of other functional domains by replacing the targeting sequence with any of the above described nucleolar targeting signals or the RS domain of, e.g., the splicing regulator *tra.*

## VI. Concluding Remarks

After decades of breathtaking progress at all fronts of science and technology providing a plethora of data on an ever-increasing number of molecules, we still seem to be far from really understanding living organisms. The reductive approach of science dissecting complex problems into smaller and more manageable questions was extremely successful and led to the discovery and characterization of RNA, DNA, and proteins, the smallest building blocks of living organisms, but something still seems to be missing. At the biochemical and genetic level humans, e.g., compare to chimpanzees as, e.g., a fox to a dog, but we write poetry or fly to the moon and they do not. Likewise, putting all the known and unknown components of the nucleus together in a test tube is far from working like a nucleus *in vivo* and is likely to end up in a mess. The answer for both dilemmas seems to lie in higher order structures and the challenge for the next century is to piece all the individual building blocks together expecting that the cell and the organism is much more than just the sum of the properties of its components.

At the cellular level, one of the first steps in this direction is to examine the three-dimensional architecture of the nucleus and to determine the subnuclear localization of the cellular components during cell cycle and differentiation. Just as at present protein expression and post-translational modification profiles are established to understand their regulatory effects during, e.g., the cell cycle, it will be necessary to also take into account the subcellular localization and hence limited spatial availability of cellular components. In other words, the subcellular and subnuclear localization of proteins represents yet another variable in the overall regulatory networks. The functional organization of the nucleus changes the equations in enzyme kinetics; protein targeting and the assembly into huge protein machineries or factories appears to be one way to integrate complex biochemical processes and to achieve efficiency and coordination. The latter is by no means a biological luxury but rather can make all the difference in such "competitive" environments as living cells and organisms.

One of the most intriguing questions in this context is whether the individual components contain all the information necessary for the assembly of these higher order structures. Ribosomes, e.g., can be assembled *in vitro,* if the individual components are added in the correct order using appropriate conditions. But still a mixture of ribosomal RNA and proteins will not spontaneously assemble efficiently *in vitro* and even less so if all the other cellular proteins are present. This problem seems to be solved *in vivo* by the nucleolus, which is a subnuclear compartment dedicated to the assembly of ribosomal subunits where all the ribosomal components and participating enzymes are concentrated while other proteins are excluded. Likewise, it would make sense that interchromatin granules and coiled bodies are domains dedicated to the assembly of splicing particles (spliceosomes). Moreover, the identification of distinct targeting sequences in a variety of molecules, ranging from replication proteins to splicing regulators and nucleolar proteins, suggests that the information for the higher order structures seems to be built-in at least in some cases. On the other hand, a homogenate of all cellular components does not spontaneously assemble into living cells. It is tempting to speculate that these higher order structures are duplicated and passed on to daughter cells just like chromosomes are and possibly by association with the chromosomes themselves. We would, furthermore, like to speculate that these higher order structures evolved together with the genome in an interdependent fashion, one dependent on the other, meaning the structures cannot be maintained and propagated without the genome and vice versa.

Almost two centuries ago Goethe formulated Faust's thirst for knowledge as a desire to understand what ultimately holds the world together (". . . Daß ich erkenne, was die Welt im Innersten zusammenhält . . ."). Besides the metaphysical component of this question, the problem could be rephrased for today's scientists as how "simple" molecules are assembled to form higher order structures and complex systems like nuclei, cells, organs, organisms, ecosystems, and galaxies.

## References


Adachi, Y., and Laemmli, U. K. (1992). Identification of nuclear pre-replication centers poised for DNA synthesis in *Xenopus* egg extracts: Immunolocalization study of replication protein A. *J. Cell Biol.* **119,** 1 15.

Adachi, Y., Copeland, T. D., Hatanaka, M., and Oroszlan, S. (1993). Nucleolar targeting signal of Rex protein of human T-cell leukemia virus type I specifically binds to nucleolar shuttle protein B-23. *J. Biol. Chem.* **268,** 13930–13934.

Alberts, A. S., Thorburn, A. M., Shenolikar, S., Mumby, M. C., and Feramisco, J. R. (1993). Regulation of cell cycle progression and nuclear affinity of the retinoblastoma protein by protein phosphatases. *Proc. Natl. Acad. Sci. U. S. A.* **90,** 388–392.

Alsner, J., Svejstrup, J. Q., Kjeldsen, E., Sorensen, B. S., and Westergaard, O. (1992). Identification of an N-terminal domain of eukaryotic DNA topoisomerase I dispensable for catalytic activity but essential for in vivo function. *J. Biol. Chem.* **267,** 12408–12411.

Andrade, L. E., Chan, E. K., Raska, I., Peebles, C. L., Roos, G., and Tan, E. M. (1991). Human autoantibody to a novel protein of the nuclear coiled body: Immunological characterization and cDNA cloning of p80-coilin. *J. Exp. Med.* **173,** 1407–1419.

Ascoli, C. A., and Maul, G. G. (1991). Identification of a novel nuclear domain. *J. Cell Biol.* **112,** 785–795.

Aten, J. A., Bakker, P. J., Stap, J., Boschman, G. A., and Veenhof, C. H. (1992). DNA double labelling with IdUrd and CldUrd for spatial and temporal analysis of cell proliferation and DNA replication. *Histochem. J.* **24,** 251–259.

Banfalvi, G., Wiegant, J., Sarkar, N., and van Duijn, P. (1989). Immunoflourescent visualization of DNA replication sites within nuclei of Chinese hamster ovary cells. *Histochemistry* **93,** 81–86.

Barnes, D. E., Johnston, L. H., Kodama, K., Tomkinson, A. E., Lasko, D. D., and Lindahl, T. (1990). Human DNA ligase I cDNA: Cloning and functional expression in *Saccharomyces cerevisiae. Proc. Natl. Acad. Sci. U.S.A.* **87,** 6679–6683.

Barnes, D. E., Tomkinson, A. E., Lehmann, A. R., Webster, A. D., and Lindahl, T. (1992). Mutations in the DNA ligase I gene of an individual with immunodeficiencies and cellular hypersensitivity to DNA-damaging agents. *Cell (Cambridge, Mass.)* **69,** 495–503.

Belenguer, P., Caizergues-Ferrer, M., Labbé, J. C., Dorée, M., and Amalric, F. (1990). Mitosis-specific phosphorylation of nucleolin by p34cdc2 protein kinase. *Mol. Cell. Biol.* **10,** 3607–3618.

Bell, P., Dabauvalle, M. C., and Scheer, U. (1992). In vitro assembly of prenucleolar bodies in *Xenopus* egg extract. *J. Cell Biol.* **118,** 1297–1304.

Berezney, R., and Coffey, D. S. (1975). Nuclear protein matrix: Association with newly synthesized DNA. *Science* **189,** 291–293.

Bestor, T. H., and Ingram, V. M. (1983). Two DNA methyltransferases from murine erythroleukemia cells: Purification, sequence specificity, and mode of interaction with DNA. *Proc. Natl. Acad. Sci. U.S.A.* **80,** 5559–5563.

Bestor, T. H., Laudano, A., Mattaliano, R., and Ingram, V. M. (1988). Cloning and sequencing of a cDNA encoding DNA methyltransferase of mouse cells. The carboxyl-terminal domain of the mammalian enzymes is related to bacterial restriction methyltransferases. *J. Mol. Biol.* **203,** 971–983.

Bjornsti, M. A., and Wang, J. C. (1987). Expression of yeast DNA topoisomerase I can complement a conditional-lethal DNA topoisomerase I mutation in *Escherichia coli. Proc. Natl. Acad. Sci. U.S.A.* **84,** 8971–8975.

Blencowe, B. J., Nickerson, J. A., Issner, R., Penman, S., and Sharp, P. A. (1994). Association of nuclear matrix antigens with exon-containing splicing complexes. *J. Cell Biol.* **127,** 593–607.

Borer, R. A., Lehner, C. F., Eppenberger, H. M., and Nigg, E. A. (1989). Major nucleolar proteins shuttle between nucleus and cytoplasm. *Cell (Cambridge, Mass.)* **56,** 379–390.

Bravo, R., and Macdonald-Bravo, H. (1985). Changes in the nuclear distribution of cyclin (PCNA) but not its synthesis depend on DNA replication. *EMBO J.* **4,** 655–661.

Bravo, R., and Macdonald-Bravo, H. (1987). Existence of two populations of cyclin/proliferating cell nuclear antigen during the cell cycle: Association with DNA replication sites. *J. Cell Biol.* **105,** 1549–1554.

Cardoso, M. C., and Leonhardt, H. (1995). Immunofluorescence techniques in cell cycle studies. *In* "Cell Cycle: Materials and Methods" (M. Pagano, ed.), pp. 15–28. Springer-Verlag, Heidelberg (in press).

Cardoso, M. C., Leonhardt, H., and Nadal-Ginard, B. (1993). Reversal of terminal differentiation and control of DNA replication: Cyclin A and Cdk2 specifically localize at subnuclear sites of DNA replication. *Cell (Cambridge, Mass.)* **74,** 979–992.

Cardoso, M. C., Joseph, C., Nadal-Ginard, B., and Leonhardt, H. (1995). Targeting of human DNA Ligase I to nuclear sites of DNA replication. Submitted for publication.

Carmo-Fonseca, M., Tollervey, D., Pepperkok, R., Barabino, S. M., Medes, A., Brunner, C., Zamore, P. D., Green, M. R., Hurt, E., and Lamond, A. I. (1991). Mammalian nuclei contain foci which are highly enriched in components of the pre-mRNA splicing machinery. *EMBO J.* **10,** 195–206.

Carmo-Fonseca, M., Pepperkok, R., Carvalho, M. T., and Lamond, A. I. (1992). Transcription-dependent colocalization of the U1, U2, U4/U6, and U5 snRNPs in coiled bodies. *J. Cell Biol.* **117,** 1–14.

Celis, J. E., and Celis, A. (1985). Cell cycle-dependent variations in the distribution of the nuclear protein cyclin proliferating cell nuclear antigen in cultured cells: Subdivision of S phase. *Proc. Natl. Acad. Sci. U.S.A.* **82,** 3262–3266.

Celis, J. E., Madsen, P., Nielsen, S., and Celis, A. (1986). Nuclear patterns of cyclin (PCNA) antigen distribution subdivide S-phase in cultured cells—some applications of PCNA antibodies. *Leuk. Res.* **10,** 237–249.

Chalfie, M., Tu, Y., Euskirchen, G., Ward, W. W., and Prasher, D. C. (1994). Green fluorescence protein as a marker for gene expression. *Science* **263,** 802–805.

Cochrane, A. W., Perkins, A., and Rosen, C. A. (1990). Identification of sequences important in the nucleolar localization of human immunodeficiency virus Rev: Relevance of nucleolar localization to function. *J. Virol.* **64,** 881–885.

Cook, P. R. (1988). The nucleoskeleton: Artefact, passive framework or active site? *J. Cell Sci.* **90,** 1–6.

Créancier, L., Prats, H., Zanibellato, C., Amalric, F., and Bugler, B. (1993). Determination of the functional domains involved in nucleolar targeting of nucleolin. *Mol. Biol. Cell.* **4,** 1239–1250.

Dang, C. V., and Lee, W. M. (1989). Nuclear and nucleolar targeting sequences of c-erb-A, c-myb, N-myc, p53, HSP70, and HIV tat proteins. *J. Biol. Chem.* **264,** 18019–18023.

D'Arpa, P., Machlin, P. S., Ratrie, H., 3rd, Rothfield, N. F., Cleveland, D. W., and Earnshaw, W. C. (1988). cDNA cloning of human DNA topoisomerase I: Catalytic activity of a 67.7-kDa carboxyl-terminal fragment. *Proc. Natl. Acad. Sci. U.S.A.* **85,** 2543–2547.

de Bruyn Kops, A., and Knipe, D. M. (1988). Formation of DNA replication structures in herpes virus-infected cells requires a viral DNA binding protein. *Cell (Cambridge, Mass.)* **55,** 857–868.

DeCaprio, J. A., Ludlow, J. W., Figge, J., Shew, J. Y., Huang, C. M., Lee, W. H., Marsilio, E., Paucha, E., and Livingston, D. M. (1988). SV40 large tumor antigen forms a specific complex with the product of the retinoblastoma susceptibility gene. *Cell (Cambridge, Mass.)* **54,** 275–283.

Din, S., Brill, S. J., Fairman, M. P., and Stillman, B. (1990). Cell-cycle-regulated phosphorylation of DNA replication factor A from human and yeast cells. *Genes Dev.* **4,** 968–977.

Dryja, T. P., Rapaport, J., McGee, T. L., Nork, T. M., and Schwartz, T. L. (1993). Molecular etiology of low-penetrance retinoblastoma in two pedigrees. *Am. J. Hum. Genet.* **52,** 1122–1128.

Dundr, M., and Raska, I. (1993). Nonisotopic ultrastructural mapping of transcription sites within the nucleolus. *Exp. Cell Res.* **208,** 275–281.

Durfee, T., Mancini, M. A., Jones, D., Elledge, S. J., and Lee, W. H. (1994). The amino-terminal region of the retinoblastoma gene product binds a novel nuclear matrix protein that co-localizes to centers for RNA processing. *J. Cell Biol.* **127,** 609–622.

Dutta, A., and Stillman, B. (1992). cdc2 family kinases phosphorylate a human cell DNA replication factor, RPA, and activate DNA replication. *EMBO J.* **11,** 2189–2199.

Elledge, S. J., Richman, R., Hall, F. L., Williams, R. T., Lodgson, N., and Harper, J. W. (1992). CDK2 encodes a 33-kDa cyclin A-associated protein kinase and is expressed before CDC2 in the cell cycle. *Proc. Natl. Acad. Sci. U.S.A.* **89,** 2907–2911.

Endo, S., Kubota, S., Siomi, H., Adachi, A., Oroszlan, S., Maki, M., and Hatanaka, M. (1989). A region of basic amino-acid cluster in HIV-1 Tat protein is essential for trans-acting activity and nucleolar localization. *Virus Genes* **3,** 99–110.

Fakan, S. (1994). Perichromatin fibrils are in situ forms of nascent transcripts. *Trends Cell Biol.* **4,** 86–90.

Fakan, S., and Hancock, R. (1974). Localizaion of newly-synthesized DNA in a mammalian cell as visualized by high resolution autoradiography. *Exp. Cell Res.* **83,** 95–102.

Fakan, S., and Nobis, P. (1978). Ultrastructural localization of transcription sites and of RNA distribution during the cell cycle of synchronized CHO cells. *Exp. Cell Res.* **113,** 327–337.

Fakan, S., and Puvion, E. (1980). The ultrastructural visualization of nucleolar and extranucleolar RNA synthesis and distribution. *Int. Rev. Cytol.* **65,** 255–299.

Fakan, S., Puvion, E., and Sphor, G. (1976). Localization and characterization of newly synthesized nuclear RNA in isolate rat hepatocytes. *Exp. Cell Res.* **99,** 155–164.

Fakan, S., and Puvion, E. (1980). The ultrastructural visualization of nucleolar and extranucleolar RNA synthesis and distribution. *Int. Rev. Cytol.* **65,** 255–299.

Fotedar, R., and Roberts, J. M. (1991). Association of p34cdc2 with replicating DNA. *Cold Spring Harbor Symp. Quant. Biol.* **56,** 325–333.

Fox, M. H., Arndt-Jovin, D. J., Jovin, T. M., Baumann, P. H., and Robert-Nicoud, M. (1991). Spatial and temporal distribution of DNA replication sites localized by immunofluorescence and confocal microscopy in mouse fibroblasts. *J. Cell Sci.* **99,** 247–253.

Franke, W. W., Kleinschmidt, J. A., Spring, H., Krohne, G., Gründ, C., Trendelenburg, M. F., Stoehr, M., and Scheer, U. (1981). A nucleolar skeleton of protein filaments demonstrated in amplified nucleoli of *Xenopus laevis*. *J. Cell Biol.* **90,** 289–299.
Fu, X. D., and Maniatis, T. (1990). Factor required for mammalian spliceosome assembly is localized to discrete regions in the nucleus. *Nature (London)* **343,** 437–441.
Gallagher, I. M., Alfa, C. E., and Hyams, J. S. (1993). p63cdc13, a B-type cyclin, is associated with both the nucleolar and chromatin domains of the fission yeast nucleus. *Mol. Biol. Cell* **4,** 1087–1096.
Girard, J. P., Lehtonen, H., Caizergues-Ferrer, M., Amalric, F., Tollervey, D., and Lapeyre, B. (1992). GAR1 is an essential small nucleolar RNP protein required for pre-rRNA processing in yeast. *EMBO J.* **11,** 673–682.
Girard, J. P., Bagni, C., Caizergues-Ferrer, M., Amalric, F., and Lapeyre, B. (1994). Identification of a segment of the small nucleolar ribonucleoprotein-associated protein GAR1 that is sufficient for nucleolar accumulation. *J. Biol. Chem.* **269,** 18499–18506.
Granboulan, N., and Granboulan, P. (1965). Cytochimie ultrastructurale du nucléole. II. Etude des sites de synthèse du RNA dans le nucléole et le noyau. *Exp. Cell Res.* **38,** 604–619.
Gratzner, H. G. (1982). Monoclonal antibody to 5-bromo- and 5-iododeoxyuridine: A new reagent for detection of DNA replication. *Science* **218,** 474–475.
Gruenbaum, Y., Cedar, H., and Razin, A. (1982). Substrate and sequence specificity of a eukaryotic DNA methylase. *Nature (London)* **295,** 620–622.
Gui, J. F., Lane, W. S., and Fu, X. D. (1994a). A serine kinase regulates intracellular localization of splicing factors in the cell cycle. *Nature (London)* **369,** 678–682.
Gui, J. F., Tronchere, H., Chandler, S. D., and Fu, X. D. (1994b). Purification and characterization of a kinase specific for the serine- and arginine-rich pre-mRNA splicing factors. *Proc. Natl. Acad. Sci. U.S.A.* **91,** 10824–10828.
Hassan, A. B., and Cook, P. R. (1993). Visualization of replication sites in unfixed human cells. *J. Cell Sci.* **105,** 541–550.
Hatanaka, M. (1990). Discovery of the nucleolar targeting signal. *BioEssays* **12,** 143–148.
Hogg, A., Bia, B., Onadim, Z., and Cowell, J. K. (1993). Molecular mechanisms of oncogenic mutations in tumors from patients with bilateral and unilateral retinoblastoma. *Proc. Natl. Acad. Sci. U.S.A.* **90,** 7351–7355.
Hozák, P. (1995). Catching RNA polymerase I in flagranti-ribosomal genes are transcribed in the dense fibrillar component of the nucleolus. *Exp. Cell Res.* **216,** 285–289.
Hozák, P., Hassan, A. B., Jackson, D. A., and Cook, P. R. (1993). Visualization of replication factories attached to nucleoskeleton. *Cell (Cambridge, Mass.)* **73,** 361–373.
Hozák, P., Cook, P. R., Schofer, C., Mosgoller, W., and Wachtler, F. (1994). Site of transcription of ribosomal RNA and intranucleolar structure in HeLa cells. *J. Cell Sci.* **107,** 639–648.
Huang, S., and Spector, D. L. (1992). U1 and U2 small nuclear RNAs are present in nuclear speckles. *Proc. Natl. Acad. Sci. U.S.A.* **89,** 305–308.
Huberman, J. A., Tsai, A., and Deich, R. A. (1973). DNA replication sites within the nuclei of mammalian cells. *Nature (London)* **241,** 32–36.
Hughes, J. M., and Ares, M., Jr. (1991). Depletion of U3 small nucleolar RNA inhibits cleavage in the 5′ external transcribed spacer of yeast pre-ribosomal RNA and impairs formation of 18S ribosomal RNA. *EMBO J.* **10,** 4231–4239.
Imai, Y., Matsushima, Y., Sugimura, T., and Terada, M. (1991). Purification and characterization of human papillomavirus type 16 E7 protein with preferential binding capacity to the underphosphorylated form of retinoblastoma gene product. *J. Virol.* **65,** 4966–4972.
Jackson, D. A., and Cook, P. R. (1986). Replication occurs at a nucleoskeleton. *EMBO J.* **5,** 1403–1410.
Jiménez-Garcia, L. F., Segura-Valdez, M. D., Ochs, R. L., Rothblum, L. I., Hannan, R., and Spector, D. L. (1994). Nucleologenesis: U3 snRNA-containing prenucleolar bodies move to sites of active pre-rRNA transcription after mitosis. *Mol. Biol. Cell.* **5,** 955–966.

Kass, S., Tyc, K., Steitz, J. A., and Sollner-Webb, B. (1990). The U3 small nucleolar ribnucleoprotein functions in the first step of preribosomal RNA processing. *Cell (Cambridge, Mass.)* **60,** 897–908.

Kill, I. R., Bridger, J. M., Campbell, K. H., Maldonado-Codina, G., and Hutchinson, C. J. (1991). The timing of the formation and usage of replicase clusters in S-phase nuclei of human diploid fibroblasts. *J. Cell Sci.* **100,** 869–876.

Kodama, K., Barnes, D. E., and Lindahl, T. (1991). In vitro mutagenesis and functional expression in *Escherichia coli* of a cDNA encoding the catalytic domain of human DNA ligase I. *Nucleic Acids Res.* **19,** 6093–6099.

Kubota, S., Siomi, H., Satoh, T., Endo, S., Maki, M., and Hatanaka, M. (1989). Functional similarity of HIV-I rev and HTLV-I rex proteins: Identification of a new nucleolar-targeting signal in rev protein. *Biochem. Biophys. Res. Commun.* **162,** 963–970.

Lamond, A. I., and Carmo-Fonseca, M. (1993). The coiled body. *Trends Cell Biol.* **3,** 198–204.

Leonhardt, H., and Bestor, T. H. (1993). Structure, function and regulation of mammalian DNA methyltransferase. *In* "DNA Methylation: Molecular Biology and Biological Significance" (J. P. Jost and H. P. Saluz, eds.), pp. 109–119. Birkhäuser, Basel.

Leonhardt, H., Page, A. W., Weier, H. U., and Bestor, T. H. (1992). A targeting sequence directs DNA methyltransferase to sites of DNA replication in mammalian nuclei. *Cell (Cambridge, Mass.)* **71,** 865–873.

Leonhardt, H., Galceran, J., Suchyna, T., Berezney, R., and Cardoso, M. C. (1995). In preparation.

Li, E., Bestor, T. H., and Jaenisch, R. (1992). Targeted mutation of the DNA methyltransferase gene results in embryonic lethality. *Cell (Cambridge, Mass.)* **69,** 915–926.

Li, E., Beard, C., and Jaenisch, R. (1993). Role for DNA methylation in genomic imprinting. *Nature (London)* **366,** 362–265.

Li, H., and Bingham, P. M. (1991). Arginine/serine-rich domains of the su(wa) and tra RNA processing regulators target proteins to a subnuclear compartment implicated in splicing. *Cell (Cambridge, Mass.)* **67,** 335–342.

Li, J. J., and Kelly, T. J. (1984). Simian virus 40 DNA replication in vitro. *Proc. Natl. Acad. Sci. U.S.A.* **81,** 6973–6977.

Ludlow, J. W., DeCaprio, J. A., Huang, C. M., Lee, W. H., Paucha, E., and Livingston, D. M. (1989). SV40 large T antigen binds preferentially to an underphosphorylated member of the retinoblastoma susceptibility gene product family. *Cell (Cambridge, Mass.)* **56,** 57–65.

Maeda, Y., Hisatake, K., Kondo, T., Hanada, K., Song, C. Z., Nishimura, T., and Muramatsu, M. (1992). Mouse rRNA gene transcription factor mUBF requires both HMG-box1 and an acidic tail for nucleolar accumulation: Molecular analysis of the nucleolar targeting mechanism. *EMBO J.* **11,** 3695–3704.

Malkas, L. H., Hickey, R. J., Li, C., Pedersen, N., and Baril, E. F. (1990). A 21S enzyme complex from HeLa cells that functions in simian virus 40 DNA replication in vitro. *Biochemistry* **29,** 6362–6374.

Mancini, M. A., Shan, B., Nickerson, J. A., Penman, S., and Lee, W. H. (1994). The retinoblastoma gene product is a cell cycle-dependent, nuclear matrix-associated protein. *Proc. Natl. Acad. Sci. U.S.A.* **91,** 418–422.

Manders, E. M., Stap, J., Brakenhoff, G. J., van Driel, R., and Aten, J. A. (1992). Dynamics of three-dimensional replication patterns during the S-phase, analysed by double labelling of DNA and confocal microscopy. *J. Cell Sci.* **103,** 857–862.

Mattaj, I. W. (1994). RNA processing. Splicing in space. *Nature (London)* **372,** 727–728.

Mazzotti, G., Rizzoli, R., Galanzi, A., Papa, S., Vitale, M., Falconi, M., Neri, L. M., Zini, N., and Maraldi, N. M. (1990). High-resolution detection of newly synthesized DNA by anti-bromodeoxyuridine antibodies identifies specific chromatin domains. *J. Histochem. Cytochem.* **38,** 13–22.

Mears, W. E., Lam, V., and Rice, S. A. (1995). Identification of nuclear and nucleolar localization signals in the herpes simplex virus regulatory protein icp27. *J. Virol.* **69,** 935–947.

Meier, U. T., and Blobel, G. (1992). Nopp140 shuttles on tracks between nucleolus and cytoplasm. *Cell* (*Cambridge, Mass.*) **70,** 127–138.

Meier, U. T., and Blobel, G. (1994). Nap57, a mammalian nucleolar protein with a putative homolog in yeast and bacteria. *J. Cell Biol.* **127,** 1505–1514.

Messmer, B., and Dreyer, C. (1993). Requirements for nuclear translocation and nucleolar accumulation of nucleolin of *Xenopus laevis. Eur. J. Cell Biol.* **61,** 369–382.

Mills, A. D., Blow, J. J., White, J. G., Amos, W. B., Wilcock, D., and Laskey, R. A. (1989). Replication occurs at discrete foci spaced throughout nuclei replicating in vitro. *J. Cell Sci.* **94,** 471–477.

Mittnacht, S., and Weinberg, R. A. (1991). G1/S phosphorylation of the retinoblastoma protein is associated with an altered affinity for the nuclear compartment. *Cell* (*Cambridge, Mass.*) **65,** 381–393.

Mittnacht, S., Lees, J. A., Desai, D., Harlow, E., Morgan, D. O., and Weinberg, R. A. (1994). Distinct sub-populations of the retinoblastoma protein show a distinct pattern of phosphorylation. *EMBO J.* **13,** 118–127.

Moir, R. D., Montag-Lowy, M., and Goldman, R. D. (1994). Dynamic properties of nuclear lamins: Lamin B is associated with sites of DNA replication. *J. Cell Biol.* **125,** 1201–1212.

Moroianu, J., and Riordan, J. F. (1994). Identification of the nucleolar targeting signal of human angiogenin. *Biochem. Biophys. Res. Commun.* **203,** 1765–1772.

Mougey, E. B., Pape, L. K., and Sollner-Webb, B. (1993). A U3 small nuclear ribonucleoprotein-requiring processing event in the 5′ external transcribed spacer of *Xenopus* precursor rRNA. *Mol. Cell. Biol.* **13,** 5990–5998.

Nakamura, H., Morita, T., and Sato, C. (1986). Structural organizations of replicon domains during DNA synthetic phase in the mammalian nucleus. *Exp. Cell Res.* **165,** 291–297.

Nakayasu, H., and Berezney, R. (1989). Mapping replicational sites in the eucaryotic cell nucleus. *J. Cell Biol.* **108,** 1–11.

Noguchi, H., Prem veer Reddy, G., and Pardee, A. B. (1983). Rapid incorporation of label from ribonucleoside disphosphates into DNA by a cell-free high molecular weight fraction from animal cell nuclei. *Cell* (*Cambridge, Mass.*) **32,** 443–451.

Ochs, R. L., Lischwe, M. A., Spohn, W. H., and Busch, H. (1985). Fibrillarin: A new protein of the nucleolus identified by autoimmune sera. *Biol. Cell.* **54,** 123–133.

Ookata, K., Hisanaga, S., Okano, T., Tachibana, K., and Kishimoto, T. (1992). Relocation and distinct subcellular localization of p34cdc2–cyclin B complex at meiosis reinitiation in starfish oocytes. *EMBO J.* **11,** 1763–1772.

Peculis, B. A., and Gall, J. G. (1992). Localization of the nucleolar protein NO38 in amphibian oocytes. *J. Cell Biol.* **116,** 1–14.

Pelham, H. R. (1984). Hsp70 accelerates the recovery of nucleolar morphology after heat shock. *EMBO J.* **3,** 3095–3100.

Peter, M., Nakagawa, J., Dorée, M., Labbé, J. C., and Nigg, E. A. (1990). Identification of major nucleolar proteins as candidate mitotic substrates of cdc2 kinase. *Cell* (*Cambridge, Mass.*) **60,** 791–801.

Pfeifle, J., Boller, K., and Anderer, F. A. (1986). Phosphoprotein pp135 is an essential component of the nucleolus organizer region (NOR). *Exp. Cell Res.* **162,** 11–22.

Pinol-Roma, S., and Dreyfuss, G. (1993). hnRNP proteins: Localization and transport between the nucleus and the cytoplasm. *Trends Cell Biol.* **3,** 151–155.

Prigent, C., Lasko, D. D., Kodama, K., Woodgett, J. R., and Lindahl, T. (1992). Activation of mammalian DNA ligase I through phosphorylation by casein kinase II. *EMBO J.* **11,** 2925–2933.

Quinlan, M. P., Chen, L. B., and Knipe, D. M. (1984). The intranuclear location of a herpes simplex virus DNA-binding protein is determined by the status of viral DNA replication. *Cell* (*Cambridge, Mass.*) **36,** 857–868.

Raska, I., Reimer, G., Jarnik, M., Kostrouch, Z., and Raska, K., Jr. (1989). Does the synthesis of ribosomal RNA take place within nucleolar fibrillar centers or dense fibrillar components? *Biol. Cell.* **65,** 79–82.

Raska, I., Ochs, R. L., Andrade, L. E., Chan, E. K., Burlingame, R., Peebles, C., Gruol, D., and Tan, E. M. (1990). Association between the nucleolus and the coiled body. *J. Struct. Biol.* **104,** 120–127.

Raska, I., Andrade, L. E., Ochs, R. L., Chan, E. K., Chang, C. M., Roos, G., and Tan, E. M. (1991). Immunological and ultrastructural studies of the nuclear coiled body with autoimmune antibodies. *Exp. Cell Res.* **195,** 27–37.

Reed, S. I. (1991). G1-specific cyclins: In search of an S-phase-promoting factor. *Trends Genet.* **7,** 95–99.

Rendon, M. C., Rodrigo, R. M., Goenechea, L. G., Garcia-Herdugo, G., Valdivia, M. M., and Moreno, F. J. (1992). Characterization and immunolocalization of a nucleolar antigen with anti-NOR serum in HeLa cells. *Exp. Cell Res.* **200,** 393–403.

Rikkonen, M., Peranen, J., and Kaariainen, L. (1992). Nuclear and nucleolar targeting signals of Semliki Forest virus nonstructural protein nsP2. *Virology* **189,** 462–473.

Riley, D. J., Lee, E., and Lee, W. H. (1994). The retinoblastoma protein—more than a tumor suppressor. *Annu. Rev. Cell Biol.* **10,** 1–29.

Roberts, J. M., and D'Urso, G. (1988). An origin unwinding activity regulates initiation of DNA replication during mammalian cell cycle. *Science* **241,** 1486–1489.

Roussel, P., Andre, C., Masson, C., Geraud, G., and Hernandez-Verdun, D. (1993). Localization of the RNA polymerase I transcription factor hUBF during the cell cycle. J. Cell Sci. **104,** 327–337.

Saunders, W. S., Cooke, C. A., and Earnshaw, W. C. (1991). Compartmentalization within the nucleus: Discovery of a novel subnuclear region. *J. Cell Biol.* **115,** 919–931.

Savino, R., and Gerbi, S. A. (1990). In vivo disruption of *Xenopus* U3 snRNA affects ribosomal RNA processing. *EMBO J.* **9,** 2299–2308.

Scheer, U., and Benavente, R. (1990). Functional and dynamic aspects of the mammalian nucleolus. *BioEssays* **12,** 14–21.

Scheer, U., and Rose, K. M. (1984). Localization of RNA polymerase I in interphase cells and mitotic chromosomes by light and electron microscopic immunocytochemistry. *Proc. Natl. Acad. Sci. U.S.A.* **81,** 1431–1435.

Scheer, U., Thiry, M., and Goessens, G. (1993). Structure, function and assembly of the nucleolus. *Trends Cell Biol.* **3,** 236–241.

Schmidt-Zachmann, M. S., and Nigg, E. A. (1993). Protein localization to the nucleolus: A search for targeting domains in nucleolin. *J. Cell Sci.* **105,** 799–806.

Schmitz, M. L., Henkel, T., and Baeuerle, P. A. (1991). Proteins controlling the nuclear uptake of NF-κB, Rel and dorsal. *Trends Cell Biol.* **1,** 130–137.

Siomi, H., Shida, H., Nam, S. H., Nosaka, T., Maki, M., and Hatanaka, M. (1988). Sequence requirements for nucleolar localization of human T cell leukemia virus type I pX protein, which regulates viral RNA processing. *Cell (Cambridge, Mass.)* **55,** 197–209.

Smith, H. C., and Berezney, R. (1980). DNA polymerase alpha is tightly bound to the nuclear matrix of actively replicating liver. *Biochem. Biophys. Res. Commun.* **97,** 1541–1547.

Sobczak-Thepot, J., Harper, F., Florentin, Y., Zindy, F., Brechot, C., and Puvion, E. (1993). Localization of cyclin A at the sites of cellular DNA replication. *Exp. Cell Res.* **206,** 43–48.

Spector, D. L. (1990). Higher order nuclear organization: Three-dimensional distribution of small nuclear ribonucleoprotein particles. *Proc. Natl. Acad. Sci. U.S.A.* **87,** 147–151.

Spector, D. L. (1993). Macromolecular domains within the cell nucleus. *Annu. Rev. Cell Biol.* **9,** 265–315.

Spector, D. L. (1994). RNA processing. Cycling splicing factors. *Nature (London)* **369,** 604.

Spector, D. L., and Smith, H. C. (1986). Redistribution of U-snRNPs during mitosis. *Exp. Cell Res.* **163,** 87–94.

Spector, D. L., Fu, X. D., and Maniatis, T. (1991). Associations between distinct pre-mRNA splicing components and the cell nucleus. *EMBO J.* **10,** 3467–3481.

Stillman, B. W., and Gluzman, Y. (1985). Replication and supercoiling of simian virus 40 DNA in cell extracts from human cells. *Mol. Cell. Biol.* **5,** 2051–2060.

Sun, J., Salem, H. H., and Bird, P. (1992). Nucleolar and cytoplasmic localization of annexin V. *FEBS Lett.* **314,** 425–429.

Templeton, D. J. (1992). Nuclear binding of purified retinoblastoma gene product is determined by cell cycle-regulated phosphorylation. *Mol. Cell. Biol.* **12,** 435–443.

Templeton, D. J., Park, S. H., Lanier, L., and Weinberg, R. A. (1991). Nonfunctional mutants of the retintoblastoma protein are characterized by defects in phosphorylation, viral oncoprotein association, and nuclear tethering. *Proc. Natl. Acad. Sci. U.S.A.* **88,** 3033–3037.

Tollervey, D., Lehtonen, H., Carmo-Fonesca, M., and Hurt, E. C. (1991). The small nucleolar RNP protein NOP1 (Fibrillarin) is required for pre-rRNA processing in yeast. *EMBO J.* **10,** 573–583.

Trimbur, G. M., and Walsh, C. J. (1993). Nucleolus-like morphology produced during the in vitro reassociation of nucleolar components. *J. Cell Biol.* **122,** 753–766.

Tubo, R. A., and Berezney, R. (1987). Identification of 100 and 150 S DNA polymerase alpha-primase megacomplexes solubilized from the nuclear matrix of regenerating rat liver. *J. Biol. Chem.* **262,** 5857–5865; erratum: p. 9921.

van Dierendonck, J. H., Keyzer, R., van de Velde, C. J., and Cornelisse, C. J. (1989). Subdivision of S-phase by analysis of nuclear 5-bromodeoxyuridine staining patterns. *Cytometry* **10,** 143–150.

Waga, S., Bauer, G., and Stillman, B. (1994). Reconstitution of complete SV40 DNA replication with purified replication factors. *J. Biol. Chem.* **269,** 10923–10934.

Walton, T. H., Moen, P., Jr., Fox, E., and Bodnar, J. W. (1989). Interactions of minute virus of mice and adenovirus with host nucleoli. *J. Virol.* **63,** 3651–3660.

Wang, J. Y. J., Knudsen, E. S., and Welch, P. J. (1994). The retinoblastoma tumor suppressor protein. *Adv. Cancer Res.* **64,** 25–85.

Warner, J. R. (1990). The nucleolus and ribosome formation. *Curr. Opin. Cell Biol.* **2,** 521–527.

Waseem, N. H., and Lane, D. P. (1990). Monoclonal antibody analysis of the proliferating cell nuclear antigen (PCNA). Structural conservation and the detection of a nucleolar form. *J. Cell Sci.* **96,** 121–129.

Wilcock, D., and Lane, D. P. (1991). Localization of p53, retinoblastoma and host replication proteins at sites of viral replication in herpes-infected cells. *Nature (London)* **349,** 429–431.

Wobbe, C. R., Dean, F., Weissbach, L., and Hurwitz, J. (1985). In vitro replication of duplex circular DNA containing the simian virus 40 DNA origin site. *Proc. Natl. Acad. Sci. U.S.A.* **82,** 5710–5714.

Wu, Z. A., Murphy, C., Wu, C.-H. H., Tsvetkov, A., and Gall, J. G. (1993). Snurposomes and coiled bodies. *Cold Spring Harbor Symp. Quant. Biol.* **58,** 747–754.

Wu, Z. A., Murphy, C., and Gall, J. G. (1994). Human p80-coilin is targeted to sphere organelles in the amphibian germinal vesicle. *Mol. Biol. Cell* **5,** 1119–1127.

Xing, Y., Johnson, C. V., Dobner, P. R., and Lawrence, J. B. (1993). Higher level organization of individual gene transcription and RNA splicing. *Science* **259,** 1326–1330.

Yan, C., and Mélèse, T. (1993). Multiple regions of NSR1 are sufficient for accumulation of a fusion protein within the nucleolus. *J. Cell Biol.* **123,** 1081–1091.

Yang, L., Wold, M. S., Li, J. J., Kelly, T. J., and Liu, L. F. (1987). Roles of DNA topoisomerases in simian virus 40 DNA replication in vitro. *Proc. Natl. Acad. Sci. U.S.A.* **84,** 950–954.

Yanishevsky, R. M., and Prescott, D. M. (1978). Late S phase cells (Chinese hamster ovary) induce early S phase DNA labeling patterns in G1 phase nuclei. *Proc. Natl. Acad. Sci. U.S.A.* **75,** 3307–3311.

Zamore, P. D., and Green, M. R. (1991). Biochemical characterization of U2 snRNP auxiliary factor: An essential pre-mRNA splicing factor with a novel intranuclear distribution. *EMBO J.* **10,** 207–214.

Zhang, G., Taneja, K. L., Singer, R. H., and Green, M. R. (1994). Localization of pre-mRNA splicing in mammalian nuclei. *Nature* (*London*) **372,** 809–812.

Zini, N., Santi, S., Ognibene, A., Bavelloni, A., Neri, L. M., Valmori, A., Mariani, E., Negri, C., Astaldi-Ricotti, G. C., and Maraldi, N. M. (1994). Discrete localization of different DNA topoisomerases in HeLa and K562 cell nuclei and subnuclear fractions. *Exp. Cell Res.* **210,** 336–348.

# Nuclear Matrix Acceptor Binding Sites for Steroid Hormone Receptors: A Candidate Nuclear Matrix Acceptor Protein

Andrea H. Lauber, Nicole P. Sandhu, Mark Schuchard, M. Subramaniam, and Thomas C. Spelsberg
Department of Biochemistry and Molecular Biology, Mayo Clinic and Mayo Foundation, Rochester, Minnesota 55904

Steroid/nuclear–hormone receptors are ligand-activated transcription factors that have been localized to the nuclear matrix. The classic model of hormone action suggests that, following activation, these receptors bind to specific "steroid response elements" on the DNA, then interact with other factors in the transcription initiation complex. However, evidence demonstrates the existence of specific chromatin proteins that act as accessory factors by facilitating the binding of the steroid receptors to the DNA. One such protein, the "receptor binding factor (RBF)-1", has been purified and shown to confer specific, high-affinity binding of the progesterone receptor to the DNA. Interestingly, the RBF-1 is localized to the nuclear matrix. Further, the RBF-1 binds specifically to a sequence of the c-*myc* proto-oncogene that has the appearance of a nuclear matrix attached region (MAR). These results, and other findings reviewed here, suggest that the nuclear matrix is involved intimately in steroid hormone-regulated gene expression.

**KEY WORDS:** Progesterone receptor, Steroid hormone, Receptor binding factor-1, Acceptor protein, Matrix-attached region, c-*myc* proto-oncogene.

## I. Introduction

The nuclear matrix is involved not only in the structural organization of interphase and metaphase chromatin, but also in DNA replication and gene transcription. The latter function suggests extensive interactions between nuclear matrix and transcription-regulating proteins such as nuclear receptors.

Steroid hormones acting at cognate receptors can initiate a cascade of events that influence cellular function and signaling. The receptors for estrogen, progesterone, androgens, and glucocorticoid have been localized within the cell nucleus and are ligand-activated transcription factors (Carson-Jurica *et al.,* 1990; Gronemeyer, 1991). Findings demonstrating that steroid hormone receptors are localized to the nuclear matrix (Berezney and Coffey, 1977; Barrack and Coffey, 1980, 1982; Barrack, 1983, 1987a,b,c) have led to questions concerning the interactions between receptors and the nuclear matrix, especially with regard to molecular mechanisms of hormone-regulated gene transcription.

A nuclear matrix protein involved in the chromatin binding of the avian oviduct-progesterone receptor (PR) has been identified. This protein, called "receptor binding factor" (RBF), acts as a chromatin/DNA acceptor protein for PR. The RBF has been purified, characterized and functional studies have been initiated. Specific DNA-binding sites for RBF have been identified and found to resemble matrix attachment region (MAR)-like sequences (Lauber *et al.,* 1995). Proteins homologous to the RBF but specific to other steroid receptor systems have been identified and some of their properties have been described here for comparison with the PR-RBF.

It has become apparent that there are interactions between steroid receptors and nuclear matrix factors. Together, these elements regulate gene transcription. Examples of how steroid receptors and nuclear matrix might interact to influence cellular functions are cited here.

## II. Mechanisms of Steroid-Regulated Gene Expression

### A. Steroid Receptors as Transcription Factors

Steroid hormones act throughout the biological system to modulate cellular processes. Glucocorticoids have profound effects on metabolism, development, inflammatory, and stress responses and are necessary for life. The gonadal steroid hormones, estrogen, progesterone, and androgens, regulate and modulate numerous functions during development, such as sexual differentiation of the brain and periphery. In adulthood, these hormones are necessary for reproductive function, maintenance of bone mass, and muscle integrity. However, in combination with other factors, these hormones become potent facilitators of cancers. Thus, molecular mechanisms of these agents are topics of intense study and debate. Since in-depth reviews of molecular mechanisms of steroid hormone action are available (Freedman and Luisi, 1993; Truss and Beato, 1993; Glass, 1994), known mechanisms of hormone action will be reviewed here briefly.

Mechanisms of steroid hormone action and regulation of gene expression have received enormous attention as a result of advances in understanding structure/function relationship of steroid receptors (see reviews cited above). The steroid hormone receptors are ligand-dependent transacting factors, composed of modular functional domains. There are regions for transactivation function, for ligand binding, and a region comprised of two zinc-finger units that bind to DNA. Nuclear localization and dimerization sequences have been identified (Green and Chambon, 1991).

Steroid hormones passively diffuse into all cells and are preferentially retained only in target cells that contain specific receptor proteins. The hormones form stable complexes with specific intracellular receptor proteins (R) that are steroid- and cell-type specific. Each ligand binds reversibly, but with high affinity ($10^{-10}$ to $10^{-8}$ $M$) to its cognate receptor. Ligand binding results in an "activation" of the receptor, which is still not completely understood. Receptor activation requires dissociation of an initial oligomeric receptor complex, which includes a multitude of factors including heat shock proteins (Green and Chambon, 1991). Evidence suggests that receptor homodimerization is required for transactivation function and that conformational and post-translational changes in the steroid receptor may be implicated as well. Together, these processes enable the receptor to bind with high affinity to specific sites on chromatin. This may be followed by a change in chromatin/nucleosome structure (Hager and Archer, 1991) and interactions with other transcription factors. While it is unclear how this occurs, it seems that these complexes interact with the transcription initiation complex, which includes the RNA polymerase II (Bagchi *et al.*, 1992).

A major question concerning the mechanisms of steroid hormone action is the role of DNA sequences that bind the receptors. In the case of glucocorticoids and gonadal steroid hormones, the specific DNA-binding sequences known as "consensus" or "response elements" consist of two inverted repeats, separated by three interspacing nucleotides. These elements act as traditional enhancers of transcription. The sequences of the binding elements differ between the estrogen receptor and the rest of the steroid receptors, but the DNA consensus elements for androgen, progestin, and glucocorticoid receptors are identical (Green and Chambon, 1991; Truss and Beato, 1993). This raises the question of how different steroid hormone systems can have separate specific functions. Some evidence suggests that receptor–DNA binding and specific subsequent transcriptional events may be dependent upon receptor-specific amino acids making contact with different base pairs of the same DNA sequence. Also, *in vivo,* hormone specificity may be conferred by cells expressing only one species of steroid receptor that can bind to the glucocorticoid, androgen, progestin response element. Undoubtedly, some combination of factors is important for specific hormone/receptor action.

Additional questions pertain to precise mechanisms of interaction between steroid receptors and DNA in the nuclear environment. In the native state, DNA is wound around nucleosomal structures and covered with chromatin proteins and, thus, may not be accessible directly to transacting proteins. Moreover, it is well documented that steroid receptor–DNA binding is characterized by low affinity, nonsaturable interactions, while receptor–chromatin binding has a saturable, high-affinity, receptor-specific binding (Spelsberg *et al.,* 1987a,b,c).

Emerging patterns of data suggest that additional nuclear chromatin factors confer specificity to steroid hormone–DNA interactions. In some cases these proteins function as "chromatin acceptor sites." Existence of such factors does not negate the model of direct steroid receptor–DNA interactions. These factors may help position the receptor appropriately on the chromatin/DNA or perhaps assist in chromatin unwinding, allowing for direct interactions between steroid receptors and DNA (Hager and Archer, 1991). In fact, such scenarios are plausible, as steroid receptors interact with other transcription factors in the regulation of gene expression (Renkawitz *et al.,* 1982,1989; Cordingley *et al.,* 1987a,b; Jost *et al.,* 1987; Martinez *et al.,* 1987; Schüle *et al.,* 1988a,b; Tora *et al.,* 1988; Diamond *et al.,* 1990; Gaub *et al.,* 1990; Yang-Yen *et al.,* 1990; Muller *et al.,* 1991; Tsai *et al.,* 1991). A major focus of this chapter is the role of nuclear acceptor proteins that confer specific binding sites for the steroid hormone receptor on DNA. The role of other (accessory) factors that participate in regulation of steroid hormone action are described as well.

## B. Direct versus Cascade Models of Steroid Hormone Action

### 1. Ligand-Bound Receptors Regulate Gene Expression

The direct action of steroid receptors on regulation of gene expression, described above, accounts for the general mechanism of action of steroid hormones and does not differ greatly from the one originally proposed by Jensen *et al.* (1968) and Gorski *et al.* (1968) and widely reviewed (Katzenellenbogan, 1980; Spelsberg *et al.,* 1983, 1987a,b, 1989; Moudgil, 1987; Rories and Spelsberg, 1989; Hager and Archer, 1991; Muller *et al.,* 1991; Tsai *et al.,* 1991; Landers and Spelsberg, 1992). The major tenant of this model is that once the receptor binds to chromatin, the activated steroid–receptor

(SR) complex functions as transcription factor, modulating the expression of target genes and/or, perhaps altering post-transcriptional steps, resulting in altered steady-state levels of specific mRNAs and proteins.

Important to understanding the physiological and mechanistic aspects of steroid-regulated gene expression is the chronology in which these events occur. Entrance of steroids into the cell and the formation of the steroid–receptor complex is rapid, occurring within minutes of injection into an animal. Following steroid–receptor complex formation (1–4 min post-treatment), this complex binds to nuclear acceptor sites (2–5 min), affects changes in RNA synthesis (several hours for "late" genes), and ultimately alters levels of proteins resulting from alterations in protein synthesis and turnover (4–8 hr post-treatment). Indeed, some of the major physiological effects of steroids in cells are not apparent until 12 to 24 hr following steroid treatment. Thus, the period elapsing between the SR binding to the nuclear acceptor sites and the late gene transcriptional response is somewhat of a mechanistic paradox. This has been termed the "lag" phase or period during which obscure undefined molecular events occur (Palmiter *et al.,* 1976).

A question that arises from this steroid action pathway has to do with the function of the lag phase. What are the roles of the early versus late responding genes in the mediation of hormone action? What accounts for the action of steroids that occur at the level of mRNA half-life and processing, protein processing, and membrane transport? These questions are addressed by the cascade model of steroid action (Spelsberg, 1976; Spelsberg *et al.,* 1987a,b,c, 1989; Landers and Spelsberg, 1992).

## 2. The Cascade Model: Direct Hormonal Regulation of Proteins That Alter Expression of Other Genes

In *Drosophila,* the steroid "ecdysone" induces the expression of early genes (puffs) followed by appearance of late puffs, which are dependent on protein synthesis (Clever and Karlson, 1960; Ashburner, 1972; Landers and Spelsberg, 1992). A cascade model was suggested (Clever, 1964; Clever and Romball, 1966; Britten and Davidson, 1969) wherein the protein products of the early regulated genes, in turn, regulate later genes. Subsequently, two of the early genes were cloned and sequenced; one had high homology to the steroid receptor family of transcription factors, and the other, a high homology to the "ets" nuclear proto-oncogene family (Burtis *et al.,* 1990; Segraves and Hogness, 1990). It was believed, then, that these early gene protein products would regulate other genes.

The steroid-target cell systems in animals show a striking similarity to the *Drosophila* ecdysone system. There is a wide variation in time elapsed between steroid administration and measurable changes in transcription among many of the regulated genes (Rories and Spelsberg, 1989; Spelsberg *et al.,* 1989; Landers and Spelsberg, 1992). In an almost identical fashion to the ecdysone-induced early and late puffing in the fly polytene chromosomes, there are early and late steroid hormone responsive genes in animal systems. The early genes show immediate changes in gene transcription and mRNA levels. Interestingly, these changes in relative protein levels are independent of hormone-induced protein synthesis.

There are several examples of early responding "regulatory" genes. The early induced genes of rat uterus (DeAngelo and Gorski, 1970), the pS2 gene of MCF-7 human breast cancer cells (Brown *et al.,* 1984), the c-*myc* gene in breast cancer cells (Dubik *et al.,* 1987), and the N-*myc* in rat uterus (Murphy *et al.,* 1987; Travers and Knowler, 1987) show rapid changes in mRNA levels following estrogen administration. Some steroidal effects are inhibitory, as estrogen rapidly inhibits c-*jun* gene expression (Lau *et al.,* 1990; Webb *et al.,* 1990), while the antiestrogen, Tamoxifen, induces c-*jun* gene expression in the avian oviduct within the early time interval. There are other examples of steroids regulating the expression of nuclear proto-oncogenes (Schuchard *et al.,* 1993). Glucocorticoids can rapidly inhibit the expression of the c-*myc* gene in avian oviduct within 5 min of hormone exposure (Rories *et al.,* 1989). Glucocorticoids also can inhibit c-*myc* gene expression in a lymphosarcoma cell line, in addition to repressing cell proliferation (Forsthoefer and Thompson, 1987).

Interestingly, proto-oncogenes may be dually regulated by different steroids. Although estrogen can increase c-*myc* and N-*myc* gene expression, progesterone appears to decrease c-*myc* mRNA levels in the avian oviduct within 5 min of administration (Fink *et al.,* 1988). The c-*myc* gene codes for a nuclear protein now thought to positively regulate the genes involved in the initiation of DNA replication (Gonda *et al.,* 1982; Kelley and Siebenlist, 1986; Studzinski *et al.,* 1986). Thus, there exists a possible role for c-*myc* and other nuclear proto-oncogenes in the steroid regulation of cell proliferation (Hyder *et al.,* 1994). This system also could serve as an example of a steroid-modulation of a regulatory *myc* gene with the *Myc* protein itself acting as a transcription factor. In fact, all nuclear proto-oncogenes could fit this model.

Some of the known structural genes that are regulated by steroids in mammalian cells represent a second class of steroid-responsive genes that we have termed "late" genes; they may be under "secondary control." These late genes show significant changes in mRNA levels 1–6 hr after steroid administration. These changes *require* protein synthesis. One example of a late steroid response is the estradiol induction of *Xenopus* liver vitellogenin mRNA, which requires a lag time of 90 min or more after estrogen administration (Baker and Shapiro, 1978). Another example occurs in avian oviduct, where there is a 3- to 4-hr lag period between the time progesterone or estrogen binds to nuclear acceptor sites, and the increase in the mRNA levels of genes coding for the ovalbumin egg white proteins (Harris *et al.,* 1975; Palmiter *et al.,* 1976; Schutz *et al.,* 1978). Similarly, estrogen up-regulates the progesterone receptor in numerous target tissues such as uterus, brain, and breast cancer cells, but there is an approximately 12-hr lapse between estrogen exposure and PR induction (Lauber and Pfaff, 1990). Hence, from a chronological view it seems likely that the rapidly responding early genes code for proteins that can regulate the late responses.

We have formulated a cascade model for ovarian steroid action based on the existence of early (regulatory) versus late (structural) genes (Fig. 1). This model was supported by the identification of rapidly responding nuclear proto-oncogenes whose steroid regulation is independent of protein

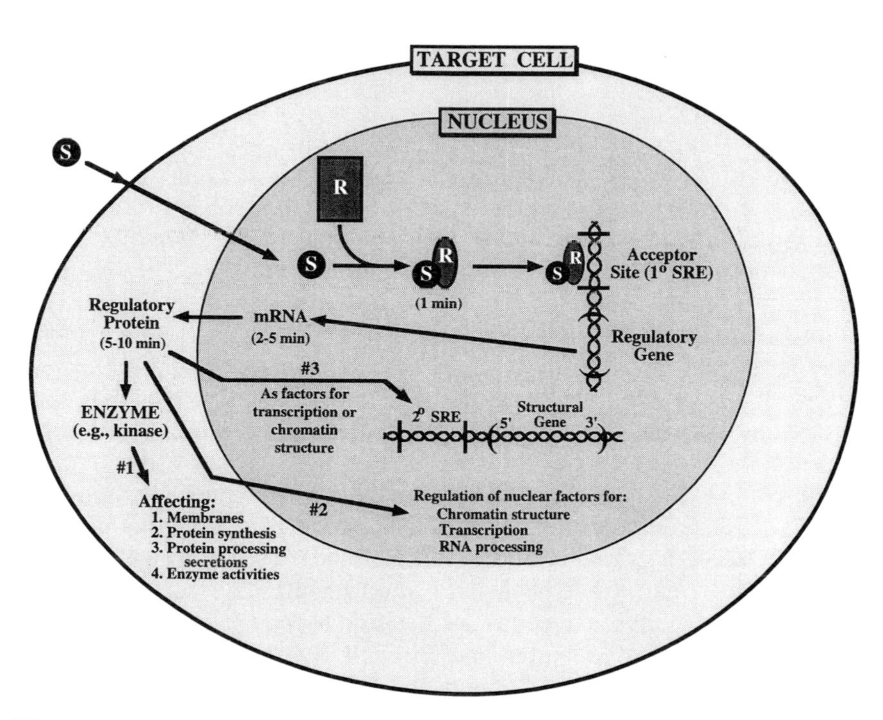


FIG. 1 Cascade model for steroid action on gene expression. Activated steroid hormone receptors bind to the nuclear acceptor sites located on primary (1°) steroid regulatory elements in the 5′ flanking domain of a regulatory gene (early or 1° gene). This binding modulates transcription of the regulatory gene within minutes after the steroid enters the target cell. Within a given target cell there may be several regulatory genes responding to a particular steroid whose protein products participate in a multitude of events within the cell. This includes the gene product that regulates the transcription of a structural gene (late or 2° gene). R, receptor; S, steroid; RS, activated receptor; 1° SRE, primary steroid regulatory element; 2° SRE, secondary SRE (modified with permission from Rories and Spelsberg, 1989).

synthesis (Schuchard *et al.,* 1993). Thus, the protein products of these early responding regulatory genes could subsequently regulate the expression of late structural genes at the transcriptional or post-transcriptional level.

Additionally, evidence for the existence of steroid-regulated proteins involved in the regulation of mRNA half-life has been reported (Cochran and Deeley, 1989; Leibold *et al,* 1990; Nielsen and Shapiro, 1990). Steroid-induced regulation of chromatin structure, including nucleosome positioning and generation of DNase-sensitive domains, have been reported, as well (Cordingley *et al.,* 1987a; Rories and Spelsberg, 1989; Hager and Archer, 1991).

## III. Role of the Nuclear Matrix in Steroid Hormone Action

Evidence suggests that the nuclear matrix is intimately involved in the regulation of several key cellular processes (Barrack, 1987a,b). The nuclear matrix is a complex subcomponent of the nucleus that is present in all eukaryotic cells. It is composed of a peripheral pore-complex lamina, an internal fibrogranular network containing both protein and RNA, and a residual nucleolus. Although the details of the nature, organization, and interaction of these various components are still not clear, there is evidence that the nuclear matrix is involved in DNA organization and replication, mRNA synthesis and processing, DNA loop attachment sites, DNA polymerase $\alpha$ activity and topoisomerase activity (Barrack, 1987a,b). In a practical sense, the nuclear matrix represents the residual structure after detergent treatment, DNase I digestion, and high salt extraction of nuclei. It reportedly consists of 7%, or less, of total nuclear DNA and much of the heterogeneous nuclear RNA (Barrack and Coffey, 1982; Berezney, 1984; Barrack, 1987a,b).

The nuclear matrix has been implicated in hormone action. This has been suggested by steroid receptor localization to the nuclear matrix following steroid treatment (Agutter and Birchell, 1979; Barrack and Coffey, 1982, 1983; Berezney, 1984; Colvard and Wilson, 1984; Hora *et al.,* 1986; Alexander *et al.,* 1987; Barrack, 1987c). Steroid receptors have been identified in the nuclear matrix of several target tissues in hormone-stimulated animals including rat liver, uterus, prostate, prostate cancer cell lines, hen liver, and human prostate (Barrack, 1987a,b,c). These matrix-associated steroid receptor sites display a high-affinity, saturable, tissue-specific binding. The majority of the receptors are associated with the internal ribonucleoprotein network (i.e., the internal nuclear matrix) of the total nuclear matrix structure (Barrack and Coffey, 1980, 1982, 1983; Barrack, 1987a,b,c).

Evidence suggests that the nuclear matrix of steroid target cells contains intact nuclear acceptor sites for steroid receptor complexes, and that these sites show binding properties similar to the chromatin acceptor sites. In addition, the nuclear matrix appears to be associated with transcriptionally active genes, and with DNase I-hypersensitive sites (Barrack and Coffey, 1982; Berezney, 1984; Barrack, 1987a,b,c). Thus, the nuclear matrix is associated with many of the biological functions that are regulated by steroid hormones. Further, nuclear matrix binding sites have many of the properties identified with the chromatin acceptor sites for steroid receptors (Webster *et al.,* 1976; Hora *et al.,* 1986; Rories and Spelsberg, 1989; Landers and Spelsberg, 1992).

Spelsberg and co-workers found that the chromatin acceptor sites for the avian oviduct PR are maintained not only after histones are removed by high salt solutions (Webster *et al.,* 1976), but also after extensive nuclease (DNase I) digestion (Hora *et al.,* 1986). The resulting protein/DNA particles, generated by the nuclease digestion of dehistonized avian oviduct chromatin, resemble the nuclear matrix in composition of RNA and protein, representing about 5–10% of the nuclear protein, and 2–5% of the nuclear DNA. These sites display specific, saturable, and high-affinity binding ($K_D \sim 10^{-10} M$) of the PR (Hora *et al.,* 1986). Since these chromatin acceptor site complexes have the same properties as those isolated as by methods used to isolate the nuclear matrix, they are now presumed to be one and the same.

A variety of steroid receptors and their nuclear binding sites are associated with nuclear matrix: these include estrogen receptors (Barrack and Coffey, 1980) and nuclear binding (acceptor) sites (Metzger and Korach, 1990), androgen receptors (Barrack and Coffey, 1980) and their nuclear acceptor sites (Colvard and Wilson, 1984), mineralocorticoid receptors (van Stensel *et al.,* 1991), and the progesterone receptor acceptor sites that are discussed below (Hora *et al.,* 1986; Schuchard *et al.,* 1991a,b). Thus, while the intact (native) chromatin displays highly specific binding sites for a variety of steroid receptors, the nuclear matrix also is implicated in mechanisms of steroid hormone action.

## IV. Nuclear Acceptor Sites for Steroid Receptors and Acceptor Proteins Bound to DNA

Using a different approach than that employed for evaluating interactions between steroid receptors and response elements, we investigated the direct binding of steroid receptors to the native nuclei, nuclear chromatin, or chromatin fractions, and the nuclear matrix (Spelsberg *et al.,* 1987c, 1989;

Landers and Spelsberg, 1992). Results of these studies revealed nuclear acceptor sites that display a saturable, specific binding of steroid receptors with high-affinity binding ranging from $K_D \sim 10^{-10}$ to $10^{-8}$ *M*. Indeed, each species of steroid receptor appears to associate with a unique class of nuclear acceptor sites. These nuclear sites for a variety of steroid receptors, residing in the chromatin or in the nuclear matrix, involve specific nuclear binding proteins (T. S. Ruh *et al.,* 1981; DeBoer *et al.,* 1984; Jost *et al.,* 1984; Chuknyiska *et al.,* 1985; Cushing *et al.,* 1985; Frankel and Senior, 1986; Pavlik *et al.,* 1986; M. F. Ruh *et al.,* 1987; Ogle, 1987; Spelsberg *et al.,* 1987c; Edwards *et al.,* 1989; Horton *et al.,* 1991; Rejman *et al.,* 1991; Schuchard *et al.,* 1991a,b). An acceptor protein has also been identified for the thyroid receptor (Murray and Towle, 1989).

Chromatin acceptor sites for the avian oviduct PRs have been studied extensively. These sites display high-affinity PR binding *in vivo* and *in vitro* (Fig. 2) and appear to consist of specific chromatin proteins bound to DNA. Also, there is evidence that specific DNA sequences are involved (Toyoda *et al.,* 1985; Rories and Spelsberg, 1989). Initially, the binding of [$^3$H]progesterone in the avian oviduct *in vivo* was characterized, and a cell-free binding assay was developed to mimic the former. The cell-free binding of [$^3$H]PR to avian-oviduct chromatin and to a dehistonized chromatin preparation, nucleoacidic protein (NAP), is saturable, has high affinity ($K_D \sim 10^{-9}$ *M*), and is receptor specific (Spelsberg *et al.,* 1987a; Rories and Spelsberg, 1989). In addition to the similarity in *in vivo*-like binding properties, the cell-free

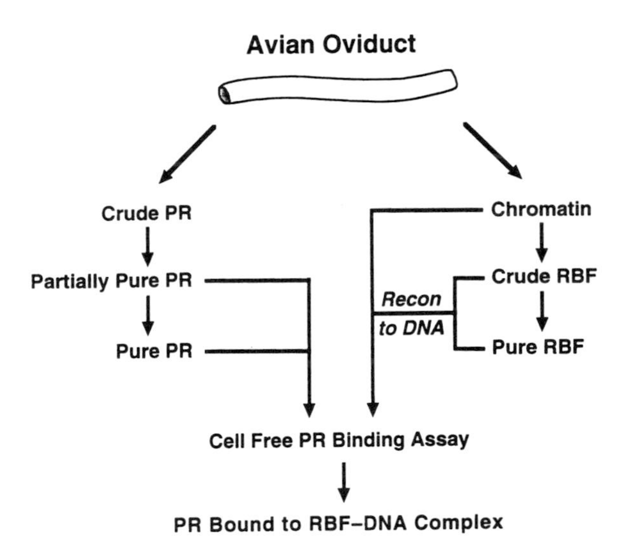


FIG. 2 Cell-free binding of PR to nuclear acceptor sites. This diagram represents the general scheme for determining interactions between the PR and RBV *in vitro.*

binding has yeilded similar numbers of receptor molecules bound per cell as measured *in vivo* (Spelsberg *et al.,* 1987a,b,c). Furthermore, the progesterone receptor must be ligand-bound and activated to display specific binding to these chromatin acceptor sites *in vitro.* Chuknyiska *et al.* (1984) and Spelsberg *et al.* (1987b, 1989), using two separate steroid target systems, showed specific binding to nuclear acceptor sites *in vivo* and *in vitro* in isolated nuclei. There are many similarities between binding sites on chromatin and those associated with the nuclear matrix (Rories and Spelsberg, 1989; Spelsberg *et al.,* 1989). The chromatin acceptor sites for PR and the nuclear matrix binding sites for PR have recently been shown to be one and the same class of acceptor sites (Schuchard *et al.,* 1991a,b). Common properties of these sites are their DNase resistance, their high-affinity binding of steroid receptors, and their composition of proteins tightly bound to DNA.

Other laboratories have examined intact (whole) nuclei and chromatin for specific binding sites associated with other steroid receptors and target cell systems. These laboratories reported similar properties as described above for the avian oviduct PR. Systems in which specific chromatin acceptor sites for PR have been identified are the sheep hypothalamus (Perry and Lopez, 1978) and the hamster uterus (Cobb and Leavitt, 1987). Chromatin acceptor sites that specifically bind estrogen receptors were found in the avian oviduct (Ruh and Spelsberg, 1983; DeBoer *et al.,* 1984), calf uteri (Ruh *et al.,* 1981; Ross and Ruh, 1984; Cushing *et al.,* 1985), rat uteri (Chuknyiska *et al.,* 1985), mouse uteri (Pavlik *et al.,* 1986), shark testes (Ruh *et al.,* 1986), human breast cancer cell lines (Sun *et al.,* 1983; Frankel and Senior, 1986), rodent malignant breast tissue (Shyamala *et al.,* 1986; Klinge *et al.,* 1987), and in mouse and sheep brain (Perry and Lopez, 1978; Lopez *et al.,* 1985). Further, specific chromatin acceptor sites for androgen receptors were reported for the rat prostate and Sertoli cells (Klyzsejko-Stefanowicz *et al.,* 1976; Wang, 1978), and for glucocorticoid receptors in the rat liver, mouse mammary, human leukemia cells (Hamana and Iwai, 1978; Ruh *et al.,* 1987), and in the MCF-7 cell human breast cancer cell line (Sun *et al.,* 1983). These chromatin acceptor sites were shown to have receptor-specific, saturable, high-affinity SR binding in cell-free binding assays, as described above for the PR–avian oviduct system. In some cases, acceptor sites were found to be masked in the intact chromatin.

Chromatin acceptor sites consist of nonhistone proteins tightly bound to DNA. Removal of these DNA-bound proteins causes a loss in the specific binding sites. One group has reported the reconstitution of the acceptor sites for the estrogen receptor using nonhistone protein fractions and genomic DNA as described below for the avian oviduct PR system (Ross and Ruh, 1984). The exact role of acceptor sites and acceptor proteins in the steroid receptor binding to the chromatin–gene structures is unclear, but conceivably, they help guide the receptors to specific nuclear binding sites in neighboring steroid-regulated genes.

## V. Acceptor Proteins for Steroid/Thyroid Hormone Receptors: Receptor Binding Factors (RBFs)

### A. Progesterone RBF in Avian Oviduct

The chromatin acceptor sites for the avian oviduct PR can be reconstituted by reannealing a specific fraction of chromatin proteins to hen genomic DNA (Spelsberg *et al.,* 1983, 1984; Toyoda *et al.,* 1985; Goldberger *et al.,* 1986, 1987; Goldberger and Spelsberg, 1988). This is the same fraction that results in a loss of specific binding sites when removed from the chromatin. Figure 2 outlines the cell-free binding assay used to assess the PR binding to the nuclear sites. Table I outlines the native-like properties of cell-free binding compared to *in vivo* (native) binding. The reconstituted sites display the same specific PR-binding properties as the native acceptor sites (Spelsberg *et al.,* 1984, 1989). As outlined in Fig. 3, the reconstituted sites display a specific, saturable PR binding and little or no binding when a nonfunctional PR preparation is used (Spelsberg and Halberg, 1980; Spelsberg *et al.,* 1984). Further, efficacy of the reconstituted sites depends upon the ratio of chromatin fraction reannealed to hen DNA and specific DNA sequences

TABLE I
Comparisons of *in Vivo* and Cell-Free Nuclear Binding by Avian PR

1. Cell-free binding levels in isolated nuclei competed by *in vivo* injections of PG but not estrogen.
2. *In vitro* and *in vivo* nuclear binding require
   - Intact receptor
   - Activated receptor
   - Pg-bound receptor
   - Functional receptors
3. *In vitro* and *in vivo* nuclear bindings show
   - High-affinity binding, saturable binding
   - Similar levels of binding (sites per nucleus)
   - The same seasonal and developmental patterns of PR binding
4. Cell-free PR binding to native acceptor site chromatin and reconstituted acceptor sites [crude RBF (CP-3)–DNA complexes] displays the above characteristics in both cell-free and *in vivo* assays.

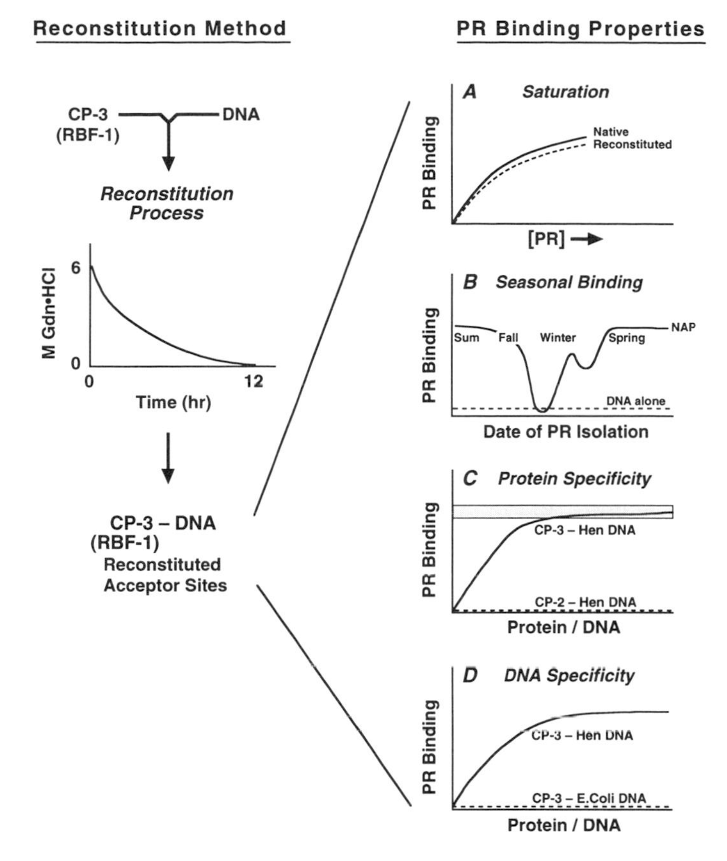


**A.** **Generates high affinity, saturable PR binding on avDNA. Activity is titratable on avDNA and generates a limited number of sites.**

**B.** **RBF-DNA complexes do not bind nonfunctional PR, only functional PR, i.e., the cell-free binding to native acceptor sites displays the same pattern of binding with seasonal receptors as does the nuclear binding *in vivo.***

**C.** **Activity is found in one fraction (CP-3) of nonhistone, chromosomal proteins. The activity has been found to be localized to the nuclear matrix.**

**D.** **Significant PR binding activity was generated with mammalian, fish, avian, and amphibian DNA. No PR binding activity was generated with E. coli, insect, viral, or plant DNA.**

FIG. 3 Chromatin proteins enhance specific PR binding to DNA. These graphs illustrate that reconstituted chromatin/DNA fractions enhance specific binding of progesterone receptor to DNA. Chromatin proteins (CP-3) extracted with guanidine, then reconstituted together with hen genomic DNA, generate specific high-affinity PR binding sites (see Fig. 4). Note, in panel B, that functional ability of the chicken PR varies across the season (Boyd and Spelsberg, 1979).

in the hen DNA (Spelsberg *et al.,* 1984; Toyoda *et al.,* 1985; Goldberger and Spelsberg, 1988). As discussed later, similar results have been reported by Ruh and co-workers (Ross and Ruh, 1984) for the reconstitution of the estrogen receptor acceptor sites in mammalian systems.

Recently, we have purified a specific chromatin acceptor protein, termed receptor binding factor-1 (RBF-1), for the avian oviduct PR (Rejman *et al.,* 1991; Schuchard *et al.,* 1991a). Reconstitution of chromatin DNA complexes was used to monitor the PR "acceptor activity" (i.e., acceptor protein) during the fractionation of avian oviduct chromatin proteins using molecular sieve, chromatofocusing, hydroxylapatite, and hydrophobic chromatographies. As a result, two, and possibly other, candidate chromatin acceptor proteins (RBF-2, RBF-3) for PR have been identified (Rejman *et al.,* 1991). The procedure to purify this acceptor activity over 100,00-fold to apparent homogeneity is outlined in Fig. 4. Table II outlines the chemical and biological properties of this acceptor protein. When bound to avian DNA, both impure and pure RBF-1 fractions of these chromatin proteins generate high-affinity, saturable, receptor-specific binding of the avian PR to DNA (Schuchard *et al.,* 1991a).

RBF-1 is a unique 10-kDa, hydrophobic that is tissue specific and appears to be a phosphoprotein (Rejman *et al.,* 1991). Using antibodies against this protein, a 10-kDa species was detected by Western immunoblots (Goldberger and Spelsberg, 1988). The 10-kDa protein species cofractionated with the highest levels of PR acceptor activity and it generated specific PR binding when complexed to hen genomic DNA (Rejman *et al.,* 1991; Schuchard *et al.,* 1991a). Additional studies have supported a biological role for this protein as a PR acceptor protein. Antibodies prepared against a reconstituted 10-kDa acceptor protein–DNA complex were capable of inhibiting the specific PR, but not estrogen receptor (ER), binding to undissociated (native) acceptor sites in oviduct chromatin (Goldberger *et al.,* 1987). These data supported earlier studies demonstrating that ER and PR had distinct nuclear acceptor sites in the avian oviduct (Kon and Spelsberg, 1982; Ruh and Spelsberg, 1983). These results, together with the fact that this 10-kDa protein can generate a receptor-dependent, saturable binding of PR to hen genomic DNA, supports a role for this protein as an acceptor site for the PR in the avian oviduct.

Immunohistochemical studies have shown that RBF-1 and PR are localized within the nuclei of epithelial cells of the undifferentiated avian oviduct (Zhuang *et al.,* 1993). However, the PR and RBF-1 immunoreactivity also were evident in epithelial, glandular, and stromal cells of the oviduct following estrogen treatment and cellular differentiation. Western blotting showed anti-RBF-1 activity to be high in the avian oviduct tissues, in addition to heart, kidney, and brain. However, lung, pancreas, and bone osteoclasts and giant cells displayed little activity (Landers *et al.,* 1994). Since many

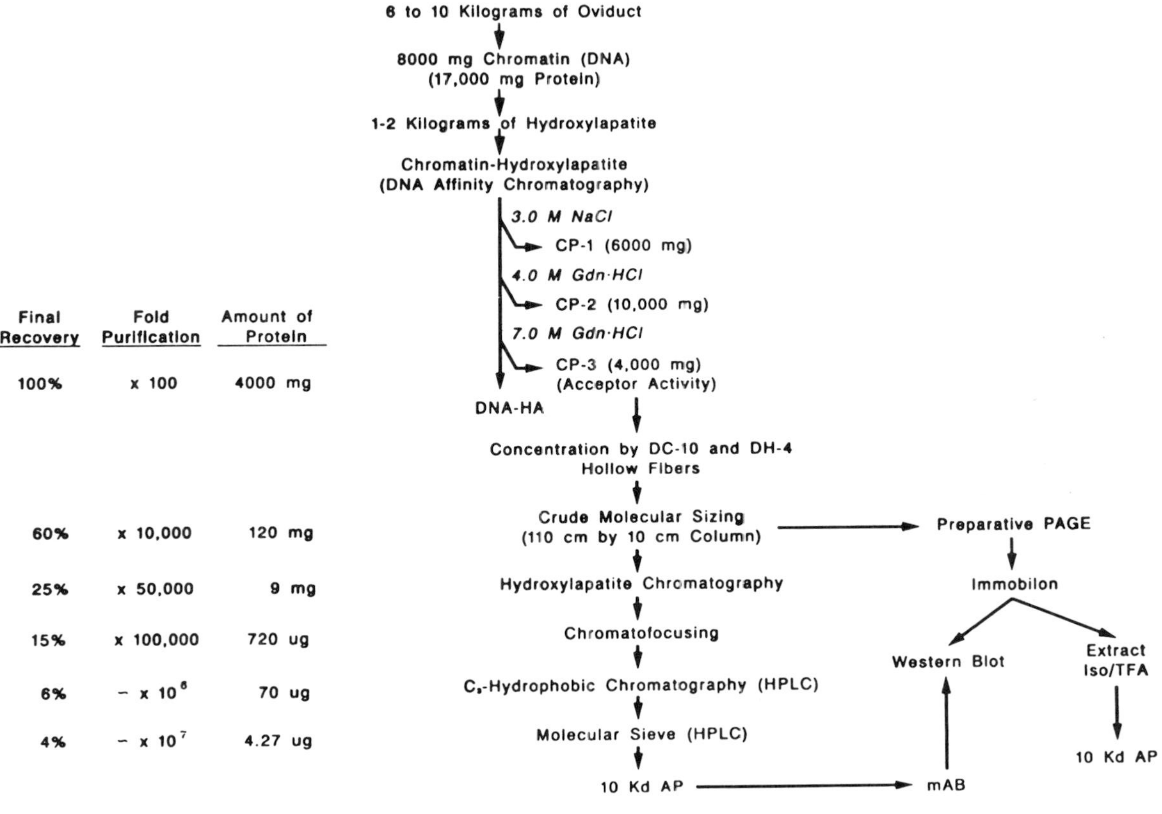


FIG. 4 Isolation and purification of the RBF-1. Various guanidine concentrations were utilized to strip chromatin proteins; PR acceptor activity was found in fraction CP-3. The purified RBF-1 was isolated and purified from this fraction and now is isolated from Western blots utilizing anti-RBF-1 antibodies.

TABLE II

Properties of the Avian RBF-1 Protein

| | |
|---|---|
| Size | 10 kDa |
| Sequence | N-terminal, hydrophobic B structure<br>C-terminal, hydrophilic, $\alpha$ helical |
| Homologies | ~70% to ATPase subunit<br>~70% uv-induced nuclear protein (metallothinien protein) |
| Conjugated (sugars, etc.) | No |
| Phosphorylated | Yes |
| Isoelectric point | ~pH 6, determined<br>~pH 9 theoretical by amino acid sequence |
| Solubility | Poor, due to N-terminal hydrophobic domain |
| Cellular localization | Nuclear matrix |
| DNA binding | High-affinity, specific binding to sequences of the avian c-*myc* gene |
| Tissue specificity | Avian oviduct, heart, liver, liver; rat ovary, uterus, prostate |

of the tissues in which RBF-1 immunoreactivity has been localized do not use progesterone or contain detectable levels of the PR, it seems likely that RBF-1 might also function in conjunction with the other steroid hormone receptors, such as glucocorticoid receptor which is found in high levels in heart. This would not be unexpected, in fact, as PR and GR bind the same steroid-response element and, at times, the PR binds glucocorticoid when present in relatively high concentrations.

The RBF-1 and the PR nuclear binding sites also have been localized to the oviduct nuclear matrix using MAb in Western blot analyses and cell-free binding assays (Schuchard *et al.,* 1991a). Fig. 5 depicts a model of the nuclear matrix with the RBF-1 protein bound by the avian oviduct PR. Early studies using Southwestern blotting indicated that these RBF-1 DNA binding elements may reside in or near the nuclear matrix that borders the steroid-regulated c-*myc* gene. Southern blot analyses suggest that nuclear matrix DNA that binds to the RBF contains sequences similar to those found in the promoter of the c-*myc* gene, and possibly the c-*jun* and c-*fos* genes as well (Schuchard *et al.,* 1991b).

The saturable but lower affinity PR binding retained on the chromatin following removal of RBF-1 was determined to contain distinct RBF-like activity (RBF-2). This partially deproteinized chromatin fraction was termed NAP. The monoclonal antibodies against RBF-1 do not recognize any proteins in the crude RBF-2 preparations. The PR binding to the RBF-2

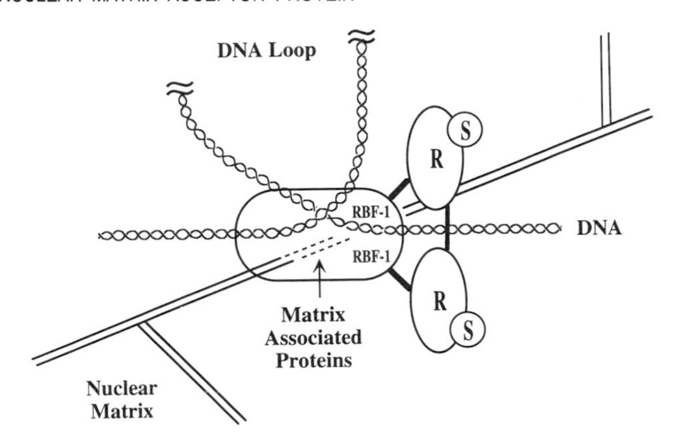


FIG. 5 This diagram represents a model of nuclear matrix as the site of steroid receptor interaction with chromatin acceptor proteins. RBF-1 is bound tightly to nuclear matrix-associated DNA, and may function to bring DNA into proximity of the activated steroid receptor. RBF-1 = receptor binding factor 1; SR = activated steroid receptor complex.

class of sites has been shown to be saturable, high affinity, receptor-specific, and DNase-resistant (Hora *et al.,* 1986; Rories and Spelsberg, 1989; Horton *et al.,* 1991). However, the binding to the RBF-2-DNA complexes appears to have an affinity at least one order of magnitude less than the RBF-1 sites. Further, the DNA sequences involved in the binding of RBF-2 appear to be largely extranuclear matrix (data not published). Efforts have focused on the RBF-1, however, since it has been purified and evidence suggests that it is associated with the highest affinity PR binding activity in whole chromatin.

In summary, the acceptor proteins generate specific SR/DNA binding sites that display specificity in the cell-free binding assays similar to those found in the avian oviduct and in other steroid-target tissues. When RBF-1 and associated proteins (less than 10% of chromatin) are selectively removed from the whole chromatin (nuclear matrix), the highest affinity class of sites measured in oviduct chromatin is abolished. When RBF is reannealed to the DNA, the specific binding sites are regenerated. The binding of [$^3$H]PR to the RBF-1 (chromatin) class of sites can be competed with PR, but not with ER complexes in cell-free and intact cell-binding assays. Further, the RBF binds only functional PR. The precise actions of the two classes of PR acceptor sites in avian oviduct chromatin are not known. The RBF-1 class of sites may be completely localized to the nuclear matrix while the RBF-2 may be less associated with the nuclear matrix. It is not yet known if both RBEs are functional and/or if both proteins are associated with the same genes.

## B. Progesterone RBF-1 in Rat Tissues

Antiavian RBF-1 immunoreactivity found in the nuclei of cells in rat uterus, ovary, and prostate has been identified by immunohistochemistry. These findings suggest that an RBF-like protein may be found in mammalian species (Zhuang *et al.*, 1993). In all rat tissues, the epithelial cells showed the greatest staining, with some in stromal and nonmuscle cells. Western blot analysis has revealed a 15-kDa RBF-1-like protein in the uterine chromatin. Thus, an RBF-1-like avian factor is also found in the nuclei of a mammalian species. The extent of homology between the 15-kDa rat uterine protein and that of the 10-kDa avian RBF-1 remains to be determined.

## C. Estrogen Receptor RBF-1 in Avian Oviduct

Studies from this laboratory have identified an estrogen receptor acceptor-like protein in avian oviduct. The estrogen receptor was shown to have a limited number of high-affinity nuclear acceptor sites similar to, but distinct, from the PR acceptor sites (Kon and Spelsberg, 1982; Ruh and Spelsberg, 1983). Similar to the RBF activity for PR, specific ER activity could be regenerated by reconstitution of the chromatin protein to genomic DNA. The exact size and nature of this ER-type RBF are unknown, but none of the antibodies that recognize the avian PR-RBF displayed cross-reactivity to the ER-specific acceptor protein.

## D. Estrogen Receptor RBF in Rabbit Endometrium

Similar candidate chromatin acceptor proteins have been identified in conjunction with other steroid receptor systems. Additional proteins have been shown to be important for specific binding of the estrogen receptor to the DNA (Spelsberg, 1983; Ross and Ruh, 1984; Singh *et al.*, 1984, 1986; Ruh and Spelsberg *et al.*, 1988) and for the androgen receptors (Rennie *et al.*, 1987).

The 17-kDa candidate acceptor protein for ER in rabbit uterine chromatin has similar properties to the 10-kDa RBF-1 for avian oviduct PR and the avian oviduct ER identified by this laboratory (Ruh and Spelsberg, 1983; Spelsberg *et al.*, 1988). The same methods used to isolate avian PR acceptor sites were used to isolate the ER acceptor protein in the rabbit uteri (Ruh and Spelsberg, 1983; Singh *et al.*, 1984, 1986; Spelsberg *et al.*, 1988). Importantly, the 17-kDa protein binds only ER and not PR (T. S. Ruh and M. F. Ruh, personal communication). Similar to the properties of the RBF-1, this candidate acceptor protein and others identified by Ruh and co-workers (Singh *et al.*, 1986) can be recombined with genomic DNAs to regenerate the specific steroid binding to these DNAs. The similarity between the avian RBF for PR and the RBFs for avian ER and rabbit uterine ER have been described previously (Spelsberg *et al.*, 1988).

### E. Thyroid Hormone RBF-like Proteins

Other laboratories have presented evidence for the role of acceptor-like proteins for the DNA (chromatin) binding of thyroid hormone receptor (Murray and Towle, 1989; Burnside *et al.*, 1990; Lazar and Berrodin, 1990). The carboxyl terminal end of the thyroid receptor, a member of the steroid/nuclear receptor family, appears to interact with the nuclear protein factors (Lazar and Berrodin, 1990). These nuclear acceptor proteins may be implicated in the DNA binding of the thyroid receptor in a manner similar to that described for the chromatin acceptor proteins affiliated with the steroid receptors.

## VI. Binding of Avian Progesterone Receptor RBF to the Promoter Region of c-*myc*

Specific, high-affinity binding of the avian oviduct progesterone receptor to target-cell nuclei is dependent upon association of DNA with specific chromatin acceptor proteins. Studies also have suggested that these acceptor sites are functional only when bound to specific DNA sequences (Spelsberg *et al.*, 1984; Toyoda *et al.*, 1985). Specifically, the purified RBF, when reconstituted along with hen genomic DNA, generates a specific PR-binding complex (Schuchard *et al.*, 1991a). The next step was to find progesterone-responsive genes, identify the specific DNA sequences that bind to the RBF, then elucidate the precise function of this apparent acceptor protein.

Studies utilizing Southwestern blot analysis have revealed that the avian PR-RBF binds to isolated avian oviduct nuclear matrix DNA, suggesting that this site may be the location of interactions among the RBF, PR, and steroid hormone-regulated genes. These findings, in conjunction with discovery that progesterone rapidly down-regulates the c-*myc* proto-oncogene mRNA levels (Fink *et al.*, 1988), prompted investigations to determine whether the RBF acceptor protein might influence PR binding and transactivation of the c-*myc* gene. Southern blotting had shown homology between selective regions of the c-*myc* gene and nuclear matrix DNA suggesting that, like RBF and PR proteins, this gene might interact directly with the matrix (Schuchard *et al.,* 1991b); see Fig. 6). Specifically, nuclear matrix DNA displayed homology to a 1021-bp fragment ("E") which encompasses the two transcription start sites, a putative PRE (see Strahle *et al.,* 1987), and Exon 1. Thus, it was necessary to determine if the PR-RBF would bind directly and specifically to the same 1021-bp segment of the upstream region of the c-*myc* gene which has homology to nuclear matrix DNA sequences.

A modified Southwestern assay was employed to assess RBF binding to the c-*myc* gene. Initially, c-*myc* DNA was digested with restriction endonucleases and fragments were purified and used to probe purified, reconstituted RBF that was slotted onto nitrocellulose. The $^{32}P$-labeled 1021-bp fragment E bound to RBF better than the other labeled DNA fragment regions indicated in the "D," "F," or "G" regions shown in Fig. 6b. These data revealed that RBF interacts with the same region of the c-*myc* gene that has homology to nuclear matrix DNA (Schuchard *et al.,* 1991b). Recently, the interaction between RBF and 1021-bp c-*myc* fragment has been dissected further. Specific fragments of DNA from different areas of E section were made by PCR amplification and labeled with $^{32}P\gamma$ ATP. The RBF binds preferably to the I and K fragments. Notably, there is a 50% homology in a 116-bp stretch of the I and K elements that bind the RBF. Interestingly, RBF binding to the c-*myc* DNA was abolished when the incubation buffer contained $Zn^{2+}$, while binding was highest in the presence of 1 m*M* CaCl and 50 m*M* K glutamate. The balance of salt and divalent cations is crucial for obtaining optimal RBF-c-*myc* DNA binding as has been noted for other protein–DNA interactions.

## VII. Role of MAR Sequences in Gene Transcription and Potential for RBF Binding to MAR-like Elements

### A. Role of MAR Sequences in Gene Transcription

Recent studies have examined specific conditions for nuclear matrix–DNA interactions (Hakes and Berezney, 1991a,b). Binding of the nuclear matrix complex to DNA is temperature dependent and once binding has occurred, the complex is predominantly salt resistant (0.5–4.0 KCl). Interestingly, the nuclear matrix complex binds preferentially to ssDNA compared to binding to dsDNA or to RNA. The nuclear matrix complex displays higher affinity binding to matrix DNA and to MAR sequences (Cockerill and Garrard, 1986) compared to that of total genomic DNA.

MARs represent the sites of chromosomal DNA that bind the nuclear matrix (Gasser and Laemmli, 1986a, 1987). MARs have been found to be

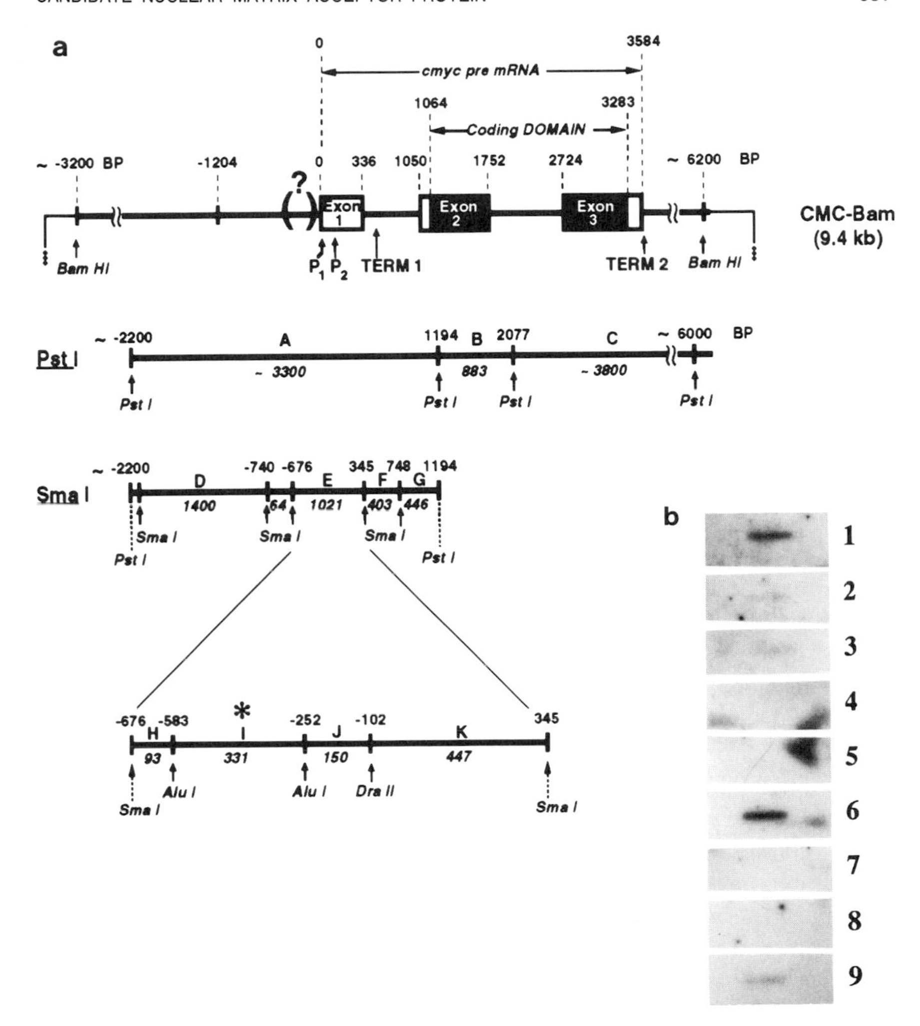


FIG. 6 Restriction map of the genomic clone of the avian c-*myc* gene cloned into a Bam H1 site of pBR-322. (CMC-BAM, American Type Culture Collection received from Rockville, MD; from Schuchard *et al.*, 1991b). DNA regions labeled A–K have been assessed for RBF binding, in some cases for homology with nuclear matrix DNA, as shown in b. (b) Southern blot of c-Myc fragment binding to nuclear matrix DNA. DNA fragments and sequences from cDNAs were slot blotted to filters and probed with $^{32}$P-labeled nuclear matrix DNA and autoradiographed. Instances of hybridization suggest DNA sequence homology with the matrix DNA (from Schuchard *et al.*, 1991b). Some fragments were included as controls. (1) Region A of the c-*myc* gene (−2200 to +1194 bp); (2) Region C of the c-*myc* gene (+2078 to ~6000 bp); (3) Region B of the c-*myc* gene (+1194 to +2077 bp); (4) Upstream region of c-*myc* gene (−3200 to −2200 bp); (5) Ovalbumin cDNA (control); (6) Region E of the c-*myc* gene (−676 to +345 bp); (7) Progesterone receptor response element (control); (8) Estrogen receptor response element (control); (9) GAPDH cDNA (control).

A-T-rich regions of DNA that contain a significant number of ATTA and ATTTA motifs (Comings and Wallack, 1978; Boulikas, 1993a,b). Studies suggest that part of the biological significance of the MAR sequences may be regulation of gene expression. For example, deletions of MAR sequences decrease transcription levels (Blasquez *et al.,* 1989; Xu *et al.,* 1989). Conversely, insertion of the MAR sequence from the chicken lysozyme gene into a CAT reporter gene enhanced enzyme activity relative to transfected constructs without the MAR (Stief *et al.,* 1989). These studies show the potential influence of MAR sequences on regulation of gene transcription.

Other studies have demonstrated further the relationship between MARs and transcription. MAR sites have been localized at the borders of different genes, close to 5′ and 3′ ends of *cis*-regulatory sequences or DNase/hypersensitive sites (DHSs) (Gasser and Laemmli, 1986a,b). DHSs mark the position of short 50 to 400-bp nucleosome-free chromatin regions where transcription factors have access to DNA recognition sequences (Emerson and Felsenfeld, 1984; Bonifer *et al.,*1991). Enhanced DNase 1 sensitivity is a well-documented characteristic of actively transcribed chromatin (Weintraub and Groudine, 1976; Elgin *et al.,* 1989). A MAR was also found between 8.85 and 11.1 kb upstream of the chicken lysozyme gene (Phi-Van and Strätling, 1988), and an other group found a DHS region of the chicken lysozyme gene between −10 kb and +10 kb relative to the transcription start site (Jantsen *et al.,* 1986; Bonifer *et al.,* 1991). Thus, MARs have been mapped to chromatin domain boundaries for the lysozyme gene and appear to border the DHS domain. These findings are of particular interest in the chicken lysozyme gene since it is regulated by estrogen and progesterone, as are some of the DSH sites.

Many biological processes have been found associated with nuclear matrix and MARs including *in vivo* replicating DNA (Vogelstein *et al.,* 1980; Berezney and Buckholtz, 1981), DNA synthesis, and transcriptionally active genes (Ciejek *et al.,* 1982, 1983; Robinson *et al.,* 1982; Thorburn *et al.,* 1988). Specific enzyme activities associated with these processes have also been characterized. Topoisomerase II activity has also been found to be an important component of the nuclear matrix (Berrios *et al.,* 1985; Gasser and Laemmli, 1986a), which may suggest that the matrix prepares individual chromatin domains for replication and/or transcription by incorporating torsional stress in defined regions (Cockerill and Garrard, 1986). The sequence ATATTT corresponds to the "invariable core" of the 15-bp topoisomerase II recognition sequence (Sander and Hsieh, 1985). Earlier studies showed that mouse nuclear matrix proteins displayed a preference for binding DNA that was A-T rich (Comings and Wallack, 1978). Nearly all MARs that have been sequenced contain this topoisomerase II consensus sequence (Cockerill and Garrard, 1986; Gasser and Laemmli, 1986b; Kas and Chasin, 1987; Levy-Wilson and Fortier, 1989). Several MARs also have been shown to contain specific binding sites and cleavage sites for topoisomerase II (Sperry *et al.,* 1989).

### B. Interactions between Steroid Receptors and RBF Involving a MAR-like Sequence

As mentioned earlier, many laboratories have reported that steroid receptors and their nuclear binding sites are associated with nuclear matrix (Barrack and Coffey, 1982) including those for estrogen receptors (Barrack and Coffey, 1980; Metzger and Korach, 1990), androgen receptors (Barrack and Coffey, 1980; Colvard and Wilson, 1984), progesterone receptor binding sites (Hora *et al.,* 1986; Schuchard *et al.,* 1991a), and mineralocorticoid receptors (van Stensel *et al.,* 1991).

In our laboratory, we have utilized Southern blotting with nuclear matrix DNA and genomic sequences to show sequence homologies among upstream regions of three genes, c-*myc,* c-*jun,* and alpha-actin (Schuchard *et al.,* 1991a,b). The 5′-flanking regions of these genes were shown to have homology to the hen matrix DNA, suggesting they contain MAR-like sequences.

Employing optimal ion concentrations and binding conditions, we have used Southwestern blotting to identify a specific 64-bp region of the c-*myc* that yields the highest level of specific binding to the purified RBF. The sequence has a string of 16 thymines followed by an area rich in A's and T's and then ending with a stretch of 8 A's. Indeed, the preponderance of A's and T's in this region is reminiscent of a MAR. However, specificity of RBF binding to the c-*myc* gene is not determined indescriminantly by stretches of A's and T's as a region of the retinoblastoma gene having a 70% homology to the c-*myc* 64-bp sequence does not bind RBF.

Together, these data suggest that RBF binds to a region of the c-*myc* gene that contains a sequence reminiscent of a MAR site situated within the 64-bp fragment. Nuclear matrix DNA has homologous sequences to this general region of the c-*myc* gene. The exact function of RBF binding to the c-*myc* gene 5′-flanking domain remains unclear. The fact that progesterone rapidly represses c-*myc* expression could implicate this complex in transcriptional regulation of the gene.

## VIII. Role of Steroid Receptor Accessory Factors in Nuclear/Chromatin Binding

### A. Definition of Accessory Factors

The DNA binding elements for steroid and unclear hormone receptors are similar. This would suggest that simple SR/response element complexes would be insufficient for exerting a varied molecular response. There must be mechanisms for conferring receptor-specific genomic events. One way this could occur is through protein–protein interactions at the receptor/DNA level.

The steroid receptor acceptor proteins or receptor binding factors discussed previously represent one class of factors that could modulate response specificity by allowing only one species of receptor to bind DNA. Another protein that alters hormone action may be the among-cell and tissue-specific transcription factors. Interactions among steroid receptors and transcription factors, such as Jun and Fos, may occur prior to and during the interaction of receptor with target DNA (Renkawitz *et al.*, 1989; Tsai *et al.*, 1991; Landers and Spelsberg, 1992).

## B. Accessory Factors for Nuclear and Steroid Hormone Receptors

Another class of nuclear binding proteins includes the "accessory" or "associate" factors. These are proteins that may function in the binding of a receptor to DNA and, perhaps, have a role in modulating subsequent transcriptional effects. These factors may be important for fine-tuning the regulatory action of the steroid receptors.

The relationship of the accessory factors to the acceptor proteins is not well established. In some instances it seems the accessory factors interact with the receptor prior to binding DNA. However, distinction is neither simplistic nor clearly defined (e.g., Hall *et al.*, 1995). Perhaps the two classes of proteins represent the same family of proteins but some bind to steroid response elements while others associate with MAR-like domains.

### 1. Thyroid Hormone Receptor Accessory Factors

Receptors for thyroid hormones are located within the cell nucleus and are classified as nuclear hormone receptors. The binding interactions of receptor to DNA response elements are analogous to those of the estrogen system. However, the thyroid hormone receptor can homodimerize or heterodimerize with other proteins prior to binding DNA (see Glass, 1994). Nonetheless, in the most general sense, this nuclear hormone receptor species is subject to the same sorts of regulatory systems as steroid hormone receptors.

The findings of Murray and Towle (1989) suggested that *in vitro* binding of either the $\alpha$ or $\beta$ forms of the T3 receptor to a modified rat GH gene TRE (thyroid hormone response element) required the addition of liver nuclear extracts to the reactions subjected to gel shift analysis. This activity was determined to be heat labile, suggesting the existence of a protein that is necessary for T3 receptor/DNA interaction. Analysis of nuclear extracts from a variety of tissues demonstrated that a protein/DNA complex was formed in the presence of all extracts, but a different complex was predominant in the presence of extract from liver.

Burnside and co-workers (1990) described a protein called "TRAP" (T3 receptor auxiliary protein), a 65,000-Da nuclear protein. Studies demonstrated that the binding of an *in vitro* synthesized T3 receptor species, Hc-erbA-$\beta$, to a TRE could be enhanced three- to fourfold by the addition of GH3 cell nuclear extract to the binding assay, presumably through TRAP in the extract (Burnside *et al.,* 1990). The TRAP factor also enhanced the binding of the rat erbA$\alpha$-1 and the pituitary specific erbA$\beta$-2 forms of the T3 receptor, and nuclear extracts from a variety of cell types were found to contain this activity was not generated by the addition of histones, BSA, or cytosol rather than nuclear extract. Data (Beebe *et al.,* 1991) suggest that TRAP interacts directly with the TRE, perhaps to stabilize the binding of the TR to the DNA.

King *et al.* (1993) reported that an accessory protein(s) modifies TR-induced bending of specific DNA sequences. Using various TRE sequences within sets of circularly permeated flanking sequences, they observed that partially purified and *in vitro* translated TR were able to induce DNA bending; however, the bend angles and bend centers differed with different sources of TR. When a receptor-depleted fraction from a rat liver TR preparation was added to *in vitro* translated TR, receptor binding to DNA was stimulated and heterodimer formation with TR appeared to occur. Changes in bend center and angle resulted that more closely resembled those produced by the native receptor. Thus, changes in DNA bending induced by the TRAP or other TR accessory proteins could be one way in which these proteins modulate accessibility of *cis*-acting DNA elements within the chromatin and/or bring together transacting factors that are not in close proximity in the absence of hormones.

## 2. Glucocorticoid Receptor Accessory Factors

Cavanaugh and Simons (1990) observed that maximal binding of a subpopulation of glucocorticoid receptor–steroid complexes to calf thymus DNA is mediated by a small molecular weight factor. The factor appeared to be most abundant in nuclear fractions, and this protein (approximately 700–3000 Da) does not appear to bind alone to DNA. This protein(s) is present in nuclear extracts of rat liver and kidney, and in human HeLa and MCF-7 nuclear extracts (Cavanaugh and Simons, 1994). Findings also suggest that a similar, low molecular weight factor is required for the DNA binding of distinct subpopulations of both the progesterone and estrogen receptor complexes.

Recent evidence suggests that COUP-TF and HNF-4 serve as accessory factors for glucocorticoid-regulated gene expression. These proteins interact with DNA sequences within a glucocorticoid-responsive unit which is composed in part of two response elements (Hall *et al.,* 1995).

### 3. Androgen Receptor Accessory Factors

Kupfer *et al.* (1993) recently described a receptor accessory factor (RAF) that was found to enhance the specific binding of the rat androgen (AR) and glucocorticoid receptors (GR) to DNA as demonstrated by the formation of hetermeric complexes demonstrated by gel shift analysis (Kupfer *et al.,* 1993). RAF has a predicted molecular mass of about 130 kDa. This factor appears to interact directly with the AR.

Another protein called ASTP, or ATP-stimulated translocation promoter, was described by Okamoto *et al.* (1988). This nuclear protein increases the nuclear binding of partially purified and activated GR complexes in the presence of ATP. ASTP was purified from the liver of adrenalectomized rats and was shown to have a molecular mass of 93 kDa and to be composed of two identical subunits of 48 kDa. Functional activity requires physiological concentrations of ATP but does not bind ATP–agarose; neither ADP nor AMP affect the activity. In the presence of ASTP, the binding of GR to chromatin and nuclei is increased, but the binding to DNA–cellulose is not affected. These findings suggest that ASTP may affect chromosomal structure, which in turn alters the binding of GR to the rat liver nuclear acceptor sites.

### 4. Progesterone Receptor Accessory Factors

Edwards *et al.* (1991) have reviewed some of the mechanisms involved in regulating binding of steroid receptor to specific DNA sequences. The model system they used involved PR from human T47D breast cancer cells and the progesterone response element (PRE) from the mouse mammary tumor virus (MMTV). They demonstrated that the ability of mammalian progesterone receptor to bind specific DNA sequences is substantially greater in the nuclei of intact cells, than to those receptors that have been activated in cell-free cytosol. Data from these studies also suggested that the dissociation of hsp90 from PR is necessary for maximal activation of receptor–DNA binding function, but it is not sufficient. Nuclear extract lacking PR and mixed with activated cytosol PR displayed a dose-dependent increase in the amount of cytosol PR that binds the DNA, implicating the necessity for other proteins.

The activity responsible for the increase in receptor binding to DNA was found to be both heat and trypsin sensitive, with a mass range of 50–75 kDa suggesting that the activity is proteinaceous. The activity was also demonstrated to be present in both steroid-responsive and steroid-nonresponsive cells suggesting that the factor may be ubiquitous and might function similarly in other systems.

Onate *et al.* (1994) recently published studies demonstrating that nuclear extracts from a variety of cell types are able to enhance the binding of baculovirus-produced purified human PR to PREs. The enhanced DNA binding was found to be dependent on the amount of nuclear extract added but the increased levels were saturable. Gel shift analysis demonstrated that HMG-1, a DNA-bending protein, suffices for producing the stimulatory activity. The effect appeared to be very selective for HMG-1 since BSA, insulin, gelatin, and RNase in the same concentrations failed to enhance binding of hPR to the PRE. It was determined that HMG-1 did not alter receptor dimerization or the steroid-binding activity of the receptor. Coimmunoprecipitation studies found both the PR and HMG-1 to be present in the enhanced complex. HMG-1 was found not only to bind PR on a PR affinity column, but also to the DNA whereby it enhanced PR binding. Whether the HMG-1 can be classified as a true accessory protein that interacts with the receptor or its response element is unclear at present.

Prendergast *et al.* (1994) suggests that the HMG-1 protein consists of two conserved DNA binding domains called A and B boxes. The two domains that recognize DNA structure rather than the base sequence were expressed as peptides, purified, and found to stimulate the binding of PR to PRE. The B box alone substituted for the full protein in enhancing binding. It is suggested that HMG-1 facilitates the binding of PR by inducing a structural change in the target DNA, but this remains to be elucidated.

## 5. Estrogen Receptor Accessory Forms

Feavers *et al.* (1987) identified two nonhistone proteins, called NHP-1 and NHP-2, that bind specifically to the concensus estradiol response element in gel retardation assays. NHP-1, a 70-kDa protein, binds to the half dyads of the ERE. The protein did not bind $E_2$, nor did it cross-react with an antibody directed against the ER. The authors hypothesize that NHP-1 may stabilize binding of the ER complex to the ERE. The authors suggest that ER might not bind directly to the ERE. Rather, it is suggested that NHP-1 binds to the ERE to exert its transcriptional effect, despite much information to the contrary. NHP-2, a 60-kDa protein that binds to the DNA 3′ of the dyad, is a region demonstrated to be implicated in estradiol-induced transcription of the *Xenopus* vitellogenin gene (Seiler-Tuyns *et al.*, 1986). Both NHP proteins bind only to double-stranded DNA. *In vitro* reconstitution experiments demonstrated that the two proteins increase the efficiency of binding of the ER complex to ERE, although neither protein was found to be either tissue or species specific.

Mukherjee and Chambon (1990) demonstrated that purified human ER does not form a detectable complex with an ERE in a gel retardation assay under the same conditions as extracts from ER-expressing yeast cells. This suggested that an additional factor (or factors) is required for the binding of ER to an ERE. A 45-kDa factor, termed "DNA binding stimulatory factor" (DBSF) was subsequently purified from yeast. DBSF is a single-stranded DNA binding protein that facilitates binding of ER to the ER response element. The DBSF was shown to favor binding to one of the two strands of the ERE, as well as to a mutated ERE, demonstrating specificity of binding was not due entirely to the palindromic structure of the ERE. As a control, an *Escherichia coli* single-stranded binding protein (SSB) did not promote binding of purified ER to the ERE. Therefore, the effect of DBSF on ER binding is not a property of all single-stranded DNA binding proteins, rather it is unique to this factor.

Recently, Halachmi *et al.* (1994) reported the identification of a 160-kDa estrogen receptor-associated protein (ERAP160) that exhibits estradiol-dependent binding to the ER. Mutation studies of the ER demonstrated that the ability of ER to transactivate parallels the ability of ER to bind ERAP160. Moreover, antiestrogens were not able to promote binding of ER to ERAP160, and blocked estrogen-dependent association of ER and ERAP160 in a dose-dependent manner. The authors postulate that ERAP160 is involved in mediating estradiol-dependent transcriptional activation by the estrogen receptor.

## C. Summary

There is evidence for the existence of accessory proteins that may function as steroid receptor acceptor proteins (receptor binding factors). The latter are defined as those nuclear proteins that bind to DNA and generate high affinity, saturable, steroid-specific binding of the nuclear receptors to the nuclear matrix or genomic DNA. Specific DNA sequences may be the sites of actions of these acceptor proteins. The accessory proteins may (1) bind to the receptor to enhance binding of the steroid/thyroid response elements; and/or (2) enhance the hormone regulation of transcription via interactions with other sequences. In general, it appears that the acceptor proteins/RBF are more closely involved in altering the chromatin/DNA structure, while the accessory factors act at the receptor level. Both of these processes enhance the receptor or DNA sequence, specificity, binding, affinity, or interactions with the transcription apparatus.

## IX. Model for Interactions among Transcriptional Regulators

We have attempted also to design a general model to describe how steroids might directly impact gene transcription, as shown in Fig. 7. The acceptor may be located at a site proximal or distal to the steroid response element enhancer sequence, as shown in Fig. 7-1. The activated steroid receptor complex binds to the acceptor site and becomes associated with the nuclear matrix (Fig. 7-2). Alternatively, the active receptor may bind an acceptor site that is already attached to the nuclear matrix. The DNA binding domain of the matrix-bound receptor could then interact with the steroid response

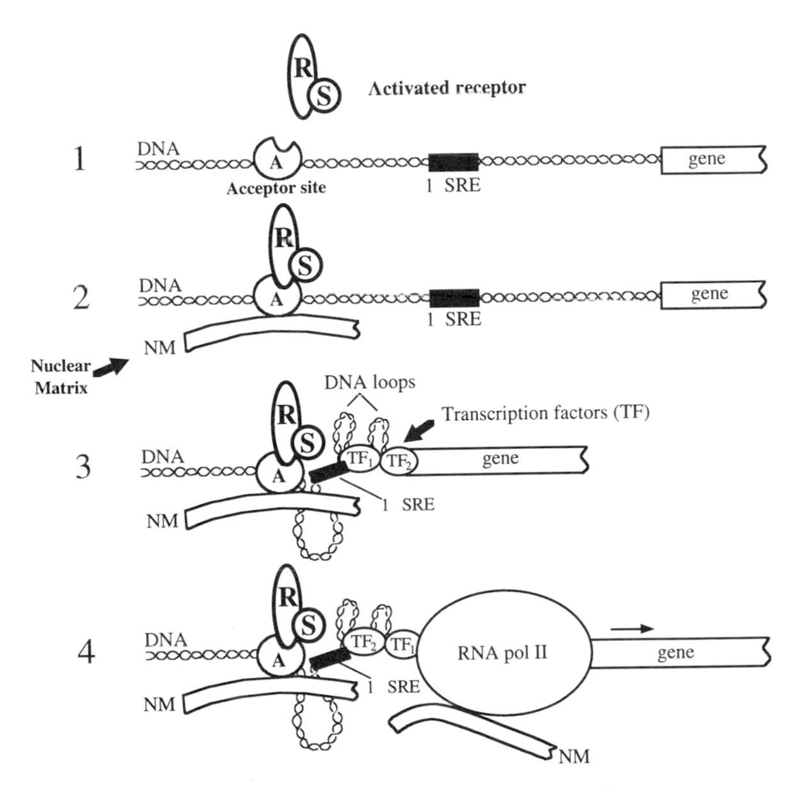


FIG. 7 Model for steroid regulation of gene expression. Steroid-induced initiation of gene transcription may include activated steroid–receptor complexes and steroid-response elements, acceptor proteins, additional transcription factors, and the nuclear matrix. Protein–DNA complexes at the promoter interact with the RNA polymerase II to regulate gene expression. These complexes are suspected to be associated with the nuclear matrix. RS, activated receptor; A, acceptor protein; 1° SRE, primary steroid response element (from Landers and Spelsberg, 1992).

element, possibly displacing a negative regulatory factor (Cordingley *et al.*, 1987b) and/or facilitating the binding of essential transcription factors to specific sites in the promoter (Fig. 7-3).

It should be noted that transcription factors can bind steroid receptors and their DNA binding sites, and show tissue specificity. Particular combinations of these may bind DNA in a cooperative fashion (Landers and Spelsberg, 1992). Since regulatory DNA elements for these factors may be a distance from the genes, the DNA may loop elements into opposition (Schleif, 1987; Ptashne, 1986, 1988) (Figs. 7-3, 7-4). The protein–DNA complex formed at the promoter is then recognized by RNA polymerase II (Fig. 7-4), and additional transcription factors bind to give rise to an active "transcription complex," with transcription of the gene ensuing (Dynan and Tjian, 1985). The nuclear matrix is shown in contact with RNA polymerase to reflect the evidence that actively transcribed genes are associated with nuclear matrix (Jackson and Cook, 1985; Zehnbauer and Vogelstein, 1985). The active receptor–acceptor complex would be in continuous contact with the SRE.

## X. Summary and Conclusions

Rapid advances in the structure and function of steroid receptors have been described. The existence and role of "early genes" in the steroid regulation of structural gene expression has also been discussed. In the cascade model, the regulatory genes code for nuclear proteins that subsequently regulate the expression of the structural genes. This model accounts for many of the known characteristics of steroid regulation of gene expression in target tissue.

Since steroids and their receptors are highly conserved over evolution, the nuclear acceptor sites surely are also conserved. It appears that each steroid species may have its own nuclear acceptor site. Several classes of nuclear acceptor sites for steroid receptors have been reported. The chromatin and matrix acceptor sites for steroid receptors have similar properties. Interactions among these proteins and relationships to the steroid response elements remain unclear. The composition, structure, and function of the chromatin/matrix acceptor sites, although unknown, are crucial for elucidating the mechanism of action of steroid regulation of gene expression.

The simplistic model of steroid hormone action suggests that the steroid receptor binds ligand, becomes activated and dimerizes, then binds to a response element on DNA to participate in the regulation of gene expres-

**Class A: Acceptor Protein–MAR Element (Nuclear Matrix)**

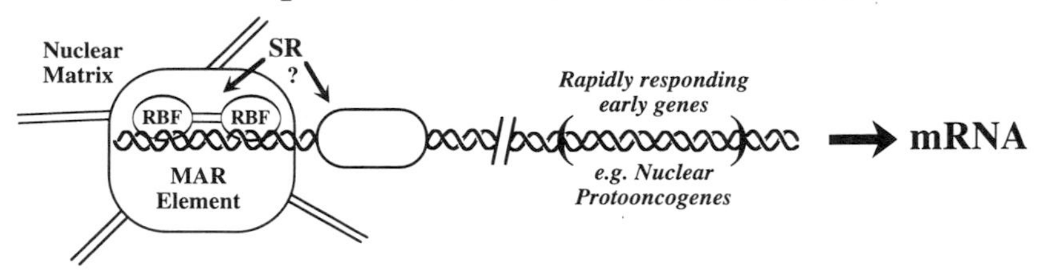


**Class B: Accessory (Associated) Protein – SRE**

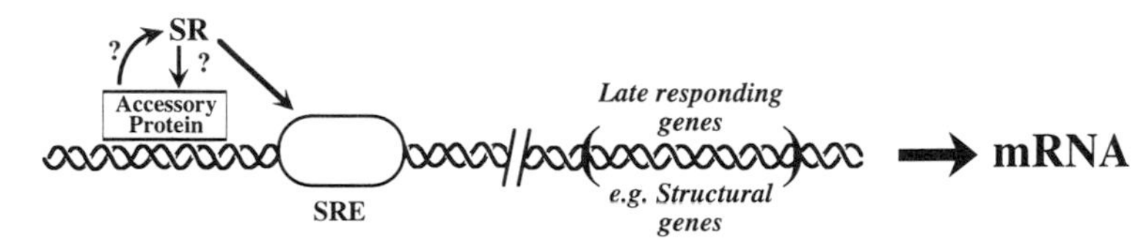


FIG. 8 Model depicting possible classes of nuclear acceptor sites: Role of acceptor versus accessory proteins. This model shows two tentative classes of nuclear binding "acceptor" sites for steroid receptors (SR) based on literature. Two classes of sites have been detected *in vivo* and *in vitro*; one has a $K_D$ of $10^{-9}$ and the other $10^{-8}$ (see Pikler *et al.*, 1976; Spelsberg, 1976; Spelsberg *et al.*, 1976). Class A sites appear to involve the acceptor protein bound to an A-T-rich MAR-like sequence. These sites may represent the higher affinity sites for SR binding. Class B may involve the accessory (associated) proteins bound to or near steroid elements. See text for comparisons of acceptor and accessory proteins.

sion. It remains unclear, however, if multiple proteins and the nuclear matrix are involved in these processes.

Evidence suggests that isolated acceptor proteins, e.g., RBFs, may be structural components of the chromatin acceptor sites for steroid receptors. The RBF for the progesterone receptor is one protein that may participate in progesterone action by interacting with the progesterone receptor and a nuclear MAR-like site. One role of the RBF, then, might be to confer specificity to the PR–DNA binding, since the PR–DNA response element is recognized similarly by androgen, glucocorticoid, and progesterone receptors.

It is interesting to speculate on the distinctions and potential role of acceptor and accessory factors in the regulation of steroid hormone action. A model shown in Fig. 8 directs attention to the possible functions of such proteins.

While precise mechanisms remain to be elucidated, it is evident that the genomic effects of steorids are intertwined with the nuclear matrix function. These findings add further complexity and excitement to the increasingly complicated task of unraveling mechanisms of steroid hormone action.

## References

Agutter, P. S., and Birchall, K. (1979). Functional differences between mammalian nuclear protein matrices and pore–lamina complex laminae. *Exp. Cell Res.* **124,** 453–460.

Alexander, R. B., Greene, G. L., and Barrack, E. R. (1987). Estrogen receptors in the nuclear matrix: Direct demonstration using monoclonal antireceptor antibody. *Endocrinology* (*Baltimore*) **120,** 1851–1857.

Ashburner, M. (1972). Patterns of puffing activity in the salivary glands of *Drosophila.* VI. Induction of ecdysone in salivary glands of *D. melanogaster* cultures *in vitro. Chromosoma* **38,** 255.

Bagchi, M. K., Tsai, M. J., O'Malley, B. W., and Tsai, S. Y. (1992). Analysis of the mechanism of steroid hormone receptor-dependent gene activation in cell-free systems. *Endocr. Rev.* **13,** 525.

Baker, H., and Shapiro, D. (1978). Rapid accumulation of vitellogenin messenger RNA during secondary estrogen stimulation of *Xenopus laevis. J. Biol. Chem.* **253,** 4521.

Barrack, E. R. (1983). The nuclear matrix of the prostate contains acceptor sites for androgen receptors. *Endocrinology* (*Baltimore*) **113,** 430–432.

Barrack, E. R. (1987a). Localization of steroid hormone receptors in the nuclear matrix. *In* "Steroid Hormone Receptors: Their Intracellular Localization" (C. R. Clark, ed.), pp. 86–127. Horwood, Chichester, England.

Barrack, E. R. (1987b). Specific association of androgen receptors and estrogen receptors with the nuclear matrix: Summary and perspectives. *In* "Recent Advances in Steroid Hormone Action" (V. K. Moudgil, ed.), pp. 85–107. de Gruyter, New York.

Barrack, E. R. (1987c). Steroid hormone receptor localization in the nuclear matrix: Interaction with acceptor sites. *J. Steroid Biochem.* **27,** 115.

Barrack, E. R., and Coffey, D. S. (1980). The specific binding of estrogens and androgens to the nuclear matrix of sex hormone responsive tissues. *J. Biol. Chem.* **255,** 7265–7275.

Barrack, E. R., and Coffey, D. S. (1982). Biological properties of the nuclear matrix: Steroid hormone binding. *Recent Prog. Horm. Res.* **38,** 133.

Barrack, E. R., and Coffey, D. S. (1983). The role of the nuclear matrix in steroid hormone action. *Biochem. Actions Horm.* **10,** 23–90.

Beebe, J. S., Darling, D. S., and Chin, W. W. (1991). 3,5,3′-triiodothyroinine receptor auxillary protein (TRAP) enhances receptor binding by interactions within the thyroid hormone response element. *Mol. Endocrinol.* **5,** 85–93.

Berezney, R. (1984). Organization and functions of the nuclear matrix in chromosomal nonhistone proteins. *In* "Chromosomal Nonhistone Proteins" (L. S. Hnilica, ed.), Vol. 4, pp. 120–180. CRC Press, Boca Raton, FL.

Berezney, R., and Buckholtz, L. A. (1981). Dynamic association of replicating DNA fragments with the nuclear matrix of regenerating liver. *Exp. Cell Res.* **312,** 1–13.

Berezney, R., and Coffey, D. S. (1977). Nuclear matrix isolation and characterization of a framework structure from rat liver nuclei. *J. Cell Biol.* **73,** 616–637.

Berrios, M., Osheroff, N., and Fischer, P. (1985). In situ localization of DNA topisomerase II, a major polypeptide component of the *Drosophila* nuclear matrix. *Proc. Natl. Acad. Sci. U.S.A.* **82,** 4142–4146.

Blasquez, V. C., Xu, M., Moses, S. C., and Garrard, W. T. (1989). Immunoglobin κ gene expression after stable integration. *J. Biol. Chem.* **264,** 21183–21189.

Bonifer, C., Hecht, A., Saueressia, H., Winter, D. M., and Sippel, A. E. (1991). Dynamic chromatin: The regulatory domain organization of eukaryotic gene loci. *J. Cell. Biochem.* **47,** 99–108.

Boulikas, T. (1993a). Nature of DNA sequences at the attachment regions of genes to the nuclear matrix. *J. Cell. Biochem.* **52,** 14–22.

Boulikas, T. (1993b). Homeodomain protein binding sites, inverted repeats, and nuclear matrix attachment regions along the human ß-γλobin gene complex. *J. Cell. Biochem.* **52,** 23–36.

Boyd, P. A., and Spelsberg, T. C. (1979). Seasonal changes in the molecular species and nuclear binding of the chick oviduct progesterone receptor. *Biochemistry* **18,** 3685.

Britten, R. J., and Davidson, E. H. (1969). Gene regulation for higher cells: A theory. *Science* **165,** 349.

Brown, M., Jeltsch, J.-M., Roberts, M., and Chambon, P. (1984). Activation of pS2 gene transcription is a primary response to estrogen in the human breast cancer cell line MCF-7. *Proc. Natl. Acad. Sci. U.S.A.* **81,** 6344.

Burnside, J., Darling, D., and Chin, W. (1990). A nuclear factor that enhances binding of thyroid hormone receptors to thyroid hormone response elements. *J. Biol. Chem.* **265,** 2500.

Burtis, K. C., Thummel, C. S., Jones, C. W., Karin, F. D., and Hogness, D. S. (1990). The Drosophilia 74Ef early puff contains E74, a complex ecdysone-inducible gene that encodes two ets-related proteins. *Cell (Cambridge, Mass.)* **61,** 85.

Carson-Jurica M. A., Schrader, W. T., and O'Malley, B. W. (1990). Steroid receptor family: Structure and function. *Endocr. Rev.* **11,** 201–220.

Cavanaugh, A. H., and Simons, S. S. (1990). Glucocorticoid receptor binding to calf thymus DNA. 1. Identification and characterization of a macromolecular factor involved in receptor–steroid complex binding to DNA. *Biochemistry* **29,** 989–995.

Cavanaugh, A. H., and Simons, S. S. (1994). Factor-assisted DNA binding as a possible general mechanism for steroid receptors. Functional heterogeneity among activated receptor–steroid complexes. *J. Steroid Biochem. Molc. Biol.* **48,** 433–446.

Chuknyiska, R. S., Haji, M., Foote, R. H., and Roth, G. S. (1984). Effects of in vivo estradiol administration of availability of rat uterine nuclear acceptor sites measured in vitro. *Endocrinology (Baltimore)* **115,** 836–838.

Chuknyiska, R. S., Haji, M., Foote, R. H., and Roth, G. S. (1985). Age associated changes in nuclear binding of rat uterine estradiol receptor complexes. *Endocrinology (Baltimore)* **116,** 547–551.

Ciejek, E. M., Nordstrom, J. L., Tsai, M.-J., and O'Malley, B. W. (1982). Ribonucleic acid precursors are associated with the chick oviduct nuclear matrix. *Biochemistry* **21,** 4945–4953.

Ciejek, E. M., Tsai, M. J., and O'Malley, B. W. (1983). Actively transcribed genes are associated with the nuclear matrix. *Nature (London)* **306,** 607–609.

Clever, U. (1964). Actinomycin and puromycin: Effects on sequential gene activation by ecdysone. *Science* **146,** 794–795.

Clever, U., and Karlson, P. (1960). Induction of puff changes in salivary gland chromosomes. *Chironomus tentans* by ecdysone. *Exp. Cell Res.* **20,** 623.

Clever, U., and Romball, C. G. (1966). RNA and protein synthesis in the cellular response to a hormone, ecdysone. *Proc. Natl. Acad. Sci. U.S.A.* **56,** 1470.

Cobb, A. D., and Leavitt, W. W. (1987). Characterization of nuclear acceptor sites for mammalian progesterone receptor: Comparison with the chick oviduct system. *Gen. Comp. Endocrinol.* **67,** 214–220.

Cochran, A., and Deeley, R. G. (1989). Detection and characterization of degradative intermediates of avian apo very low density lipoprotein II mRNA present in estrogen-treated birds and following destabilization by hormone withdrawal. *J. Biol. Chem.* **264,** 6495.

Cockerill, P. N., and Garrard, W. T. (1986). Chromosomal loop anchorage of the kappa immunoglobulin gene occurs next to the enhancer in a region containing topoisomerase II sites. *Cell (Cambridge, Mass.)* **44,** 273–282.

Colvard, D. S., and Wilson, E. M. (1984). Androgen receptor binding to nuclear matrix in vitro and its inhibition by 8S androgen receptor binding factor. *Biochemistry* **23,** 3479–3486.

Comings, D. E., and Wallack, A. S. (1978). DNA-binding properties of nuclear matrix proteins. *J. Cell Sci.* **34,** 233–246.

Cordingley, M. G., Riegel, A. T., and Hager, G. L. (1987a). Steroid-dependent interaction of transcription factors with the inducible promoter of mouse mammary tumor virus *in vivo*. *Cell* (*Cambridge, Mass.*) **48,** 261–270.

Cordingley, M. G., Richard-Foy, H., Lichter, A., and Hager, G. L. (1987b). The hormone response element of the MMTV LTR: A complex regulatory region. *In* "DNA: Protein Interactions and Gene Regulation" (E. B. Thompson and J. Papaconstantinou, eds.), pp. 233–243. Univ. of Texas Press, Austin.

Cushing, C. L., Bambara, R. A., and Helf, R. (1985). Interactions of estrogen-receptor and anti-estrogen-receptor complexes with nuclei in vitro. *Endocrinology* (*Baltimore*) **116,** 2419–2429.

DeAngelo, A., and Gorski, J. (1970). Role of RNA synthesis in the estrogen induction of a specific uterine protein. *Proc. Natl. Acad. Sci. U.S.A.* **66,** 693.

DeBoer, W., Snippe, L., Ab, G., and Gruber, M. (1984). Interaction of calf uterine estrogen receptors with chicken target cell nuclei. *J. Steroid Biochem.* **20,** 387–390.

Diamond, M. I., Miner, J. N., Yoshinaga, S. K., and Yamamoto, K. R. (1990). Transcription factor interactions: Selectors of positive or negative regulation from a single DNA element. *Science* **249,** 1266–1272.

Dubik, D., Dembinski, T. C., and Shiu, R. P. C. (1987). Stimulation of c-*myc* oncogene expression associated with estrogen-induced proliferation of human breast cancer cells. *Cancer Res.* **47,** 6517–6521.

Dynan, W. S., and Tjian, R. (1985). Control of eukaryotic messenger RNA synthesis by sequence-specific DNA-binding proteins. *Nature* (*London*) **316,** 774–778.

Edwards, D. P., Kuhnel, B., Estes, P. A., and Nordeen, S. K. (1989). Human progesterone receptor binding to mouse mammary tumor virus deoxyribonucleic acid: Dependence on hormone and non-receptor nuclear factors. *Mol. Endocrinol.* **3,** 381– 391.

Edwards, D. P., DeMarzo, A. M., Onate, S. A., Beck, C. A., Estes, P. A., and Nordeen, S. K. (1991). Mechanisms controlling steroid receptor binding to specific DNA sequences. *Steroids* **56,** 271–278.

Elgin, S. C. R., Cartwright, I. L., Fleischmann, G., Gilmour, D. S., and Thomas, G. H. (1989). Alterations in chromatin structure associated with gene activation. *In* "DNA-Protein Interactions in Transcription" pp. 287–296. Alan R. Liss, New York.

Emerson, B. M., and Felsenfeld, G. (1984). Specific factor conferring nuclease hypersensitivity at the 5′ end of the chicken adult ß-globin gene. *Proc. Natl. Acad. Sci. U.S.A.* **81,** 95–99.

Feavers, I. M., Jiricny, J., Moncharmont, B., Saluz, H. P., and Jost, J. P. (1987). Interaction of two nonhistone proteins with the estradiol response element of the avian vitellogenin gene modulates the binding of estradiol-receptor complex. *Proc. Natl. Acad. Sci. U.S.A.* **84,** 7453–7457.

Fink, K. L., Wieben, E. D., Woloschak, G. E., and Spelsberg, T. C. (1988). Rapid regulation of c-*myc* proto-oncogene expression by progesterone in the avian oviduct. *Proc. Natl. Acad. Sci. U.S.A.* **85,** 1796–1800.

Forsthoefer, A., and Thompson, E. A. (1987). Glucocorticoid regulation of transcription of the c-*myc* cellular proto-oncogene in P1798 cells. *Mol. Endocrinol.* **1,** 899–907.

Frankel, F. R., and Senior, M. B. (1986). The estrogen receptor complex is bound at unusual chromatin regions. *J. Steroid Biochem.* **24,** 983–988.

Freedman, L. P., and Luisi, B. F. (1993). On the mechanism of DNA binding by nuclear hormone receptors: A structural and functional perspective. *J. Cell. Biochem.* **51,** 140–150.

Gasser, S. M., and Laemmli, U. K. (1986a). The organization of chromatin loops: Characterization of a scaffold attachment site. *EMBO J.* **5,** 511–518.

Gasser, S. M., and Laemmli, U. K. (1986b). Cohabitation of scaffold binding regions with upstream/enhancer elements of three developmentally regulated genes of *D. melanogaster*. *Cell* (*Cambridge, Mass.*) **46,** 521–530.

Gasser, S. M., and Laemmli, U. K. (1987). A glimpse at chromosomal order. *Trends Genet.* **3,** 16–22.

Gaub, M. P., Bellard, M., Scheuer, I., Chambon, P., and Sassone-Corsi, P. (1990). Activation of the ovalbumin gene by the estrogen receptor involves the fos–jun complex. *Cell (Cambridge, Mass.)* **63,** 1267–1276.

Glass, C. K. (1994). Differential recognition of target genes by nuclear receptor monomers, dimers, and heterodimers. *Endocr. Rev.* **15,** 391–407.

Goldberger, A., and Spelsberg, T. C. (1988). Partial purification and preparation of polyclonal antibodies against candidate chromatin acceptor proteins for the avian oviduct progesterone receptor. *Biochemistry* **27,** 2103–2109.

Goldberger, A., Littlefield, B. A., Katzman, J., and Spelsberg, T. C. (1986). Monoclonal antibodies to the nuclear binding sites of the avian oviduct progesterone receptor. *Endocrinology (Baltimore)* **118,** 2235–2241.

Goldberger, A., Horton, M., Katzmann, J., and Spelsberg, T. C. (1987). Characterization of the nuclear acceptor sites for the avian oviduct progesterone receptor using monoclonal antibodies. *Biochemistry* **26,** 5811–5816.

Gonda, T. J., Sheiness, D. K., and Bishop, J. M. (1982). Transcripts from the cellular homologs of retroviral oncogenes: Distribution among chicken tissues. *Mol. Cell. Biol.* **2,** 617–624.

Gorski, J., Toft, D. O., Shyamala, G., Smith, D., and Notides, A. (1968). Hormone receptors: Studies on the interaction of estrogens with the uterus. *Recent Prog. Horm. Res.* **24,** 45–80.

Green, S., and Chambon, P. (1991). The estrogen receptor. *In* "Nuclear Hormone Receptors" (M. G. Parker, ed.), pp. 15–38. Academic Press, San Diego, CA.

Gronemeyer, H. (1991). Transcription activation by estrogen and progesterone receptors. *Annu. Rev. Genet.* **25,** 89–123.

Hager, G. L., and Archer, T. K. (1991). The interaction of steroid receptors with chromatin. *In* "Nuclear Hormone Receptors" (M. G. Parker, ed.), pp. 217–234. Academic Press, San Diego, CA.

Hakes, D. J., and Berezney, R. (1991a). DNA binding properties of the nuclear matrix and individual nuclear matrix proteins. *J. Biol. Chem.* **266,** 11131–11140.

Hakes, D. J., and Berezney, R. (1991b). Molecular cloning of matrin F/G: A DNA binding protein of the nuclear matrix that contains putative zinc finger motifs. *Proc. Natl. Acad. Sci. U.S.A.* **88,** 6186–6190.

Halachmi, S., Marden, E., Martin, G., MacKay, H., Abbondanza, C., and Brown, M. (1994). Estrogen receptor-associated proteins: Possible mediators of hormone-induced transcription. *Science* **264,** 1455–1458.

Hall, R. K., Sladek, F. M., and Granner, D. K. (1995). The orphan receptors COUP-TF and HNF-4 serve as accessory factors required for induction of phosphoenolpyruvate carboxykinase gene transcription by glucocorticoids. *Proc. Natl. Acad. Sci. U.S.A.* **92,** 412–416.

Hamana, K., and Iwai, K. (1978). Glucocorticoid receptor complex binds to nonhistone protein and DNA in rat liver chromatin. *J. Biochem. (Tokyo)* **83,** 279–286.

Harris, S. E., Rosen, J. M., Means, A. R., and O'Malley, B. W. (1975). Use of a specific probe for ovalbumin messenger RNA to quantitate estrogen-induced gene transcripts. *Biochemistry* **14,** 2072.

Hora, J., Horton, M. J., Toft, D. O., and Spelsberg, T. C. (1986). Nuclease resistance and the enrichment of native nuclear acceptor sites for the avian oviduct progesterone receptor. *Proc. Natl. Acad. Sci. U.S.A.* **83,** 8839–8843.

Horton, M., Landers, J., Subramaniam, M., Goldberger, A., Toyoda, H., Littlefield, B., Hora, J., Gosse, B., and Spelsberg, T. C. (1991). Enrichment of a second class of native acceptor sites for the avian oviduct progesterone receptor as intact chromatin fragments. *Biochemistry* **30,** 9523–9530.

Hyder S. M., Stancel, G. M., and Loose-Mitchell, D. S. (1994). Steroid hormone-induced expression of oncogene encoded nuclear proteins. *CRC Crit. Rev. Eukaryotic Gene Express.* **4,** 55–116.

Jackson, D. A., and Cook, P. R. (1985). Transcription occurs at a nucleo-skeleton. *EMBO J.* **4,** 919–925.

Jantsen, K., Fritton, H. P., and Igo-Kemenes, T. (1986). The DNAse I sensitive domain of the chicken lysozyme gene spans 24 kb. *Nucleic Acids Res.* **14,** 6085–6099.

Jensen, E. V., Suzuki, T., Kawashima, T., Stumpf, W. E., Jungblut, P. W., and DeSombre, E. R. (1968). A two step mechanism for the interaction of estradiol with rat uterus. *Proc. Natl. Acad. Sci. U.S.A.* **59,** 632–638.

Jost, J.-P., Seldram, M., and Geiser, M. (1984). Preferential binding of estrogen-receptor complexes to a region containing the estrogen-dependent and hypomethylation site preceding the chicken vitellogenin II gene. *Proc. Natl. Acad. Sci. U.S.A.* **81,** 429–433.

Jost, J.-P., Saluz, H., Jirleny, J., and Moncharmont, B. (1987). Estradiol-dependent transacting factor binds preferentially to a dyad-symmetry structure within the third intron of the avian vitellogenin gene. *J. Cell. Biochem.* **35,** 69–82.

Kas, E., and Chasin, L. A. (1987). Anchorage of the Chinese hamster dihydrofolate reductase gene to the nuclear scaffold occurs in an intragenic region. *J. Mol. Biol.* **198,** 677–692.

Katzenellenbogan, B. S. (1980). Dynamics of steroid hormone receptor action. *Annu. Rev. Physiol.* **42,** 17–35.

Kelley, K., and Siebenlist, U. (1986). The regulation and expression of c-*myc* in normal and malignant cells. *Annu. Rev. Immunol.* **4,** 317–338.

King, I. N., de Soyza, T., Catanzaro, D. F., and Lavin, T. N. (1993). Thyroid hormone receptor-induced bending of specific DNA sequences is modified by an accessory factor. *J. Biol. Chem.* **268,** 495–501.

Klinge, C. M., Bambara, R. A., Zain, S., and Helf, R. (1987). Estrogen receptor binding to nuclei from normal and neoplastic rat mammary tissues in vitro. *Cancer Res.* **47,** 2852–2859.

Klyzsejko-Stafanowicz, L., Chui, J. F., Tsai, Y. H., and Hnilica, L. S. (1976). Acceptor proteins in rat androgenic tissue chromatin. *Proc. Natl. Acad. Sci. U.S.A.* **73,** 1954–1958.

Kon, O. L., and Spelsberg, T. C. (1982). Acceptor sites for the estrogen receptor in hen oviduct chromatin. *Biochem. J.* **210,** 905–912.

Kupfer, S. R., Marschke, K. B., Wilson, E. M., and French, F. S. (1993). Receptor accessory factor enhances specific DNA binding of androgen and glucocorticoid receptors. *J. Biol. Chem.* **268,** 17519–17527.

Landers, J. P., and Spelsberg, T. C. (1992). New concepts in steroid hormone action: Transcription factors, proto-oncogenes and the cascade model for steroid regulation and gene expression. *CRC Crit. Rev. Eukaryotic Gene Express.* **2**(1), 19–63.

Landers, J. P., Subramaniam, M., Gosse, B., Weinshilboum, R., and Spelsberg, T. C. (1994). The ubiquitous nature of the progesterone receptor binding factor-1 (RBF-1) in avian tissues. *J. Cell. Biochem.* **55,** 241–251.

Lau, C. K., Subramaniam, M., Rasmussen, K., and Spelsberg, T. C. (1990). Rapid inhibition of the c-*jun* proto-oncogene expression in avian oviduct by estrogen. *Endocrinology* (*Baltimore*) **127,** 2595–2597.

Lauber, A. H., and Pfaff, D. W. (1990). Estrogen regulation of mRNAs in the brain and relationship to lordosis behavior. *Curr. Top. Neuroendocrinol.* **10,** 115–147.

Lauber, A. H., Schuchard, M., Subramaniam, S., and Spelsberg, T. C. (1995). In preparation.

Lazar, M. A., and Berrodin, T. J. (1990). Thyroid hormone receptor from distinct nuclear protein dependent and independent complexes with a thyroid hormone response element. *Mol. Endocrinol.* **4,** 1627.

Leibold, E. A., Landino, A., and Yu, Y. (1990). Structural requirements of iron-responsive elements for binding of the protein involved in both transferrin receptor and ferritin mRNA post-transcriptional regulation. *Nucleic Acids Res.* **18,** 1819–1824.

Levy-Wilson, B., and Fortier, C. (1989). The limits of the DNase-I-sensitive domain of the human apolipoproteins B gene coincide with the locations of chromosomal anchorage loops and define the 5′ and 3′ boundaries of the gene. *J. Biol. Chem.* **264,** 21196–21204.

Lopez, A., Burgos, J., and Ventanas, J. (1985). The binding of [$^3$H]oestradiol receptor complex to hypothalamic chromatin of male and female mice. *Int. J. Biochem.* **17,** 1207–1211.

Martinez, E., Givel, F., and Wahli, W. (1987). The estrogen-responsive element as an inducible enhancer: DNA sequence requirements and conversion to a glucocorticoid responsive element. *EMBO J.* **6,** 3719.

Metzger, D. A., and Korach, K. S. (1990). Cell-free interaction of the estrogen receptor with mouse uterine nuclear matrix: Evidence of saturability, specificity, and resistance to KCl extraction. *Endocrinology* (*Baltimore*) **126,** 2190–2195.

Moudgil, V. K., ed. (1987). "Recent Advances in Steroid Hormone Action." de Gruyter, New York.

Mukherjee, R., and Chambon, P. (1990). A single-stranded DNA-binding protein promotes the binding of the purified estrogen receptor to its responsive element. *Nucleic Acids Res.* **18,** 5713–5716.

Muller, M., Baniahmad, C., Kaltschmidt, C., Schule, R., and Renkawitz, R. (1991). Co-operative transactivation of steroid receptors. *In* "Nuclear Hormone Receptors" (M. G. Parker, ed.), pp. 155–174. Academic Press, San Diego, CA.

Murphy, L. J., Murphy, L. C., and Friesen, H. G. (1987). Estrogen induction of n-myc and c-*myc* proto-oncogene expression in the rat uterus. *Endocrinology* (*Baltimore*) **120,** 1882–1888.

Murray, M. B., and Towle, H. C. (1989). Identification of nuclear factors that enhance binding of the thyroid hormone receptor to a thyroid hormone response element. *Mol. Endocrinol.* **3,** 1434–1442.

Nielsen, D. A., and Shapiro, D. J. (1990). Estradiol and estrogen receptor-dependent stabilization of a mini-vitellogenin mRNA lacking 5,100 nucleotides of coding sequence. *Mol. Cell. Biol.* **10,** 371–376.

Ogle, T. F. (1987). Nuclear acceptor sites for progesterone-receptor complexes in rat placenta. *Endocrinology* (*Baltimore*) **121,** 28–35.

Okamoto, K., Isohasi, F., Ueda, K., Kokufu, I., and Sakamoto, Y. (1988). Purification and characterization of an adenosine triphosphate-stimulated factor that enhances the nuclear binding of activated glucocorticoid-receptor complex from rat liver. *Endocrinology* (*Baltimore*) **123,** 2752–2761.

Onate, S. A., Prendergast, P., Wagner, J. P., Nissen, M., Reeves, R., Pettijohn, D. E., and Edwards, D. P. (1994). The DNA-bending protein HMG-1 enhances progesterone receptor binding to its target DNA sequences. *Mol. Cell. Biol.* **14,** 3376–3391.

Palmiter, R., Moore, P., Mulvihill, E., and Emtage, S. (1976). A significant lag in the induction of ovalbumin messenger RNA by steroid hormones: A receptor translocation hypothesis. *Cell* (*Cambridge, Mass.*) **8,** 557.

Pavlik, E. J., Van Nagell, J. R., Nelsen, K., Gallion, H., and Donaldson, E. S. (1986). Antagonism to estradiol in the mouse: Reduced entry of receptors complexed with 4-hydroxytamoxifen into $Mg^+$ + soluble chromatin fraction. *Endocrinology* (*Baltimore*) **118,** 1924–1934.

Perry, B. N., and Lopez, A. (1978). The binding of [$^3$H]-labelled oestradiol- and progesterone-receptor complexes to hypothalamic chromatin of male and female sheep. *Biochem. J.* **176,** 873–883.

Phi-Van, L., and Strätling, W. H. (1988). The matrix attachment regions of the chicken lysozyme gene co-map with the boundaries of the chromatin domain. *EMBO J.* **7,** 655–664.

Pikler, G. B., Webster, R. A., and Spelsberg, T. C. (1976). Nuclear binding of progesterone in hen oviduct: Binding to multiple sites in vitro. *Biochem. J.* **156,** 399–408.

Prendergast, P., Onate, S. A., Christensen, K., and Edwards, D. (1994). Nuclear accessory factors enhance the binding of protesterone receptor to specific target DNA. *J. Steroid Biochem. Mol. Biol.* **48,** 1–13.

Ptashne, M. (1986). Gene regulation by proteins acting nearby and at a distance. *Nature* (*London*) **322,** 697–701.

Ptashne, M. (1988). How eukaryotic transcriptional activators work. *Nature* (*London*) **335,** 683–689.

Rejman, J., Landers, J. P., Goldberger, A., McCormick, D., Gosse, B., and Spelsberg, T. C. (1991). Purification of a nuclear protein (Receptor Binding Factor -1) associated with the chromatin acceptor sites for the avian oviduct progesterone receptor. *J. Protein Chem.* **10,** 651–667.

Renkawitz, R., Beug, H., Graf, T., Matthias, P., and Grez, M. (1982). Expression of a chicken lysozyme recombinant gene is regulated by progesterone and dexamethasone after microinjection into oviduct cells. *Cell* (*Cambridge, Mass.*) **31,** 167–176.

Renkawitz, R., Schüle, R., Kaltschmidt, C., Baniahmad, A., Altschmied, J., Steiner, C., and Muller, M. (1989). Clustered arrangement and interaction of steroid hormone receptors with other transcription factors. *In* "Proceedings of the 40th Colloquium Mosbach," pp. 21–28. Springer-Verlag, Heidelberg.

Rennie, P. S., Bowden, J.-F., Brunchovsky, N., Frenette, P.-S., Foekens, J. A., and Chen, H. (1987). DNA and protein components of nuclear acceptor sites for androgen receptors in the rat prostate. *J. Steroid Biochem.* **27,** 513–520.

Robinson, S. I., Nelkin, B. D., and Vogelstein, B. (1982). The ovalbumin gene is associated with the nuclear matrix of chicken oviduct cells. *Cell* (*Cambridge, Mass.*) **28,** 99–106.

Rories, C., and Spelsberg, T. C. (1989). Ovarian steroid action on gene expression: Mechanisms and models. *Annu. Rev. Physiol.* **51,** 653–681.

Rories, C., Lau, C. K., Fink, K., and Spelsberg, T. C. (1989). Rapid inhibition of *c-myc* gene expression by a glucocorticoid in the avian oviduct. *Mol. Endocrinol.* **3,** 991–1001.

Ross, P., and Ruh, T. S. (1984). Binding of the estradiol-receptor complex to reconstituted nucleoacidic protein from calf uterus. *Biochem. Biophys. Acta* **782,** 18–25.

Ruh, M. F., Singh, R. K., Mak, P., and Callard, G. V. (1986). Tissue and species specificity of unmasked nuclear acceptor sites for the estrogen receptor of squalus testes. *Endocrinology* (*Baltimore*) **118,** 811–817.

Ruh, M. F., Singh, R. K., Ruh, T. S., and Shyamala, G. (1987). Binding of glucocorticoid receptors to mammary chromatin acceptor sites. *J. Steroid Biochem.* **28,** 581–586.

Ruh, T. S., and Spelsberg, T. C. (1983). Acceptor sites for the estrogen receptor in hen oviduct chromatin. *Biochem. J.* **210,** 905–912.

Ruh, T. S., Ross, P., Wood, D. M., and Keene, J. L. (1981). The binding of [$^3$H]oestradiol receptor complexes to calf uterine chromatin. *Biochem. J.* **200,** 133–142.

Sander, M., and Hsieh, T. S. (1985). *Drosophila* topisomerase II double-stranded DNA cleavage: Analysis of DNA sequence homology at the cleavage site. *Nucleic Acids Res.* **13,** 1057–1072.

Schleif, R. (1987). Why should DNA loop? (Commentary) *Nature* (*London*) **327,** 369–370.

Schuchard, M., Rejman, J. J., McCormick, D. J., Gosse, B., Ruesink, T., and Spelsberg, T. C. (1991a). Characterization of a purified chromatin acceptor protein (receptor binding factor 1) for the avian oviduct progesterone receptor. *Biochemistry* **30,** 4534–4542.

Schuchard, M., Subramaniam, M., Ruesink, T., and Spelsberg, T. C. (1991b). Nuclear matrix localization and specific matrix DNA binding factor -1 of the avian oviduct progesterone receptor. *Biochemistry* **30,** 9516–9522.

Schuchard, M., Landers, J. P., Punkay-Sandhu, N., and Spelsberg, T. C. (1993). Steroid hormone regulation of nuclear proto-oncogenes. *Endocr. Rev.* **14,** 659–669.

Schüle, R., Muller, M., Kaltschmidt, C., and Renkawitz, R. (1988a). Many transcription factors interact synergistically with steroid receptors. *Science* **242,** 1418–1420.

Schüle, R., Muller, M., Otsuka-Murakami, H., and Renkawitz, R. (1988b). Cooperativity of the glucocorticoid receptor and the CACCC-box binding factor. *Nature* (*London*) **332,** 87–90.

Schutz, G., Nguyen-Huu, M., Giesecke, K., Hynes, N., and Groner, B. (1978). Hormonal control of egg white protein messenger RNA synthesis in the chicken oviduct. *Cold Spring Harbor Symp. Quant. Biol.* **42,** 617.

Schwabe, J. W. R., Chapman, L., Finch, F. J., and Rhodes, D. (1993). The crystal structure of the estrogen receptor DNA-binding domain bound to DNA: How receptors discriminate between their response elements. *Cell (Cambridge, Mass.)* **75,** 567–578.

Segraves, W. A., and Hogness, D. S. (1990). The E75 ecdysone-inducible gene responsible for the 75B early puff in *Drosphila* encodes two members of the steroid receptor superfamily. *Genes Dev.* **4,** 204–219.

Seiler-Tuyns, A., Walker, P., Martinez, E., Merillat, A. M., Girel, F., and Wahli, W. (1986). Identification of estrogen-responsive DNA sequences by transient expression experiments in a human breast cancer cell line. *Nucleic Acids Res.* **14,** 8755–8770.

Shyamala, G., Singh, R. K., Ruh, M. F., and Ruh, T. S. (1986). Relationship between mammary ER and estrogenic sensitivity II: Binding of cytoplasmic receptor to chromatin. *Endocrinology (Baltimore)* **119,** 819–826.

Singh, R. K., Ruh, M. F., and Ruh, T. S. (1984). Binding of [$^3$H]estradiol and [$^3$H]H1285 receptor complexes to rabbit uterine chromatin. *Biochim. Biophys. Acta* **800,** 33–40.

Singh, R. K., Ruh, M. F., Butler, W. B., and Ruh, T. S. (1986). Acceptor sites on chromatin for receptor bound by estrogen versus antiestrogen in antiestrogen-sensitive and -resistant MCF-7 cells. *Endocrinology (Baltimore)* **118,** 1087–1095.

Spelsberg, T. C. (1976). Nuclear binding of progesterone in chick oviduct: Multiple binding sites in vivo and transcriptional response. *Biochem. J.* **156,** 391–398.

Spelsberg, T. C., and Halberg, F. (1980). Circannual rhythms in steroid receptor concentration and nuclear binding in chick oviduct. *Endocrinology (Baltimore)* **107,** 1234–1244.

Spelsberg, T. C., Pikler, G. M., and Webster, R. A. (1976). Progesterone binding to hen oviduct genome: Specific versus nonspecific binding. *Science* **194,** 197–199.

Spelsberg, T. C., Littlefield, B. A., Seelke, R., Martin-Dani, G., Toyoda, H., Boyd-Leinen, P., Thrall, C., and Kon, O. (1983). Role of specific chromosomal proteins and DNA sequences in the nuclear binding sites for steroid receptors. *Recent Prog. Horm. Res.* **39,** 463–517.

Spelsberg, T. C., Gosse, B., Littlefield, B. A., Toyoda, H., and Seelke, R. (1984). Reconstitution of native-like nuclear acceptor sites of the avian oviduct progesterone receptor: Evidence for involvement of specific chromatin proteins and specific DNA sequences. *Biochemistry* **23,** 5103–5112.

Spelsberg, T. C., Goldberger, A., Horton, M., and Hora, J. (1987a). Nuclear acceptor sites for sex steroid hormone receptors in chromatin. *J. Steroid Biochem.* **27,** 133–147.

Spelsberg, T. C., Hora, J., Horton, M., Goldberger, A., and Littlefield, B. A. (1987b). Specific DNA binding proteins and DNA sequences involved in steroid hormone regulation of gene expression. *In* "DNA: Protein Interactions and Gene Regulation" (E. B. Thompson and J. Papaconstantinous, eds.), pp. 259–267. Univ. of Texas Press, Austin.

Spelsberg, T. C., Horton, M., Fink, K., Goldberger, A., Rories, C., Gosse, B., and Rasmussen, K. (1987c). A new model for steroid regulation of gene transcription using chromatin acceptor sites and regulatory genes and their products. *In* Recent Advances in Steroid Hormone Action" (V. K. Moudgil, ed.), pp. 31–48. de Gruyter, Berlin.

Spelsberg, T. C., Ruh, T., Ruh, M., Goldberger, A., Horton, M., Hora, J., and Singh, R. (1988). Nuclear acceptor sites for steroid hormone receptors: Comparisons of steroids and antisteroids. *J. Steroid Biochem.* **31,** 579–592.

Spelsberg, T. C., Rories, C., Rejman, J., Goldberger, A., Fink, K., Lau, C. K., Colvard, D., and Wiseman, G. (1989). Steroid action on gene expression: Possible roles of regulatory genes and nuclear acceptor sites. *Biol. Reprod.* **40,** 54–69.

Sperry, A. O., Blasques, V. C., and Garrard, W. T. (1989). Dysfunction of chromosomal loop attachment sites: Illegitimate recombination linked to matrix association regions and topisomerase II. *Proc. Nat. Acad. Sci. U.S.A.* **86,** 5496–5501.

Stief, A., Winter, D. M., Strätling, W. H., and Sippel, A. E. (1989). A nuclear DNA attachment element mediates elevated and position-independent gene activity. *Nature (London)* **341,** 343–345.

Strahle, U., Klock, G., and Schutz, G. (1987). A DNA sequence of 15 base pairs is sufficient to mediate both glucocorticoid and progesterone induction of gene expression. *Proc. Natl. Acad. Sci. U.S.A.* **84,** 7871–7875.

Studzinski, G. P., Brelvi, Z. S., Feldman, S. C., and Watt, R. A. (1986). Participation of c-*myc* protein in DNA synthesis of human cells. *Science* **234,** 467–470.

Sun, L.-H. K., Pfendner, E. G., Senior, M. B., and Frankel, F. R. (1983). Progesterone, glucocorticoid and estradiol receptors in MCF-7 cells bind to chromatin. *Mol. Cell. Endocrinol.* **30,** 267–278.

Thorburn, A., Moore, R., and Knowland, J. (1988). Attachment of transcriptionally active DNA sequences to the nucleoskeleton under isotonic conditions. *Nucleic Acids Res.* **16,** 7183.

Tora, L., Gronemeyer, H., Turcotte, B., Gaub, M.-P., and Chambon, P. (1988). The N-terminal region of the chicken progesterone receptor specifies target gene activation. *Nature (London)* **333,** 185–188.

Toyoda, T., Seelke, R., Littlefield, B. A., and Spelsberg, T. C. (1985). Evidence for specific DNA sequences in the nuclear acceptor sites of the avian oviduct progesterone receptor. *Proc. Natl. Acad. Sci. U.S.A.* **82,** 4722–4726.

Travers, M. T., and Knowler, J. T. (1987). Oestrogen-induced expression of oncogenes in the immature rat uterus. *FEBS Lett.* **211,** 27–30.

Truss, M., and Beato, M. (1993). Steroid hormone receptors: Interaction with deoxyribonucleic acid and transcription factors. *Endocr. Rev.* **14,** 459–479.

Tsai, S. Y., Tsai, M.-J., and O'Malley, B. W. (1991). The steroid receptor superfamily: Transactivators of gene expression. *In* "Nuclear Hormone Receptors" (M. G. Parker, ed.), pp. 103–124. Academic Press, San Diego, CA.

van Stensel, B., van Haarst, A. D., deKloet, E. R., and van Driel, R. (1991). Binding of corticosteroid receptors to rat hippocampus nuclear matrix. *FEBS Lett.* **292,** 229–231.

Vogelstein, B., Pardoll, D. M., and Coffey, D. S. (1980). Supercoiled loops and eukaryotic DNA replication. *Cell (Cambridge, Mass.)* **22,** 79–85.

Wang, T. Y. (1978). The role of nonhistone chromosomal proteins in the interaction of prostate chromatin with androgen receptor complex. *Biochim. Biophys. Acta* **518,** 8188.

Webb, D. K., Moulton, B. C., and Khan, S. (1990). Estrogen induced expression of the c-jun proto-oncogene in the immature and mature rat uterus. *Biochem. Biophys. Res. Commun.* **168,** 721.

Webster, R. A., Pikler, G. M., and Spelsberg, T. C. (1976). Nuclear binding of progesterone in hen oviduct: Role of acidic chromatin proteins in high-affinity binding. *Biochem. J.* **156,** 409–418.

Weintraub, H., and Groudine, M. (1976). Chromosome subunits in active genes have an altered conformation. *Science* **193,** 848–856.

Xu, M., Hammer, R. E., Blasquez, V. C., Jones, S. L., and Garrard, W. T. (1989). Immunoglobin k gene expression after stable integration. *J. Biol. Chem.* **264,** 21190–21195.

Yang-Yen, H.-F., Chambard, J.-C., Sun, Y.-L., Smeal, T., Schmidt, T. J., Drouin, J., and Karin, M. (1990). Transcriptional interference between c-jun and the glucocorticoid receptor. Mutual inhibition of DNA binding due to direct protein-protein interaction. *Cell (Cambridge, Mass.)* **62,** 1205–1215.

Zehnbauer, B. A., and Vogelstein, B. (1985). Supercoiled loops and the organization of replication and transcription in eukaryotes. *BioEssays* **2,** 52–54.

Zhuang, Y.-H., Landers, J. P., Schuchard, M. D., Syvala, H., Gosse, B., Ruesink, R., Spelsberg, T. C., and Tuohimaa, P. (1993). Immunohistochemical localization of the avian progesterone receptor and its candidate receptor binding factor (RBF-1). *J. Cell. Biochem.* **53,** 383–393.

# Nuclear Matrix and the Cell Cycle

Peter Loidl and Anton Eberharter
Department of Microbiology, University of Innsbruck-Medical School, A-6020 Innsbruck, Austria

The facts that the nuclear matrix represents a structural framework of the cell nucleus and that nuclear events, such as DNA replication, transcription, and DNA repair, are associated with this skeletal structure suggest that its components are subject to cell cycle-regulatory mechanisms. Cell cycle regulation has been shown for nuclear lamina assembly and disassembly during mitosis and chromatin reorganization. Little attention has so far been paid to internal nuclear matrix proteins and matrix-associated proteins with respect to the cell cycle. This survey attempts to summarize available data and presents experimental evidence that important metabolic functions of the nucleus are regulated by the transient, cell cycle-dependent attachment of enzymes and regulatory proteins to the nuclear matrix. Results on thymidine kinase and RNA polymerase during the synchronous cell cycle of *Physarum polycephalum* demonstrate that reversible binding to the nuclear matrix represents an additional level of regulation for nuclear processes.

**KEY WORDS:** Nuclear matrix, Lamins, Cell cycle, Thymidine kinase, RNA polymerase, c-*myc, Physarum polycephalum,* Phosphorylation, Histone acetylation, Chromatin.

## I. Introduction

> What are called structures are slow processes of long duration, functions are quick processes of short duration. (von Bertalanffy, 1952).

The chromatin of eukaryotic cells undergoes dramatic structural changes during the cell cycle. Before mitosis, dispersed interphase chromatin is induced to condense to form metaphase chromosomes. After proper interaction with the mitotic spindle and correct chromosome segregation, chromatin has to decondense after mitotic division and adapts various structural conformations that are compatible with processes like DNA replication, transcription, or DNA repair in a highly ordered, cell cycle-dependent manner. During the past decade it became evident that these dramatic structural transitions occur on an underlying skeletal element, termed nuclear matrix in interphase and chromosome scaffold during mitosis. The nuclear matrix (Berezney and Coffey, 1974) or scaffold has been operationally defined as a structure that can be isolated from interphase cells or metaphase chromosomes after removal of soluble proteins and histones, and digestion of the majority of DNA. The fact that newly synthesized DNA (Berezney and Coffey, 1975) as well as transcribed sequences (Robinson *et al.,* 1982) were found to be associated with the nuclear matrix considerably stimulated the scientific interest in this subnuclear structure. It was conceptually attractive that the cell nucleus, like the cytoplasm and some extracellular tissue compartments, should contain structural elements to determine its form and organization and to guide regulatory signals to their site of action. During the past 10 years experimental evidence has been presented that most metabolic functions of the cell nucleus are at least transiently associated with the nuclear matrix. Therefore, this subnuclear structure represents a functional compartment to which nuclear processes are actually associated. Bertalanffy's equation of structure and function (von Bertalanffy, 1952) is particularly applicable to the nuclear matrix in the sense that the concept of the nuclear matrix as an independent nuclear structure affecting the organization and function of chromatin might be equivalent to the alternative, sometimes discussed concept of the matrix being a reflection of nuclear processes without an independent existence in the absence of such processes.

The nuclear matrix mainly consists of the nuclear lamina, the nucleolar remnant, and an internal nuclear framework. Morphologically, the internal nuclear matrix contains filaments that can be identified by whole-mount electron microscopy (Fey *et al.,* 1984). At least some of these nuclear fibers have ultrastructural features reminiscent of intermediate filaments (Jackson and Cook, 1988). However, the protein components of these nucleoplasmic filaments are not known. Part of the filaments may be composed of NuMa, a 236-kDa protein that is able to polymerize into filaments (Compton *et al.,* 1992; Yang *et al.,* 1992; Compton and Cleveland, 1993). Moreover, there is immunohistochemical evidence that the nucleoplasm and its internal meshwork may contain intermediate filament-type proteins or lamin-related proteins (Beven *et al.,* 1991; Fidlerova *et al.,* 1992; Frederick *et al.,* 1992; Grabher *et al.,* 1992; Bridger *et al.,* 1993; Eberharter *et al.,* 1993; Lang and Loidl, 1993), actin, and actin-binding protein (Correas, 1991).

Adjacent of the inner membrane of the nuclear envelope is the nuclear lamina, which is composed of three major proteins, lamins A, B, and C (Franke, 1987). Lamin B is constitutively expressed in all mammalian somatic cell types, whereas lamins A and C are not expressed in early development, but appear later in differentiated cells (Lebel *et al.,* 1987; Rober *et al.,* 1989). In fact, lamins A and C arise from the same gene by alternative splicing (McKeon *et al.,* 1986). The morphology of the nuclear lamina as revealed by electron microscopy is different when *Xenopus* oocytes are compared with mammalian cell nuclei. Whereas the nuclear lamina of *Xenopus* contains a meshwork of orthogonally oriented filaments (Weber and Geisler, 1985; Aebi *et al.,* 1986), the lamina of mammalian nuclei represents an amorphous fibrillar meshwork (Dwyer and Blobel, 1976).

The role of nuclear lamins in nuclear envelope assembly has been extensively studied *in vitro* using *Xenopus* cell-free extracts and CHO mitotic extracts (Burke and Gerace, 1986; Newport *et al.,* 1990); depletion of lamins from CHO mitotic extracts resulted in inhibition of nuclear envelope assembly, whereas depletion of lamin LIII from *Xenopus* nuclear extract did not block nuclear envelope assembly around a chromatin substrate. The regulation of lamin expression has been studied in mouse embryos and in embryonal carcinoma stem cell lines (Lebel *et al.,* 1987; Stewart and Burke, 1987); in preimplantation embryos and teratocarcinoma stem cells a single B-type lamin is expressed, whereas lamins A and C are not expressed before the initiation of organogenesis.

The interconnection of the nuclear lamina with the nuclear envelope is at least partly mediated by lamin B and a lamin B receptor polypeptide that has been shown to be part of the internal membrane of the nuclear envelope in avian erythrocytes (Worman *et al.,* 1988, 1990). The association of lamin B with the lamin B receptor protein is dependent upon phosphorylation of the lamin receptor (Appelbaum *et al.,* 1990). Moreover, the lamin B receptor has DNA-binding motifs and could therefore have a function in chromatin organization. Recently, it has been shown that lamin B is involved in the organization of replicating chromatin during S-phase (Moir *et al.,* 1994). Transport of the cdc2-cyclin B complex has been shown to precede nuclear lamina breakdown (Pines and Hunter, 1991); the p34 cdc2–cyclin B complex can phosphorylate lamins on mitosis-specific sites resulting in lamin disassembly *in vitro* (Peter *et al.,* 1990; Dessev *et al.,* 1991). Purified lamin A and C proteins have also been demonstrated to bind chromatin in a specific and saturable way (Burke, 1990; Yuan *et al.,* 1991; Glass *et al.,* 1993). Paddy *et al.* (1990) showed that lamins do not always form a continuous layer around the nuclear periphery, but are also organized as a discontinuous network. Altogether, the data suggest that lamins might act as a substrate for the formation of replication complexes and are therefore subject to stringent cell cycle control mechanisms.

In contrast to the nuclear lamins, little is known about the nature and function of internal nuclear matrix proteins. This may partly be due to the fact that it is difficult to decide whether a protein is transiently bound to the nuclear matrix or represents an essential nuclear matrix constituent, in the sense of a real building block. In any case, proteins of the nuclear matrix have to be regulated in a cell cycle-specific way in order to fulfill distinct functions during different nuclear processes. One has to consider several possibilities of cell cycle-dependent regulation: (1) synthesis or degradation of nuclear matrix constituents, (2) assembly or disassembly of nuclear matrix structures, (3) secondary association of proteins to the nuclear matrix, (4) binding of DNA and associated structural and regulatory components to the nuclear matrix, and (5) post-translational modification of nuclear matrix components. The present review will focus on the cell cycle-dependent synthesis of matrix components and the transient association of proteins with the nuclear matrix.

## II. Cell Cycle Regulation

### A. Current Concepts

The eukaryotic cell comprises a series of distinct events, the most dramatic of which is mitosis. Most cells have four basic tasks during their cell cycle. They must grow ($G_1$- and $G_2$-period), replicate their DNA (S-phase), segregate their chromosomes into two identical sets, and divide into two independent daughter cells (mitosis). To ensure that cells divide and faithfully segregate copies of the genetic information to the daughter cells, it is essential that mitosis occurs only after DNA replication has been completed. Furthermore, the cell has to repair DNA damage and grow to a certain size before it can initiate mitosis. Therefore multiple checkpoints exist during the cell cycle that act to ensure that each event can only be started once previous events have been terminated. When a single cell cycle event is disturbed, these feedback control mechanisms arrest the further progression through the cycle. The most important function of such dependency pathways is to prevent irreversible damage to the cell that results from a temporary loss of a distinct cell cycle function.

The key components of cell cycle regulation are members of the p34 protein kinase family and cyclins (Murray and Hunt, 1993). Protein p34 was identified as the product of the cell division cycle genes, cdc2 of *Schizosaccharomyces pombe,* and CDC28 of *Saccharomyces cerevisiae.* Cyclins were discovered in fertilized sea urchin eggs. The cell cycle is basically driven by the interaction of cyclins, the stability of which fluctuates during the cell cycle, and p34-cdc2, which is invariably expressed during the entire cycle. Cyclins are divided into classes based on sequence similarity and physiological function: mitotic cyclins (cyclin B), S-phase cyclins (cyclin A), and G1 cyclins (Cln in yeast, cyclins C, D, E, F in vertebrates). Mechanisms and structural correlates have to exist which, on one hand perceive signals of the cell's competence to pass certain cell cycle transition points, and on the other hand convert cell cycle signals into distinct structural changes. The nuclear matrix might represent such a structural correlate.

There are few data on a possible relation between cell cycle control and nuclear matrix, although it would appear attractive to postulate a regulation level where cell cycle processes are triggered by transient association of regulatory components to a subnuclear scaffold. Few direct links exist between cell cycle control proteins and the nuclear matrix. (1) The cyclin B/p34-cdc2 complex phosphorylates nuclear lamins and is therefore involved in nuclear envelope breakdown. (2) Human cyclins A and B1 are differentially located in the cell and undergo cell cycle-dependent nuclear transport with cyclin B1 being associated with condensed chromosomes during pro- and metaphase (Pines and Hunter, 1991). Moreover, mitosis-specific proteins are located at the core or axis of mitotic chromatin and are likely to be directly involved in the assembly of the chromsome scaffold structure (Hirano and Mitchison, 1991). (3) Regulatory proteins are transiently associated to the nuclear matrix in a cell cycle-dependent manner, as shown for the c-*myc* homologous protein of *Physarum* (Waitz and Loidl, 1991).

## B. The Synchronous Cell Cycle of *Physarum polycephalum*

*Physarum polycephalum* is an acellular myxomycete that grows as a multinucleate plasmodium in its diploid stage of the life cycle and as unicellular amoebae in the haploid phase (Burland *et al.,* 1993). The cell cycle of plasmodia is distinguished from that of other organisms by the natural synchrony of nuclear division. Up to $10^{10}$ nuclei in a common cytoplasm divide with perfect synchrony. Not only mitosis itself occurs synchronously, but also events during interphase, facilitating biochemical analysis. The plasmodial cell cycle consists of an S-phase, which starts immediately after mitosis without an intervening $G_1$-period. S-phase occupies 2–3 hr, $G_2$-period approximately 7 hr; nuclear division itself lasts for about 30 min. The plasmodial mitotic cycle differs from the amoebal cycle in its absence of cytokinesis. The nuclear membrane persists during the entire duration of mitosis. The occurrence of plasmodial mitosis without dissolution of the nuclear membrane is reminiscent of fungi and some animal embryos. However, chromatin condensation and decondensation and the assembly and disassembly of the mitotic spindle do occur at mitosis.

Despite its natural synchrony, the use of *Physarum* for cell cycle studies is sometimes questioned due to the lack of a $G_1$-period, the multinucleated state, and the intranuclear mitosis. However, these features are not only shared by numerous other protists, but also by early embryos of many animals. A major advantage of the plasmodial stage of *Physarum* for cell cycle studies is the possibility to prepare giant macroplasmodia (up to $10^{10}$ nuclei) and to remove multiple pieces, even large, at different time points during the cycle without disturbing the synchronous progression of the cell cycle. Moreover, the cell cycle stage can be simply monitored by phase-contrast microscopy of tiny explants of the plasmodium without time-consuming staining procedures. The fact of the closed mitosis enables one to isolate nuclei during all mitotic stages of nuclear division. These features allow the accurate and reproducible analysis of multiple events at multiple stages of unperturbed mitotic cycles with a high amount of material for biochemical analysis.

## III. Cell Cycle Dependence of Nuclear Matrix Constituents

When nuclear matrix preparations from mouse cells of different origin were compared by two-dimensional gel electrophoresis, a limited set of proteins was found to be common to all cell types, apart from tissue-specific polypeptides, and the majority of these proteins could even be identified in cells of other mammalian species (Stuurman *et al.,* 1990); the authors termed this set of proteins "minimal matrix." The only protein that was identified among these matrix proteins was lamin B. After careful electrophoretic analysis of nuclear matrix preparations from different human cell lines, Kallajoki and Osborn (1994) pointed out that it is difficult to decide whether proteins are true matrix constituents, associated polypeptides, or even contaminating proteins. Berezney and co-workers (Nakayasu and Berezney, 1991) identified a set of major nuclear matrix proteins that they termed "matrins." Among 12 consistently found abundant matrix proteins were lamins A, B, and C, the nuclear protein B-23, and ribonucleoproteins. Eight matrin proteins were separated and characterized; for most of these proteins DNA binding properties were demonstrated (Hakes and Berezney, 1991). Except for lamins, only very few experimental data on cell cycle dependence of matrix constituents are available. For future research, experimental criteria will be essential to discriminate internal matrix proteins from those proteins that are tightly bound to the nuclear matrix.

### A. Lamins

Due to the dramatic structural changes of the nuclear envelope and lamina during mitosis, the lamin proteins are obvious candidates to be involved in cell cycle-regulatory pathways. It was reported by different research groups that lamins were phosphorylated by protein kinases that had been recognized as important cell cycle regulators. Bacterially expressed human lamin C, assembled into filaments *in vitro,* showed increased phosphorylation on specific sites in response to maturation-promoting factor from a frog egg extract; phosphorylation was accompanied by disassembly of the lamin filaments (Ward and Kirschner, 1990). The cdc2 protein kinase was shown to phosphorylate chicken B-type lamins *in vitro* on sites that are specifically phosphorylated during mitosis *in vivo;* concomitantly cdc2 kinase induced nuclear lamina depolymerization (Peter *et al.,* 1990; Lüscher *et al.,* 1991). The same was shown with chicken lamin B2 introduced into *Schizosaccharomyces pombe* (Enoch *et al.,* 1991). When lamin head-to-tail polymers reconstituted from bacterially expressed chicken lamin B2 *in vitro* were phosphorylated by cdc2 kinase at physiological sites, polymer disassembly was observed; the subsequent dephosphorylation allowed reformation of lamin polymers (Peter *et al.,* 1991). These data clearly demonstrate that phosphorylation/dephosphorylation is sufficient to control the assembly of lamin B dimers. Moreover, the lamin B receptor too has been shown to undergo mitosis-specific phosphorylation by p34cdc2 kinase (Courvalin *et al.,* 1992), which results in altered binding of lamin B (Appelbaum *et al.,* 1990). Using immunopurified avian lamin B receptor protein (p58), five proteins could be shown to be specifically associated with p58; two of these proteins could be identified as lamins A and B, apart from a kinase activity that phosphorylates p58 itself *in vivo* and *in vitro* at serine residues (Simos and Georgatos, 1992). p58 kinase is not identical with protein kinase A and cdc2 kinase. The B-type lamins have been shown to remain associated with p58 during mitosis (Meier and Georgatos, 1994). In the same report an early association of lamin B and p58 with the chromosomes has been observed, supporting the idea that this chromosome-bound lamin and its receptor serve as a "seed" for further homotypic interactions at the end of cell division. In mammalian cells additional proteins (lamina-associated polypeptides, LAPs) exist that specifically bind lamins (Foisner and Gerace, 1993); at least one of these proteins, LAP 2, also binds to mitotic chromosomes. LAPs are phosphorylated during mitosis, and phosphorylation inhibits binding to lamins and chromosomes. Again these proteins may have a key role in events of nuclear envelope reassembly after mitosis. It is important in this context that lamins A and C contain a chromatin binding site (Burke, 1990; Glass and Gerace, 1990; Yuan *et al.,* 1991; Glass *et al.,* 1993), and matrix-attached regions (MARs) interact specifically with lamin B1 in rat liver cells (Luderus *et al.,* 1992).

Lamins and lamin receptor analogs have also been described in lower eukaryotes. In yeast, analogues of lamins A and B, as well as lamin B receptor have been identified (Georgatos *et al.,* 1989). In the myxomycete

*Physarum polycephalum,* a lamin B homolog has been characterized (Lang and Loidl, 1993). Synthesis and total amount of the *Physarum* lamin B does not significantly fluctuate during the cell cycle, indicating that regulation either affects modification of the protein or the assembly and disassembly events (Lang *et al.,* 1993).

Apart from being substrate to cdc2 kinase, lamins are subject to phosphorylation by protein kinase C. This was shown for lamin B2 in mouse keratinocytes (Kasahara *et al.,* 1991) and lamin B in human leukemic cells (Hocevar *et al.,* 1993). In chicken lamin B2, four protein kinase C phosphorylation sites have been identified; two sites were phosphorylated throughout interphase, whereas two additional serine residues became phosphorylated specifically in response to activation of protein kinase C by phorbol ester (Hennekes *et al.,* 1993). The consequence of phosphorylation was a strong inhibition of transport to the nucleus. Some of the residues phosphorylated by cdc2 kinase or protein kinase C have been changed by constructed point mutations in human lamin A in order to study the functional consequences (Heald and McKeon, 1990; Haas and Jost, 1993). The studies revealed that certain phosphorylation sites were related to mitotic lamin A disassembly, whereas others affected nuclear localization.

Another important indication that nuclear lamins are not only constituents of the nuclear periphery comes from morphological studies of nucleoplasmic lamin foci during interphase. A colocalization of lamin B with sites of bromodeoxyuridine incorporation and proliferating cell nuclear antigen decoration was reported during S-phase of mouse 3T3 cells (Moir *et al.,* 1994). Therefore lamins may mediate interactions between the nuclear matrix, chromatin, and the replicational machinery.

Taken together, the data indicate more and more that nuclear lamins are assembled and reorganized during various cell cycle events. This dynamic picture of the nuclear lamina differs considerably from the view of a static protein network that only alters its structure during disassembly at mitosis; it is, however, suggestive, since lamins are affected by different components of the cellular signal transduction network. It is attractive to consider lamin proteins as a molecular routing system originating from cytoskeletal elements to the internal nuclear matrix, a route that mediates signals involved in fundamental cell cycle events, like mitosis or DNA replication. Similar disassembly processes triggered by p34-cdc2 protein kinase, as reported for nuclear lamins, occur at the cytoplasmic vimentin-containing intermediate filament network during mitosis (Chou *et al.,* 1990, 1991). It has been recently demonstrated that microinjection of antivimentin antibodies into prometaphasic 3T3 cells inhibited nuclear lamin reassembly at the end of cell division (Kouklis *et al.,* 1993). This observation emphasizes that intermediate filament networks of the cytoplasm and nuclear lamins should be regarded as a functional continuum.

## B. Other Proteins

In contrast to our detailed knowledge about nuclear lamins and their regulation during the cell cycle, we know little about the cell cycle-dependent synthesis, assembly, modification, and decay of other proteins that represent integral components of the nuclear matrix. Berezney and co-workers (Nakayasu and Berezney, 1991) have identified a set of internal nuclear matrix proteins, the matrins, none of which has been investigated in detail with respect to cell cycle regulation. The few remaining reports in the literature are difficult to compare, since they differ considerably in terms of methodology, the employed experimental system, and the developmental stage.

We have recently investigated the protein composition and synthesis of nuclear matrix proteins during the synchronous cell cycle of *Physarum polycephalum* (Lang *et al.,* 1993). The protein composition was found to be invariant when analyzed at eight cell cycle time points by one-dimensional SDS polyacrylamide gel electrophoresis, clearly indicating the lack of cell cycle phase-specific matrix proteins; approximately 50 polypeptides could be visualized by Coomassie-blue staining, irrespective of the nuclear matrix preparation protocol (Eberharter *et al.,* 1993; Lang *et al.,* 1993). When the synthesis of these nuclear matrix proteins was measured by incorporation of [$^{35}$S]methionine–cysteine, none of the detected proteins was synthesized in a strictly cell cycle-dependent mode. Only slight quantitative changes in the rate of synthesis or assembly could be observed. Four proteins exhibited a cell cycle-dependent synthesis pattern; two proteins ($M_r$ 16 and 21 kDa) had a biphasic pattern with maxima in late S-phase and late $G_2$ period; a 46-kDa protein had a clear maximum of synthesis at the S–$G_2$ transition, but extremely low levels during late $G_2$-period and mitosis, whereas a 48-kDa polypeptide had a clear maximum in mitosis with very low levels during the rest of interphase. Both proteins were phosphorylated in a cell cycle-dependent manner. However, the identity of these proteins is still obscure.

Similar results with nuclear matrix proteins of unknown function and indentity have been obtained in other experimental systems too. Human fibroblasts at different cell cycle stages were analyzed for possible differences in their synthesis rate for matrix polypeptides. Only 10 of approximately 150 proteins exhibited some cell cycle dependence. One of these proteins with $M_r$ 47 kDa exhibited the most significant cell cycle fluctuation with maximum synthesis during S-phase (Dell'Orco and Whittle, 1994). Milavetz and Edwards (1986) demonstrated that mouse 3T3 cells did not differ significantly in the nuclear matrix composition from African green monkey kidney cells; irrespective of serum stimulation, matrix proteins exhibited slow turnover in contrast to lamin A. Three stress proteins localized in the nuclear matrix and three other unidentified proteins ($M_r$ 52, 55, and 120 kDa) were synthesized and phosphorylated with different rates during the cell cycle in HL-60 cells after treatment with retinoic acid (Pipkin *et al.,* 1991). Mitotin, a 125-kDa nuclear matrix protein, is synthesized continuously during the cell cycle, but exhibits a characteristic accumulation at the $G_2$/mitosis transition that is due to the appearance of phosphorylated forms with enhanced metabolic stability (Zhelev *et al.,* 1990); analysis of mitotin mRNA also demonstrated invariant levels during the synchronized cell cycle of Raji and WISH cells (Todorov *et al.,* 1991).

During the synchronous cell cycle of *Physarum polycephalum* we have observed significant differences in the chromatin assembly of newly synthesized core histones between the nuclear matrix and bulk nuclear chromatin (Lang *et al.,* 1993). We always found approximately 5% of the total nuclear core histones tightly bound to the nuclear matrix in *Physarum* (Waitz and Loidl, 1988; Grabher *et al.,* 1992; Lang *et al.,* 1993) as well as in maize (P. Loidl, unpublished results) regardless of the cell cycle stage. In bulk nuclear chromatin assembly of newly synthesized core histones started before mitosis and continued until the end of S-phase (3 hr past mitosis); this synthesis pattern has been already reported earlier (Loidl and Gröbner, 1987a). In contrast, the assembly of newly synthesized core histones into nuclear matrix structures was restricted to a much shorter time period starting at the onset of mitosis until approximately 1 hr after mitosis; at later time points of S-phase, assembly of newly synthesized histones into the nuclear matrix was negligible, though the total synthesis and assembly of histones into bulk nuclear chromatin was still high (Lang *et al.,* 1993). Since nuclear matrix core histones are most likely those assembled into DNA of MARs, the results indicate that MARs replicate early during S-phase and are extremely tightly bound to the nuclear matrix. Surprisingly, we did not observe incorporation of newly synthesized H2A and H2B into nuclear matrix-bound chromatin during the G2-period, although there is almost continuous synthesis and assembly of these histones into bulk nuclear chromatin during $G_2$-period, when most of the transcription occurs in *Physarum* (P. Loidl, unpublished results; Loidl and Gröbner, 1987a,b). One would expect assembly of newly synthesized H2A and H2B into the nuclear matrix during $G_2$-period in the light of numerous experiments that demonstrate a tight association of transcribed chromatin with the nuclear matrix.

One could still argue that nuclear matrix-bound core histones do not represent a distinct subpopulation, but rather the artificial association of part of the total nuclear histones to a subnuclear structure. However, the specific pattern of post-translational acetylation of core histones in the nuclear matrix strongly suggests that the nuclear matrix-bound chromatin is functionally distinct. In a detailed cell cycle study during the naturally synchronous *Physarum* cycle, we could demonstrate that the small percentage of matrix-bound histones is distinguished by an extraordinarily high acetylation state and a specific distribution of acetylated histone subspecies (Lang *et al.*, 1993). Bulk nuclear histone H4 has its highest acetate content during early S-phase (Loidl *et al.*, 1983), reaching values around 1.1, whereas the values in the nuclear matrix-bound H4 are 1.8, irrespective of the cell cycle stage (Lang *et al.*, 1993). Although at present we do not understand the functional significance of this extremely high acetylation state, it indicates a general functional difference between bulk chromatin and nuclear matrix-attached chromatin.

Apart from the cell cycle-dependent phosphorylation and acetylation of nuclear matrix polypeptides, there are data on other post-translational modifications of nuclear matrix components in the literature. Prenylation of unidentified nuclear matrix proteins was reported during the cell cycle of HepG2 cells (Sepp-Lorenzino *et al.*, 1991); prenylation peaked during mid-S-phase and declined until the cells reached mitosis. We have previously shown that 10% of total nuclear ADP-ribosylation affects proteins of the nuclear matrix of *Physarum polycephalum* (Golderer *et al.*, 1991). Although total nuclear ADP-ribosylation was twice as high in $G_2$-period as compared to S-phase, the amount of ADP-ribosylated nuclear matrix proteins was twice as high in S-phase; moreover the matrix-associated ADP-ribosylation was characterized by a higher proportion of poly(ADP-ribose). The nuclear matrix-associated ADP-ribosylation was mainly due to modification of H2A, H2B, and actin (P. Loidl, unpublished results).

There is one example of an internal nuclear matrix protein that is regulated through the cell cycle and to which a function has been attributed. Numatrin is a tightly bound nuclear matrix protein of 40 kDa that is identical to the nucleolar protein B23 (Feuerstein *et al.*, 1988a). In a study using human and murine B and T lymphocytes it was demonstrated that numatrin abundance had a clear peak during S-phase in response to mitogenic stimulation (Feuerstein *et al.*, 1988b); moreover, numatrin was abundant in a variety of malignant cells. Like lamins, numatrin is also subject to post-translational phosphorylation by p34-cdc2 protein kinase (Feuerstein, 1991; Feuerstein and Randazzo, 1991). For this reason and due to its DNA-binding activity, numatrin is thought to be involved in DNA-associated processes during proliferation in a highly cell cycle-dependent way (Feuerstein *et al.*, 1990).

## IV. Cell Cycle Dependence of Nuclear Matrix-Associated Proteins

If one asumes that the nuclear matrix represents a functional compartment of the cell nucleus where important nuclear events take place, it is conceivable that many regulatory proteins, enzymes, nucleic acids, and nonproteinacious molecules bind transiently to this nuclear structure. Such transient association with the matrix at distinct stages of the cell cycle may represent an additional level of regulation of nuclear processes that until now has not been seriously considered. Association with the nuclear matrix has been frequently reported for a huge number of different proteins, although only very few reports tried to answer the question whether this association was transient or stable. For this reason we will briefly overview proteins that have been demonstrated to associate with the nuclear matrix and focus on those rare examples where a cell cycle-dependent interrelation has been established.

## A. Enzymes

It is now well accepted that DNA replication and transcription occur at the nucleoskeleton (Cook, 1989, 1991, 1994). One therefore expects the enzymatic apparatus for both processes to be attached to the nuclear matrix. Since transcription is a selective process, affecting only a small percentage of the cells total genome, the coupling of transcription to the nuclear scaffold has mainly been verified by demonstrating actively transcribed genes to be associated to the matrix (Robinson *et al.,* 1982; Andreeva *et al.,* 1992; Thorburn and Knowland, 1993). In contrast, for DNA replication numerous results indicate that the necessary enzymes are firmly attached to the nuclear matrix (Cook, 1991).

We will specifically address enzymes involved in DNA replication and transcription during the naturally synchronous cell cycle of the myxomycete *Physarum polycephalum* with respect to association with the nuclear matrix.

### 1. Thymidine Kinase and Thymidylate Synthase

It has been demonstrated that isolated nuclear matrices from *Physarum polycephalum* catalyzed DNA synthesis without exogenous templates. The activity of these matrix preparations changed according to the *in vivo* DNA replication in macroplasmodia during the cell cycle (Shioda and Murakami-Murofushi, 1989; Shioda *et al.,* 1989). Corresponding results have been obtained in other experimental systems (Collins and Chu, 1987; Tubo and Berezney, 1987). We wanted to address the question whether, apart from DNA polymerase activity, other enzymatic activities involved in the production of precursor molecules for DNA replication are associated with the nuclear matrix during the synchronous cell cycle of *Physarum.* Two enzymes engaged in the production of deoxythymidine monophosphate seemed to be interesting candidates for a detailed cell cycle investigation. Thymidine kinase is a salvage pathway enzyme that phosphorylates deoxythymidine, whereas thymidylate synthase represents the main enzyme for deoxythymidine monophosphate production, that converts deoxyuridine monophosphate into deoxythymidine monophosphate. Both enzymes have been shown to fluctuate periodically during the cell cycle and life cycle of *Physarum* (Gröbner and Loidl, 1982, 1983), exhibiting a pronounced maximum of activity during early and mid-S-phase, with low levels during $G_2$-period. We analyzed both enzyme activities during all steps of nuclear matrix preparation from macroplasmodia at different time points during the cell cycle. Nuclear matrices were prepared by using encapsulated macroplasmodial pieces as described (Waitz and Loidl, 1988; Grabher *et al.*, 1994).

Figure 1 shows that thymidine kinase and thymidylate synthase behave differently with respect to subcellular location. Thymidylate synthase is not bound to the nuclear matrix, neither in S-phase nor in $G_2$-period, but is solubilized after treatment with Triton and low salt buffers. In contrast, thymidine kinase is completely insoluble during S-phase, with almost 80% of the total activity bound to the nuclear matrix. In $G_2$-period the enzyme is readily solubilized by Triton and low salt buffers, without significant activity in the final nuclear matrix preparation. Figure 2 shows a summary of a series of experiments over a time period starting from 2 hr prior to mitosis 2 until 5 hr after mitosis 3. It is evident that thymidylate synthase is not bound to the nuclear matrix, irrespective of the cell cycle stage. Thymidine kinase shows an interesting pattern, since a high percentage of the total activity is bound to the nuclear matrix during S-phase, whereas only very low matrix activity can be detected in early and mid-$G_2$-period. A similar result was obtained with another salvage pathway enzyme, deoxycytidine kinase of *Physarum* (P. Loidl, unpublished results). One has to consider that the plasmodial cell cycle of *Physarum* lacks a $G_1$-period; therefore, S-phase immediately starts after the termination of mitosis. This is the reason why S-phase-associated events (e.g., thymidine kinase activity, thymidylate synthase activity, histone biosynthesis) already are induced before the onset of mitosis in the late $G_2$-period of the preceding cell cycle. Clearly, the cell cycle pattern of nuclear matrix-bound thymidine kinase is congruent with the time period of DNA synthesis.

In line with our result of a periodic association of salvage pathway enzymes to the nuclear matrix is a recent report on channeling salvage pathway DNA precursors to nuclear matrix-associated sites of DNA replication (Panzeter and Ringer, 1993). In this study it was shown that deoxythymidine-derived nucleotides, in contrast to orotic acid-derived nucleotides, bypassed the soluble deoxythymidine triphosphate pool and engaged directly in DNA synthesis at the nuclear matrix. The authors discussed the possibility that thymidine kinase, as a membrane-bound enzyme, could effectively trap deoxythymidine, although it remains obscure how deoxy-

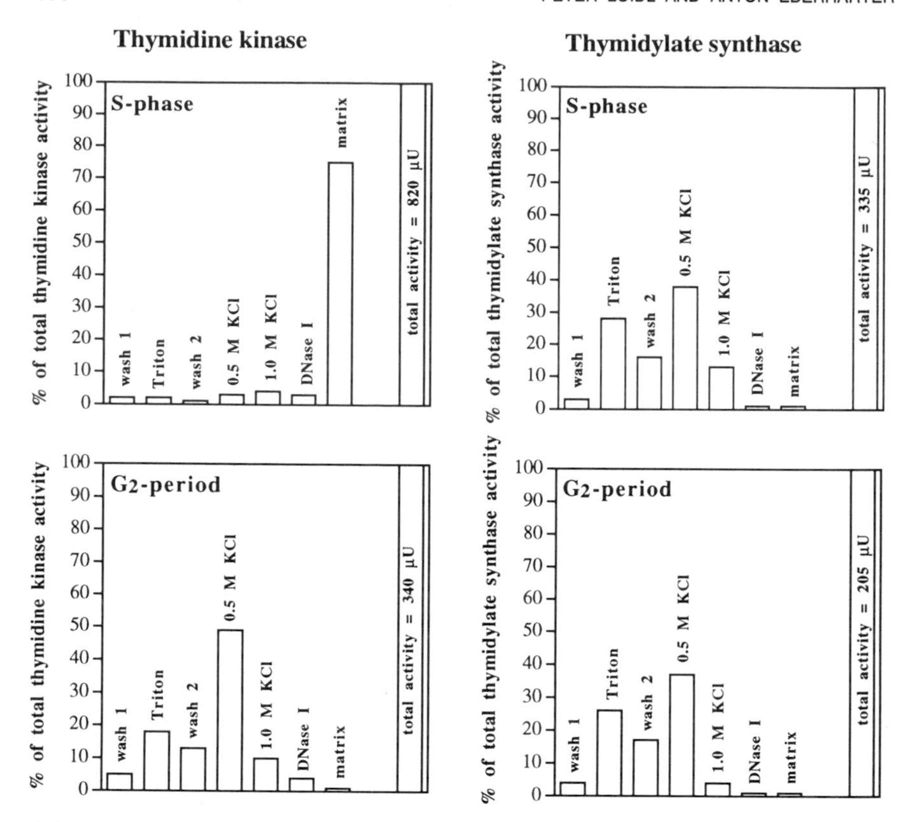


FIG. 1 Thymidine kinase and thymidylate synthase activities during different steps of nuclear matrix preparation of macroplasmodia of *Physarum polycephalum.* Two giant S-shaped macroplasmodia at 1 hr after mitosis 2 (S-phase) and 7.5 hr after mitosis 2 ($G_2$-period) were encapsulated into agarose beads and nuclear matrices were prepared as described (Waitz and Loidl, 1988), except that KCl was used for deproteinization instead of NaCl. All steps of preparation were analyzed for thymidine kinase and thymidylate synthase activities as previously described (Gröbner and Loidl, 1982). Enzyme activities at each step are expressed as percentage of the total activity.

thymidine monophosphate would then bypass cytoplasmic pools. Our results suggest that deoxythymidine is trapped and transported to the nucleus by other mechanisms, but is then directly phosphorylated by nuclear matrix-bound thymidine kinase at the sites of DNA replication. A possible explanation would be that active thymidine kinase molecules translocate from membrane sites to the nuclear matrix in a cell cycle-dependent way. Interestingly, thymidine kinase transcription is regulated at the $G_1$/S-phase transition in mammalian cells by a complex that contains a retinoblastoma-like protein and a cdc2 kinase (Dou *et al.,* 1992).

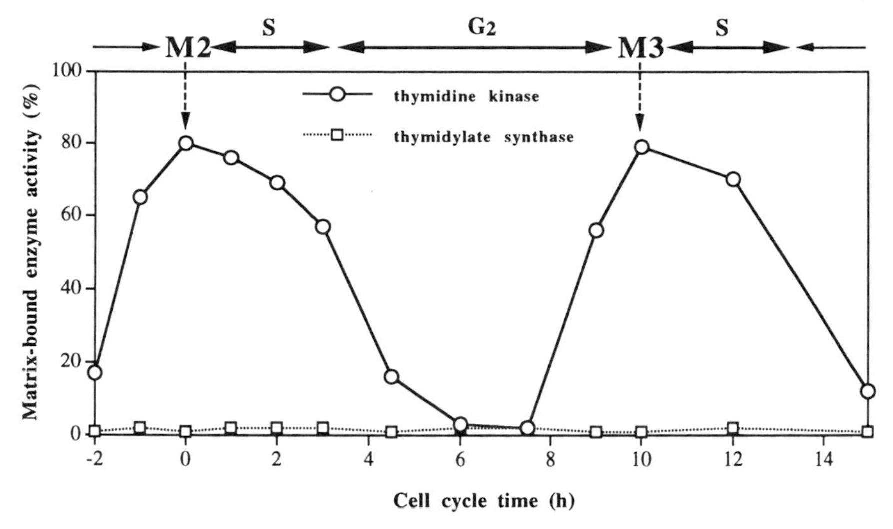


FIG. 2 Nuclear matrix-bound thymidine kinase and thymidylate synthase activities during the synchronous cell cycle of *Physarum polycephalum.* The experiment depicted in Fig. 1 was performed at 13 time points of the cell cycle (from 2 hr prior to mitosis 2 until 5 hr after mitosis 3). Matrix-bound activities are expressed as percentage of total enzyme activity. Dashed arrows mark the time of mitosis 2 and mitosis 3, respectively (telophase).

## 2. RNA Polymerase

$\alpha$-Amanitin-sensitive and -resistant RNA polymerase activities were shown to be present at almost constant levels during the *Physarum* cell cycle (Hildebrandt and Sauer, 1976). In order to see whether a proportion of RNA polymerase activity was bound to the nuclear matrix, as shown in other systems (Dickinson *et al.,* 1990), we analyzed the different steps of matrix preparation for RNA polymerase activity and compared different time points during the *Physarum* cell cycle (early S-phase, late S-phase, late $G_2$-period).

Figure 3 shows that during early and late S-phase the majority of RNA polymerase activity was solubilized by high salt buffers with low residual activity in the nuclear matrix fraction. However, in late $G_2$-period almost no enzyme activity could be solubilized by high salt solutions, but the enzyme activity was almost quantitatively retained in the nuclear matrix. To get a clear picture over the whole cell cycle we performed this type of experiment over a time period starting 2 hr before mitosis 2 until 3 hr after mitosis 3, covering more than one complete cell cycle. Figure 4 shows that approximately 90% of the total RNA polymerase activity was matrix bound in late $G_2$-period before mitosis 2. At the time of mitosis 2, RNA polymerase activity obviously abandoned the nuclear matrix, since only ~10% of the total activity remained attached to the matrix. The same was true for early

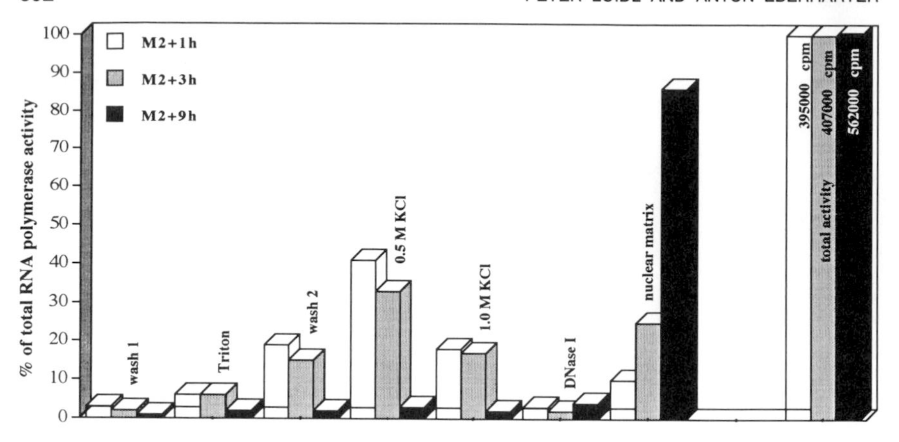


FIG. 3 RNA polymerase activity during different steps of nuclear matrix preparation of macroplasmodia of *Physarum polycephalum.* Three giant S-shaped macroplasmodia at 1 hr after mitosis 2 (S-phase), 3 hr after mitosis 2 (S–$G_2$ transition), and 9 hr after mitosis 2 (late $G_2$-period) were encapsulated into agarose beads and nuclear matrices were prepared as decribed (Waitz and Loidl, 1988), except that KCl was used for deproteinization instead of NaCl. All steps of preparation were analyzed for RNA polymerase activity essentially as described (Hildebrandt and Sauer, 1973). Enzyme activities at each step are expressed as percentage of the total activity.

S-phase, and also could be seen during S-phase of the subsequent cycle (after mitosis 3). With progression through $G_2$-period the percentage of matrix-bound RNA polymerase activity steadily increased to reach approximately 90% in late $G_2$-period, prior to the onset of mitosis 3. In contrast, total RNA polymerase activity exhibited only moderate fluctuations during the cell cycle, in accordance with previously published data (Hildebrandt and Sauer, 1976). Isotope dilution experiments during the *Physarum* cell cycle revealed that the synthesis of ribosomal RNA and transfer RNA occurs throughout the cycle. The rate of synthesis increased with progression through the cycle, which leads to elevated synthesis during later stages of $G_2$-period (Turnock, 1979). On the basis of total enzyme activities such an increasing synthesis rate cannot be explained. However, the cell cycle fluctuation of RNA polymerase activity tightly associated with the nuclear matrix closely resembles the pattern of synthesis rate. The cell cycle pattern of neither total RNA polymerase nor matrix-bound polymerase changed significantly when $\alpha$-amanitin was used to discriminate A-type and B-type enzymes (P. Loidl, unpublished results). Despite almost continuous synthesis of RNA polymerases during the cell cycle, the actual rate of RNA synthesis is obviously regulated via cell cycle-dependent, transient attachment of enzymes to the active sites at the nuclear matrix.

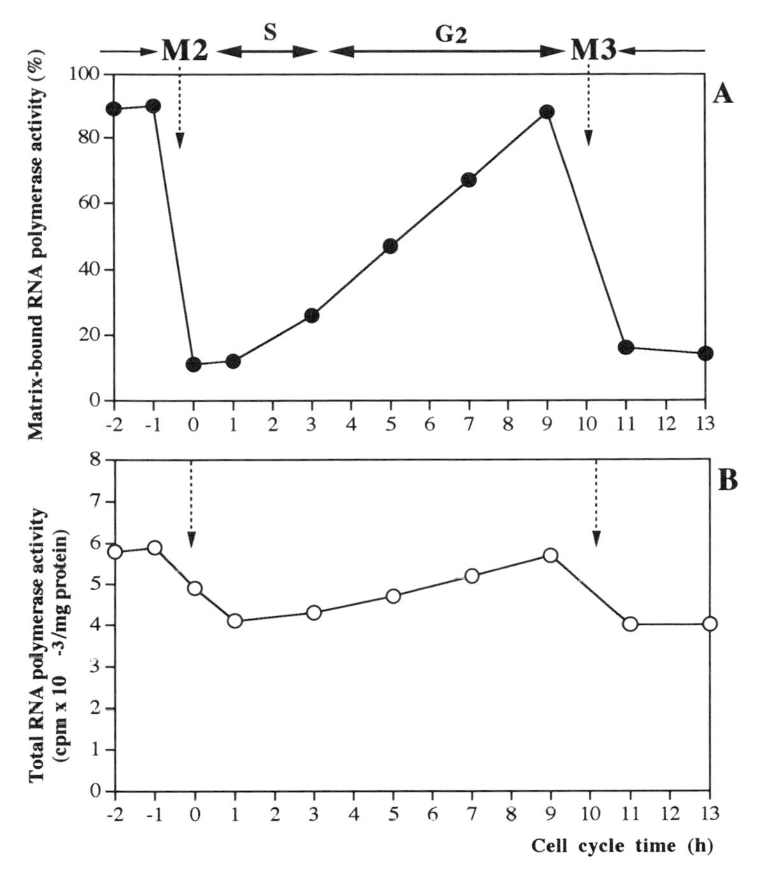


FIG. 4 Nuclear matrix-bound and total RNA polymerase activity during the synchronous cell cycle of *Physarum polycephalum.* The experiment depicted in Fig. 1 was performed at 10 time points of the cell cycle (from 2 hr prior to mitosis 2 until 3 hr after mitosis 3). A, Matrix-bound RNA polymerase activities are expressed as percentage of total enzyme activity. B, Total specific RNA polymerase activities (cpm/mg protein). Dashed arrows mark the time of mitosis 2, and mitosis 3, respectively (telophase).

### 3. Other Enzymes

A growing number of enzymes was recognized to be bound to the nuclear matrix in a variety of organisms, although no cell cycle data are available. Particularly interesting are those enzymes, especially kinases, that have regulatory functions and are involved in intracellular signal transduction pathways. A common feature to these enzymes is that they frequently change their intranuclear location. Calcium/calmodulin-dependent protein kinase II was shown to be localized among the nuclear matrix in rat and human cell culture lines during interphase. During mitosis the protein became part of the mitotic apparatus (Ohta *et al.,* 1990). Interestingly, calmodulin seems to be involved in the *in vivo* binding of a matrix-associated region in yeast (Fishel *et al.,* 1993). A casein kinase II analogue of pea is also bound to the nuclear matrix and nuclear lamina and phosphorylates a lamin A-like protein (Li and Roux, 1992). Casein kinase 2 was also found to be associated with the nuclear matrix in various tissues of the rat (Tawfic and Ahmed, 1994). There is evidence that casein kinase 2 is directly involved in the modification of nuclear matrix proteins. Several kinases involved in the phosphoinositide metabolism have been localized in the nuclear matrix of mouse and rat cells. Interestingly, phosphatidylinositol 4-kinase was found exclusively in the lamina–pore complex, whereas phosphatidylinositol 4-phosphate 5-kinase was only found in the internal nuclear matrix (Payrastre *et al.,* 1992). Diacylglycerol kinase and phospholipase C activities were also preferentially detected in the internal nuclear matrix. The differential localization of these enzymes in the matrix indicates that phosphoinositide metabolism may play a complex role in the nucleus. Isoforms of phosphoinositidase are specifically localized in the nuclear matrix, but also among the cytoskeleton of Swiss 3T3 cells (Zini *et al.,* 1993). Phosphoinositidase C-b is present at the same sites of the matrix where phospholipids and protein kinase C could be identified. From the results it is clear that elements of the phosphoinositide signal transduction system are located inside the nucleus. Unfortunately, we do not know anything about cell cycle regulation of these enzymes or a possible transient, but cell cycle-dependent nuclear matrix association.

Recently, enzymes involved in the post-translational acetylation of core histones, acetyltransferases, and deacetylases have been reported to be tightly bound to the nuclear matrix, representing true members of internal matrix proteins (Hendzel and Davie, 1992; Hendzel *et al.,* 1994). However, these results are contradictory to numerous data in the literature (e.g., López-Rodas *et al.,* 1991a,b), since the routine quantitative extraction of these enzymes is performed by moderate salt buffers. Our laboratory has analyzed nuclear matrices prepared by different protocols from yeast, *Physarum,* maize, and chicken erythrocytes, but was unable to find a significant proportion of one of the numerous enzyme forms in the nuclear matrix, regardless of the cell cycle stage (Brosch *et al.,* 1992; Grabher *et al.,* 1994).

## B. Regulatory Elements

A transient, cell cycle-dependent association with the nuclear matrix, as observed with enzymes, may also contribute to modulation of the biological function of regulatory proteins in the cell nucleus. In cultured GC cells, a rat pituitary tumor cell line, thyroid hormone nuclear receptors were found to be partly associated with the nuclear matrix and therefore the functional state of T3 receptors may be determined by a cell cycle-dependent attachment to the matrix (Kumarasiri *et al.,* 1988). During the synchronous cell cycle of *Physarum polycephalum* we could show that the c-*myc* homologous protein became periodically associated with different substructures of the nuclear matrix (Waitz and Loidl, 1991). Although the amount of c-*myc* protein was constant during the cell cycle, the proportion of nuclear matrix-bound protein was higher during S-phase as compared to several time points in $G_2$-period. Not only the proportion of matrix-bound c-*myc* protein changed, but also the distribution among the nuclear scaffold structure. Whereas during S-phase c-*myc* protein was specifically clustered at the nuclear periphery, presumably the nuclear lamina, it was uniformly distributed all over the internal nuclear matrix in $G_2$-period. Like thymidine kinase and RNA polymerase, c-*myc* also undergoes a cell cycle-dependent change of localization, despite an obviously invariant level through the cycle. We have analyzed several proto-oncogene proteins (c-*fos,* c-*jun,* c-*myb* and a tumor suppressor gene product (p53) during the *Physarum* cell cycle, but could get no evidence for possible association with the nuclear matrix (López-Rodas *et al.,* 1992; P. Loidl, unpublished results). Interestingly, the matrix association of the c-*myc* protein could be disrupted by short treatment of macroplasmodia with butyrate, an inhibitor of histone deacetylases (López-Rodas *et al.,* 1992); this butyrate-induced disruption only occurred during the $G_2$-period when c-*myc* protein was bound to the internal nuclear matrix. During S-phase when the protein was concentrated at the nuclear lamina, butyrate treatment had no effect. This indicates that c-*myc* protein binds to matrix-associated chromatin regions. This binding could be dependent on a certain degree of core histone acetylation that is disturbed by butyrate treatment (López-Rodas *et al.,* 1993; Loidl, 1994).

Another protein that is associated with the nuclear matrix in a cell cycle-dependent way, very much like c-*myc* in *Physarum,* is the retinoblastoma gene product. This protein is known as a tumor suppressor involved in cell cycle regulation, though its mechanism of action is still unclear. Isolated nuclear matrices from synchronized cultured cells contained a significant amount of hypophosphorylated retinoblastoma protein only during early $G_1$-period (Mancini *et al.,* 1994). The protein was preferentially found in the nuclear periphery and in nucleolar remnants, but not in the internal nuclear matrix filaments. The authors suggested that the interaction of retinoblastoma gene product with the nuclear matrix might be important for its ability to regulate cell cycle progression. Ozaki *et al.* (1994) found that lamin A can associate with the retinoblastoma protein *in vitro* and they could identify a distinct binding domain in lamin A. Within the retinoblastoma protein-binding domain of lamin A there is a short amino acid sequence that is also present in the retinoblastoma protein-binding region of the transcription factor E2F-1.

DNA-binding activities related to or identical to a variety of transcription factors (SP-1; ATF, CCAAT, C/EBP, OCT-1, AP-1) are present in nuclear matrix proteins of different cell types (Van Wijnen *et al.,* 1993). Comparison of the relative abundance of these binding activities in matrix and nonmatrix nuclear fractions suggests that the distribution between these two fractions is cell type specific and growth dependent. These findings indicate that transcription factors are present in the nuclear matrix and this localization may be essential to the role of the nuclear matrix in gene regulation. Together with the fact that regulatory proteins such as c-*myc* or retinoblastoma gene product are associated with the nuclear matrix in a cell cycle-dependent way, it is conceivable that the nuclear matrix somehow focuses and channels a number of signals for the regulation of growth and gene expression.

## C. Other Proteins

Among the variety of proteins associated with the nuclear matrix are only a few for which we know the exact function or the biological reason why they bind to a nuclear scaffold. Some of these proteins are potentially interesting in terms of cell cycle regulation, although most studies lack any cell cycle aspect. Human telomeres have been found to be attached to the nuclear matrix via a TTAGGG repeat (De Lange, 1992). Several viral proteins have a high affinity for the nuclear matrix and could be colocalized with the nuclear matrix (Greenfield *et al.,* 1991; Fredman and Engler, 1993; Lu *et al.,* 1993). The functional significance of viral protein matrix attachment is still unclear, but it may assist in proper positioning of DNA for DNA replication (Fredman and Engler, 1993) and may be dependent on cell cycle position of infected cells.

In order to characterize the multipe nuclear matrix proteins, Penman and co-workers have created a library of monoclonal antibodies. Most of these antibodies give a speckled immunofluorescence, since the antibodies recognize proteins of the internal nuclear matrix (Nickerson *et al.,* 1992; Wan *et al.,* 1994). The epitopes detected by these antibodies show characteristic relocations during mitosis, although identity and function remained unclear. Similar morphological studies have been performed with antisera from patients suffering from autoimmune diseases that frequently detect nuclear matrix proteins (Casiano *et al.,* 1993).

## V. Conclusions

The huge amount of DNA within the cell nucleus has to be dynamically organized and therefore requires a structural system to coordinate the different processes occurring among chromatin. Experimental evidence has accumulated during the past decade that the nuclear matrix as an underlying nuclear framework fulfills such coordinative functions. The fact that processes occurring at the chromatin site are under a stringent cell cycle control implicates that the structure and function of the matrix should also change in a cell cycle-dependent manner. Cell cycle control mechanisms can affect the nuclear matrix and its complex constituents at several levels. (1) Synthesis and degradation of its components; (2) assembly and disassembly of complex structures from monomeric elements; (3) secondary association of molecules to the matrix; (4) binding of DNA to the nuclear matrix; and (5) post-translational modification of matrix constituents. For one well-defined set of nuclear matrix proteins, the nuclear lamins, experimental data covering these aspects exist, and we are just beginning to understand the contribution of lamins to nuclear architecture and function in a cell cycle context. For components of the internal nuclear matrix only few data with respect to cell cycle regulation are available, a fact partly due to the lack of unambiguous identification and characterization of internal nuclear matrix proteins. This situation is further complicated by the difficulty of distinguishing real matrix proteins from associated proteins whose association is dependent on cell cycle and development stage. However, there is increasing evidence that the transient cell cycle phase-specific attachment of enzymes and regulatory proteins to the nuclear matrix represents an important, so far underrated level of cellular regulation.

## Acknowledgments

The authors thank their colleagues G. Brosch, M. Goralik-Schramel, T. Lechner, A. Loidl, A. Lusser, and G. Stöffler for their help and interest during preparation of this manuscript. Experimental work was supported by the Austrian Science Foundation (Project 9223) and the Dr. Legerlotz-Foundation.

## References

Aebi, U., Cohn, J., Buhle, L., and Gerace, L. (1986). The nuclear lamina is a meshwork of intermediate-type filaments. *Nature (London)* **323,** 560–564.

Andreeva, M., Markova, D., Loidl, P., and Djondjurov, L. (1992). Intranuclear compartmentalization of transcribed and nontranscribed c-*myc* sequences in Namalva-S cells. *Eur. J. Biochem.* **207,** 887–894.

Appelbaum, J., Blobel, G., and Georgatos, S. D. (1990). In vivo phosphorylation of the lamin B receptor. *J. Biol. Chem.* **265,** 4181–4184.

Berezney, R., and Coffey, D. S. (1974). Identification of a nuclear protein matrix. *Biochem. Biophys. Res. Commun.* **60,** 1410–1417.

Berezney, R., and Coffey, D. S. (1975). Nuclear protein matrix: Association with newly synthesized DNA. *Science* **189,** 291–293.

Beven, A., Guan, Y., Peart, J., Cooper, C., and Shaw, P. (1991). Monoclonal antibodies to plant nuclear matrix reveal intermediate filament-related components within the nucleus. *J. Cell Sci.* **98,** 293–302.

Bridger, J. M., Kill, I. R., O'Farell, M., and Hutchinson, C. J. (1994). Internal lamin structures within G1 nuclei of human dermal fibroblasts. *J. Cell Sci.* **104,** 297–306.

Brosch, G., Georgieva, E. I., López-Rodas, G., Lindner, H., and Loidl, P. (1992). Specificity of *Zea mays* histone deacetylase is regulated by phosphorylation. *J. Biol. Chem.* **267,** 20561–20564.

Burke, B. (1990). On the cell-free association of lamins A and C with metaphase chromosomes. *Exp. Cell Res.* **186,** 169–176.

Burke, B., and Gerace, L. (1986). A cell free system to study reassembly of the nuclear envelope at the end of mitosis. *Cell (Cambridge, Mass.)* **44,** 639–652.

Burland, T. G., Solnica-Krecel, L., Bailey, J., Cunningham, D. B., and Dove, W. F. (1993). Patterns of inheritance, development and the mitotic cycle in the protist *Physarum polycephalum. Adv. Microb. Physiol.* **35,** 1–69.

Casiano, C. A., Landberg, G., Ochs, R. L., and Tan, E. M. (1993). Autoantibodies to a novel cell cycle-regulated protein that accumulates in the nuclear matrix during S phase and is localized in the kinetochores and spindle midzone during mitosis. *J. Cell Sci.* **106,** 1045–1056.

Chou, Y.-H., Bischoff, J. R., Beach, D., and Goldman, R. D. (1990). Intermediate filament reorganization during mitosis is mediated by p34-cdc2 phosphorylation of vimentin. *Cell (Cambridge, Mass.)* **62,** 1063–1071.

Chou, Y.-H., Ngai, K.-L., and Goldman, R. (1991). The regulation of intermediate filament reorganization in mitosis. *J. Biol. Chem.* **266,** 7325–7328.

Collins, J. M., and Chu, A. K. (1987). Binding of the DNA polymerase alpha-DNA primase complex to the nuclear matrix in HeLa cells. *Biochemistry* **26,** 5600–5607.

Compton, D. A., and Cleveland, D. W. (1993). NuMA is required for the proper completion of mitosis. *J. Cell Biol.* **120,** 947–957.

Compton, D. A., Szilak, I., and Cleveland, D. W. (1992). Primary structure of NuMa, an intranuclear protein that defines a novel pathway for segregation of proteins at mitosis. *J. Cell Biol.* **116,** 1395–1408.

Cook, P. R. (1989). The nucleoskeleton and the topology of transcription. *Eur. J. Biochem.* **185,** 487–501.

Cook, P. R. (1991). The nucleoskeleton and the topology of replication. *Cell (Cambridge, Mass.)* **66,** 627–635.

Cook, P. R. (1994). RNA polymerase: Structural determinant of the chromatin loop and the chromosome. *BioEssays* **16,** 425–430.

Correas, I. (1991). Characterization of isoforms of protein 4.1 present in the nucleus. *Biochem. J.* **279,** 581–585.

Courvalin, J.-C., Segil, N., Blobel, G., and Worman, H. J. (1992). The lamin B receptor of the inner nuclear membrane undergoes mitosis-specific phosphorylation and is a substrate for p34cdc2-type protein kinase. *J. Biol. Chem.* **267,** 19035–19038.

De Lange, T. (1992). Human telomeres are attached to the nuclear matrix. *EMBO J.* **11,** 717–724.

Dell'Orco, R. T., and Whittle, W. L. (1994). Nuclear matrix composition and in vitro cellular senescence. *Exp. Gerontol.* **29,** 139–149.

Dessev, G., lovcheva-Dessev, C., Bischoff, J. R., Beach, D., and Goldman, R. (1991). A complex containing p34 cdc2 and cyclin B phosphoryltes the nuclear lamin and disassembles nuclei of clam oocytes in vitro. *J. Cell Biol.* **112,** 523–533.

Dickinson, P., Cook, P. R., and Jackson, D. A. (1990). Active RNA polymerase I is fixed within the nucleus of HeLa cells. *EMBO J.* **9,** 2207–2214.

Dou, Q.-P., Markell, P. J., and Pardee, A. B. (1992). Thymidine kinase transcription is regulated at G1/S phase by a complex that contains retinoblastoma-like protein and a cdc2 kinase. *Proc. Natl. Acad. Sci. U.S.A.* **89,** 3256–3260.

Dwyer, N., and Blobel, G. (1976). A modified procedure for the isolation of a pore complex-lamina fraction from rat liver nuclei. *J. Cell Biol.* **70,** 581–591.

Eberharter, A., Grabher, A., Gstraunthaler, G., and Loidl, P. (1993). Nuclear matrix of the lower eukaryote *Physarum polycephalum* and the mammalian epithelial LLC-PK1 cell line. *Eur. J. Biochem.* **212,** 573–580.

Enoch, T., Peter, M., Nurse, P., and Nigg, E. A. (1991). p34cdc2 acts as a lamin kinase in fission yeast. *J. Cell Biol.* **112,** 797–807.

Feuerstein, N. (1991). Phosphorylation of numatrin and other nuclear proteins by cdc2 containing CTD kianse cdc2/p58. *J. Biol. Chem.* **266,** 16200–16208.

Feuerstein, N., and Randazzo, P. A. (1991). In vivo and in vitro phosphorylation studies of numatrin, a cell cycle regulated nuclear protein, in insulin-stimulated NIH 3T3 HIR cells. *Exp. Cell Res.* **194,** 289–296.

Feuerstein, N., Chan, P. K., and Mond, J. J. (1988a). Identification of numatrin, the nuclear matrix protein associated with induction of mitogenesis, as the nucleolar protein B23. *J. Biol. Chem.* **263,** 10608–10612.

Feuerstein, N., Spiegel, S., and Mond, J. J. (1988b). The nuclear matrix protein, numatrin (B23), is associated with growth-factor-induced mitogenesis in Swiss 3T3 fibroblasts and with T lymphocyte proliferation stimulated by lectins and anti-T cell antigen receptor antibody. *J. Cell Biol.* **107,** 1629–1642.

Feuerstein, N., Mond, J. J., Kinchington, P. R., Hickey, R., Karjalainen Lindsberg, M.-L., Hay, I., and Ruyechan, W. T. (1990). Evidence for DNA binding activity of numatrin (B23), a cell-cycle regulated nuclear matrix protein. *Biochem. Biophys. Acta* **1087,** 127–136.

Fey, E. G., Wan, K. M., and Penman, S. (1984). Epithelial cytoskeletal framework and nuclear matrix-Intermediate filament scaffold: Three-dimensional organization and protein composition. *J. Cell Biol.* **98,** 1973–1984.

Fidlerova, H., Sovova, V., Krekule, I., Viklicky, V., and Levan, G. (1992). Immunofluorescence detection of the vimentin epitope in chromatin structures of cell nuclei and chromsomes. *Hereditas* **117,** 265–273.

Fishel, B. R., Sperry, A. O., and Garrard, W. T. (1993). Yeast calmodulin and a conserved nuclear protein participate in the in vivo binding of a matrix association region. *Proc. Natl. Acad. Sci. U.S.A.* **90,** 5623–5627.

Foisner, R., and Gerace, L. (1993). Integral membrane proteins of the nuclear envelope interact with lamins and chromosomes, and binding is modulated by mitotic phosphorylation. *Cell (Cambridge, Mass.)* **73,** 1267–1279.

Franke, W. W. (1987). Nuclear lamins and cytoplasmic intermediate filament proteins: A growing multigene family. *Cell (Cambridge, Mass.)* **48,** 3–4.

Frederick, S. E., Mangan, M. E., Carey, J. B., and Gruber, P. J. (1992). Intermediate filament antigens of 60 and 65 kDa in the nuclear matrix of plants: Their detection and localization. *Exp. Cell Res.* **199,** 213–222.

Fredman, J. N., and Engler, J. A. (1993). Adenovirus precursor to terminal protein interacts with the nuclear matrix in vivo and in vitro. *J. Virol.* **67,** 3384–3395.

Georgatos, S. D., Maroulakou, I., and Blobel, G. (1989). Lamin A, lamin B, and lamin B receptor analogues in yeast. *J. Cell Biol.* **108,** 2069–2082.

Glass, C. A., Glass, J. R., Taniura, H., Hasel, K. W., Blevitt, J. M., and Gerace, L. (1993). The $\alpha$-helical rod domain of human lamins A and C contains a chromatin binding site. *EMBO J.* **12,** 4413–4424.

Glass, J. R., and Gerace, L. (1990). Lamins A and C bind and assemble at the surface of mitotic chromosomes. *J. Cell Biol.* **111,** 1047–1057.

Golderer, G., Loidl, P., and Gröbner, P. (1991). Cell cycle dependent ADP-ribosylation of the nuclear matrix. *Eur. J. Cell Biol.* **55,** 183–185.

Grabher, A., Eberharter, A., Gstraunthaler, G., and Loidl, P. (1992). Characterization of nuclear matrix proteins of *Physarum polycephalum* and mammalian cells. *Cell Biol. Int. Rep.* **16,** 1151–1157.

Grabher, A., Brosch, G., Sendra, R., Lechner, T., Eberharter, A., Georgieva, E., López-Rodas, G., Franco, L., Dietrich, H., and Loidl, P. (1994). Subcellular location of enzymes involved in histone acetylation. *Biochemistry* **33,** 14887–14895.

Greenfield, I., Nickerson, J., Penman, S., and Stanley, M. (1991). Human papillomavirus 16 E7 protein is associated with the nuclear matrix. *Proc. Natl. Acad. Sci. U.S.A.* **88,** 11217–11221.

Gröbner, P., and Loidl, P. (1982). Thymidylate synthetase during synchronous nuclear division cycle and differentiation of *Physarum polycephalum. Biochim. Biophys. Acta* **697,** 83–88.

Gröbner, P., and Loidl, P. (1983). Response of the dTMP-synthesizing enzymes to differentiation processes in *Physarum polycephalum. Exp. Cell Res.* **144,** 385–391.

Haas, M., and Jost, E. (1993). Functional analysis of phosphorylation sites in human lamin A controlling lamin disassembly, nuclear transport and assembly. *Eur. J. Cell Biol.* **62,** 237–247.

Hakes, D. J., and Berezney, R. (1991). DNA binding properties of the nuclear matrix and individual nuclear matrix proteins. *J. Biol. Chem.* **266,** 11131–11140.

Heald, R., and McKeon, F. (1990). Mutations of phosphorylation sites in lamin A that prevent nuclear lamina disassembly in mitosis. *Cell* (*Cambridge, Mass.*) **61,** 579–589.

Hendzel, M. J., and Davie, J. R. (1992). Nuclear distribution of histone deacetylase: A marker enzyme for the internal nuclear matrix. *Biochim. Biophys. Acta* **1130,** 307–313.

Hendzel, M. J., Sun, J.-M., Chen, H. Y., Rattner, J. B., and Davie, J. R. (1994). Histone acetyltransferase is associated with the nuclear matrix. *J. Biol. Chem.* **269,** 22894–22901.

Hennekes, H., Peter, M., Weber, K., and Nigg, E. A. (1993). Phosphorylation on protein kinase C sites inhibits nuclear import of lamin B2. *J. Cell Biol.* **120,** 1293–1304.

Hildebrandt, A., and Sauer, H. (1973). DNA dependent RNA polymerases from *Physarum polycephalum. FEBS Lett.* **35,** 41–44.

Hildebrandt, A., and Sauer, H. W. (1976). Levels of RNA polymerases during the mitotic cycle of *Physarum polycephalum. Biochim. Biophys. Acta* **425,** 316–321.

Hirano, T., and Mitchison, T. J. (1991). Cell cycle control of higher-order chromatin assembly around naked DNA in vitro. *J. Cell Biol.* **115,** 1479–1489.

Hocevar, B. A., Burns, D. J., and Fields, A. P. (1993). Identification of protein kinase C (PKC) phosphorylation sites on human lamin B. *J. Biol. Chem.* **268,** 7545–7552.

Jackson, D. A., and Cook, P. R. (1988). Visualization of a filamentous nucleoskeleton with a 23 nm axial repeat. *EMBO J.* **7,** 3667–3677.

Kallajoki, M., and Osborn, M. (1994). Gel electrophoretic analysis of nuclear matrix fractions isolated from different human cell lines. *Electrophoresis* **15,** 520–528.

Kasahara, K., Chida, K., Tsunenaga, M., Kohno, Y., Ikuta, T., and Kuroki, T. (1991). Identification of lamin B2 as a substrate of protein kinase C in BALB/MK-2 mouse keratinocytes. *J. Biol. Chem.* **266,** 20018–20023.

Kouklis, P. D., Merdes, A., Papamarcaki, T., and Georgatos, S. D. (1993). Transient arrest of 3T3 cells in mitosis and inhibition of nuclear lamin reassembly around chromatin induced by anti-vimentin antibodies. *Eur. J. Cell Biol.* **62,** 224–236.

Kumarasiri, M. H., Shapiro, L. E., and Surks, M. I. (1988). Cell cycle dependence of thyroid hormone nuclear receptors in cultured GC cells: Relationship to nuclear matrix. *Endocrinology (Baltimore)* **122,** 1897–1904.

Lang, S., and Loidl, P. (1993). Identification of proteins immunologically related to vertebrate lamins in the nuclear matrix of the myxomycete *Physarum polycephalum. Eur. J. Cell Biol.* **61,** 177–183.

Lang, S., Decristoforo, T., Waitz, W., and Loidl, P. (1993). Biochemical and morphological characterization of the nuclear matrix during the synchronous cell cycle of *Physarum polycephalum. J. Cell Sci.* **105,** 1121–1130.

Lebel, S., Lampron, C., Royal, C. A., and Raymond, Y. (1987). Lamins A and C appear during retinoic acid-induced differentiation of mouse embryonal carcinoma cells. *J. Cell Biol.* **105,** 1099–1104.

Li, H., and Roux, S. J. (1992). Casein kinase II protein kinase is bound to lamina-matrix and phosphorylates lamin-like protein in isolated pea nuclei. *Proc. Natl. Acad. Sci. U.S.A.* **89,** 8434–8438.

Loidl, P. (1994). Histone acetylation: Facts and questions. *Chromosoma* **103,** 441–449.

Loidl, P., and Gröbner, P. (1987a). Histone synthesis during the cell cycle of *Physarum polycephalum. J. Biol. Chem.* **262,** 10195–10199.

Loidl, P., and Gröbner, P. (1987b). Postsynthetic acetylation of histones during the cell cycle: A general function for the displacement of histones during chromatin rearrangements. *Nucleic Acids Res.* **15,** 8351–8366.

Loidl, P. , Loidl, A., Puschendorf, B., and Gröbner, P. (1983). Lack of correlation between H4 acetylation and transcription during the *Physarum* cell cycle. *Nature (London)* **305,** 446–448.

López-Rodas, G., Tordera, V., Sanchez del Pino, M., and Franco, L. (1991a). Subcellular location and nucleosome specificity of yeast histone acetyltransferases. *Biochemistry* **30,** 3728–3732.

López-Rodas, G., Georgieva, E. I., Sendra, R., and Loidl, P. (1991b). Histone acetylation in *Zea mays* I. Activities of histone acetyltransferases and histone deacetylases. *J. Biol. Chem.* **266,** 18745–18750.

López-Rodas, G., Lang, S., Loidl, A., Fasching, B., Greil, R., and Loidl, P. (1992). Nuclear proto-oncogene homologous proteins during the cell cycle of *Physarum polycephalum. Cell Biol. Int. Rep.* **16,** 1185–1191.

López-Rodas, G., Brosch, G., Georgieva, E. I., Sendra, R., Franco, L., and Loidl, P. (1993). Histone deacetylase—A key enzyme for the binding of regulatory proteins to chromatin. *FEBS Lett.* **317,** 175–180.

Lu, Y.-L., Spearman, P., and Ratner, L. (1993). Human immunodeficiency virus type 1 viral protein R localization in infected cells and virions. *J. Virol.* **67,** 6542–6550.

Luderus, M. E. E., de Graaf, A., Mattia, E., den Blaauwen, J. L., Grande, M. A., de Jong, L., and van Driel, R. (1992). Binding of matrix attachment regions to lamin B1. *Cell (Cambridge, Mass.)* **70,** 949–959.

Lüscher, B., Brizuela, L., Beach, D., and Eisenman, R. N. (1991). A role for the p34cdc2 kinase and phosphatases in the regulation of phosphorylation and disassembly of lamin B2 during the cell cycle. *EMBO J.* **10,** 865–875.

Mancini, M. A., Shan, B., Nickerson, J. A., Penman, S., and Lee, W.-H. (1994). The retinoblastoma gene product is a cell cycle-dependent, nuclear matrix-associated protein. *Proc. Natl. Acad. Sci. U.S.A.* **91,** 418–422.

McKeon, F. D., Kirschner, M. W., and Caput, D. (1986). Homologies in both primary and secondary structure between nuclear envelope and intermediate filament proteins. *Nature (London)* **319,** 463–468.

Meier, J., and Georgatos, S. D. (1994). Type B lamins remain associated with the internal nuclear envelope protein p58 during mitosis: Implications for nuclear assembly. *EMBO J.* **13,** 1888–1898.

Milavetz, B. I., and Edwards, D. R. (1986). Synthesis and stability of nuclear matrix proteins in resting and serum-stimulated Swiss 3T3 cells. *J. Cell. Physiol.* **127,** 388–396.

Moir, R. D., Montag-Lowy, M., and Goldman, R. D. (1994). Dynamic properties of nuclear lamins: Lamin B is associated with sites of DNA replication. *J. Cell Biol.* **125,** 1201–1212.

Murray, A., and Hunt, T. (1993). "The Cell Cycle: An Introduction." Freeman, New York.

Nakayasu, H., and Berezney, R. (1991). Nuclear matrins: Identification of the major nuclear matrix proteins. *Proc. Natl. Acad. Sci. U.S.A.* **88,** 10312–10316.

Newport, J. W., Wilson, K. L., and Dunphy, W. G. (1990). A lamin-independent pathway for nuclear envelope assembly. *J. Cell Biol.* **111,** 2247–2259.

Nickerson, J. A., Krockmalnic, G., Wan, K. M., Turner, C. D., and Penman, S. (1992). A normally masked nuclear matrix antigen that appears at mitosis on cytoskeleton filaments adjoining chromosomes, centrioles, and midbodies. *J. Cell. Biol.* **116,** 977–987.

Ohta, Y., Ohba, T., and Miyamoto, E. (1990). $Ca^{2+}$/calmodulin-dependent protein kinase II: Localization in the interphase nucleus and the mitotic apparatus of mammalian cells. *Proc. Natl. Acad. Sci. U.S.A.* **87,** 5341–5345.

Ozaki, T., Saijo, M., Murakami, K., Enomoto, H., Taya, Y., and Sakiyama, S. (1994). Complex formation between lamin A and the retinoblastoma gene product: Identification of the domain on lamin A required for its interaction. *Oncogene* **9,** 2649–2653.

Paddy, M. R., Belmont, A. S., Saumweber, H., Agard, D. A., and Sedat, J. W. (1990). Interphase nuclear envelope lamins form a discontinuous network that interacts with only a fraction of the chromatin in the nuclear periphery. *Cell (Cambridge, Mass.)* **62,** 89–106.

Panzeter, P. L., and Ringer, D. P. (1993). DNA precursors are channelled to nuclear matrix DNA replication sites. *Biochem. J.* **293,** 775–779.

Payrastre, B., Nievers, M., Boonstra, J., Breton, M., Verkleij, A. J., and van Bergen en Henegouwen, P. M. P. (1992). A differential location of phosphoinositide kinases, diacylglycerol kinase, and phospholipase C in the nuclear matrix. *J. Biol. Chem.* **267,** 5078–5084.

Peter, M., Nakagawa, J., Dorée, M., Labbé, J. C., and Nigg, E. A. (1990). In vitro disassembly of the nuclear lamina and M phase-specific phosphorylation of lamins by cdc2 kinase. *Cell (Cambridge, Mass.)* **61,** 591–602.

Peter, M., Heitlinger, E., Häner, M., Aebi, U., and Nigg, E. A. (1991). Disassembly of in vitro formed lamin head-to-tail polymers by cdc2 kinase. *EMBO J.* **10,** 1535–1544.

Pines, J., and Hunter, T. (1991). Human cyclins A and B1 are differentially located in the cell and undergo cell cycle-dependent nuclear transport. *J. Cell Biol.* **115,** 1–17.

Pipkin, J. L., Hinson, W. G., Anson, J. F., Lyn-Cook, L. E., Burns, E. R., and Casciano, D. A. (1991). Synthesis and phosphorylation of nuclear matrix proteins following toxic dose of retinoic acid in cycling and differentiating HL-60 cells. *Appl. Theor. Electrophoresis* **2,** 17–29.

Rober, R. A., Weber, K., and Osborn, M. (1989). Differential timing of nuclear lamin A/C expression in the various organs of the mouse embryo and the young animal: A development study. *Development (Cambridge, UK)* **105,** 365–378.

Robinson, S. I., Nelkin, B. D., and Vogelstein, B. (1982). The ovalbumin gene is associated with the nuclear matrix of chicken oviduct cells. *Cell (Cambridge, Mass.)* **28,** 99–106.

Stepp-Lorenzino, L., Rao, S., and Coleman, P. S. (1991). Cell cycle dependent, differential prenylation of proteins. *Eur. J. Biochem.* **200,** 579–590.

Shioda, M., and Murakami-Murofushi, K. (1989). Changes in activity and subcellular localization of alpha-like DNA polymerase during cell cycle of *Physarum polycephalum. Biochem. Int.* **18,** 581–588.

Shioda, M., Matsuzawa, Y., Murakami-Murofushi, K., and Ohta, J. (1989). DNA synthesis by the isolated nuclear matrix from synchronized plasmodia of *Physarum polycephalum. Biochim. Biophys. Acta* **1007,** 254–263.

Simos, G., and Georgatos, S. D. (1992). The inner nuclear membrane protein p58 associates in vivo with a p58 kinase and the nuclear lamins. *EMBO J.* **11,** 4027–4036.

Stewart, C., and Burke, B. (1987). Teratocarcinoma stem cells and early mouse embryos contain only a single major lamin polypeptide closely resembling lamin B. *Cell (Cambridge, Mass.)* **51,** 383–392.

Stuurman, N., Meijne, A. M. L., van der Pol, A. J., de Jong, L., van Driel, R., and van Renswoude, J. (1990). The nuclear matrix from cells of different origin. *J. Biol. Chem.* **263,** 5460–5465.

Tawfic, S., and Ahmed, K. (1994). Association of casein kinase 2 with nuclear matrix. Possible role in nuclear matrix protein phosphorylation. *J. Biol. Chem.* **269,** 7489–7493.

Thorburn, A., and Knowland, J. (1993). Attachment of vitellogenin genes to the nucleoskeleton accompanies their activation. *Biochem. Biophys. Res. Commun.* **191,** 308–313.

Todorov, I. T., Lavigne, J., Sakr, F., Kaneva, R., Foisy, S., and Bibor-Hardy, V. (1991). Nuclear matrix protein mitotin messenger RNA is expressed at constant levels during the cell cycle. *Biochem. Biophys. Res. Commun.* **177,** 395–400.

Tubo, R. A., and Berezney, R. (1987). Pre-replicative association of multiple replicative enzyme activities with the nuclear matrix during rat liver regeneration. *J. Biol. Chem.* **262,** 1148–1154.

Turnock, G. (1979). Patterns of nucleic acid synthesis in *Physarum polycephalum. Prog. Nucleic Acid Res. Mol. Biol.* **23,** 53–104.

Van Wijnen, A. J., Bidwell, J. P., Fey, E. G., Penman, S., Lian, J. B., Stein, J. L., and Stein, G. S. (1993). Nuclear matrix association of multiple sequence-specific DNA binding activities related to SP-1, ATF, CCAAT, C/EBP, OCT-1, and AP-1. *Biochemistry* **32,** 8397–8402.

von Bertalanffy, L. (1952). "Problems of Life." Harper, New York.

Waitz, W., and Loidl, P. (1988). In situ preparation of the nuclear matrix of *Physarum polycephalum:* Ultrastructural and biochemical analysis of different matrix isolation procedures. *J. Cell Sci.* **90,** 621–628.

Waitz, W., and Loidl, P. (1991). Cell cycle dependent association of c-*myc* protein with the nuclear matrix. *Oncogene* **6,** 29–35.

Wan, K. M., Nickerson, J. A., Krockmalnic, G., and Penman, S. (1994). The B1C8 protein is in the dense assemblies of the nuclear matrix and relocates to the spindle and pericentriolar filaments at mitosis. *Proc. Natl. Acad. Sci. U.S.A.* **91,** 594–598.

Ward, G. E., and Kirschner, M. W. (1990). Identification of cell cycle-regulated phosphorylation sites on nuclear lamin C. *Cell (Cambridge, Mass.)* **61,** 561–577.

Weber, K., and Geisler, N. (1985). Intermediate filaments: Structural conservation and divergence. *Ann. N.Y. Acad. Sci.* **455,** 126–143.

Worman, H. J., Yuan, J., Blobel, G., and Georgatos, S. D. (1988). A lamin B receptor in the nuclear envelope. *Proc. Natl. Acad. Sci. U.S.A.* **85,** 8531–8534.

Worman, H. J., Evans, C. D., and Blobel, G. (1990). The lamin B receptor of the nuclear envelope inner membrane: A polytopic protein with eight potential transmembrane domains. *J. Cell Biol.* **111,** 1535–1542.

Yang, C. H., Lambie, E. J., and Snyder, M. (1992). NuMa: An unusually long coiled-coil related protein in the mammalian nucleus. *J. Cell Biol.* **116,** 1303–1317.

Yuan, J., Simos, G., Blobel, G., and Georgatos, S. D. (1991). Binding of lamin A to polynucleosomes. *J. Biol. Chem.* **266,** 9211–9215.

Zhelev, N. Z., Todorov, I. T., Philipova, R. N., and Hadjiolov, A. A. (1990). Phosphorylation-related accumulation of the 125K nuclear matrix protein mitotin in human mitotic cells. *J. Cell Sci.* **95,** 59–64.

Zini, N., Martelli, A. M., Cocco, L., Manzoli, F. A., and Maraldi, N. M. (1993). Phosphoinositidase C isoforms are specifically localized in the nuclear matrix and cytoskeleton of Swiss 3T3 cells. *Exp. Cell Res.* **208,** 257–269.

# Specificity and Functional Significance of DNA Interaction with the Nuclear Matrix: New Approaches to Clarify the Old Questions

Sergey V. Razin,*,† Irina I. Gromova,*,‡ and Olga V. Iarovaia *,†
*Institute of Gene Biology of the Russian Academy of Sciences, 117342 Moscow,Russia, †International Centre for Genetic Engineering and Biotechnology, Padriciano 99, Italy, and ‡University of Aarhus, Department of Molecular Biology, Aarhus C, Denmark

In this chapter the specificity of chromosomal DNA partitioning into topological loops is discussed. Different experimental approaches used for the analysis of the above problem are critically reviewed. This discussion is followed by presentation of a novel approach for mapping the DNA loop anchorage sites that we have developed. This approach, based on the excision of the whole DNA loops by topoisomerase II-mediated DNA cleavage at matrix attachment sites, seems to constitute a unique tool for the analysis of topological organization of chromosomal DNA in living cells. We also discuss experimental results indicating that the DNA-loop anchorage sites form "weak points" in chromosomes that are preferentially sensitive to cleavage with both endogenous and exogenous nucleases. In connection with this discussion, rationales for the supposition that DNA loops constitute basic units of eukaryotic genome organization and evolution are considered. The chapter concludes by suggesting a new model of spatial organization of eukaryotic genome within the cell nucleus that resolves apparent contradictions between different data on the specificity of DNA interaction with the nuclear matrix.

**KEY WORDS:** DNA loops, Nuclear matrix DNA, Topoisomerase II, Long-range genome fragmentation, Nucleolar genes, c-*myc* amplicone.

## I. Introduction

About 20 years ago it was found that after extraction of permeabilized cells with 2 *M* NaCl solution DNA remained constrained by periodic attachments to the insoluble nuclear remnants (Cook and Brazell, 1975, 1976; Cook *et al.,* 1976). Earlier, several other research teams independently studied structure of residual nuclei (Georgiev and Chentsov, 1960, 1962; Zbarsky *et al.,* 1962; Smetana *et al.,* 1963; Narayan *et al.,* 1967; Berezney and Coffey, 1974). The most extensive analysis of the residual nuclear structures was carried out by Berezney and Coffey (1974, 1975, 1977) who suggested the term "nuclear matrix" (NM) for these structures. One of the most important initial observations of these authors was that the residual nuclear structures retained the shape and some morphological features of nuclei after enzymatic digestion of almost all nucleic acids and extraction of chromatin proteins with a concentrated salt solution. The standard procedure for isolation of nuclear matrices described by Berezney and Coffey (1974, 1975) is still used by many researchers either in its original form or with slight modifications.

During mitosis some part of the NM seems to undergo spatial reorganization giving rise to the central scaffolding element of metaphase chromosomes. Packaging of DNA into constrained domains (loops) is preserved during mitosis (Warren and Cook, 1978). DNA loops attached to the high salt-insoluble metaphase chromosome scaffold can be directly visualized by electron microscopy (EM) of spread histone-depleted chromosomes (Paulson and Laemmli, 1977). Similarly, DNA loops attached to the NM can be observed on the spreads of high salt-extracted nuclei (Hancock and Hughes, 1982). The radial loop model of eukaryotic chromosomes has been proposed to explain the above observations (Marsden and Laemmli, 1979). According to this model, the 30-nm chromatin fibril is arranged into loops due to regular attachments to a proteinous scaffolding structure. Transitions between interphase and metaphase are thought to be achieved through a certain reorganization (folding–unfolding) of the scaffolding element without disturbing the loops. The average size of the loops seems to be the same in interphase nuclei and metaphase chromosomes constituting from 25 to 200 kb (see below; Hancock, 1982). It was reasonable to suggest that with this size there could be a correlation between the topological organization of DNA within the nuclei and the genome organization into functional units, such as replicons and transcriptions. The specificity of the eukaryotic DNA organization into loops has been intensively studied but results and conclusions of different authors are controversial. Recently, we elaborated a new approach to study chromosomal DNA loops (Razin *et al.,* 1993). Instead of characterizing loop basements (NM DNA) we concentrated our efforts on the characterization of individual loops after their excision by topoisomerase II-mediated DNA cleavage at matrix attachment sites.

In this chapter we shall reconsider all data related to the problem of specificity of the genomic DNA organization into loops. This analysis will be concluded by a suggested new model of spatial organization of the eukaryotic genome that resolves apparent contradictions between the above data. We shall also discuss rationales for the supposition that chromosomal DNA loops constitute basic units of the eukaryotic genome organization and evolution.

## II. Determination of the Size of DNA Loops

Apparent similarity between the average size of chromosomal DNA loops (Table I) and the average sizes of replicons and transcriptons always stimulated the development of models postulating that loops are structural–functional blocks of the genome. Hence, it seems reasonable to start our present discussion with a critical analysis of data on the determination of the average size of DNA loops. The approaches used to estimate average size of chromosomal DNA loops, although different in details, can be easily divided into four major groups discussed below (Fig. 1).

### A. Distance between Nicks Introduced into DNA of Nucleoids for Complete Relaxation of Superhelical Tension

The term "nucleoids" was suggested by Cook *et al.* (1976) to designate high salt-extracted nuclei, or in other words, the nuclear matrices containing all nuclear DNA. As long as chromosomal DNA is arranged into topologically closed domains (loops), it will be negatively supercoiled after disruption of nucleosomes by high salt extraction. The density of negative supercoiling of closed DNA domains can be decreased by treating nucleoids with ethidium bromide that intercalates into DNA and causes partial unwinding of the DNA double helix (Cook and Brazell, 1975, 1976; Benyajati and Worcel, 1976). By titration of nucleoids with increasing concentration of ethidium bromide, it is possible to eliminate all negative supercoils and, furthermore, to induce positive supercoiling (Fig. 1A). Changes in the superhelical density influence the sedimentation properties of nucleoids. When DNA loops are completely relaxed, the sedimentation coefficient is minimal because this is the least compact configuaration of nucleoid. The same relaxed configuration may be obtained by a completely different approach, namely by introducing a single-stranded DNA (ssDNA) scission (nick) in each loop. Once this is done, it is necessary simply to determine average distance between the nicks. It is easy to see that in the simplest case (i.e., when all DNA is arranged into loops of relatively similar sizes)

TABLE I

The Sizes of DNA Loops in Some Eukaryotic Cells Determined by Different Experimental Approaches

| Approach | Experimental object | Average size of DNA loops | Reference |
|---|---|---|---|
| Relaxation of superhelical domains in DNA | HeLa cells (human) | 660 kb | Mullenders *et al.* (1983) |
| | Mouse carcinoma cell line FM 3A | 200 kb | Nakane *et al.* (1978) |
| | Chinese hamster cell line (CHO) | 140 kb | Hartwig (1982) |
| | *Drosophila melanogaster* cultured cells | 85 kb | Benyajati and Worcel (1976) |
| Determination of the diameter of DNA loop halo in nucleoids | Mouse cell line 3T3 | 90 kb | Vogelstein *et al.* (1980) |
| % of DNA in the nuclear matrix and average size of matrix DNA fragments | Rat liver | 80 kb | Berezney and Buchholtz (1981b) |
| | Cultured mouse fibroblast (L cells) | 60 kb | Razin *et al.* (1979) |
| | HeLa cells (human) | 80 kb | Mullenders *et al.* (1983) |
| | Chicken erythrocyte | 45 kb | Ganguly *et al.* (1983) |
| | *Drosophila melanogaster* cultured cells | 25 kb | Iarovaia *et al.* (1985) |
| Measurement of DNA loop length on EM pictures | HeLa cells (human) | 30–90 kb | Paulson and Laemmli (1977) |

this distance must be equal to the double size of an average loop. If, however, the sizes of DNA loops vary significantly, or if there are populations of loops differing in their sizes, the sedimentation properties of nucleolids will be determined by relaxation of the largest loops. Hence, the average loop size may be easily overestimated by the above described approach.

## B. Diameter of DNA Loop Halo in Relaxed Nucleoids

This is possibly the most widely used procedure (Buongiorno-Nardelli *et al.,* 1982; Vogelstein *et al.,* 1985). It may be considered as a technical

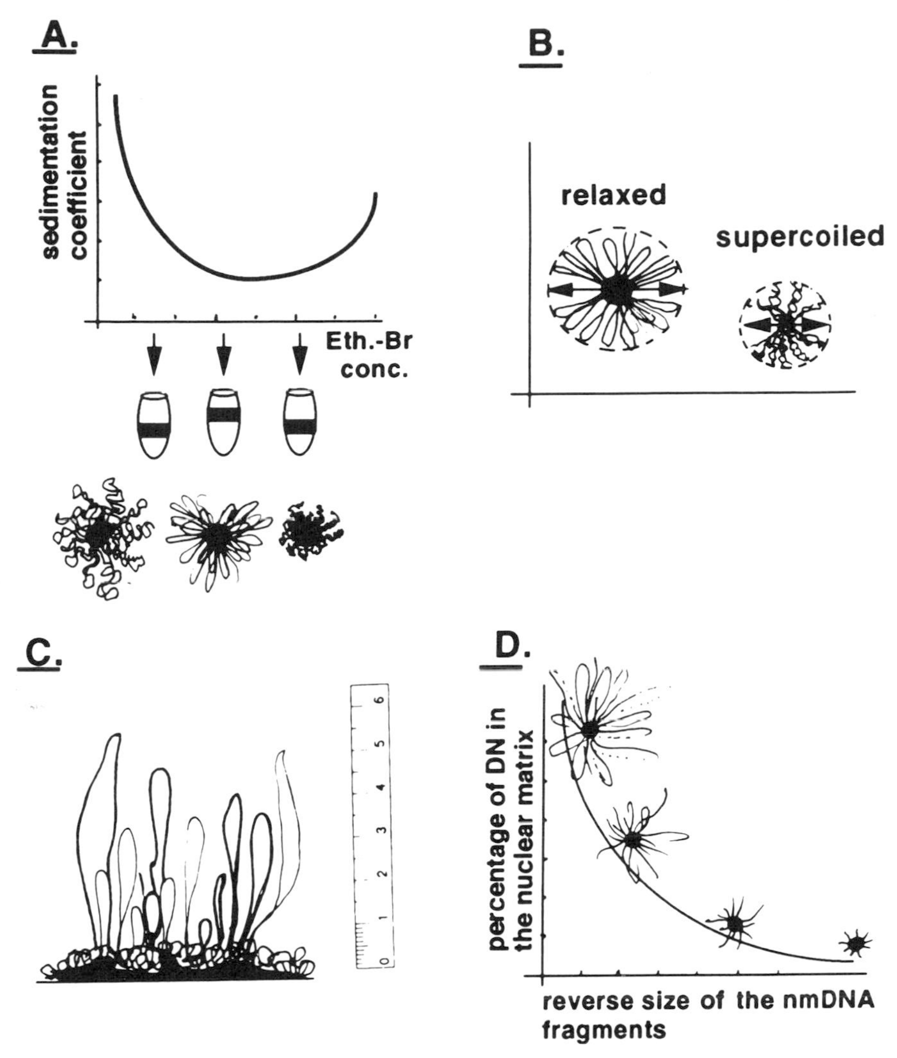


FIG. 1 Determination of the average DNA loop size by different experimental approaches. (A) A curve showing the typical dependence of nucleoid sedimentation properties on the concentration of intercalating drug in the media. The drawings beyond the curve represent schematically (from left to right) a nucleoid with negatively supercoiled loops (fast sedimenting), a nucleoid with relaxed loops (slowly sedimenting), and a nucleoid with positively supercoiled loops (fast sedimenting). (B) A scheme illustrating a possibility to determine the average DNA loop size by measuring the diameter of halo of relaxed DNA loops in nucleoids. (C) A scheme illustrating a possibility to measure sizes of relaxed DNA loops on the EM pictures. (D) A scheme showing the basis of interdependence between the average size of nuclear matrix DNA fragments and the percentage of DNA in the nuclear matrix.

modification of the above-described nucleoid relaxation protocol (Fig. 1B). An obvious advantage of this modification is the simplicity of experiments. Cells grown on coverslips are extracted with 2 *M* NaCl solution containing also a nonionic detergent. After this treatment nuclei remain attached to the coverslips through cytoskeletal elements and histone-depleted DNA forms a halo of supercoiled loops attached to the NM. After relaxation of the loops by introducing ssDNA scissions or by treatment with topoisomerase I, the diameter of the halo (which can be measured under a fluorescence microscope) will be equal to the average loop size (Fig. 1B). Again, it is necessary to keep in mind that the largest loops will determine the halo diameter and that a portion of DNA nonorganized into loops (if any) cannot be accurately estimated by this approach. According to the data of Vogelstein *et al.* (1985), this portion (the DNA that resides within confines of the NM) may constitute as much as 20% of the whole nucleoid DNA.

### C. Loop Size Based on Correlation between the Percentage of DNA and Size of Nuclear Matrix DNA Fragments

Treatment of nucleoids with nucleases producing double-stranded DNA (ds-DNA) breaks results in a gradual release of DNA from the nuclear matrix. Within the 2- to 50-kb area the size of DNA fragments remaining bound to the nuclear matrix after limited DNA cleavage turned out to correlate linearly with the percentage of DNA recovered in the nuclear matrix (Razin *et al.*, 1979; Jackson *et al.*, 1990). On the basis of this correlation (Fig. 1C) it is possible to estimate an average size of DNA loops by approximation to the starting point of digestion. Clearly, the interpretation of the above correlation is based on the assumption that all DNA is organized in more or less uniform loops and that the attachment regions are either small enough or are protected from nucleases. At least the second assumption may be incorrect because no linear correlation between the percentage of DNA in nuclear matrices and the average size of nuclear matrix DNA fragments was observed when less than 2% of nuclear DNA remained in the nuclear matrices (S. V. Razin and V. L. Mantieva, unpublished results).

### D. Loop Length of Spread Nucleoids and Histone-Depleted Chromosomes

Visualization of DNA loops in histone-depleted metaphase chromosomes (Paulson and Laemmli, 1977) and interphase nuclei (Hancock and Hughes, 1982) provides a possibility to measure directly their length (Fig. 1D). In HeLa cells loops ranging in size from 30 to 90 kb were detected (Paulson and Laemmli, 1977). One should, however, keep in mind that during spreading the DNA may be mechanically broken and that the probability of losing the longer loops will be relatively higher. Hence, the average loop size determined by direct measurements of DNA loop length by EM pictures may be underestimated. Again, the portion of DNA residing within the scaffolding elements (if any) cannot be estimated.

From the above discussion we may underline two points:

1. Whatever approach was used to determine an average size of DNA loops in the above studies, the interpretation of experimental results was based on the simplest assumption that the whole genome is organized into loops of more or less similar sizes. If a portion of the genome (even as large as 10 or 20%) were organized in a completely different fashion, this portion would be automatically excluded from the analysis by any of the experimental procedures. This is especially true in the case of the first two procedures (nucleoid loop relaxation analysis and measurement of the diameter of the DNA loop halo) which were actually used in most of the studies to estimate the DNA loop size in different organisms. Hence, what was actually estimated in all these studies is not the average loop size, but the average size of the majority of uniformly organized loops. Due to this, it became possible to observe a correlation between the DNA loop size and the replicon size in different organisms (Buongiorno-Nardelli *et al.*, 1982), although the attachment to the NM of transcribing and potentially transcribing DNA sequences (see Section III,C,1) might probably disturb this correlation.
2. The percentage of DNA nonorganized into loops cannot be estimated by any of the approaches used for determination of the average loop size.

## III. Analysis of Sequence Specificity of Nuclear Matrix DNA

### A. Introduction to the Problem

Different approaches used so far to study the specificity of genomic DNA organization into loops are conceptually similar as they all were designed to isolate and characterize DNA sequences located at the loop ends that were presumably identified as sequences residing in close proximity to the NM (NM DNA). In most of these studies DNA loops were cut by nucleases (most frequently by restriction nucleases) and DNA sequences remaining bound to the NM were isolated by differential centrifugation.

The specificity of thus obtained matrix-bound DNA was then analyzed to answer two different, although related, questions: (1) Are there specific DNA sequences located at the basements of DNA loops? (2) Do the loop ends have fixed positions in the genome (even if these positions are not defined by any DNA sequence consensus)?

To answer the first question, it seemed reasonable to solubilize from the nuclear matrix as much DNA as possible and then to study the remainder. One may expect that the DNA fraction that cannot be solubilized from nuclear matrices by nuclease treatment is a set of relatively short DNA fragments representing the DNA loop basements immersed into the nuclear matrix and hence protected from nucleases. It would be logical to use the term "nuclear matrix DNA" for designation of this protected DNA fraction. However, the term nuclear matrix DNA is actually used more widely to denote any DNA fraction remaining bound to the nuclear matrix after cleavage and solubilization of a prominent portion of loop DNA.

The specificity of the distribution of loop ends in the genome (the second of the above indicated questions) was studies by analysis of the distribution of individual DNA sequences between matrix-bound and cleaved-off DNA fractions in the course of progressive detachment of DNA from the nuclear matrix. To make all DNA sequences equally accessible for nucleases and to ensure subsequent extraction of cleaved-off portions of DNA loops, histones were removed in experiments of the above type. Initially, this was achieved by preextraction of nuclei with a concentrated salt solution (Hancock, 1982). More recently, treatment with an ionic detergent, lithium diiodosalicylate (LIS), was used for the same purpose (Mirkovich *et al.*, 1984).

The results obtained with the two above-mentioned procedures of histone removal turned out to be quite different (see below). To resolve the contradiction, it was supposed that high salt extraction caused an artificial precipitation of DNA bound to transcriptional complexes along with nascent RNA chains (Kirov *et al.*, 1984; Mirkovich *et al.*, 1984). This problem was elucidated by the elaboration of a new protocol of nuclear matrix DNA isolation with the aid of electroelution in physiological ionic strength of cleaved-off pieces of DNA loops arranged in chromatin (Jackson and Cook, 1985, 1986). In the following sections (III,B; III,C) we shall discuss the properties of nuclear matrix DNA isolated by different experimental approaches.

### B. Search for DNA Sequence Elements Essential for the Attachment of DNA

After extensive treatment with DNase I or micrococcal nuclease of either interphase nuclei or metaphase chromosomes followed by the extraction with a concentrated salt solution, only about 0.1% of DNA remains in the nuclear matrix/chromosomal scaffold (Bowen, 1981; Razin *et al.,* 1985a). The same portion of DNA can be recovered in nuclear matrices treated with either of the above-mentioned nucleases after histone removal by 2 *M* NaCl extraction (Goldberg *et al.,* 1983; Jackson *et al.,* 1984). In both cases the nuclear matrix DNA fraction protected from nucleases was found to be composed for the most part of fragments ranging in size from 50 to 200 bp. These fragments could not be removed from the matrices by repeated extractions with 2 *M* NaCl solution or even by density centrifugation in a CsCl gradient (Jackson *et al.,* 1984). Hence, they were assumed to represent loop ends immersed into the nuclear matrix.

Analysis of sequence specificity of short nuclear matrix DNA fragments gave controversial results. Some authors have found that the reassociation kinetics of these DNA samples is similar, if not identical, to that of total DNA (Basler *et al.,* 1981; Bowen, 1981). An enrichment of similarly prepared nuclear matrix DNA samples in very fast-renaturating sequences was reported by other investigators (Opstelten *et al.,* 1989). As has been shown by the hybridization analysis, some classes of repetitive sequences are overrepresented and some are underrepresented in the nuclear matrix DNA as compared to the total DNA (Jackson *et al.,* 1984). Possible reasons for the contradictions in the data on the sequence organization of nuclear matrix DNA have been discussed previously (Razin *et al.,* 1985b; Razin, 1987). It should be stressed that even those scientists who have found that the sequence organization of nuclear matrix DNA is not random also have failed to identify a particular DNA sequence element responsible for the attachment to the nuclear matrix. This failure may have several reasons. One of them is that such an element could not be identified by the analytical approaches used (namely, reassociation kinetics analysis and cross-hybridization experiments). Indeed, conserved recognition sites for sequence-specific DNA-binding proteins in most cases are very short (4 to 8 bp). If a similar recognition site or a combination of such recognition sites is necessary for matrix binding, they cannot be identified by cross-hybridization.

Furthermore, it was also speculated that a matrix attachment element (if any) was not a distinct DNA sequence, but rather a kind of sequence motive; for example, a DNA sequence capable of forming (under certain conditions) a noncanonical secondary structure. To check the above-mentioned possibilities, it seemed reasonable to clone randomly taken nuclear matrix DNA fragments and to compare their nucleotide sequences. This work was done by several research teams with essentially similar results (Goldberg *et al.,* 1983; Kalandadze *et al.,* 1988; Opstelten *et al.,* 1989; Boulikas and Kong, 1993). In all cases a significant portion or even the majority of nuclear matrix DNA fragments were found to possess a degenerated sequence organization as followed from the presence of short perfect and imperfect repeats and simple motives. Some of the nucleotide sequences of chicken erythrocyte nuclear matrix DNA fragments cloned in our lab are shown in Fig. 2. It should be mentioned that the strategy of cloning and sequencing of nuclear matrix DNA fragments can be compromised if a population of these fragments includes replication forks temporarily associated with the nuclear matrix (Berezney and Coffey, 1975); Berezney and Buchholtz, 1981a). That is why nuclear matrix DNA fragments for cloning and sequencing were isolated either from synchronized cells at $G_1$/S border (Goldberg *et al.,* 1983) or in G2 phase (Opstelten *et al.,* 1989) or from mature chicken erythrocytes that are not active in replication (Kalandadze *et al.,* 1988). In the last case the contamination of loop ends with transcribed sequences was also excluded (see below, Section III,C,1).

Our description of different attempts to identify the matrix-attachment consensus would not be complete without mentioning the *in vitro* binding assay, which has become extremely popular in the past few years. The initial observation related to the above approach was made by Dessev and collaborators (Krachmarov *et al.,* 1986) who found that an isolated nuclear lamina contained DNA binding sites capable of detaining preferentially AT-rich DNA sequences. The binding specificity was not, however, properly studied. Later on, Garrard and collaborators found that a subset of cloned DNA sequences can be detained by some affinity sites on isolated nuclear matrices (Cockerill and Garrard, 1986a,b; Sperry *et al.,* 1989). Binding of these DNA sequences to the nuclear matrices was not tissue- or even species-specific (Cockerill and Garrard, 1986b; Amati and Gasser, 1990). Although most, if not all, DNA fragments found in complexes with nuclear matrices were relatively rich in AT pairs, some other DNA fragments with a similar nucleotide composition were not detained by the nuclear matrices (Garrard, 1990; Das *et al.,* 1993). Hence, it was logical to assume that DNA fragments able to interact with the affinity sites on the nuclear matrix possessed some common property determined at the level of nucleotide sequence. The DNA sequence element essential for affinity binding to an isolated nuclear matrix was called matrix attachment region or MAR. Comparison of nucleotide sequences of several cloned DNA fragments containing MARs did not reveal any prominent homology (Mielke *et al.,* 1990; see also Section III,C,2 of this chapter). Recent evidence suggests that the essential feature of MARs is the ability to melt under relatively mild conditions (Bode *et al.,* 1992). MARs frequently colocalize with important functional elements in the genome such as promoters and enhancers (Cockerill and Garrard, 1986a; Cockerill *et al.,* 1987; Garrard, 1990). In yeast cells MARs seem to be an integral part of replication origins (Amati and Gasser, 1990). They also have been reported to confer the ability of position-independent expression to genes integrated into new genomic positions

**CO 326 245 b.p. A+T - 33%**

GACCGCTTCG GGTGTCCTTC GGGTCCGTTC GGCGGCGTTC GGGTCACTGC AGATCACTTC
GGGTGTCCTC GGGTCCGGTT CGGGTCCGTT GGCCGGCGTT CGGGTGACCG CAGAGAGCTT
CGGGTCTCTT CGGGTCCGTT CGGCGGCGTT CGGGTCACTG CAGACCGCTT CGAGTCTTTT
CGGGTCCGGT TCGGGTCCGT TCGGCGGCGT TCGGGTCACT GCAGACAACT TCGGAACTCT
TCGGG

**CO 62 220 b.p. A+T - 20%**

GGAACGGAAC GGAACGGAAC GGAACGGAGC GGAGCGGAGC AGAGATGCCA CCGCCGCCGT
CGGCCCGCCG TGACGCCGTG CGCCGGCGGG GGCGGCGGGG AGAGCCGGGG GGGGAACCCG
AGTGCCGCAG CTCCGCCGGT ACCGGAACCC CCGGAGAGGG TGCGTGGCAC GGGGGGGTGC
GGGAGGCGAG GGGGTGGCGG GAGGGCACGG CAGCAGGGGG

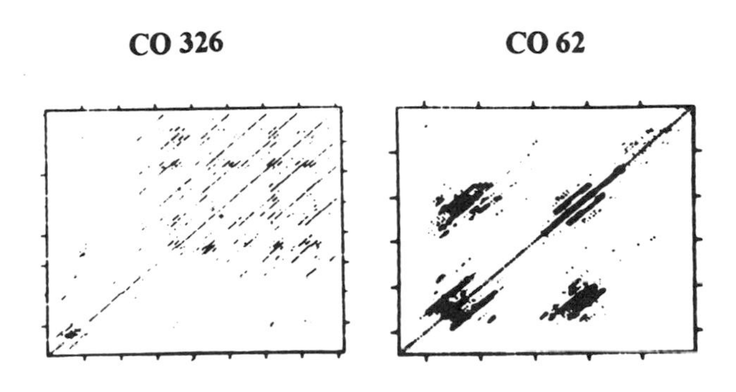


**CO 30 160 b.p. A+T - 26%**

ATCCGGGGGG GGTCGTGGGG GGGGGGAGTC CCCAATTCCA AGGGGGGGGT CGGAGGGTGG
GGGGGGTCCC CATTCCAAGG GGGGGGCCGG AAGGGGGAGG GTCCCCATTT CTCAGGGGGG
GGTCGGAGGG GGGGGTCCCC ATTCCCAACG GGGGAGGGGG

**CO 73 517 b.p. A+T - 42%**

CCCCCCAGTG GGGAGGTCCT GTGGGGGGGT TTGGGGGGGT CCTAGGGAGG TTTTGGGGGG
GATAGTGGGG TTTGGGGGGG TCCTATGGGG GATGGGGGGG TTCTGGAGGG GTCCAAGGGG
GGATACTGGG GTTTGGGGGG GTCCTATGGG GAATGGGGGG GTCCCTGAGG GGTCCAAGGG
GGGATAGTGG GGTTTGGGGG GGCTGGGTCC GTTGGGGATG GGTAGGAATG CGGCCCACTG
TGAGGTCACG GGGTGACTGT GACTCTTCGG TGCCATCACG AAGCACACAG ATAGGTGGAG
GTTTGCTGCC CCGTGGCTAT TGCGTCAGTA TTTTGGGCAT TTGCCTGCCA GAGTGCACCC
GAGGACTTTC AGTGGGCCAT TTTCCCAAGC AGTGACAAAT AAAGGATTCT TGAAGCAAGC
AGTAGAGCAG CTTTGGTGTT GCCGTTGATC AGTGCAGTAG CTGTGACCGT GCCGGTATTT
GATAGTAGTG TCAGCAAAGT AAATGCTGTG CTGGGTA

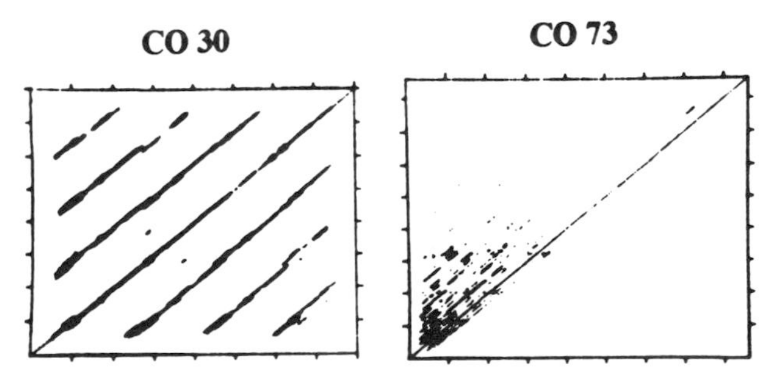


FIG. 2 Sequences of several cloned nuclear matrix DNA fragments with the dot-matrix maps of internal homologies (Kalandadze *et al.*, 1988).

(Stief *et al.*,1989; Klehr *et al.*, 1991). Hence, it is likely that MARs represent a new class of regulatory elements of the eukaryotic genome. A detailed analysis of functional significance of MARs is beyond the scope of the present chapter. The reader will find this analysis in other chapters of this volume.

For our discussion the key question is whether MARs are the sites of chromosomal loop anchorage. The answer to this question is not at all clear. It is, however, obvious that the organization of nuclear DNA into topological loops is not affected by high salt extraction (see above), while complexes of MAR elements with an isolated nuclear matrix do not endure even 0.5 *M* NaCl treatment (Vassetzky *et al.*, 1989). Thus, at best they may constitute only a part of the high salt-resistant nuclear matrix attachment sites normally stabilized by some other interactions. It may happen, however, that MARs are not at all located at the basement of DNA loops in nucleoids. Preferential localization of the kappa immunoglobulin MAR on the nuclear scaffold after histone removal by LIS treatment (Cockerill and Garrard, 1986a) may be explained by rebinding in the course of the treatment of LIS-extracted nuclei with restriction nucleases (see Section III,C,2) and therefore cannot be regarded as conclusive proof. Hence, the possibility of participation of MAR elements in the anchorage DNA loops on the nuclear matrix is not presently supported by solid experimental evidence and should be treated with caution. The question can be eventually clarified by high-resolution fluorescence *in situ* hybridization of MAR-specific probes to DNA of nucleoids fixed at coverslips. This approach has been already successfully used for analysis of the distribution of some unique DNA sequences within DNA loops of nucleoids (Gerdes *et al.*, 1994).

## C. Analysis of the Specificity of Distribution of Unique Genes within Chromosomal DNA Loops

### 1. High Salt Extraction Procedure

The experimental approach for mapping locations of individual DNA sequences within topological loops was designed by Cook and Brazell (1980). They suggested to determine an approximate distance between a DNA sequence under study and a matrix attachment site by analyzing relative representation of this DNA sequence in nuclear matrix DNA fractions of different sizes that could be collected in the course of progressive detachment of DNA from the nuclear matrix (Fig. 3). According to the simplest model of DNA loop organization, any given DNA sequence may be found preferentially in nuclear matrix DNA as long as the average size of nuclear matrix DNA fragments exceeds more than twice the distance between this

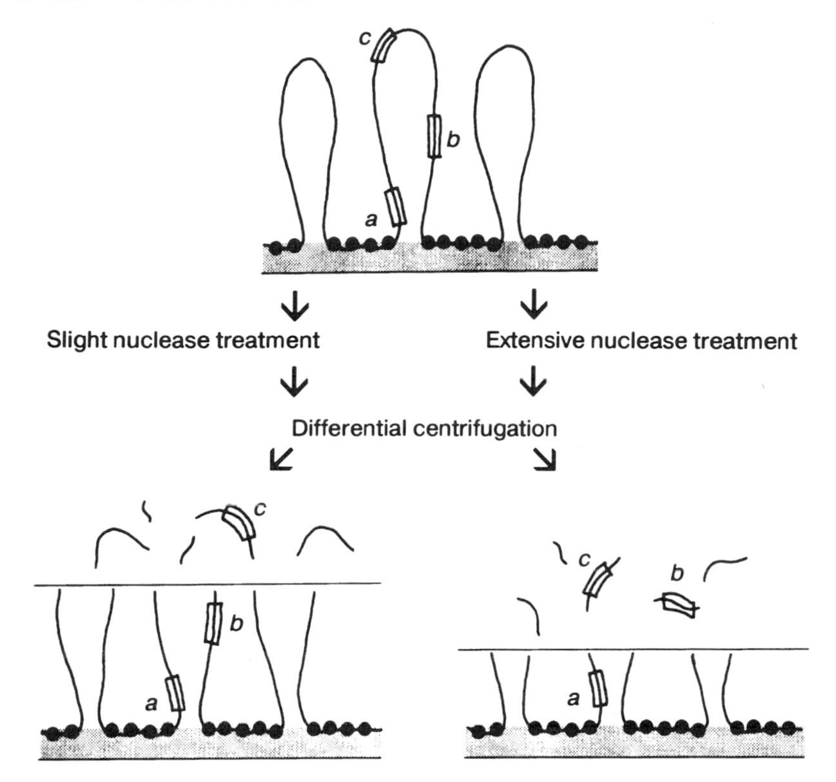


FIG. 3 A scheme illustrating the possibility of determining an approximate distance between a DNA sequence under study and a matrix attachment site by analyzing relative representation of this DNA sequence in different size fractions of nuclear matrix DNA. The design represents a hypothetical DNA loop with three regions of interest located close to matrix attachment site ("a"), at the middle of the loop ("c"), and at the half-distance from the matrix attachment site to the middle of the loop ("b"). A brief nuclease treatment releases into the supernatant only the sequence element c, while a more extensive treatment releases both b and c. It is easy to see that any given DNA sequence element will be released into supernatant when the average size of nuclear matrix DNA fragments is less than double distance between this sequence element and the matrix attachment site.

DNA sequence and the matrix attachment site (Fig. 3). Then (when the average size of nuclear matrix DNA fragments decreases beyond the above point) the same DNA sequence may move from the matrix-bound to the cleaved-off DNA fraction. It is hardly necessary to say that for correct estimation of a distance between a sequence of interest and a matrix attachment site all loop DNA should be equally accessible for a nuclease used for the detachment of DNA from the nuclear matrix. Furthermore, the nuclease should cut all DNA sequences with an equal probability. To meet the first of the above conditions, the nuclease treatment was carried out after histone removal by high salt extraction (Cook and Brazell, 1980; Ciejek *et al.,* 1983; Robinson *et al.,* 1983; Jost and Seldran, 1984; Small and Vogelstein, 1985; Small *et al.,* 1985; Vogelstein *et al.,* 1985). The second condition was usually disregarded, as sequence-specific restriction endonucleases were used for DNA cleavage from the nuclear matrix by most of the researchers (Ciejek *et al.,* 1983; Robinson *et al.,* 1983; Small *et al.,* 1985; Small and Vogelstein, 1985). This disregard of the second condition resulted in a decreased resolution of the approach. Consequently, the distance between the DNA sequence under study and the nuclear matrix attachment site was not actually established. Instead of this, it was only possible to find out whether any given sequence was located close to the matrix attachment site or somewhere within the loop. The results of the mapping experiments of the above type were rather unexpected. Most of the researchers observed a correlation between the transcriptional status of a gene and its attachment to the nuclear matrix. In other words, active genes were generally found in nuclear matrix DNA, while nonactive ones were found in cleaved-off DNA fractions (Cook *et al.,* 1982; Robinson *et al.,* 1982, 1983; Ciejek *et al.,* 1983; Hentzen *et al.,* 1984; Jost and Seldran, 1984; Small *et al.,* 1985; Small and Vogelstein, 1985). The most convincing results were probably obtained in experiments with hormone-induced premature cell differentiation. Cell type-specific genes that started their expression in differentiated cells were associated with the nuclear matrix in these cells and were not associated with the nuclear matrix in maternal (nondifferentiated) cells (Robinson *et al.,* 1983; Jost and Celdran, 1984). A positive correlation between the expression of viral genomes integrated into the host cell DNA and their attachment to the nuclear matrix was also observed (Nelkin *et al.,* 1980; Cook *et al.,* 1982). Furthermore, the elongation complexes of RNA polymerase II were found to be concentrated close to matrix attachment sites (Razin and Iarovaia, 1985), and nascent RNA chains were found to be bound to the nuclear matrix (Jackson *et al.,* 1981; Ciejek *et al.,* 1982). In view of the above-mentioned results it is reasonable to discuss two questions. The first one is whether the specificity of nuclear matrix DNA revealed in experiments with high salt-extracted nuclei can be explained by rebinding of MAR sequences to the affinity sites on the nuclear matrix. This rebinding could occur in the course of nucleoid treatment with restriction nucleases. The important question is therefore whether the samples were reextracted with 2 *M* NaCl solution after the cleavage of loop DNA with nucleases. In one of the initial publications of Vogelstein and collaborators it is clearly stated that this reextraction was carried out (Robinson *et al.,* 1982) and some other researchers reproduced exactly the experimental procedure described in this publication (Jost and Seldran, 1984). However, in other cases (Robinson *et al.,* 1983; Small and Vogelstein, 1985) the reextraction step was possibly omitted (at least it is not mentioned in "Methods" sections). Hence, the nuclear matrix DNA fractions studied in some of the above-discussed experiments could actually include DNA fragments rebound to the nuclear matrix after high salt extraction (i.e., MAR sequences). It should be stressed, however, that all basic conclusions concerning the association of active genes with the nuclear matrix are supported by the data obtained in experiments with reextraction of nuclease-treated nucleoids with 2 *M* NaCl solution (Robinson *et al.,* 1982; Jost and Seldran, 1984). Furthermore, recent observations made in experiments on high-resolution *in situ* hybridization of different DNA probes to the DNA loop halo of immobilized nucleoids support the conclusion that active genes are bound to the nuclear matrix (Gerdes *et al.,* 1994).

The second question deserving discussion is whether active genes are bound to the nuclear matrix through transcriptional complexes or, indirectly, through nascent RNA chains. It is also necessary to consider a formal possibility that the whole phenomenon (association of active genes with the nuclear matrix) is an experimental artifact attributed to the precipitation of either nascent RNA chains or both nascent RNA chains and transcription complexes in the course of high salt extraction. Several lines of experimental evidence show that none of the above suppositions is correct. First, neither the organization of DNA into loops (in nucleoids), nor the association of active genes with the nuclear matrix is affected by RNase treatment (Cook and Brazell, 1978; Iarovaia *et al.,* 1985; Vogelstein *et al.,* 1985). Second, the inhibition of transcription by either $\alpha$-amanitine or by actinomycin D does not cause the detachment of active genes from the nuclear matrix (Iarovaia *et al.,* 1985; Small *et al.,* 1985). Third, the arrest of transcription of most normally active genes under heat shock conditions (and the dissociation of transcriptional complexes from these genes) also does not cause the detachment from the nuclear matrix of normally transcribed, although temporarily silent, genes (Small *et al.,* 1985). Fourth, in the course of hormone-induced cell differentiation an association of cell type-specific genes with the nuclear matrix occurs before transcription is actually started (Jost and Celdran, 1984). Taken together, all these data support the conclusion that transcription per se is not essential for the association of active genes with the nuclear matrix. This association is likely to constitute a condition of transcriptional activation rather than a consequence of transcription.

It should be mentioned here that not only transcriptionally active genes, but also replicating DNA sequences were found in close association with the nuclear matrix (Berezney and Coffey, 1975; Pardoll *et al.,* 1980; Berezney and Buchholtz, 1981a).

The functional dependence and heterogeneity of DNA–nuclear matrix interactions could be hardly assimilated by the simple model of DNA organization into relatively uniform loops anchored to the nuclear matrix.

On the other hand, the assumption that all specific DNA interactions with the nuclear matrix are transcription related (Small and Vogelstein, 1985) also failed to explain some experimental observation, for example a correlation between an average size of DNA loops and an average size of replicons in cells of different species (Buongiorno-Nardelli *et al.*, 1982). To solve the problem, we have suggested the existence of a special group of "structural" or "permanent" attachment sites that delimit the borders of supercoiled DNA loops in nucleoids and are not related to transcription (Razin *et al.*, 1985c; Razin, 1987). A similar suggestion also was discussed by other authors (Getzenberg *et al.*, 1991; Pienta *et al.*, 1991; Jackson *et al.*, 1992). If this suggestion is correct, transcription-related associations of DNA with the nuclear matrix may be responsible for the anchorage of the majority of nuclear matrix DNA fragments without affecting the average size of DNA loops estimated by the above-discussed (Section II) experimental approaches. In order to test out supposition, we have mapped the preferential positions of DNA attachment to the nuclear matrix in the domain of chicken $\alpha$-globin genes in chicken erythroblasts, mature erythrocytes, and mature sperm nuclei (Razin *et al.*, 1985c; Farache *et al.*, 1990; Kalandadze *et al.*, 1990). The rationales for this experiment were that in terminally differentiated transcriptionally silent mature erythrocytes and mature sperm nuclei, transcription-dependent association of DNA with the nuclear matrix might disappear while permanent attachment sites were likely to be preserved. Indeed, while in erythroblast nuclei the whole transcriptionally active area was bound to the nuclear matrix, in erythrocyte and sperm nuclei only two major attachment sites framing the domain from the upstream and the downstream were detected. The same DNA fragments were also attached to the nuclear matrix of cultured chicken fibroblasts (Kalandadze *et al.*, 1990). Hence, at least within the domain of chicken $\alpha$-globin genes, it was possible to discriminate permanent (perserved in nonactive nuclei) and transcription-dependent interactions of DNA with the nuclear matrix. It is important to mention that in the above experiments the nuclear matrices were reextracted with 2 *M* NaCl solution after the nuclease treatment (Razin *et al.*, 1985c; Farache *et al.*, 1990; Kalandadze *et al.*, 1990). Hence, the permanent interactions of DNA with the nuclear matrix identified in erythrocyte and sperm nuclei could not be artificially caused by rebinding of MAR elements to an isolated nuclear matrix.

An important question was whether permanent sites of DNA attachment to the nuclear matrix also could be found in other genomic areas. In order to clarify this question, we have compared the sequence complexities of nuclear matrix DNA samples prepared from erythroblasts and mature erythrocytes (Razin *et al.*, 1986). Trace amounts of either total DNA or nuclear matrix DNA from erythroblasts or erythrocytes were reannealed in reactions driven by a vast excess of either erythrocyte or erythroblast nuclear matrix DNA (Fig. 4). The most important observation made in these experiments was that in both renaturation reaction with $C_0t$ values sufficient for complete renaturation of homologous nuclear matrix DNA probes total DNA was reannealed only partially (25% of unique sequences in the reaction driven by erythroblast nuclear matrix DNA and 10% of unique sequences in the reaction driven by erythrocyte nuclear matrix DNA). This mean that both nuclear matrix DNA fractions contain only subsets of unique DNA sequences present in total DNA. Consequently, one may conclude that matrix attachment sites have specific positions in the genome. In the other case all unique sequences present in total DNA are to be found in the vicinity of matrix attachment sites (i.e., in nuclear matrix DNA). Similar considerations made it possible to conclude that erythrocyte nuclear matrix DNA is a subfraction of erythroblast nuclear matrix DNA. Indeed, the latter was not reannealed completely in the reaction driven by erythrocyte nuclear matrix DNA while both nuclear matrix DNA samples were reannealed completely in the reaction driven by erythroblast nuclear matrix DNA. Hence, the erythroblast nuclear matrix DNA is composed of two subfractions: permanently attached to nuclear matrix DNA fragments (i.e., erythrocyte nuclear matrix DNA) and DNA

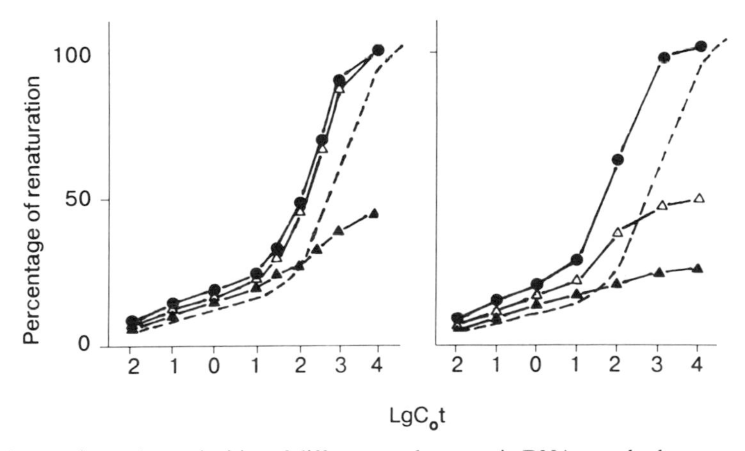


FIG. 4 Comparison of complexities of different nuclear matrix DNA samples by corenaturation analysis. The curves show kinetics of reannealing of trace amounts of total DNA (filled triangles), nuclear matrix DNA from chicken erythroblasts (open triangles), and nuclear matrix DNA from mature chicken erythrocytes (filled circles) in renaturation reactions driven by vast excess of either erythroblast nuclear matrix DNA (left) or erythrocyte nuclear matrix DNA (right). Broken lines show the profile of self-renaturation of total chicken DNA. For details of the experiment see Razin *et al.* (1986).

fragments interacting with the nuclear matrix in accordance with the functional processes.

Recent evidence indicates that replication origins are permanently bound to the nuclear matrix (Aelen *et al.*, 1983; Van der Velden *et al.*, 1984; Carri *et al.*, 1986; Dijkwel *et al.*, 1986; Razin *et al.*, 1986). It seems therefore likely that the topological organization of DNA into loops is directly related to the organization of replication units.

### 2. LIS Extraction Procedure

As has been already mentioned (Section III,C,1), the validity of the high salt extraction protocol for nuclear matrix DNA isolation was questioned in connection with an enrichment of the resulting DNA preparation with transcriptionally active sequences. It has been proposed that high salt extraction can cause the precipitation of transcription complexes onto the nuclear matrix (Kirov *et al.*, 1984; Mirkovich *et al.*, 1984). The results of experiments demonstrating that an association of active genes with the nuclear matrix does not depend on transcription (see Section III,C,1) were not completely convincing because opposite observations were made by other authors (Mirkovich *et al.*, 1984). In order to avoid possible artifacts of high salt extraction, it has been proposed to disrupt chromatin by treatment with an ionic detergent LIS, in a low ionic strength buffer (Mirkovich *et al.*, 1984). This method permitted the identification of specific genomic elements bound to the nuclear remnants that were called nuclear scaffolds in order to avoid possible confusion with nuclear matrices isolated using high salt extraction. Correspondingly, genomic elements found in complex with the nuclear scaffold were called scaffold-attached regions or SARs (Mirkovich *et al.*, 1984). The same SARs have been found to be bound to scaffolds of interphase nuclei and metaphase chromosomes (Mirkovich *et al.*, 1988). Most surprisingly, SARs turned out to be indistinguishable from MARs identified by *in vitro* binding assay. Nuclear (chromosomal) scaffolds were found to possess the same affinity sites capable of binding *in vitro* MAR/SAR elements, as a nuclear matrix isolated by the high salt-extraction procedure (Cockerill and Garrard, 1986a; Izaurralde *et al.*, 1988; Mirkovich *et al.*, 1988; Phi-van and Strätling, 1988). Transcriptionally active DNA sequences are not preferentially associated with scaffolds prepared by the LIS extraction protocol (Mirkovich *et al.*, 1984). Genomic distribution of SARs has been extensively studied by many research teams. The results of the above studies have been reviewed by Gasser *et al.* (1989), Roberge and Gasser (1992), and Laemmli *et al.* (1992). Hence it is hardly necessary to review them again in the present chapter. What should be stressed once again is that SARs/MARs undoubtedly play an important function in the genome organization. They usually comap with regulatory genomic sequences and at least in some cases (Bode and Maass, 1988; Phi-van and Strätling, 1988; Levy-Wilson and Fortier, 1989) seem to frame the DNase I-sensitive genomic domains. Different SARs/MARs do not possess extensive sequence homology, but usually are AT-rich and contain multiple recognition sites for *in vitro* cleavage with topoisomerase II (Gasser *et al.,* 1989; Roberge and Gasser, 1992). Identification of several nuclear proteins that bind specifically to MAR/SAR elements (Von-Kries *et al.,* 1991; Dickinson *et al.,* 1992; Ludérus *et al.,* 1992; Romig *et al.,* 1992) presents an additional, although indirect, argument for the functional significance of SARs/MARs.

The question that is not at all clear is whether SARs identified as "*in vivo* bound" to the nuclear/chromosomal scaffold are DNA loop anchorage sites. It is easy to see that the whole experimental strategy of SAR mapping is based on the model of topological DNA organization in high salt-extracted nuclei (nucleoids). In particular, the authors have assumed that the treatment of isolated nuclei with LIS removes histones but does not disturb the organization of DNA into loops (Mirkovich *et al.,* 1984). In this case one can expect to find in proximity to the nuclear scaffold (after cleavage of loop DNA by restriction nucleases) DNA sequences located at the loop basements. One may be, however, surprised to realize that nobody has ever checked whether DNA is indeed organized into constrained domains (closed loops) in LIS-extracted nuclei. For example, experiments on titration of LIS-extracted nuclei with ethidium bromide never have been done. Hence, it may as well happen that the treatment of isolated nuclei with LIS in hypotonic buffer destroys all specific interactions of DNA with the nuclear matrix (scaffold). Even in this case the DNA will coprecipitate with the nuclear matrix (scaffold) because it is difficult to separate the extremely long DNA chain from the matrix network. This separation may of course occur after mild nuclease digestion. On the other hand, if LIS does not destroy preexisting DNA associations with the nuclear matrix, a mild nuclease digestion followed by LIS extraction would not cause a significant release of DNA from the high-sedimenting fraction. Hence, analysis of the dependence between the average size of scaffold-bound DNA fragments and the percentage of scaffold-bound DNA in the course of progressive nuclease treatment (i.e., the experiments performed many times with high salt-extracted nuclei (Igo-Kemenes and Zachau, 1977; Razin *et al.,* 1979) could help to clarify what happens with scaffold/matrix attachment sites in the course of LIS extraction. Surprisingly, these experiments also have never been done with LIS-extracted nuclei or metaphase chromosomes. The results of experiments of Jackson and collaborators (1990) on determination of DNA loop size in LIS-extracted nuclei can hardly be considered as conclusive evidence for the preservation of DNA loop organization in the course of LIS extraction. Indeed, in these experiments the cleavage of loop DNA and elution of cleaved-off DNA fragments were done when LIS had been washed out from the agarose blocks with embedded cells. Hence the rebinding of MARs to the exposed affinity sites on the nuclear matrix could occur.

It is also worth mentioning that complexes of DNA with the nuclear scaffold resistant to LIS extraction in nuclei may be expected to remain resistant to this extraction after the removal of DNA loops by the treatment with restriction nucleases. However, complexes of SARs with nuclear scaffolds are not resistant to reextraction with LIS. That is why they are normally separated from cleaved-off DNA simply by several washes in a restriction buffer (Mirkovich *et al.,* 1984, 1988; Izaurralde *et al.,* 1988).

Hence, there is no evidence that the LIS extraction procedure makes it possible to identify preexisting *in vivo* interactions of DNA with scaffolding structures. On the contrary, it is likely that specific complexes of SARs with LIS-extracted nuclear (chromosomal) scaffolds represent a result of binding *in vitro.* As has been discussed previously (Izaurralde *et al.,* 1988) SARs have plenty of possibilities to interact with the corresponding affinity sites at the step of treatment of LIS-extracted nuclei with restriction nucleases. Indeed, MAR/SAR binding sites (but not their complexes with target DNA sequences) have been shown to be resistant to LIS treatment (Izaurralde *et al.,* 1988; Mirkovich *et al,* 1988). Incubation of isolated matrices with a mixture of DNA sequences under study in a buffer suitable for DNA treatment with restriction nucleases is normally used for *in vitro* binding of MARs to the affinity sites on the nuclear matrix (Cockerill and Garrard, 1986a,b). Genomic DNA fragments that do not contain SARs can easily substitute a nonspecific competitor used in the classical (Cockerill and Garrard, 1986a,b) *in vitro* binding assay. The LIS extraction procedure therefore may be considered as a technical variant of the *in vitro* binding assay for identification of DNA sequences capable of interacting with the affinity sites on an isolated nuclear matrix. Further investigations are necessary to find out whether these affinity sites and the target DNA sequences (MARs/SARs) are involved somehow in DNA loop anchorage to the nuclear matrix. As has been already mentioned (Section III,B), the high resolution *in situ* hybridization of MAR/SAR-containing probes with nuclear DNA halos (Gerdes *et al.,* 1994) may serve as an important tool for elucidating a possible function of MARs/SARs in the topological organization of eukaryotic DNA.

### 3. Electroelution Procedure

In order to avoid possible artifacts of either high salt extraction or treatment with detergents in hypotonic buffer solutions, it has been suggested to prepare nuclear matrix DNA by electroelution of cleaved-off pieces of DNA loops in physiological ionic strength (Jackson and Cook, 1985, 1986).

An advantage of this procedure is that it is not necessary to disrupt the nucleosomal organization. Long pieces of DNA arranged in chromatin, which are normally not soluble at physiological ionic strength, still may be electroeluted from nuclei. A disadvantage is that the organization of loop DNA in chromatin modulates the accessibility of different DNA sequences and hence the cleavage of loop DNA by nucleases proceeds in a nonrandom fashion. Nuclear matrix-bound DNA isolated by the electroelution procedure was found to be essentially similar to that isolated by the high salt-extraction method. In particular, preferential associations of both transcribing (Jackson and Cook, 1985) and replicating (Jackson and Cook, 1986) DNA sequences with the nuclear matrix have been demonstrated. It is clear that in these experiments such associations could not be attributed to the artificial precipitation of transcriptional complexes and replication forks. Importantly, it was also found that the extraction of nuclei with hypotonic buffers may give rise to different artifacts (Jackson and Cook, 1985, 1986). The same observations were also made when the properties of nuclear matrix DNA isolated using either high salt extraction alone or a combination of low salt and high salt extractions were compared (Razin *et al.,* 1985b).

## IV. Excision of DNA Loops by Selective DNA Cleavage at Matrix Attachment Sites

The previous section of this chapter could be finished with a remark that the 15-year history of the research on the specificity of DNA organization into loops is full of contradictions and unsettled questions. To clarify at least some of these questions, it seemed reasonable to think about a completely different approach for characterization of chromosomal DNA loops. Until now the attention of all scientists interested in the above problem has been concentrated on the isolation and characterization of the loop ends. In this section the recently developed methods of excision and characterization of whole DNA loops will be discussed.

### A. Excision by Topoisomerase II-Mediated DNA Cleavage

DNA topoisomerase II has been identified as a component of the nuclear matrix/chromosomal scaffold isolated by the high salt extraction procedure (Berrios *et al.,* 1985; Earnshaw *et al.,* 1985). Topoisomerase II induces temporary double-stranded scissions in DNA in the course of a relaxation reaction normally catalyzed by this enzyme. Intermediate cleavage complexes can be accumulated by the inhibition of the religation step of the topoisomerase II-mediated cleavage-reunion reaction with a variety of antitumor drugs, such as VM-26, VP16, or m-AMSA (Chen *et al.,* 1984). As a component of the nuclear matrix, DNA topoisomerase II may be expected to have preferential access to the DNA loop ends. Hence it seemed reasonable to try to use topoisomerase II-mediated DNA cleavage for excision of genomic DNA loops. The idea was supported by the results of mapping of topoisomerase II cleavage sites in the chicken domain of $\alpha$-globin genes (Razin *et al.,* 1991). Inhibition of the activity of topoisomerase II in living chicken cells resulted in the excision of this domain as a DNA fragment of unique size (~25 kb long). At least the upstream end of this DNA fragment was mapped within the area known to be permanently attached to the nuclear matrix isolated by the high salt-extraction procedure (Razin *et al.,* 1991). It was clear, however, that the nuclear matrix attachment sites cannot be unique targets for *in vivo* DNA cleavage with topoisomerase II. Indeed, it has been reported that the enzyme can introduce scissions at all accessible sites in chromatin including DNase I-hypersensitive sites (Reitman and Felsenfeld, 1990) and even internucleosomal spacers (Udvardy and Schedl, 1991). To decrease (or even to eliminate) the probability of topoisomerase II-mediated DNA cleavage outside the matrix attachment sites we have suggested extracting permeabilized cells with 2 *M* NaCl solution and then treating them with topoisomerase II inhibitors (Razin *et al.,* 1993; Gromova *et al.,* 1995a; Iarovaia *et al.,* 1995a). We have reasoned that the extraction of nuclei with concentrated salt solutions does not affect the nuclear matrix (chromosomal scaffold) integrity and does not release DNA loops from the nuclear matrix (see Section II). At the same time, this extraction may remove soluble topoisomerase II capable of interacting with DNA loops outside the matrix-attached regions. Furthermore, the preferential accessibility of DNase I-hypersensitive regions for the soluble topoisomerase II will be lost with the disruption of nucleosomes by 2 *M* NaCl extraction. It is also important that topoisomerase II is not inactivated by incubation in a concentrated salt solution. In a preliminary experiment we have shown that the high salt-insoluble topoisomerase II of the nuclear matrix is indeed able to introduce scissions in DNA (Vassetzky *et al.,* 1989). On the basis of the above-mentioned facts we have elaborated a procedure for excising individual loops from DNA (Razin *et al.,* 1993; Gromova *et al.,* 1995; Iarovaia *et al.,* 1995a). Cells were permeabilized with a nonionic detergent and extracted with 2 *M* NaCl in order to obtain nucleoids with extended DNA loops attached to the nuclear matrix containing DNA–topoisomerase II. The nucleoids were then incubated under conditions favoring topoisomerase II-mediated DNA cleavage [magnesium-containing buffer supplemented with VM-26, an inhibitor of topoisomerase II known to block the religation half-reaction catalyzed by the enzyme (Chen *et al.,*

1984)]. After termination of the incubation by addition of SDS and digestion of proteins with proteinase K, the excised loops and their oligomers were separated by pulsed field gel electrophoresis (PFGE) and subjected to Southern blot analysis.

To test the above protocol, we have analyzed the specificity of cleavage ribosomal RNA genes by high salt-soluble topoisomerase II (Razin *et al.*, 1993; Iarovaia *et al.*, 1993). Hybridization of Southern filters with a probe representing encoding sequences for 18S and 28S rRNA revealed a regular pattern of bands with the sizes divisible by the size of rDNA repeats (Fig. 5C). The size distribution of the bands showed a clear dependence on the concentration of VM-26 at the step of incubation of nucleoids in a $Mg^{2+}$-containing buffer. At a high VM-26 concentration virtually all hybridizing material was converted into a single band having the size of a rDNA repeated unit.

We have proposed that each repeated unit of ribosomal DNA is organized into a loop attached to the nuclear matrix through a DNA sequence element that has a specific location within the repeat (Fig. 6). High-resolution analysis performed using indirect end-labeling of topoisomerase II cleavage products additionally cut by different restriction enzymes made it possible to conclude that cleavages occurred in a nontranscribed spacer within about a 4-kb area located just upstream to the start of the 45S rRNA precursor.

It is of interest that a regular pattern of bands divisible in their size by the rDNA repeat size was also observed when living cells were treated with inhibitors of topoisomerase II, although the background was significantly higher in this case (Fig. 5B). This pattern was especially clearly seen when metaphase cells were treated with VM-26 (Fig. 5D). We have concluded that matrix attachment sites constitute preferential targets for topoisomerase II-mediated DNA cleavage even in intact nuclei. That is why the regular pattern could still be seen in the above experiments. The high background observed in experiments with living nonsynchronized cells could be easily explained by a less specific cleavage of DNA with soluble topoisomerase II reported to occur at all accessible places in chromatin (Reitman and Felsenfeld, 1990; Udvardy and Schedl, 1991). In agreement with the proposed model (Fig. 6), the high salt extraction eliminated or at least reduced significantly the above background without affecting the regular pattern of the "loop-sized" bands. It is therefore highly unlikely that the high salt treatment causes disintegration, redistribution, or creation *de novo* of matrix attachment sites organized with the participation of topoisomerase II. It is also worth noting that the region of attachment of the ribosomal DNA repeat determined by the topoisomerase-mediated cleavage protocol is located in a nontranscribed spacer. Hence, the protocol permits discriminating loop ends from functional associations of DNA with the nuclear matrix that are routinely revealed with the aid of the high

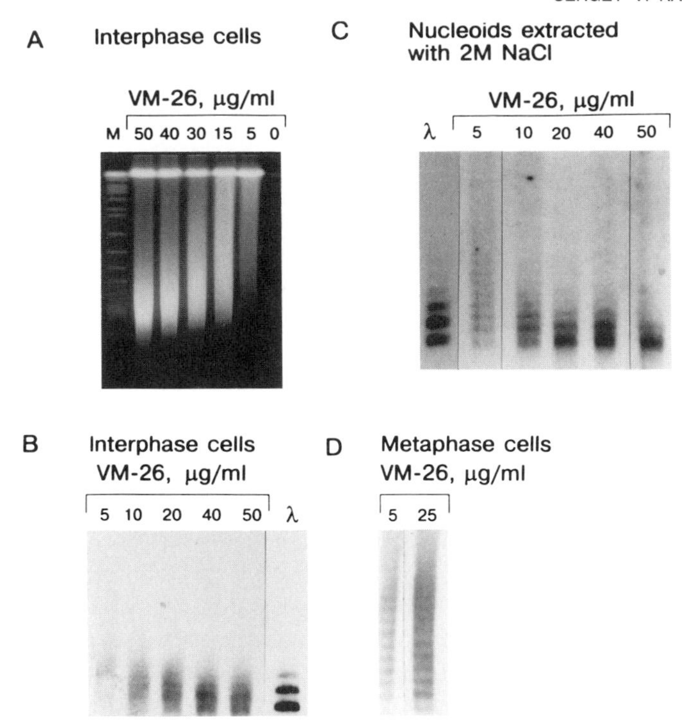


FIG. 5 Cleavage of the Chinese hamster cultured cell (CHO cell line) ribosomal genes by an endogenous topoisomerase II in living cells and in high salt-extracted nuclei. (A) Size distribution of DNA fragments released from the genome of CHO cells incubated in the media containing different amounts of VM-26 (staining with ethidium bromide after separation by PFGE). Cells embedded in agarose blocks were incubated with 0–50 $\mu$g/ml of VM-26 as indicated above the slots. After cell lysis and digestion of proteins, the released DNA fragments were separated by PFGE (see Razin *et al.,* 1993, for details of the experiment). Lane M shows distribution of a molecular weight marker (*Saccharomyces cerevisiae* chromosomal DNA). (B) After separation by PFGE (see A) the released DNA fragments were transferred to nylon filter and hybridized with the $^{32}$P-labeled probe representing coding sequences for 18S and 28S ribosomal RNA. The slot designated "λ" shows distribution of the λ phage DNA concatemers. (C) Cells embedded into agarose blocks were first extracted with 2 *M* NaCl in the presence of 0.2% of NP-40 and then were incubated in the buffer for topoisomerase II-mediated DNA cleavage supplemented with different amounts of VM-26, as indicated above the slots. After lysis, digestion of proteins and separation of released DNA fragments by PFGE, Southern blot analysis was carried out with the same probe as in the experiment shown in section B. (D) The same experiment as shown in section B, but metaphase cells were treated with VM-26.

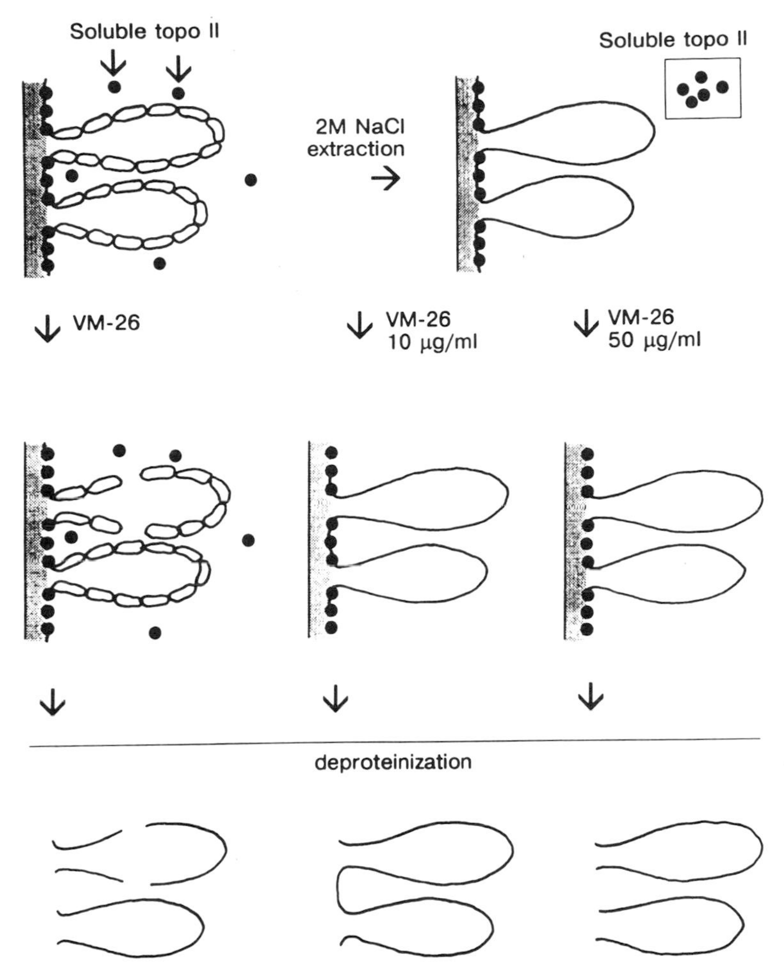


FIG. 6 Interpretation of the results represented in Fig. 5. In living cells the 30-nm chromatin fibril is arranged into loops by periodic attachment to the nuclear matrix. The latter contains topoisomerase II (filled circles) that can cut DNA at loop basements. In addition, soluble topoisomerase II can interact with DNA at a number of accessible sites creating an apparently random cleavage pattern (as far as the low-resolution analysis by PFGE is considered). The resulting cleavage pattern will represent the superimposing of this random cleavage pattern over the regular pattern of cleavages at loop basements. With the soluble topoisomerase II removed by high salt extraction, the cleavages can occur only within matrix attachment sites. Varying concentration of VM26 it is possible to cut DNA at all loop basements or less frequently so that multimers of loops are also released.

salt extraction procedure. The absence of a significant background in an experiment with *in vivo* DNA cleavage by topoisomerase II in metaphase cells can be easily explained by the removal of the most part of soluble topoisomerase II from chromosomes at the beginning of mitosis (Swedlow *et al.*, 1993).

The most important conclusion following from experiments with a ribosomal gene cluster is that an experimental protocol based on a particular model of topological organization of DNA within eukaryotic cell nuclei has permitted us to obtain exactly those results that were predicted by the model. Clearly, this not only justifies the suggested experimental approach but also corroborates the initial model. It should be stressed that the interpretation of the above-discussed data (Fig. 6) does not depend on a particular function that topoisomerase II plays in the nuclear matrix. Topoisomerase II may be an integral part of the nuclear matrix (chromosomal scaffold) as has been suggested previously (Berrios *et al.*, 1985; Earnshaw *et al.*, 1985) or it may be just one of a number of proteins interacting with the ends of DNA loops. For example, it has been shown recently that topoisomerase II may be complexed with another nuclear matrix protein, known as "scaffold protein 2" (Ma *et al.*, 1993). Further investigations will possibly clarify the function of topoisomerase II in organization of sites of DNA attachment to the nuclear matrix. Our data only demonstrate that either in living cells or in high salt-extracted nuclei topoisomerase II interacts with DNA at matrix attachment sites or, at least, is located in close proximity to DNA at these sites. Otherwise the excision of DNA loops by topoisomerase II-mediated DNA cleavage would not be possible.

After checking the validity of the topoisomerase II-mediated cleavage protocol for DNA loop excision in experiments with a cluster of ribosomal genes, we have applied this protocol to several other genomic areas including an amplified human c-*myc* gene locus (Gromova *et al.*, 1995a) and 14B–15B locus of *Drosophila melanogaster* X-chromosome (Iarovaia *et al.*, 1995b). In all cases the long-range distribution of DNA cleavages induced by high salt-insoluble DNA–topoisomerase II was specific and possessed certain common features. Cleavages occurred within distinct but relatively long (~5 kb) regions. We suggest calling these regions "cleavage areas." It is likely that multiple and relatively weak interactions of DNA with the nuclear matrix scattered over the cleavage area constitute together a high salt-resistant attachment site or, speaking more precisely, a "matrix attachment area." Although the sites of DNA interaction with nuclear matrix proteins may be protected from nucleases, the spacer regions (still constituting a part of the matrix attachment area) may as well be accessible or even hypersensitive (see the scheme in Fig. 7). This complex organization of at least permanent matrix attachment sites can explain the failure to identify any matrix attachment consensus DNA sequence by analysis of short nu-

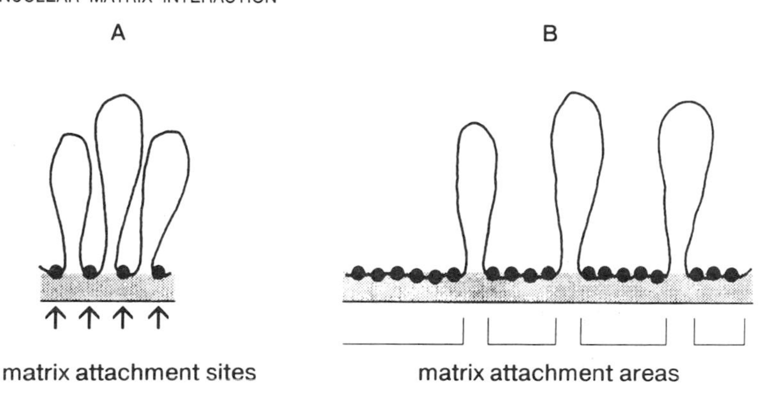


FIG. 7 Two models of organization of DNA loop anchorage on the nuclear matrix. (A) The simplest model suggesting that loops are separated by matrix attachment sites organized with participation of relatively short DNA fragments immersed into the nuclear matrix. (B) A model postulating that interloop regions are relatively long. Within these regions ("matrix attachment areas") the DNA fragments directly involved in interaction with the nuclear matrix proteins (filled circles) may be protected from nucleases, while the neighboring regions (i.e., DNA sequences located between multiple attachment sites within the attachment area) may happen to be sensitive or even hypersensitive to nucleases (due to the disturbance of regular nucleosomal organization within the whole attachment area.)

clear matrix DNA fragments isolated after limit nuclease digestion (see Section III,B). It should be stressed that an involvement of relatively long DNA fragments in anchorage of DNA loops at the nuclear matrix (i.e., matrix attachment areas instead of matrix attachment sites) is not in contradiction with the previously published data, as the approaches used for estimation of DNA loop size were not suitable for determination of a portion of DNA located at the nuclear matrix (see discussion in Section II).

The cleavage areas are separated by loop-sized (20–100 kb) domains of lower sensitivity to the topoisomerase II-mediated DNA cleavage. At first sight, the existence of the long regions of diminished accessibility for DNA topoisomerase II is contradictory to the previously reported observations. Indeed, a detailed mapping of topoisomerase II cleavage sites in some genomic regions showed that in living cells the enzyme could introduce DNA scissions in all accessible places, even in internucleosomal spacers (Rowe *et al.,* 1986; Udvardy and Schedl, 1991). These results hardly permitted one to expect that the long-range pattern of DNA cleavage by topoisomerase II would be nonrandom. However, the above observations were all made when the distribution of topoisomerase II-mediated DNA scissions in short genomic areas accessible to cleavage by topoisomerase II in living cells was studied. It was also reported that some genes are much more accessible to topoisomerase II cleavage as compared to other ones (Riou *et al.*, 1989). The attempts to relate the above accessibility with the transcriptional status of the corresponding genes were not successful (Riou *et al.*, 1989). Our present results suggest that drastic variations in the efficiency of cleavage of different genes by topoisomerase II may reflect the pattern of DNA interaction with the nuclear matrix.

No clear correlation between the distribution of matrix attachment areas and the distribution of transcription units was observed in our experiments. In some cases these areas partially overlap the transcription units. For example, the first two exons of the human c-*myc* gene are located within the matrix attachment area (Gromova *et al.*, 1995a). Some long genes, such as the human muscle dystrophy gene, are clearly arranged in more than one loop (M. Lagarkova and R. Hancock, unpublished) and hence several cleavage areas fall within a single transcription unit.

When information about the positions of replication origins within the regions under study was available, they turned out to be located within the matrix attachment areas (Razin *et al.*, 1993; Gromova *et al.*, 1995a) in agreement with the previous observations indicating a permanent association of replication origins with the nuclear matrix (see Section III,C,1). In connection with these observations, it seems interesting that, similarly to matrix attachment areas, the replication origins of higher eukaryotes also seem to present relatively large areas in which the replication initiation occurs with a preferential probability (Hamlin, 1992).

### B. Excision by Endogenous and Exogenous Nucleases

Recent evidence suggests that the degradation of genomic DNA in the course of apoptosis starts with an excision of 50 to 300-kb DNA fragments (Filipski *et al.*, 1990; Brown *et al.*, 1993; Oberhammer *et al.*, 1993). Products of similar size also accumulate as a result of chromosomal DNA cleavage with some exogenous nucleases (Clark *et al.*, 1987). The pattern of large-scale fragmentation of chromosomal DNA by nucleases has been suggested to reflect a certain periodicity of 30-nm chromatin fibril organization into loops attached to the nuclear scaffold (Filipski *et al.*, 1990; Oberhammer *et al.*, 1993). The only reason for this supposition was, however, a certain similarity in sizes. To gain further insights into the mechanisms of large-scale fragmentation of the eukaryotic genome by nucleases, it seemed reasonable to compare the pattern of this fragmentation with the pattern of long-range DNA fragmentation by nuclear matrix-bound topoisomerase II. Recently, we have done this kind of analysis using as a model an amplified human c-*myc* gene locus (Gromova *et al.*, 1995b). Surprisingly, different nucleases (DNase I, mung bean nuclease, S1 nuclease, endogenous $Ca^{2+}Mg^{2+}$-dependent nuclease) were found to cleave nuclear DNA preferentially within the matrix attachment areas generating the same sets of large DNA fragments as those generated by nuclear matrix-bound DNA topoisomerase II (Fig. 8). In agreement with the data it was also found that DNA located within the permanent matrix attachment site in the upstream part of the domain of chicken $\alpha$-globin genes is preferentially susceptible for S1 cleavage in permeabilized cells (Recillas-Targa, 1994). Furthermore, exogenous Bal 31 nuclease generated the same specific pattern of long-range fragmentation of human ribosomal genes as nuclear matrix-bound DNA–topoisomerase II did (Iarovaia *et al.,* 1995c). None of the above results could be reproduced when either high salt-extracted nuclei or free DNA were treated with exogenous nucleases. Hence, the preferential accessibility of the matrix-attached regions in permeabilized nuclei is determined by the mode of chromatin packaging in the nuclei. For some reasons the matrix-bound DNA turned out to be preferentially accessible for nucleases. A model explaining this preferential accessibility will be discussed in the final section of the present chapter (Section VI).

### C. DNA Interaction with the Nuclear Matrix and the Cell Proliferation Status

The experiments described in Sections IV,A and IV,B, as well as most of other experiments on characterization of the specificity of DNA loop organization were carried out with cultured cells of higher eukaryotes. We were concerned over the possibility that the topological organization of DNA may be different in resting cells (i.e., in the majority of cells of adult multicellular organisms). To clarify the question, we have studied the specificity of nucleolar DNA organization into loops in normal and activated to proliferation human lymphocytes using the two above discussed (Sections VI,A and IV,B) procedures of DNA loop excision (Iarovaia *et al.,* 1995a). In activated lymphocytes nucleolar genes were found to be organized into loops of the same size as the size of individual rDNA repeats. The loops could be excised from the genome by DNA cleavage at matrix attachment sites with either endogenous topoisomerase II or exogenous nuclease Bal 31. In contrast, in normal lymphocytes none of these enzymes generated any specific pattern of long-range fragmentation of nucleolar genes (Fig. 9, and Iarovaia *et al.,* 1995a). Hence, it is likely that the proliferation arrest correlates with a certain reorganization at higher levels of DNA packaging. Further studies are necessary to understand the biological significance of the dependence of the mode of DNA interaction with the nuclear matrix on the cell proliferation status. One of the possibilities is that in proliferating cells the active replication origins are attached to the nuclear matrix (Aelen *et al.,* 1983; Van der Velden *et al.,* 1984; Carri *et al.,* 1986; Dijkwel *et al.,*

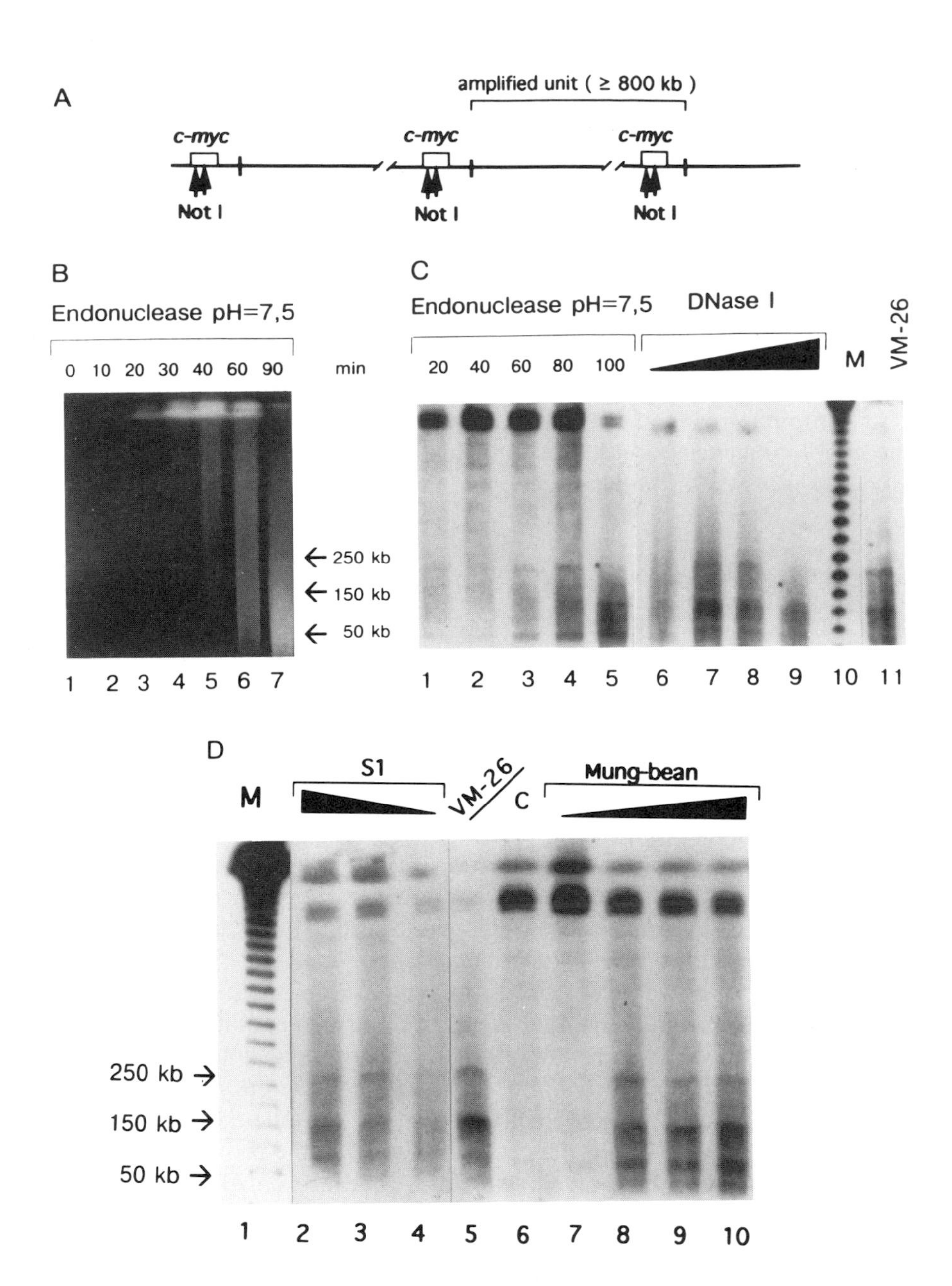
A
amplified unit ( ≥ 800 kb )
c-myc
c-myc
c-myc
Not I
Not I
Not I
B
Endonuclease pH=7,5
0 10 20 30 40 60 90 min
250 kb
150 kb
50 kb
1 2 3 4 5 6 7
C
Endonuclease pH=7,5
DNase I
20 40 60 80 100
M
VM-26
1 2 3 4 5 6 7 8 9 10 11
D
S1
VM-26
Mung-bean
M
C
250 kb
150 kb
50 kb
1 2 3 4 5 6 7 8 9 10

1986; Razin *et al.*, 1986; Brylawski *et al.*, 1993) and that the proliferation arrest is maintained via distraction of these attachments. The above supposition is corroborated by recent findings that in cells of developing *Xenopus laevis* embryos the number of attachment sites decrease with the reduction in a number of active replication origins (Micheli *et al.*, 1993).

## V. DNA Loops as Basic Units of the Genome Organization and Evolution

It has long been known that chromosomal rearrangements frequently occur at some specific places ("hot spots") in the genome. As shown by analysis

FIG. 8 Analysis of specificity of the human c-*myc* amplicone long-range fragmentation by endogenous and exogenous nucleases. (A) A map of the organization of the c-*myc* amplicone in human OC-NYH-VM cells (see Razin *et al.*, 1993, for the details). The positions of the gene are shown by open rectangles. The third exon of the c-*myc* gene, which was used as a probe in hybridization experiments, is located just downstream of the pair of *Not* I cleavage sites indicated by arrows. (B) Size distribution of DNA fragments released from the genome of permeabilised OC-NYH-VM cells incubated for different time intervals (from 0 to 90 min as indicated above the slots) at 37°C in a neutral buffer supplemented with $Mg^{2+}$ and $Ca^{2+}$ ions (digestion with endogenous $Ca^{2+}/Mg^{2+}$-dependent nucleases). After separation by PFGE the DNA fragments were stained with ethidium bromide. (C) Permeabilized cells were either incubated for different time intervals (from 0 to 100 min, as indicated above the slots) at 37°C in a neutral buffer favoring the digestion of DNA with endogenous nucleases (lanes 1–5) or were treated for 30 min at 0°C with different amounts of exogenous DNase I in a $Mg^{2+}$-containing neutral buffer. The DNA samples from cells treated with 2, 10, 25, and 50 $\mu$g/ml of DNase I are shown, respectively, in lanes 6–9. After termination of incubations and digestion of proteins, all DNA samples were additionally digested with NotI (to permit indirect end-labeling). Thereafter they were separated by PFGE and subjected to Southern blot analysis using $^{32}$P-labeled third exon of the c-*myc* gene as the hybridization probe. Lane 11 demonstrates the distribution of DNA products of topoisomerase II-mediated DNA cleavage (excised DNA loops). The distribution of $\lambda$ phage DNA concatemers is shown in lane 10. (D) Permeabilized cells were incubated for 10 min at 37°C in $Zn^{2+}$-containing buffer (pH 5.0) supplemented with 300, 150, or 50 units, respectively of S1 nuclease (lines 2–4) or with 60, 150, or 300 units, respectively, of mung bean nuclease (lines 8–10). After additional digestion with *Not*I, separation by PFGE and Southern blot transfer, fragments bearing the third exon of c-*myc* gene were visualized by hybridization. Lanes 6 and 7 represent DNA from nontreated cells (lane 6) and cells incubated for 10 min at 37°C in $Zn^{2+}$-containing buffer (pH 5.0) without enzymes. Lane 5 demonstrates the distribution of DNA products of topoisomerase II-mediated DNA cleavage (excised DNA loops). The distribution of $\lambda$ phage DNA concatemers is shown in lane 1. Note that either endogenous digestion or treatment of permeabilized cells with exogenous nucleases (including the single-stranded DNA-specific nucleases) resulted in accumulation of characteristic 80-, 150-, and 260-kb fragments of the c-*myc* amplicone similar (if not identical) to the fragments excised from this amplicone by the topoisomerase II-mediated cleavage.

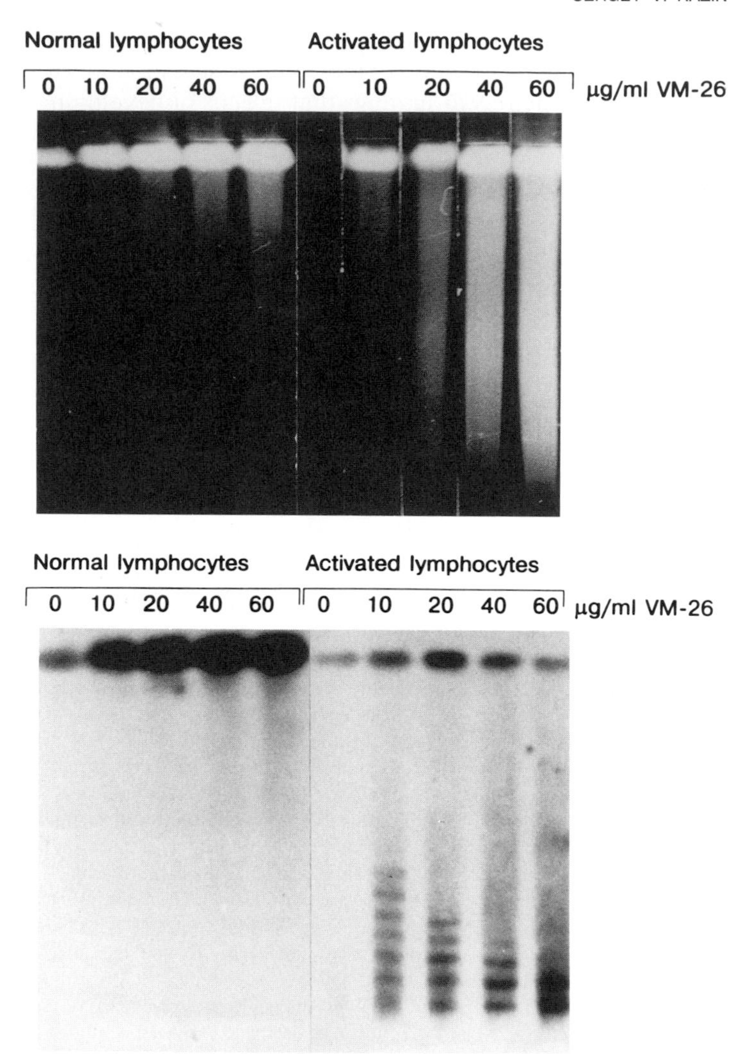


FIG. 9 Analysis of the specificity of topoisomerase II-mediated long-range fragmentation of ribosomal genes in normal and activated to proliferation human lymphocytes. (upper) Separation by PFGE of DNA fragments excised from the genome of normal and activated to proliferation human lymphocytes incubated for 30 min at 37°C in RPMI medium supplemented with VM-26 (from 0 to 60 $\mu$g/ml, as indicated above the slots). (lower) Hybridization of the $^{32}$P-labeled probe representing coding sequences for 18S and 28S ribsomal RNA with the material transferred from gels shown in A. Note the presence of a regular pattern of bands in slots containing the DNA from activated lymphocytes and absence of any bands (with the exception of the compression band) in slots containing the DNA from normal lymphocytes.

of the long-range distribution of hot spots of chromosomal DNA rearrangement they are usually separated by 20 to 100-kb pieces of DNA that are rarely involved in rearrangements (Henglein *et al.,* 1989; Geng *et al.,* 1993). A correlation between the above distances and the average size of topological DNA loops seems to be suggestive.

Little is known about the mechanisms of chromosomal DNA amplification and translocation. Recent evidence suggests, however, that a double-stranded DNA break can initiate a chain of events resulting in any of the above-mentioned rearrangements (Windle *et al.,* 1991). We have found that matrix attachment areas constitute more preferential targets for either exogenous or endogenous nucleases as compared to loop DNA (Section IV,B). Hence it is likely that the DNA chain integrity within these areas may be preferentially injured, promoting the possibility of translocations between matrix attachment areas. A DNA recombinase activity with as yet poorly studied properties has been recently found in nuclear matrix preparations (Dave *et al.,* 1991; Pandey *et al.,* 1991). The possibility of direct participation of nuclear matrix-bound topoisomerase II in DNA recombination has been also discussed in the literature (Blasques *et al.,* 1989; Sperry *et al.,* 1989; Bodley *et al.,* 1993). Indirectly, topoisomerase II-mediated DNA breaks that can accumulate under certain conditions might facilitate recombination events within matrix attachment areas. Though separated by long pieces of loop DNA, the attachment areas are likely to be located close to each other in a nuclear space. This may promote the possibility of recombination events between matrix attachment areas. Indeed, in a model experiment it has been demonstrated that association of circular DNA plasmids with the nuclear matrix facilitates the topoisomerase II-mediated incorporation of these plasmids into catenated networks (Tsutsui *et al.,* 1989). Furthermore, irrespective of the above-discussed contradictions the studies of the sequence specificity of nuclear matrix DNA still indicate that at least this DNA, which is involved in the formation of permanent attachment sites, is in a way degenerated as it frequently contains simple sequence motives, nonspecific repeats, etc. (Goldberg *et al.,* 1983; Opstelten *et al.,* 1989; Kalandadze *et al.,* 1990). This sequence organization seems to be favorable for illegitimate recombination between matrix attachment areas.

To move from all the above considerations in the direction of experimental analysis, we have compared the positions of known hot spots of DNA rearrangement around the human c-*myc* gene with the positions of matrix attachment areas and found a good correlation (I. I. Gromova and S. V. Razin, unpublished). Hence, translocations indeed frequently occur at matrix attachment sites without disrupting loop DNA. This may constitute a general principle of the genome organization and evolution as a set of loop DNA domains. One can only speculate as to how the eukaryotic chromosome was once assembled (possibly by recombination between several simple genomes that gave rise to the first domains). What seems to be clear is that not only the rearrangement of modern genomes, but also their disintegration in the course of regulated cell death (apoptosis) starts with the accumulation of breaks at the domain borders resulting in the excision of DNA loops and their oligomers (see Section IV,B). Hence, it is possible that DNA loops constitute certain basic units of the genome organization.

## VI. Nuclear Matrix as a System of Channels Used for mRNP Transport

At first sight it seems difficult to explain why matrix-bound DNA (matrix attachment areas) presents a more preferential target for exogenous nucleases as compared to loop DNA. Indeed, according to the radial loop model of chromosome organization (Marsden and Laemmli, 1979) the loop ends should be hidden within the chromosome "body" and hence be less accessible for exogenous nucleases. Similarly, it seems difficult to reconcile the data on preferential sensitivity of active genes to nucleases (Weintraub and Groudin, 1976) with the data on their attachment to the nuclear matrix (Section III,C,1). Speaking more generally, one may raise a question as to how the same structure (nuclear matrix) may function as a scaffolding element for the assembly of chromosomal DNA loops and simultaneously as a place where active functional processes occur.

Here we are proposing a new nuclear matrix model that seems to be able to solve the above paradigm and to settle some other inherent problems of the whole "nuclear matrix story." Our main supposition is that the scaffolding element providing a surface for the attachment of DNA loops is not a filament but a channel connected with the nuclear pore. Chromosomal DNA domains are assembled around this channel in the same fashion as a modern multistory building is assembled around a tower with elevators (Fig. 10). An internal nuclear matrix is therefore postulated to be a network of such channels that provides possibilities for ingoing transport of different compounds (precursors for DNA and RNA synthesis, enzymes, and regulatory proteins) from cytoplasm and for an opposite transport of messenger RNA from the sites of synthesis and splicing to cytoplasm. It indeed has been shown that DNA precursors are channeled directly to the places of DNA replication (replication foci) on the nuclear matrix (Panzeter and Ringer, 1993). Furthermore, the RNA synthesis, processing, and transport were found to be spatially organized along some distinct pathways within nuclei (Lawrence *et al.,* 1989; Spector, 1990; Carter *et al.,* 1993; Xing *et al.,* 1993). At least at some cases these pathways (channels?) could be traced

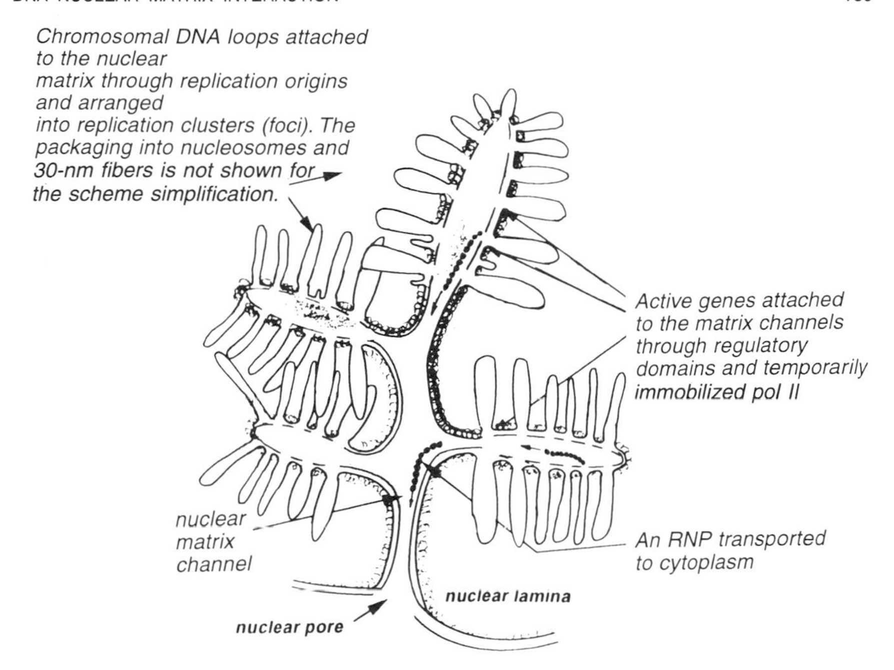


FIG. 10 The model of an interphase chromosome organization based on an assumption that the nuclear matrix is a system of internal nuclear channels.

up to the nuclear envelop (Spector, 1990). The specific distribution of mRNA pathways remains preserved after the removal of chromatin by high salt extraction (Xing and Lawrence, 1991). Hence, the above pathways constitute a part of the high salt-insoluble nuclear matrix. Once the RNA transport is blocked for any reason, the putative matrix channels will be overloaded with RNP particles that can easily aggregate (Akenstorf *et al.*, 1984) to form an RNA filament inside the channel. This may happen as a result of different treatments in the course of nuclei and/or nuclear matrix isolation and that is how an RNA matrix (Smetana *et al.*, 1963; Narayan *et al.*, 1967; Faiferman and Pogo, 1975) might be created.

The channels are not likely to have a distinct morphology or distinct sizes. On the contrary, they may branch and expand locally, giving rise to caverns around which replicon clusters (Nakayasu and Berezney, 1989; Hozák *et al.*, 1993) and/or transcription foci (Carter *et al.*, 1993; Jackson *et al.*, 1993; Xing *et al.*, 1993) can be assembled.

As one can see, the proposed model easily explains why the functional activity in the nuclei is all arranged around the nuclear matrix. Furthermore, it is possible that under certain conditions exogenous nucleases can penetrate into the nuclear matrix channels and hence preferentially digest exactly those DNA sequences that are attached to the matrix. High salt extraction and some other treatments used for histone removal may probably cause a collapse of the matrix channels and an aggregation of the internal staff. As a result, chromosomal DNA will be accessible only from the outside. This means that matrix-attached DNA sequences will become partially protected and that their solubilization is the course of nuclease digestion will occur in the last turn.

It also should be noted the proposed model suggests how different protein factors may be directly transported from the places of their synthesis in the cytoplasm to specific target sequences on chromosomal DNA. The operation of some regulatory mechanisms choosing the necessary targets by a simple selection of channels seems to be quite possible at this level. An advantage of such a mechanism would be the ability to ensure a simultaneous activation of a selected group of independent targets connected with the same channel, for example all replicons in a given cluster.

## VII. Concluding Remarks

It has long been recognized that the eukaryotic genome is partitioned into a number of relatively independent functional domains. A question of whether this partitioning correlates with a subdivision of chromosomal DNA into structural domains (DNA loops fixed at the nuclear matrix) has been addressed in a number of experiments. Surprisingly, after almost 20 years of research the question still remains unanswered. Furthermore, even the commonly held views are not always based on solid experimental data. For example, no conclusive evidence has been ever presented that the LIS-extraction procedure identifies preexisting *in vivo* interactions of DNA with the nuclear matrix. In contrast, the available experimental data strongly suggest that this protocol only permits study of the specificity of *in vitro* binding of DNA to the nuclear matrix. Nevertheless, the LIS extraction procedure is still frequently referred to as a protocol for identification of DNA sequences *in situ* bound to the nuclear matrix, i.e., is unreasonably opposed to the *in vitro* binding assay (Adom *et al.,* 1992).

In this chapter we have tried to reconsider the apparently contradictory data on the specificity of DNA organization into loops and to suggest possible explanations for at least some of the above contradictions. Further progress in the studies of structural–functional domains of eukaryotic genome will depend on the development of new experimental approaches. As has been already mentioned before, the high-resolution *in situ* hybridization technique (Carter *et al.,* 1993; Xing *et al.,* 1993) constitutes a powerful tool for investigation of different aspects of the eukaryotic genome spatial organization. This technique recently has been used for mapping the positions of unique DNA sequences in DNA loop halo of nucleoids (Gerdes *et al.*, 1994). Utilization of the same approach in combination with different methods of histone removal may be expected to clarify the so far putative function of SARs/MARs in DNA loop anchorage on the nuclear matrix.

The present progress in the studies of the long-range functional organization of eukaryotic genome became possible due to the elaboration of reliable methods for cloning and physical mapping of large genomic areas. The currently used physical maps are, however, biologically senseless because they are all based on casual distribution of recognition sequences for the rear-cutting restriction enzymes. Development of an experimental procedure for the genome fragmentation into DNA loops (Razin *et al.*, 1993; Gromova *et al.*, 1995a) raises a possibility to make domain/loop maps of large genomic areas and eventually, of whole chromosomes. These maps will reflect the internal principle of genome construction and hence will allow relating of the structural architecture of the genome to its functional organization. As has been discussed in this chapter (Section V), the DNA loops are likely to represent the basic structural units of the genome. Elucidation of general principles of organization and functioning of these units will constitute an important step toward understanding the principles of the whole genome organization and evolution.

## Acknowledgments

Research was supported by Grant 93-04-21558 from Russian Foundation for Support of Fundamental Science, Grant MKF000 from International Science Foundation, and an ICGEB grant to Sergey Razin.

## References

Adom, J. N., Gouilleux, F., and Richard-Foy, H. (1992). Interaction with the nuclear matrix of a chimeric construct containing a replication origin and a transcription unit. *Biochim. Biophys. Acta* **1171,** 187–197.

Aelen, J. M. A., Opstelten, R. J. G., and Wanka, F. (1983). Organization of DNA replication in *Physarum.* Attachment of origins of replicons and replication forks to the nuclear matrix. *Nucleic Acids Res.* **11,** 1181–1196.

Akenstorf, H., Convay, G. C., Wooley, J. C., and LeStourgeon, W. M. (1984). Nuclear matrix-like filaments form through artifactual rearrangement of hnRNP proteins. *J. Cell Biol.* **99,** (4, Pt.2), 233.

Amati, B., and Gasser, S. M. *Drosophila* scaffold-attached regions bind nuclear scaffolds and can function as ARS elements in both budding and fission yeasts. *Mol. Cell. Biol.* **10,** 5442–5454.

Basler, J., Hastie, N. D., Pietras, D., Matsui, S.-I., Sandberg, A. A., and Berezney, R. (1981). Hybridization of nuclear matrix attached deoxyribonucleic acid fragments. *Biochemistry* **20,** 6921–6929.

Benyajati, C., and Worcel, A. (1976). Isolation, characterization and structure of the folded interphase genome of *Drosophila melanogaster. Cell* (*Cambridge, Mass.*) **9,** 393–407.

Berezney, R., and Buchholtz, L. A. (1981a). Dynamic association of replicating DNA fragments with the nuclear matrix of regenerating rat liver. *Exp. Cell Res.* **132,** 1–13.

Berezney, R., and Buchholtz, L. A. (1981b). Isolation and characterization of rat liver nuclear matrices containing high-molecular weight deoxyribonucleic acid. *Biochemistry* **20,** 4995–5002.

Berezney, R., and Coffey, D. S. (1974). Identification of a nuclear protein matrix. *Biochem. Biophys. Res. Commun.* **60,** 1410–1417.

Berezney, R., and Coffey, D. S. (1975). Nuclear protein matrix: Association with newly synthesized DNA. *Science* **189,** 291–292.

Berezney, R., and Coffey, D. S. (1977). Nuclear matrix: Isolation and characterization of a framework structure from rat liver nuclei. *J. Cell Biol.* **73,** 616–637.

Berrios, M., Osheroff, N., and Fisher, P. (1985). In situ localization of DNA topoisomerase II, a major polypeptide of *Drosophila* nuclear matrix. *Proc. Natl. Acad. Sci. U.S.A.* **82,** 4142–4146.

Blasques, V. C., Sperry, A. O., Cockerill, P. N., and Garrard, W. T. (1989). Protein: DNA interactions of chromosomal loop attachment sites. *Genome* **31,** 503–509.

Bode, J., and Maass, K. (1988). Chromatin domain surrounding the human interferon-$\beta$ gene as defined by scaffold attached regions. *Biochemistry* **27,** 4706–4711.

Bode, J., Kohwi, I., Dickinson, L., Joh, T., Klehr, D., Mielke, C., and Kohwi-Shigematsu, T. (1992). Biological significance of unwinding capability of nuclear matrix-associated DNAs. *Science* **255,** 195–197.

Bodley, A. L., Huang, H. C., Yu, C. A., and Liu, L. F. (1993). Integration of simian virus-40 into cellular DNA occurs at or near topoisomerase-II cleavage hot spots induced by VM-26 (teniposide). *Mol. Cell. Biol.* **13,** 6190–6200.

Boulikas, T., and Kong, C. F. (1993). Multitude of inverted repeats characterize a class of anchorage sites of chromatin loops to the nuclear matrix. *J. Cell. Biochem.* **53,** 1–12.

Bowen, B. C. (1981). DNA fragments associated with chromosomal scaffolds. *Nucleic Acids Res.* **19,** 5093–5108.

Brown, D. G., Sun, X.-M., and Cohen, G. M. (1993). Dexamethasone-induced apoptosis involves cleavage of DNA to large fragments prior to internucleosomal fragmentation. *J. Biol. Chem.* **268,** 3037–3039.

Brylawski, B. P., Tsongalis, G. J., Cordeirostone, M., May, W. T., Comeau, L. D., and Kaufman, D. G. (1993). Association of putative origins of replication with the nuclear matrix in normal human fibroblasts. *Cancer Res.* **53,** 3865–3868.

Buongiorno-Nardelli, M., Gioacchino, M., Carri, M. T., and Marilley, M. (1982). A relationship between replicon size and supercoiled loop domains in the eukaryotic genome. *Nature* (*London*) **298,** 100–102.

Carri, M. T., Micheli, G., Graziano, E., Pace, T., and Buongiorno-Nardelli, M. (1986). The relationship between chromosomal origins of replication and the nuclear matrix during the cell cycle. *Exp. Cell Res.* **164,** 426–436.

Carter, C. C., Bowman, D., Carrington, W., Fogarty, K., McNeil, J. A., Fay, F. S., and Lawrence, J. B. (1993). A three-dimensional view of precursor messenger RNA metabolism within mammalian nucleus. *Science* **259,** 1330–1335.

Chen, G. L., Yang, L., Rowe, T. C., Halligan, B. D., Tewey, K. M., and Liu, L. (1984). Nonintercalative antitumor drugs interfere with the breakage-reunion reaction of mammalian topoisomerase II. *J. Biol. Chem.* **259,** 13560–13566.

Ciejek, E. M., Nordstrom, J. L., Tsai, M.-J., and O'Malley, B. W. (1982). Ribonucleic acid precursors are associated with the chick oviduct nuclear matrix. *Biochemistry* **21,** 4945–4953.

Ciejek, E. M., Tsai, M.-J., and O'Malley, B. W. (1983). Actively transcribed genes are associated with the nuclear matrix. *Nature (London)* **306,** 607–609.

Clark, R. W., Tseng, P. O., and Lechuga, J. M. (1987). A nuclease-derived fragments of metaphase DNA and its relationship to the replicon. *Exp. Cell Res.* **169,** 296–310.

Cockerill, P. N., and Garrard, W. T. (1986a). Chromosomal loop anchorage of the kappa immunoglobulin gene occurs next to the enhancer in a region containing topoisomerase II sites. *Cell (Cambridge, Mass.)* **44,** 273–282.

Cockerill, P. N., and Garrard, W. T. (1986b). Chromosomal loop anchorage sites appear to be evolutionary conserved. *FEBS Lett.* **204,** 5–7.

Cockerill, P. N., Yuen, M.-H., and Garrard, W. T. (1987). The enhancer of the immunoglobulin heavy chain locus is flanked by presumptive chromosomal loop anchorage elements. *J. Biol. Chem.* **262,** 5394–5397.

Cook, P. R., and Brazell, I. A. (1975). Supercoils in human DNA. *J. Cell Sci.* **19,** 261–279.

Cook, P. R., and Brazell, I. A. (1976). Conformational constrains in human DNA. *J. Cell Sci.* **22,** 287–302.

Cook, P. R., and Brazell, D. A. (1978). Spectro fluorometric measurement of the binding of ethidium bromide to superhelical DNA from cell nuclei. *Eur. J. Biochem.* **84,** 464–477.

Cook, P. R., and Brazell, D. A. (1980). Mapping sequences in loops of nuclear DNA by their progressive detachment from the nuclear cage. *Nucleic Acids Res.* **8,** 2895–2906.

Cook, P. R., Brazell, I. A., and Jost, E. (1976). Characterization of nuclear structures containing superhelical DNA. *J. Cell Sci.* **22,** 303–324.

Cook, P. R., Lang, J., Hayday, A., Lania, L., Fried, M., Chiswell, D. J., and Wyke, A. (1982). Active viral genes in transformed cells lie close to the nuclear cage. *EMBO J.* **1,** 447–452.

Das, A. T., Ludérus, M. E., and Lamers, W. H. (1993). Identification and analysis of a matrix-attachment region 5′ of the rat glutamate-dehydrogenase-encoding gene. *Eur. J. Biochem.* **215,** 777–785.

Dave, V. P., Modak, M. J., and Pandey, V. N. (1991). Nuclear matrix bound V(D)J recombination activity in rat thymuc nuclei: An in vitro system. *Biochemistry* **30,** 4763–4767.

Dickinson, L. A., Joh, T., Kohwi, Y., and Kohwi-Shigematsu, T. (1992). A tissue-specific MAR/SAR DNA-binding protein with unusual binding site recognition. *Cell (Cambridge, Mass.)* **70,** 631–645.

Dijkwel, P. A., Wenink, P. W., and Poddighe, J. (1986). Permanent attachment of replication origins to the nuclear matrix in BHK cells. *Nucleic Acids Res.* **14,** 3241–3249.

Earnshaw, W. C., Halligan, B., Cooke, C. A., Heck, M. M. S., and Liu, L. F. (1985). Topoisomerase II is a structural component of mitotic chromosome scaffolds. *J. Cell Biol.* **100,** 1706–1715.

Faiferman, I., and Pogo, A. O. (1975). Isolation of a nuclear ribonucleoprotein network that contains heterogenous RNA and is bound to the nuclear envelope. *Biochemistry* **14,** 3808–3816.

Farache, G., Razin, S. V., Rzeszowska-Wolny, J., Moreau, J., Recillas-Targa, F., and Scherrer, K. (1990). Mapping of structural and transcription-related matrix attachment sites in the $\alpha$-globin gene domain of avian erythoblasts and erythrocytes. *Mol. Cell. Biol.* **10,** 5349–5358.

Filipski, J., Leblanc, J., Youdale, T., Sikorska, M., and Walker, P. R. (1990). Periodicity of DNA folding in higher order chromatin structures. *EMBO J.* **9,** 1319–1327.

Ganguly, A., Bagchi, B., Bera, M., Ghosh, A. N., and Sen, A. (1983). Estimation of domain length of chicken erythrocyte chromatin. *Biochem. Biophys. Res. Commun.* **739,** 286–290.

Garrard, W. T. (1990). Chromosomal loop organization in eukaryotic genomes. *Nucleic Acids Mol. Biol.* **4,** 163.

Gasser, S. M., Amati, B. B., Cardenas, M. E., and Hofmann, J., F.-X. (1989). Studies on scaffold attachment sites and their relation to genome function. *Int. Rev. Cytol.* **119,** 57–96.

Geng, J. P., Tong, J. H., Dong, S., Wang, Z. Y., Chen, S. J., Chen, Z., Zelent, A., Berger, R., and Larsen, C. J. (1993). Localization of the chromosome 15 breakpoints and expression of multiple PML-RAR alpha transcripts in acute promielocytic leukemia: A study of 28 Chinese patients. *Leukemia* **7,** 20–26.

Georgiev, G. P., and Chentsov, Yu. S. (1960). On the structure of cell nuclei: Experimental electron microscopic investigation of isolated nuclei. *Dokladi Akad. Nauk SSSR* **132,** 199–202.

Georgiev, G. P., and Chentsov, Y. S. (1962). On the structural organization of nucleolo-chromosomal ribonucleoproteins. *Exp. Cell. Res.* **27,** 570–572.

Gerdes, M. G., Carter, K. C., Moen, P. T., Jr., and Lawrence, J. B. (1994). Dynamic changes in the higher-level chromatin organization of specific sequences revealed by in situ hybridization to nuclear halos. *J. Cell Biol.* **126,** 289–304.

Getzenberg, R. H., Pienta, K. J., Ward, W. S., and Coffey, D. S. (1991). Nuclear structure and the three-dimensional organization of DNA. *J. Cell. Biochem.* **47,** 289–299.

Goldberg, G. I., Collier, I., and Cassel, A. (1983). *Proc. Natl. Acad. Sci. U.S.A.* **80,** 6887–6891.

Gromova, I. I., Thomsen, B., and Razin, S. V. (1995a). Different topoisomerase II antotumour drugs direct similar specific long-range fragmentation of an amplified *c-myc* gene *loci* in living cells and in high salt-extracted nuclei. Specific DNA sequences associated with the nuclear matrix in synchronized mouse 3T3 cells. *Proc. Natl. Acad. Sci. U.S.A.* **92,** 102–106.

Gromova, I. I., Nielsen, O. F., and Razin, S. V. (1995b). Long-range fragmentation of the eukaryotic genome by exogenous and endogenous nucleases proceeds in a specific fashion via preferential DNA cleavage at matrix attachment sites. *J. Biol. Chem.* (in press).

Hamlin, J. L. (1992). Mammalian origins of replication. *BioEssays* **14,** 651–659.

Hancock, R. (1982). Topological organization of interphase DNA: The nuclear matrix and other skeletal structures. *Biol. Cell.* **46,** 105–122.

Hancock, R., and Hughes, M. E. (1982). Organization of DNA in the eukaryotic nucleus. *Biol. Cell.* **44,** 201–212.

Hartwig, M. (1982). Organization of mammalian chromosomal DNA: Supercoiled and folded circular DNA subunits from interphase cell nuclei. *Acta Biol. Med. Ger.* **37,** 421–432.

Henglein, B., Synovzik, H., Groitl, P., Bornkamm, G. W., Hartl, P., and Lipp, M. (1989). Three breakpoints of variant t(2;8) translocations in Burkitt's lymphoma cells fall within a region 140 kilobases distal from c-*myc*. *Mol. Cell. Biol.* **9,** 2105–2113.

Hentzen, P. C., Rho, J. H., and Bekhor, I. (1984). Nuclear matrix DNA from chicken erythrocytes contains $\beta$-globin gene sequences. *Proc. Natl. Acad. Sci. U.S.A.* **81,** 304–307.

Hozák, P., Hassan, A. B., Jackson, D. A., and Cook, P. R. (1993). Visualization of replication factories attached to nucleoskeleton. *Cell* (*Cambridge, Mass.*) **73,** 361–373.

Iarovaia, O. V., Gorelova, T. V., Shuppe, N. G., and Razin, S. V. (1985). A pattern of the heat shock genes attachment to the nuclear matrix does not depend on their functional state. *Genetika* (*Moscow*) **21,** 896–901.

Iarovaia, O. V., Razin, S. V., and Hancock, R. (1993). Domain organization of mammalian ribosomal gene cluster. *Dokl. Akad. Nauk* (*Russia*) **330,** 120–122.

Iarovaia, O. V., Lagarkova, M. A., and Razin, S. V. (1995a). The specificity of human lumhocyte chromosomal DNA long range fragmentation by endogenous topoisomerase II and exogenous Bal 31 nuclease depends on cell proliferation status. *Biochemistry* **34,** 4133–4137.

Iarovaia, O. V. *et al.* (1995b). *Mol. Cell. Biol.* In preparation.

Iarovaia, O. V., Lagarkova, M. A., and Razin, S. V. (1995c). Excision of chromosomal DNA loops by treatment of permeabilised cells with nuclease Bal 31. *Mol. Gen. Genet.* (in press).

Igo-Kemenes, T., and Zachau, H. G. (1977). Domains in chromatin structure. *Cold Spring Harbor Symp. Quant. Biol.* **42,** 109–118.

Izaurralde, E., Mirkovich, J., and Laemmli, U. K. (1988). Interaction of DNA with nuclear scaffolds in vitro. *J. Mol. Biol.* **200,** 111–125.

Jackson, D. A., and Cook, P. R. (1985). Transcription occurs at a nucleoskeleton. *EMBO J.* **4,** 919–925.

Jackson, D. A., and Cook, P. R. (1986). Replication occurs at a nucleoskeleton. *EMBO J.* **5,** 1403–1410.

Jackson, D. A., McCready, S. J., and Cook, P. R. (1981). RNA is synthesized at the nuclear cage. *Nature* (*London*) **292,** 552–555.

Jackson, D. A., and Cook, P. R., and Patel, S. B. (1984). Attachment of repeated sequences to the nuclear cage. *Nucleic Acids Res.* **17,** 6709–6726.

Jackson, D. A., Dickinson, P., and Cook, P. R. (1990). The size of chromatin loops in HeLa cells. *EMBO J.* **9,** 567–571.

Jackson, D. A., Dolle, A., Robertson, G., and Cook, P. R. (1992). The attachments of chromatin loops to the nucleoskeleton. *Cell Biol. Int. Rep.* **16,** 687–696.

Jackson, D. A., Hassan, A. B., Errington, R. J., and Cook, P. R. (1993). Visualization of focal sites of transcription within human nuclei. *EMBO J.* **12,** 1059–1065.

Jost, J.-P., and Seldran, M. (1984). Association of transcriptionally active vitellogenin II gene with the nuclear matrix of chicken liver. *EMBO J.* **3,** 2205–2208.

Kalandadze, A. G., Razin, S. V., and Georgiev, G. P. (1988). Clonging and analysis of DNA fragments permanently attached to the nuclear matrix. *Dokl. Akad. Nauk SSSR* **300,** 1001–1005.

Kalandadze, A. G., Bushara, S. A., Vassetzky, Y. S., and Razin, S. V. (1990). Characterization of DNA pattern in the site of permanent attachment to the nuclear matrix located in the vicinity of replication origin. *Biochem. Biophys. Res. Commun.* **168,** 9–15.

Kirov, N., Djondjurov, L., and Tsanev, R. (1984). Nuclear matrix and transcriptional activity of the mouse $\alpha$-globin gene. *J. Mol. Biol.* **180,** 601–614.

Klehr, D., Maass, K., and Bode, J. (1991). Scaffold-attached regions from the human interferon B domain can be used to enhance the stable expression of genes under the control of various promoters. *Biochemistry* **30,** 1264–1270.

Krachmarov, C., Iovcheva, C., Hancock, R., and Dessev, G. (1986). Association of DNA with the nuclear lamina in Ehrlich ascites tumor cells. *J. Cell. Biochem.* **31,** 59–74.

Laemmli, U. K., Kas, E., Poljak, L., and Adachi, Y. (1992). Scaffold-associated regions: *cis*-acting determinants of chromatin structural loops and functional domain. *Curr. Opin. Genet. Dev.* **2,** 275–285.

Lawrence, J. B., Singer, R. H., and Marselle, L. M. (1989). Highly localized tracks of specific transcripts within interphase nuclei visualized by in situ hybridization. *Cell* (*Cambridge, Mass.*) **57,** 493–502.

Levy-Wilson, B., and Fortier, C. (1989). The limits of the DNase I-sensitive domain in the homan apolipoprotein B gene coincide with the locations of chromosomal anchorage loops and define the 5′ and 3′ boundaries of the gene. *J. Biol. Chem.* **264,** 21196–21204.

Ludérus, M. E., de Graaf, A., Mattia, E., den Blaauwen, J. L., Grande, M. A., de Jong, L., and van Driel, R. (1992). Binding of matrix attachment regions to lamin B1. *Cell* (*Cambridge, Mass.*) **70,** 949–959.

Ma, X., Saitoh, N., and Curis, P, J. (1993). Purification and characterization of a nuclear DNA-binding factor complex containing topoisomerase II and chromosome scaffold protein 2. *J. Biol. Chem.* **268,** 6182–6188.

Marsden, M. P., and Laemmli, U. K. (1979). Metaphase chromosome structure: Evidence for a radial loop model. *Cell* (*Cambridge, Mass.*) **17,** 849–858.

Micheli, G., Luzzatto, A. R. C., Carri, M. T., Decapoa, A., and Pelliccia, F. (1993). Chromosome length and DNA loop size during early embryonic development of *Xenopus laevis. Chromosoma* **102,** 478–483.

Mielke, C., Kohwi, I., Kohwi-Shigematsu, T., and Bode, J. (1990). Hierarchical binding of DNA fragments derived from scaffold-attached regions: Correlation of properties in vitro and function in vivo. *Biochemistry* **29,** 7475–7485.

Mirkovich, J., Mirault, E., and Laemmli, U. K. (1984). Organization of the higher order chromatin loop: Specific DNA attachment sites on nuclear scaffold. *Cell (Cambridge, Mass.)* **39,** 323–332.

Mirkovich, J., Gasser, S. M., and Laemmli, U. K. (1988). Scaffold attachment of DNA loops in metaphase chromosomes. *J. Mol. Biol.* **200,** 101–109.

Mullenders, L. H. F., Van Zeeland, A. A., and Natarayan, A. T. (1983). Comparison of DNA loop size and supercoiled domain size in human cells. *Mutat. Res.* **112,** 245–252.

Nakane, M., Ide, T., Anzai, K., Ohara, S., and Andoh, T. (1978). Supercoiled DNA folded by nonhistone proteins in cultured mouse carcinoma cells. *J. Biochem.* (*Tokyo*) **84,** 145–157.

Nakayasu, H., and Berezney, R. (1989). Mapping replication sites in the eukaryotic cell nucleus. *J. Cell Biol.* **108,** 1–11.

Narayan, K. S., Steele, W. J., Smetana, K., and Busch, H. (1967). Ultrastructural aspects of the ribonucleoprotein network in nuclei of Walker tumor and rat livers. *Exp. Cell Res.* **46,** 65–77.

Nelkin, B. D., Pardoll, D., and Vogelstein, B. (1980). Localization of SV40 genes within supercoiled loop domains. *Nucleic Acids Res.* **8,** 5623–5633.

Oberhammer, F., Wilson, J. W., Dive, C., Morris, I. D., Hickman, J. A., Wakeling, A. E., Walker, P. R., and Sikorska, M. (1993). Apoptotic death in epithelial cells: Cleavage of DNA to 300 and/or 50 kb fragments prior to or in the absence of internucleosomal fragmentation. *EMBO J.* **12,** 3679–3684.

Opstelten, R. J. G., Clement, J. M. E., and Wanka, F. (1989). Direct repeats at nuclear matrix-associated DNA regions and their putative control function in the replicating eukaryotic genome. *Chromosoma* **98,** 422–427.

Pandey, V. N., Dave, V. P. Amrute, S. B., Rosenbach, K., and Modak, M. J. (1991). Thymic nuclear matrix assiciated activity is not V(D)J recombinase. *Biochem. Biophys. Res. Commun.* **181,** 95–99.

Panzeter, P. L., and Ringer, D. P. (1993). DNA precursors are channelled to nuclear matrix DNA replication sites. *Biochem. J.* **293,** 775–779.

Pardoll, D. M., Vogelstein, B., and Coffey, D. S. (1980). A fixed site of DNA replication in eukarryotic cells. *Cell* (*Cambridge, Mass.*) **19,** 527–536.

Paulson, J. R., and Laemmli, U. K. (1977). The structure of histone-depleted metaphase chromosomes. *Cell* (*Cambridge, Mass.*) **12,** 815–828.

Phi-van, L., and Strätling, W. H. (1988). The matrix attachment regions of the chicken lysozyme gene comap with the boundaries of the chromatin domain. *EMBO J.* **7,** 655–664.

Pienta, K. J., Getzenberg, R. H., and Coffey, D. S. (1991). Cell structure and DNA organization. *CRC Crit. Rev. Eukaryotic Gene Express.* **1,** 355–385.

Razin, S. V., Mantieva, V. L., and Georgiev, G. P. (1979). The similarity of DNA sequences remaining bound to scaffold upon nuclease treatment of interphase nuclei and metaphase chromosomes. *Nuclei Acids Res.* **17,** 1713–1735.

Razin, S. V. (1987). DNA interactions with the nuclear matrix and spatial organization of replication and transcription. *BioEssays* **6,** 19–23.

Razin, S. V., and Iarovaia, O. V. (1985). Initiated complexes of RNA polymerase II are concentrated in the nuclear skeleton associated DNA. *Exp. Cell Res.* **158,** 273–275.

Razin, S. V., Chernokhvostov, V. V., Yarovaia, O. V., and Georgiev, G. P. (1985a). Organization of the sites for DNA attachment to the nonhistone proteinaceous nuclear skeleton. *In* "Progress in Nonhistone Protein Research" (I. Bekhor, ed.), Vol. 2, pp. 91–114. CRC Press, Boka Raton, FL.

Razin, S. V., Yarovaia, O. V., and Georgiev, G. P. (1985b). Low ionic strength extraction of nuclease-treated nuclei destroys the attachment of transcriptionally active DNA to the nuclear skeleton. *Nucleic Acids Res.* **13,** 7427–7444.

Razin, S. V., Rzheshovska-Wolni, I., Moreau, F., and Scherrer, K. (1985c). Localization of DNA attachment sites to the nuclear matrix within the domain of chicken $\alpha$-globin genes in functionally active and inactive nuclei. *Mol. Biol.* (*Moscow*) **19,** 456–466.

Razin, S. V., Kekelidze, M. G., Lukanidin, E. M., Scherrer, K., and Georgiev, G. P. (1986). Replication origins are attached to the nuclear skeleton. *Nucleic Acids Res.* **14,** 8189–8207.

Razin, S. V., Petrov, P., and Hancock, R. (1991). Precise localization of the $\alpha$-globin gene cluster within one of 20 to 300 kb fragments released by cleavage of chicken chromosomal DNA at topoisomerase II sites in vivo: Evidence that the fragments are DNA loops or domains. *Proc. Natl. Acad. Sci. U.S.A.* **88,** 8515–8519.

Razin, S. V., Hancock, R., Iarovaia, O., Westergaard, O., Gromova, I., and Georgiev, G. P. (1993). Structural-functional organization of chromosomal DNA domains. *Cold Spring Harbor Symp. Quant. Biol.* **58,** 25–35.

Recillas-Targa, F., Razin, S. V., De Moura Gallo, C. V., and Scherrer, K. (1994). Excision close to matrix attachment regions of the entire chicken $\alpha$-globin genes domain by nuclease S1 and characterization of framing structures. *Proc. Natl. Acad. Sci. U.S.A.* **91,** 4422–4426.

Reitman, M., and Felsenfeld, G. (1990). Developmental regulation of topoisomerase II sites and DNase I-hypersensitive sites in the chicken $\beta$-globin locus. *Mol. Cell. Biol.* **10,** 2779–2786.

Riou, J.-F., Lefevre, D., and Riou, G. (1989). Stimulation of the Topoisomerase II induced DNA cleavage sites in the *c-myc* protooncogene by antitumor drugs is associated with gene expression. *Biochemistry* **28,** 9104–9110.

Roberge, M., and Gasser, S. M. (1992). DNA loops: Structural and functional properties of scaffold-attached regions. *Mol. Microbiol.* **6,** 419–423.

Robinson, S. I., Nelkin, B. D., and Vogelstein, B. (1982). The ovalbumin gene is associated with the nuclear matrix of chicken oviduct cells. *Cell (Cambridge, Mass.)* **28,** 99–106.

Robinson, S. I., Small, D., Idzerda, R., McKnight, G. S., and Vogelstein, B. (1983). The association of active genes with the nuclear matrix of the chicken oviduct. *Nucleic Acids Res.* **15,** 5113–5130.

Romig, H., Fackelmayer, F. O., Renz, A., Ramsperger, U., and Richter, A. (1992). Characterization of SAF-A, a novel nuclear DNA binding protein from HeLa cells with high affinity for nuclear matrix/scaffold attachment elements. *EMBO J.* **11,** 3431–3440.

Rowe, T. C., Wang, J. C., and Liu, L. F. (1986). In vivo localization of DNA topoisomerase II cleavage sites on *Drosophila* heat shock chromatin. *Mol. Cell. Biol.* **6,** 985–992.

Small, D., and Vogelstein, B. (1985). The anatomy of supercoiled loops in the *Drosphila* 7F locus. *Nucleic Acids Res.* **13,** 7703–7713.

Small, D., Nelkin, B., and Vogelstein, B. (1985). The association of transcribed genes with the nuclear matrix of *Drosphila* cells during heat shock. *Nucleic Acids Res.* **13,** 2413–2431.

Smetana, K., Steele, W. J., and Busch, H. (1963). A nuclear ribonucleoprotein network. *Exp. Cell Res.* **31,** 198–201.

Spector, D. L. (1990). Higher order nuclear organization: Three-dimensional distribution of small nuclear ribonucleoprotein particles. *Proc. Natl. Acad. Sci. U.S.A.* **87,** 147–151.

Sperry, A. O., Blasquez, V. B., and Garrard, W. T. (1989). Disfunction of chromosomal loop attachment sites: Illegitimate recombination linked to matrix associated regions and topoisomerase II. *Proc. Natl. Acad. Sci. U.S.A.* **86,** 5497–5501.

Steif, A. C., Winter, D. M., Strätling, W. H., and Sippel, A. E. (1989). A nuclear DNA attachment element mediates elevated and position independent gene activity. *Nature (London)* **341,** 343–345.

Swedlow, J. R., Sedat, J. W., and Agard, D. A. (1993). Multiple chromosomal populations of topoisomerase II detected in vivo by time-lapse, three dimensional wide-field microscopy. *Cell (Cambridge, Mass.)* **73,** 97–108.

Tsutsui, K., Tsutsui, K., and Oda, T. (1989). Incorporation of exogenous circular DNA into large catenated networks in isolated nuclei. Evidence for involvement of the nuclear scaffold. *J. Biol. Chem.* **264,** 7644–7652.

Udvardy, A., and Schedl, P. (1991). Chromatin structure, not DNA sequence specificity, is the primary determinant of topoisomerase II sites of action in vivo. *Mol. Cell. Biol.* **11,** 4973–4984.

Van der Velden, H. M. V., van Willigen, G., Wetzeis, R. H. W., and Wanka, F. (1984). Attachment of origins of replication to the nuclear matrix and chromosomal scaffold. *FEBS Lett.* **171,** 13–16.

Vassetzky, Y. S., Razin, S. V., and Georgiev, G. P. (1989). DNA fragments which specifically bind to isolated nuclear matrix in vitro interact with matrix-associated DNA topoisomerase II. *Biochem. Biophys. Res. Commun,* **159,** 1263–1268.

Vogelstein, B., Pardoll, D. M., and Coffey, D. S. (1980). Supercoiled loops and eukaryotic DNA replication. *Cell (Cambridge, Mass.)* **22,** 79–85.

Vogelstein, B., Small, D., Robinson, S., and Nelkin, B. (1985). The nuclear matrix and the organization of nuclear DNA. *In* "Progress in Nonhistone Protein Research" (I. Bekhor, ed.), Vol. 2, pp. 115–129. CRC Press, Boca Raton, FL.

Von-Kries, J. P., Buhrmester, H., and Strätling, W. H. (1991). A matrix/scaffold attachment region binding protein: Identification, purification, and mode of binding. *Cell (Cambridge, Mass.)* **64,** 123–135.

Warren, A. C., and Cook, P. R. (1978). Supercoiling of DNA and nuclear conformation during the cell cycle. *J. Cell Sci.* **30,** 211–226.

Weintraub, H., and Groudin, M. (1976). Chromosomal subunits in active genes have an altered conformation. *Science* **73,** 848–856.

Windle, B., Draper, B. W., Yin, Y., O'Gorman, S., and Wahl, G. M. (1991). A central role for chromosomal breakage in gene amplification, deletion formation, and amplicon integration. *Genes Dev.* **5,** 160–174.

Xing, Y., and Lawrence, J. B. (1991). Preservation of specific RNA distribution within the chromatin-depleted nuclear substructure demonstrated by in situ hybridization coupled with biochemical fractionation. *J. Cell. Biol.* **112,** 1055–1063.

Xing, Y., Johnson, C. V., Dobner, P. R., and Lawrence, J. B. (1993). Higher level organization of individual gene transcription and RNA splicing. *Science* **259,** 1326–1330.

Zbarsky, I. B., Dmitrieva, N. P., and Yermolaeva, L. P. (1962). On the structure of tumour cell nuclei. *Exp. Cell Res.* **27,** 573–576.

# INDEX

## A

## D

## L



## M

## N

## O



## P

## R

## S

## T

## U



## V



## Y